Probability Theory and Stochastic Modelling

Volume 91

The **Probability Theory and Stochastic Modelling** series is a merger and continuation of Springer's two well established series Stochastic Modelling and Applied Probability and Probability and Its Applications series. It publishes research monographs that make a significant contribution to probability theory or an applications domain in which advanced probability methods are fundamental. Books in this series are expected to follow rigorous mathematical standards, while also displaying the expository quality necessary to make them useful and accessible to advanced students as well as researchers. The series covers all aspects of modern probability theory including

- Gaussian processes
- Markov processes
- Random fields, point processes and random sets
- Random matrices
- Statistical mechanics and random media
- Stochastic analysis

as well as applications that include (but are not restricted to):

- Branching processes and other models of population growth
- Communications and processing networks
- Computational methods in probability and stochastic processes, including simulation
- Genetics and other stochastic models in biology and the life sciences
- Information theory, signal processing, and image synthesis
- Mathematical economics and finance
- Statistical methods (e.g. empirical processes, MCMC)
- Statistics for stochastic processes
- Stochastic control
- Stochastic models in operations research and stochastic optimization
- Stochastic models in the physical sciences

More information about this series at http://www.springer.com/series/13205

Valeriĭ V. Buldygin • Karl-Heinz Indlekofer •
Oleg I. Klesov • Josef G. Steinebach

Pseudo-Regularly Varying Functions and Generalized Renewal Processes

Springer

Valeriĭ V. Buldygin
Department of Mathematical Analysis
National Technical University of Ukraine
Kyiv, Ukraine

Karl-Heinz Indlekofer
Department of Mathematics
University of Paderborn
Paderborn, Germany

Oleg I. Klesov
Department of Mathematical Analysis
and Probability Theory
National Technical University of Ukraine
Kyiv, Ukraine

Josef G. Steinebach
Mathematical Institute
University of Cologne
Cologne, Germany

ISSN 2199-3130 ISSN 2199-3149 (electronic)
Probability Theory and Stochastic Modelling
ISBN 978-3-319-99536-6 ISBN 978-3-319-99537-3 (eBook)
https://doi.org/10.1007/978-3-319-99537-3

Library of Congress Control Number: 2018956309

Mathematics Subject Classification (2010): 60K05, 60F15, 26A12, 60H10, 60F10, 60E07

This Springer imprint is published by the registered company Springer Nature Switzerland AG
The registered company address is: Gewerbestrasse 11, 6330 Cham, Switzerland

Preface

Renewal theory is a branch of probability theory rich in fascinating mathematical problems and also in various important applications. On the other hand, regular variation of functions is a property that plays a key role in many fields of mathematics. One of the main aims of this book is to exhibit some fruitful links between these two areas via a generalized approach to both of them.

The core of renewal theory is to study (so-called) renewal processes and their probabilistic and statistical characteristics. One of the most cited examples in renewal theory is the following, which deals with the life span of a light bulb.

Renewal Process Assume one has a light bulb in a room and one turns it on, keeping the light bulb working until it fails, after which it is replaced with a new light bulb. If ξ_i denotes the life span of the ith bulb, then $S_n = \xi_1 + \cdots + \xi_n$ represents the total life span of the first n bulbs and

$$N_t = \max\{n : S_n \le t\} \tag{1}$$

is the number of bulbs needed until time t. $\{N_t\}$ is called a renewal counting process constructed from the sequence $\{S_n\}$.

Of course, the bulbs in this example can be exchanged by any other expendable resource, which makes the example more realistic and more attractive from the point of view of applications. For example, consider arriving customers waiting in a queue until one of the servers is free to serve him/her. The arrival counting process to the queue is usually assumed to be a renewal counting process $\{N_t\}$, where the $\{\xi_i\}$ are the inter-arrival times between the customers.

Next we give some further examples of stochastic processes which are related to renewal processes and which occur both in pure and applied mathematics.

Reward Process Let some (random) event occur from time to time. Imagine that someone experiences a reward at each occurrence of the event. Let r_i be the reward earned at the time of the ith occurrence of the event. Denoting by N_t the number of occurrences of the event up to time t, we again see a renewal process with ξ_i being the time between the $(i-1)$th and ith occurrence of the event. Then

$$Y_t = \sum_{i=1}^{N_t} r_i$$

is called a reward process, which is a special case of a (so-called) compound renewal process.

Treating the r_i's as penalties rather than rewards we arrive at several other settings. One of the classical examples here arises in a risk model describing the evolution of the capital of an insurance company which experiences two opposing cash flows: incoming cash premiums and outgoing claims. Premiums from the customers may arrive at a constant rate $c > 0$ (say) and claims $\{r_i\}$ occur according to a counting process $\{N_t\}$, that is, N_t is the number of claims up to time t. So, for an insurer who starts with initial surplus x,

$$X_t = x + ct - \sum_{i=1}^{N_t} r_i, \qquad t \geq 0,$$

represents the capital at time t.

General Counting Process There are many situations similar to what has been described above. Let, for example, ξ_i be the time passed between the $(i-1)$th and ith record in a sport discipline. Then N_t, again, denotes the number of records up to time t and is, in general, called a counting process. Records can also have a negative meaning, for example losses caused by catastrophic events such as hurricanes.

Such a counting process can also be seen in pure mathematics. Let (say) S_n be the nth prime number. Then N_t (usually denoted by $\pi(t)$ in this case) plays a crucial role not only in number theory, but also in applications like cryptography when estimating the security of protocols used to transmit data. An example where a renewal process explicitly occurs in cryptomachines is discussed in Gut [159].

The central object of all the above models can be reduced to an investigation of the behavior of an underlying counting process $\{N_t\}$. In probability theory, this can effectively be done under the classical assumptions that the $\{\xi_i\}$ are

(a) independent,
(b) identically distributed,
(c) positive

random variables. But what about the asymptotic behavior of N_t in cases where these assumptions fail? Such situations are not rare at all. Imagine, for example, that a catastrophic event occurs and leads to many claims for damages. Then the times $\{\xi_i\}$ between the claims become dependent, since many of the policy holders will report to an insurance company at (roughly) the same time. Another example, in which Assumption (b) fails, is related to the so-called alternating renewal process

$\{N_t\}$ constructed from a sequence

$$S_1 = \xi_1, \quad S_2 = \xi_1 + \eta_1, \quad S_3 = \xi_1 + \eta_1 + \xi_2, \quad S_4 = \xi_1 + \eta_1 + \xi_2 + \eta_2, \dots,$$

where the $\{\xi_i\}$ are independent random variables with distribution function F_ξ and the $\{\eta_i\}$ are independent random variables with distribution function F_η. This kind of renewal process occurs in chromatographic models or in describing the motion of water in a river (see, e.g., Gut [159] for a short discussion and further references). Finally, Assumption (c) may fail if, for example, $\{\xi_i\}$ describes a money flow including both expenses and incomes, so that ξ_i could attain both positive and negative values.

Formula (1) nevertheless defines a process $\{N_t\}$, called a *generalized renewal process* in this case, even if one of the Assumptions (a)–(c) fails. Moreover, one can introduce and study some other important functionals of $\{S_n\}$ here. For example, viewing $N_t + 1$ defined via (1) as the *first time* when $\{S_n\}$ exceeds the level t, one can also consider L_t, the *last passage time* across the level t. There are other natural functionals, e.g., T_t, the total time spent by $\{S_n\}$ below the level t, etc. The just mentioned three functionals coincide if Assumption (c) holds, but otherwise they are different. All these functionals as well as many others are also called generalized renewal processes. So, generalized renewal processes arise either if N_t is constructed by (1), but one of the properties (a)–(c) fails, or if other functionals of $\{S_n\}$ (similar to N_t) are considered (we shall specify later what kind of functionals we have in mind when studying generalized renewal processes).

One of the basic questions studied in all models including renewal processes concerns the asymptotic behavior of N_t as $t \to \infty$ which, in turn, is determined by the asymptotic behavior of S_n as $n \to \infty$ (we shall clarify what we mean by "asymptotic behavior" later in this book). Therefore one may expect that the asymptotic properties of the models for (generalized) renewal processes can be derived from the asymptotic behavior of $\{S_n\}$. In turn, it is also to be expected that one can argue vice versa, i.e., that the asymptotic behavior of $\{S_n\}$ is determined by that of $\{N_t\}$. In this case, one would be able to derive statistical properties of the ξ_i's by observing the counting process $\{N_t\}$. So, $\{S_n\}$ and $\{N_t\}$ may be viewed as "dual objects" in a certain sense.

Generally, we call two objects *dual* if their asymptotic properties are related to each other as indicated above, that is, if a limit result for the first object implies a corresponding one for the second object, and vice versa. The duality of $\{N_t\}$ and $\{S_n\}$ has been proved in Gut et al. [160] for the classical setting where all the Assumptions (a)–(c) hold. We would like to mention that Doob [111] predicted this duality property in 1948, but until very recently the limit theorems for renewal processes and those for sums of random variables developed independently.

Counting processes and their dual (or "renewal") processes are often observed in number theory. For example, the total number $\pi(x)$ of prime numbers up to x and the n-th prime number p_n are dual objects, since $\pi(p_n) = n$. This duality is reflected by the *prime number theorem* stating that $\pi(x) \sim x/\ln(x)$ as $x \to \infty$ and the asymptotic behavior $p_n \sim n\ln(n)$ as $n \to \infty$; note that $n\ln(n)$ is inverse to $x/\ln(x)$ in a certain sense.

Probability theory provides other examples of dual objects. One of them is the number μ_n of records until time n in a sequence of random variables and the magnitude τ_n of the n-th record. Obviously $\mu_{\tau_n} = n$ and this duality allows one to study the asymptotic behavior of μ_n and τ_n simultaneously. By the way, the second property of inverse functions fails in both cases since, in general, $\tau_{\mu_n} \neq n$ and $\tau_{\pi(n)} \neq n$.

Another example is the duality between the tail $\overline{F} = 1 - F$ of a distribution function F concentrated on the nonnegative half-line and its quantile function τ_q. Here, if F is continuous and increasing, then both properties of inverse functions hold, i.e., $\overline{F}(\tau_q) = q$ and $\tau_{\overline{F}(x)} = x$. On the other hand, if F is either discontinuous or non-increasing, then each of these may fail. Nevertheless, one can derive the asymptotic behavior of one of these dual objects, either $\overline{F}(x)$ as $x \to \infty$ or τ_q as $q \to 0$, from the other one.

A violation of any of the Assumptions (a)–(c) means in general that the duality between $\{N_t\}$ and $\{S_n\}$ disappears. Since duality properties are important in many situations, one main aim of this monograph is to study conditions under which duality retains, more precisely, under which an almost sure convergence of $S_n/a(n)$ as $n \to \infty$ results in an almost sure convergence of $N(t)/a^{-1}(t)$ as $t \to \infty$, and vice versa, where $a(\cdot)$ is a suitable normalizing function with inverse a^{-1}.

In Klesov et al. [227], for example, such dualities have been proved in a general setting and it became clear that this could be done not only for N_t defined by (1), but also for a large class of generalized renewal processes including the functionals L_t and T_t. The key observation in Klesov et al. [227] was that, in order to preserve duality, the inverse function a^{-1} should satisfy a technical condition, namely

$$\lim_{\varepsilon\downarrow 0} \limsup_{t\to\infty} \left| \frac{a^{-1}((1\pm\varepsilon)t)}{a^{-1}(t)} - 1 \right| = 0. \tag{2}$$

Property (2) is satisfied, for example, if a is a *regularly varying* (RV) function (in the Karamata sense) with nonzero index.

Now, considering (2) as the defining property of a more general class $\mathcal{PRV}$ of functions, called *pseudo-regularly varying* (PRV) functions, and developing its characteristics further, it turns out that the theory of dual objects can be extended. It will be shown that many properties of PRV functions remain the same as in the RV case, including, e.g., an integral representation and uniform convergence properties. Having then developed a theory of PRV functions one is able to study the duality of objects in a unified manner, under which the classical setting corresponds to a particular, still very important example.

The structure of the book has certainly been influenced by the direction we have taken in our investigations of PRV functions, namely, we first obtained some basic applications to classical limit theorems in renewal theory, then we understood the importance of the PRV property (2) in this field and discovered its central role for various other limit theorems, and finally we investigated a number of further applications in different fields. Before we now briefly describe the contents of the book, we would like to mention that the asymptotics for generalized renewal processes studied here are essentially based on the fact that the objects under consideration are inverse to each other (in a certain sense), e.g., $S_{N_t} \approx t$ and $N(S_n) \approx n$, where "$\approx$" has to be given a precise meaning, of course.

We should mention that we have not aimed to touch *all* the interesting aspects and important applications of regular variation theory. Our scope is to add some new material to the various monographs and texts which are already devoted to these topics (see, for example, Bingham et al. [41] for a comprehensive discussion of both the theory and applications of regular variation). For the development of the theory see also the important contributions of Seneta [324] and Geluk and de Haan [148]. An excellent presentation of Tauberian theorems which is heavily related to regularly varying functions has been given in Korevaar [237]. Resnick [300, 302], de Haan and Ferreira [167], and Borovkov and Borovkov [48] discuss various applications of the concept of regular variation in probability theory. A role of regularly varying functions for actuarial and finance mathematics is highlighted in Embrechts et al. [119] and Novak [285]. For recent texts dealing with applications to statistical problems and long range dependence see, e.g., Mikosch [272], Samorodnitsky [316], Solier [340] or Pipiras and Taqqu [292] (see also the interesting discussion in de Haan [166], where applications to currency exchange rates, life span estimation, and sea level data are given).

In Chap. 1, we first assume the classical setting and study equivalences in the strong law of large numbers and the law of the iterated logarithm for sums of independent, identically distributed (iid) random variables and their corresponding renewal processes. As mentioned above, the proofs are essentially based on the property that the renewal process $N = \{N_t\}$ is the generalized inverse function constructed from the sequence $S = \{S_n\}$ of partial sums. Not only equivalence statements can be proved in the case of sums of iid random variables, but also the corresponding moment conditions can be derived from their counterparts for partial sums. Some nontraditional limit results for sums are also studied in this chapter.

Chapter 2 is a continuation of Chap. 1 in a more general setting, where much less is known about the limit properties of the underlying sequences. The random variables studied in this chapter are neither necessarily independent, nor identically distributed, nor nonnegative. We propose some new approaches to derive limit results for generalized renewal processes from their counterparts holding for the underlying sequences. Note that several definitions of generalized renewal processes are introduced in this chapter, where each of them reflects a certain feature of the classical definitions. Among the generalized renewal processes studied in Chap. 2 are the *first exit time*, *last exit time*, and *sojourn time*.

Chapters 3–7 provide the function-theoretic foundations of the book, while the other chapters are devoted to applications and some necessary complements. The following three classes of functions play a key role when studying dual objects in this monograph, namely $\mathcal{PRV}$, the class of *pseudo regularly varying* functions, $\mathcal{SQI}$, the class of *sufficiently quickly increasing* functions, and $\mathcal{POV}$, the class of *pseudo-O-varying* functions. In fact, all three classes appear as natural generalizations of the classical class of RV functions, but other important classes of functions will be considered as well.

In Chap. 4, special attention is paid to absolutely continuous functions (including paths of compound Poisson processes in particular). An interesting link to the property of *elasticity* in mathematical economics is established when studying absolutely continuous PRV functions.

The classes $\mathcal{PRV}$, $\mathcal{SQI}$ and $\mathcal{POV}$ can be characterized in terms of *upper* and *lower limit functions* used in studying RV and (so-called) ORV functions, an extension of the notion of RV functions due to Avakumović and Karamata. A detailed treatment of functions belonging to the classes $\mathcal{PRV}$, $\mathcal{SQI}$ and $\mathcal{POV}$ as well as their quasi-inverses is given in Chaps. 3, 4 and 7.

In Chap. 7, in particular, duality properties are discussed concerning the limit behavior of the ratio of two such functions and the corresponding ratio of their asymptotic quasi-inverses. The latter duality can be established via piecewise linear interpolation. We also study conditions, important in applications, under which generalized inverses are asymptotic inverse or quasi-inverse functions. Moreover, an application of the general results to ordinary differential equations is discussed, more precisely, it is studied how to obtain asymptotic stability of solutions of the Cauchy problem with respect to initial conditions.

The main scope of Chap. 5 is to study the class of ORV functions with *non-degenerate semigroups of regular points* introduced in Buldygin et al. [60]. This class contains the functions for which the limit $\lim_{t\to\infty} f(t\lambda)/f(t)$ exists and belongs to the interval $(0, \infty)$, but not necessarily for all numbers λ. It turns out that the set of such numbers λ is a multiplicative semigroup in $\mathbf{R}_+$. All the functions discussed can be classified according to a given semigroup of regular points, in particular, RV functions correspond to the case where this semigroup coincides with $\mathbf{R}_+$. Analogues of Karamata's theorem (both direct and inverse parts) are discussed in Chap. 6.

In Chaps. 7 and 8, we are able to generalize some results of Chap. 2, which turns out to also explain the key role of condition (2). In fact, it means that the function a^{-1} there belongs to the class $\mathcal{SQI}$ of sufficiently quickly increasing functions. It is worthwhile to mention that our approach provides a unified method to study the asymptotic behavior of renewal processes in both discrete and continuous time. The results of Chap. 8 reveal the nature of SQI, PRV, and POV properties of normalizing sequences and functions, and moreover contain conditions for SQI, PRV, and POV properties that can easily be checked. In addition to general results, Chap. 8 contains a number of applications to various specific schemes for constructing renewal processes. In particular, we prove a strong law of large numbers for generalized renewal processes constructed from nonhomogeneous compound Poisson processes.

Another application of the theory of POV functions is presented in Chap. 9. We obtain the exact order of growth (to infinity) of solutions of autonomous stochastic differential equations. The same method is applied to some other stochastic differential equations. The results of this chapter are rather general extensions of the corresponding theorems in the monograph of Gihman and Skorohod [149].

In Chap. 10, the method of Chap. 2 is also successfully applied to the case of renewal processes constructed from random walks in multidimensional time. Since there is no complete ordering in the space of multi-indices, the definition of a renewal process can only be justified by using an analogue of the process $\{T_t\}$ rather than $\{N_t\}$. The asymptotic behavior of renewal processes in this case differs from the classical one and depends on the asymptotic properties of the Dirichlet function, i.e., the number of solutions of the inequality $n_1 \cdots n_r \leq x$ for positive integers $n_1, \ldots, n_d$, where x is a parameter. Some of our results require a more precise asymptotic for the Dirichlet function than those known so far. In fact, this behavior depends on a confirmation of the Riemann hypothesis concerning the zeros of the ζ-function.

The final Chap. 11 contains another application of PRV functions to the (so-called) precise asymptotics for the complete convergence of sums of stable random variables. Moreover, we give a new proof of Karamata's theorem that the function $W(t) = \sum_{s=1}^{t} w(s)/s$ is slowly varying if this holds true for w.

In a supplementary Appendix, we provide some auxiliary results, which may also be of independent interest and are partially new to the best of our knowledge.

We should finally mention that equivalent versions of the PRV property have been introduced by many authors and that PRV functions have been studied under various different names, e.g., in Korenblyum [235], Matuszewska [267], Matuszewska and Orlicz [270], Gihman and Skorohod [149], Stadtmüller and Trautner [344, 345], Berman [30, 32], Yakymiv [372], Cline [90], Djurčić [105], Djurčić and Torgašev [108], Klesov et al. [227], Buldygin et al. [61], and Rogozin [305], but the latter list is not exhaustive at all. The papers of Korenblyum [235] and Stadtmüller and Trautner [344] deal with nondecreasing PRV functions in the framework of Tauberian theorems for Laplace transforms. In particular, the Tauberian theorem for the Laplace transform of a nondecreasing positive function f holds if and only if $f \in \mathcal{PRV}$ (see [344]). A generalization of the PRV property to the multivariate case has been considered in Yakymiv [372] and, in fact, turns out to be important for the one-dimensional case as well (see also Resnick [301]).

Kyiv, Ukraine — Valeriĭ V. Buldygin
Paderborn, Germany — Karl-Heinz Indlekofer
Kyiv, Ukraine — Oleg I. Klesov
Cologne, Germany — Josef G. Steinebach

Acknowledgements

In a stimulating paper in 1930, Karamata [205] introduced the notion of *regular variation* and proved some fundamental theorems for *regularly varying* (RV) functions (see also Karamata [206]). These results together with later extensions and generalizations turned out to be very fruitful for various fields of mathematics (cf. Seneta [324] and Bingham et al. [41] for excellent surveys on this topic and for the history of its theory and applications). Among the classical generalizations there are, e.g., the (so-called) *O-regularly varying* (ORV) functions and *O-slowly varying* (OSV) functions. We further add to this list the notions of *pseudo-regularly varying* (PRV) functions and functions of *positive order of variation* (POV functions) together with other extensions which can effectively be applied in various mathematical problems.

Our interest in generalizing the notion of Karamata's regular variation was stimulated by some applications to certain asymptotic problems in renewal theory and can be traced back to the very first days of this century. Our starting point was condition (2.23) to be imposed on the required normalizing functions, whose complicated analytical form appeared unsatisfactory to us. So, in the beginning we just treated this condition as a purely technical one, being useful for deriving some equivalence theorems in renewal theory.

Later, we succeeded in improving it to a nicer form (see (3.10)) and we also learned that we were not the first to make use of this condition. In fact, it seems to be a root for several trees of studies of important classes of functions of a real variable.

Our approach to condition (3.10) has changed after an answer of Eugene Seneta to our question about his opinion and possible related results in the literature. He kindly provided us with a preliminary list of references and suggested to ask Tatjana Ostrogorski concerning further results on this topic. Unfortunately, we were not able to take advantage of a possible discussion of the topic with her due to her untimely death.

On the other hand, among other references in Prof. Seneta's list, we learned more about the origin of the topic, in particular of the work of Avakumović [19]. Vojislav Gregor Avakumović (1910–1990) was a late professor at the University of Marburg,

Germany, where some of his students and collaborators are still working. Due to this special coincidence of the topic and the place, where we were working on it, we decided to continue our investigations further. Thanks to the support given by the Deutsche Forschungsgemeinschaft (DFG) this could be realized.

Our interest in the generalization of Karamata's theory of regular variation has expanded over time and new settings and problems have come up ever since. We have worked enthusiastically on the topic and our knowledge of the field became wider and wider. However, we do not claim at all that it is comprehensive in any respect. Moreover, our continuous extension of the list of references indicates clearly that (very likely) some important results and authors are still missing.

The present book is a personal reflection of our joint research work on this subject, mainly done during numerous mutual visits of the authors to Germany and Ukraine. Our intention is to share our knowledge of the generalized Karamata theory and its various implications with all researchers and advanced students in the field. We hope it will be useful for some of them and it may contribute to further developing the theory and its potential applications. On the other hand, we clearly see that neither the theory, nor its applications, are in a final state yet, rather they are still in progress and expanding quickly. The structure of the book reflects the historical development of our approach, that is, first some applications are discussed, after which a basic theory is created, and finally further applications are provided.

We very much enjoyed and still appreciate our academic collaboration and take the opportunity to thank DFG and the Universities of Marburg, Paderborn, and Cologne in Germany and the National Technical University of Ukraine "KPI" in Kiev, Ukraine, for continuous support and help over the years.

Last but not least we should like to thank all reviewers and editors for their very constructive remarks and suggestions, which really helped us to improve an earlier version of this manuscript.

Let us work together in the beautiful field of Karamata's ideas and their extensions and applications!

Kyiv, Ukraine — Valeriĭ V. Buldygin
Paderborn, Germany — Karl-Heinz Indlekofer
Kyiv, Ukraine — Oleg I. Klesov
Cologne, Germany — Josef G. Steinebach

Contents

1 **Equivalence of Limit Theorems for Sums of Random Variables and Renewal Processes** 1
1.1 Introduction 1
1.2 Strong Limit Theorems for Partial Sums 3
1.3 Renewal Processes with Linear Drift 11
1.3.1 Results 11
1.3.2 Proofs 12
1.4 Renewal Processes Without Linear Drift 17
1.4.1 Results 17
1.4.2 Proofs 20
1.5 Comments 24

2 **Almost Sure Convergence of Renewal Processes** 27
2.1 Introduction 27
2.2 Almost Sure Limit Theorems 31
2.2.1 The Strong Law of Large Numbers 31
2.2.2 Non-equivalent Strong Laws 35
2.2.3 Rate of Convergence 38
2.3 Examples 43
2.3.1 Renewal Processes Constructed from Independent, Identically Distributed Random Variables 43
2.3.2 Nonidentically Distributed or Dependent Interarrival Times 48
2.4 Comments 51

3 **Generalizations of Regularly Varying Functions** 53
3.1 Introduction 53
3.2 RV- and ORV-Functions 54
3.2.1 Some Notation 54
3.2.2 Upper and Lower Limit Functions 56
3.2.3 RV- and SV-Functions 57

3.2.4 ORV- and OSV-Functions 59
3.2.5 OURV-Functions 60
3.3 Four Important Classes of Functions 61
3.3.1 PRV-Functions 61
3.3.1.1 Some Simple Properties of WPRV- and PRV-Functions 65
3.3.2 PI- and SQI-Functions 66
3.3.2.1 Some Simple Properties of WSQI- and SQI-Functions 68
3.3.3 POV-Functions 69
3.3.3.1 Some Simple Properties of POV-Functions 69
3.4 Functions Preserving the Asymptotic Equivalence 70
3.4.1 A Uniform Convergence Theorem 75
3.4.2 A Uniform Convergence Theorem for SV- and RV-Functions 75
3.5 Integral Representations 76
3.5.1 Integral Representations for RV- and ORV-Functions 76
3.5.2 Integral Representations for PRV-Functions 77
3.5.3 Integral Representations for POV-Functions 79
3.6 Potter's Bounds for PRV-Functions 80
3.7 Convergence to Infinity of PI-Functions 82
3.7.1 A Counterexample 89
3.8 Characterizations of POV-Functions 91
3.9 Asymptotically Equivalent Monotone Versions of POV-Functions 92
3.10 Comments 95

4 Properties of Absolutely Continuous Functions 99
4.1 Introduction 99
4.2 Some Classes of Absolutely Continuous Functions 100
4.3 Relationships Between Limit Functions 101
4.3.1 Essential Limit Functions 102
4.3.2 Inequalities Between the Limit Functions of f and Its Density θ 103
4.4 Logarithmic Density and Integral Representations 106
4.4.1 Integral Representations for Functions in the Class $\mathbb{DL}$ 106
4.4.2 Logarithmic Densities 107
4.4.3 A Characterization of Properties of Absolutely Continuous Functions 107
4.4.4 Admissible Transformations of Densities 108
4.5 Elasticity and Indices of Asymptotic Elasticity 111
4.5.1 Elasticity of Functions 111
4.5.2 Indices of Asymptotic Elasticity 113
4.5.3 Essential Indices of Asymptotic Elasticity 114

4.5.4 Examples ... 115
4.5.5 Arithmetic Properties of Indices of Asymptotic Elasticity ... 118
4.6 Relationships in Terms of the Indices of Asymptotic Elasticity ... 119
4.7 Functions with Bounded Asymptotic Elasticity ... 121
4.7.1 Admissible Transformations of Densities ... 122
4.7.2 The ER-Property of Absolutely Continuous Functions ... 123
4.7.3 Boundaries of the Indices of Asymptotic Elasticity for Almost Monotone Densities ... 125
4.8 The PRV-Property of Absolutely Continuous Functions ... 130
4.8.1 Necessary PRV-Properties ... 130
4.8.2 The Measurability of Limit Functions of Densities and the ER- (PRV)-Property of Absolutely Continuous Functions ... 134
4.9 The SQI-Property of Absolutely Continuous Functions ... 135
4.9.1 Necessary Conditions for the PI-Property ... 136
4.9.2 Sufficient SQI-Conditions ... 138
4.10 The POV-Property of Absolutely Continuous Functions ... 139
4.11 Properties of Piecewise Linear Interpolations ... 141
4.11.1 Properties of Piecewise Linear Interpolations of Positive Sequences ... 141
4.11.2 Properties of Positive Piecewise Linear Functions ... 148
4.12 Comments ... 150

5 Nondegenerate Groups of Regular Points ... 153
5.1 Introduction ... 153
5.2 Groups of Regular Points ... 154
5.2.1 Regular Points and Groups of Regular Points ... 154
5.2.2 Full Sets of Regular Points for RV- (WRV)-Functions ... 156
5.2.3 Characterization Theorems ... 156
5.2.4 Null Sets of Regular Points ... 158
5.2.5 An Arbitrary Group of Regular Points ... 160
5.2.6 A Degenerate Group of Regular Points ... 161
5.3 Regularly Periodic and Regularly Log-Periodic Functions ... 161
5.3.1 Definitions and Some Properties of Regularly Periodic and Regularly Log-Periodic Functions ... 161
5.3.2 Sets of Regular Points for Regularly Log-Periodic Functions ... 166
5.4 $*$-Invariant Limit Functions ... 168
5.5 Weak Factorization Representations for Limit Functions ... 172
5.5.1 Some Corollaries and Remarks ... 174

5.6 Strong Factorization Representations for Limit Functions 175
5.6.1 Some Corollaries and Remarks 179
5.6.2 Factorization Representations of Limit Functions for ORV-Functions with Nondegenerate Sets of Regular Points 182
5.7 Characterization Theorems 182
5.8 Factorization Representations of Functions 186
5.9 Uniform Convergence Theorems 193
5.10 Comments 198

6 Karamata's Theorem for Integrals 201
6.1 Introduction 201
6.2 Karamata's Theorem for Integrals of RV-Functions 202
6.2.1 Karamata's Theorem: Direct Part 202
6.2.2 Karamata's Theorem: Converse Part 203
6.3 Regularly Log-Periodic and Regularly Log-Bounded Functions 203
6.3.1 Regularly Log-Bounded Functions 205
6.4 Generalizations of Karamata's Theorem 206
6.4.1 Direct Part: The Case $\rho \neq -1$ 206
6.4.2 Direct Part: The Case $\rho = -1$ 207
6.4.3 Converse Part 209
6.5 Asymptotic Properties of Integrals 210
6.6 An Auxiliary Result 215
6.7 Proof of the Generalized Karamata Theorem 219
6.7.1 Proof of the Converse Part of the Generalized Karamata Theorem 219
6.7.2 The Proof of the Generalized Karamata Theorem, Direct Part 226
6.8 Comments 227

7 Asymptotically Quasi-inverse Functions 229
7.1 Introduction 229
7.2 Asymptotically Quasi-inverse and Asymptotically Inverse Functions 232
7.2.1 Generalized Inverse Functions 233
7.2.2 Quasi-inverse Functions 233
7.2.3 Asymptotically Quasi-inverse and Asymptotically Inverse Functions 235
7.2.4 Asymptotically Quasi-inverse Functions for PRV-Functions 237
7.2.5 Asymptotically Inverse Functions for POV-Functions 241
7.2.6 Asymptotically Right and Asymptotically Left Quasi-inverse Functions 243
7.3 Properties of Asymptotically Quasi-inverse Functions 244

7.4 Asymptotically Quasi-inverse Functions and the POV-Property ... 249
7.5 Properties of a Ratio of Asymptotically Quasi-inverse Functions ... 254
7.5.1 A General Approach to the Proof ... 255
7.5.2 The Limit Behavior of the Ratio of Functions Which Are Asymptotically Quasi-inverses for SQI-Functions ... 256
7.5.3 A Characterization of SQI- and PRV-Functions ... 259
7.5.4 The Limit Behavior of the Ratio of Functions Which Are Asymptotically Quasi-inverse for POV-Functions ... 260
7.5.5 Zero and Infinite Limit Points of the Ratio of Asymptotically Quasi-inverse Functions ... 264
7.5.6 Zero and Infinite Limit Points of the Ratio of Functions Which Are Asymptotically Quasi-inverse for POV-Functions ... 266
7.5.7 The Limit Behavior of the Ratio of Functions Which Are Asymptotically Quasi-inverse for RV-Functions ... 269
7.6 Limit Points of the Ratio of Asymptotically Quasi-inverse Functions ... 270
7.6.1 Necessary Conditions for Relation (7.44) ... 275
7.7 Duality Between Certain Classes of Functions ... 276
7.7.1 Some Auxiliary Results ... 276
7.7.2 The Images of Some Classes of Functions Under the Transformations $f \mapsto f^{\sim}$ and $f^{\sim} \mapsto f$ ... 278
7.7.3 Proof of Theorems 7.73–7.78 ... 279
7.7.4 Equivalence Theorems for Asymptotically Left Quasi-inverse Functions ... 286
7.8 Generalized Inverse Functions Corresponding to Sequences ... 287
7.8.1 The Limit Behavior of the Ratio of Asymptotically Quasi-inverse Functions Constructed from Sequences ... 290
7.8.2 Piecewise Linear Interpolations of Sequences and Functions ... 291
7.8.3 Generalized Inverse Functions for Piecewise Linear Interpolations of Sequences ... 294
7.8.4 The Limit Behavior of the Ratio of Asymptotically Quasi-inverse Functions for Sequences ... 295
7.9 Stability of Solutions of Differential Equations ... 301
7.9.1 Solutions of the Cauchy Problem for an Autonomous Differential Equation ... 301

7.9.2 Asymptotic Stability of a Solution of the Cauchy Problem with Respect to the Initial Condition for an Autonomous Differential Equation 303
7.9.3 Asymptotic Stability of Solutions of the Cauchy Problem with Respect to the Initial Condition for a Non-autonomous Ordinary Differential Equation 307
7.10 Comments 308

8 Generalized Renewal Processes 311
8.1 Introduction 311
8.2 Generalized Renewal Processes Constructed from Random Sequences 314
8.2.1 The SLLN for Random Sequences and the Corresponding Renewal Processes 315
8.3 Examples 320
8.3.1 Generalized Renewal Processes Constructed from Stationary Sequences.................................... 320
8.3.2 The Asymptotic Behavior of the First Exit Time from a Domain with a Curvilinear Boundary 322
8.3.3 Generalized Renewal Processes Constructed from Markovian Gaussian Sequences Under Strong Dependence... 324
8.3.4 Generalized Renewal Processes Constructed from Sums of Independent Random Variables................ 326
8.4 Renewal Processes Constructed from Stochastic Processes 327
8.4.1 The SLLN for Stochastic Processes and Their Corresponding Renewal Processes 329
8.4.2 The SLLN for Stochastic Processes and Their Corresponding Renewal Processes Under WPRV Normalizations 331
8.4.3 The SLLN for Stochastic Processes and Their Corresponding Renewal Processes Under POV Normalizations 332
8.4.4 The SLLN for Stochastic Processes and Their Corresponding Renewal Processes Under RV Normalizations 333
8.5 Further Examples 334
8.5.1 Generalized Renewal Processes Constructed from Compound Counting Processes 336
8.5.2 Generalized Renewal Processes Constructed from Compound Poisson Processes 338
8.5.3 Generalized Renewal Processes Constructed from Compound Poisson Processes with Piecewise Constant Intensity Functions 340
8.6 Comments 341

9 Asymptotic Behavior of Solutions of Stochastic Differential Equations ... 345
9.1 Introduction ... 345
9.1.1 Gihman and Skorohod's Setting and Approach ... 347
9.1.2 Organization of the Current Chapter ... 349
9.2 Order of Growth of Solutions of Stochastic Differential Equations ... 350
9.2.1 Auxiliary Functions and Their Properties ... 351
9.2.2 Main Results ... 355
9.2.3 The Exact Order of Growth of the Process $G_a \circ \tilde{X}_b$ ... 356
9.2.4 Some Auxiliary Results ... 357
9.2.5 Proof of the Main Results ... 362
9.2.6 A Discussion of the Main Results ... 364
9.2.7 Asymptotic Stability of Solutions of Stochastic Differential Equations with Respect to Initial Conditions ... 365
9.3 Renewal Processes Constructed from Solutions of Stochastic Differential Equations ... 366
9.4 Order of Growth of Solutions of Autonomous Stochastic Differential Equations ... 369
9.4.1 Renewal Processes Constructed from Solutions of Autonomous Stochastic Differential Equations ... 373
9.5 φ-Order of Growth of Solutions of Autonomous Stochastic Differential Equations ... 373
9.6 $\varphi_{1,2}$-Equivalence of Solutions of Autonomous Ordinary Differential Equations ... 378
9.6.1 Conditions for the $\varphi_{1,2}$-Asymptotic Equivalence of Solutions of Ordinary Differential Equations in Terms of the Functions $G_k^{(\varphi_k)}$ ... 380
9.6.2 Conditions for the $\varphi_{1,2}$-Asymptotic Equivalence of Solutions of Ordinary Differential Equations in Terms of the Functions $g_k^{(\varphi_k)}$ ... 381
9.6.3 An Application of Karamata's Theorem ... 382
9.7 $\varphi_{1,2}$-Order of Growth of Solutions of Autonomous Stochastic Differential Equations ... 384
9.8 $\varphi_{1,2}$-Equivalence of Solutions of Autonomous Stochastic Differential Equations ... 387
9.9 Comments ... 391

10 Asymptotics for Renewal Processes Constructed from Multi-indexed Random Walks ... 395
10.1 Introduction ... 395
10.2 Some Classical Results from Renewal Theory ... 397
10.3 Limit Theorems for Multiple Sums ... 398
10.4 The Renewal Function for Multi-indexed Random Walks ... 400

10.5 Renewal Processes Constructed from Multi-indexed Random Walks ... 402
10.6 Asymptotics for Renewal Functions Constructed from Random Walks with Multidimensional Time ... 403
10.7 Asymptotics for Renewal Processes Constructed from Random Walks with Multidimensional Time ... 405
10.8 Examples ... 408
10.9 The Leading Coefficient in the Dirichlet Divisor Problem ... 411
10.10 An Example of Kátai ... 414
10.11 Comments ... 416

11 Spitzer Series and Regularly Varying Functions ... 419
11.1 Introduction ... 419
11.2 The Asymptotic Behavior of Large Deviation Probabilities ... 423
11.3 Proof of Theorem 11.1 ... 433
11.4 Comments ... 438

A Some Auxiliary Results ... 439
A.1 Introduction ... 439
A.2 The Asymptotic Behavior of Series of Weighted Tails of Distributions ... 440
A.3 Large Deviations in the Case of Attraction to a Stable Law ... 443
A.4 Some Auxiliary Results on Slowly Varying Functions ... 449
A.5 Distributions Attracted to Stable Laws ... 452
A.6 The Asymptotic Behavior of Integrals and Sums with Slowly Varying Functions ... 458
A.7 The Parameswaran Lemma for Sums ... 459
A.8 Comments ... 460

References ... 463

Index ... 479

Chapter 1
Equivalence of Limit Theorems for Sums of Random Variables and Renewal Processes

1.1 Introduction

Many limit results are known for cumulative sums of independent identically distributed random variables and the corresponding renewal counting processes to hold under the same conditions. Despite the coincidence of conditions for both cases, the theories have been developed independently of each other and the methods are different.

The well-known Kolmogorov strong law of large numbers for sums of independent, identically distributed random variables holds if the first moment exists. Proved by Doob almost 20 years later, the strong law of large numbers for the renewal counting process constructed from the sequence of cumulative sums holds under the same condition. Considering another classical limit theorem for cumulative sums, the law of the iterated logarithm, we see a similar phenomenon, namely that both laws of the iterated logarithms for cumulative sums and the corresponding renewal process are valid under the same assumption that the second moment exists. We also recall a third classical result, the central limit theorem, where the conditions are the same for sums and renewal processes.

So, the natural question arises as to whether or not the two results in any pair are equivalent. If yes, then one can automatically extend the statements for renewal processes by adding the necessity part known for the limit theorems for cumulative sums. A general question is whether or not the whole theories of limit theorems for sums and renewal processes are equivalent.

In this chapter, we consider the problem of the equivalence of certain asymptotic results, like the strong law of large numbers or the law of the iterated logarithm, for a sequence of sums of independent, identically distributed random variables and its corresponding renewal process. In the proofs, we consistently apply a "duality" property which links these two stochastic processes, namely we make use of the fact that the renewal process is the generalized inverse function to the underlying

V. V. Buldygin et al., *Pseudo-Regularly Varying Functions and Generalized Renewal Processes*, Probability Theory and Stochastic Modelling 91, https://doi.org/10.1007/978-3-319-99537-3_1

sequence of partial sums. Theorem 1.21 below is an example of such an equivalence result, where we prove that (1.44) and (1.45) are imply each other. In further chapters, similar equivalences will be proved for more general cases, based on certain facts from the theory of regularly varying functions and their generalizations, which are developed first. In the present chapter, we not only prove equivalence statements, but also find necessary and sufficient conditions in a particular, yet important, case.

Let $\{X_n\}_{n\geq 1}$ be a sequence of *nonnegative*, independent and identically distributed (i.i.d.) random variables, and let $S_n = \sum_{k=1}^n X_k$, $n \geq 1$, $S_0 = 0$, denote their corresponding partial sums (or *renewal sequence*). The sequence $\{S_n\}_{n\geq 1}$ is also called a *random walk*. The *renewal counting process* or simply *renewal process* $\{L(t)\}_{t\geq 0}$ with respect to $\{S_n\}_{n\geq 0}$ is defined as

$$L(t) = \max\{n \geq 0\colon S_n \leq t\}, \qquad t \geq 0. \tag{1.1}$$

A possible interpretation is that X_n represents the time between the $(n-1)$-th and n-th replacement (renewal) of a machine part (say), so that $L(t)$ counts the number of replacements (renewals) up to time t.

If the definition of the renewal process is changed slightly, then the analytic relation between these stochastic processes may become clearer. Namely, if

$$S^{\leftarrow}(t) = \min\{n \geq 1\colon S_n > t\}, \qquad t \geq 0,$$

then, according to the commonly accepted terminology, $S^{\leftarrow}(t)$ is the generalized inverse function to the sequence $\{S_n\}$. Obviously, $S^{\leftarrow}(t) = L(t) + 1$, thus the question on the asymptotic behavior of the renewal process L is equivalent to that for the function $S^{\leftarrow}$.

If the distribution of X_1 is nondegenerate in 0, i.e. $\mathsf{P}(X_1 = 0) = p < 1$ (which will be assumed throughout), then it is immediate from the nonnegativity of the X_n's and the Borel–Cantelli lemma that

$$\lim_{n\to\infty} S_n = +\infty \qquad \text{a.s.}, \tag{1.2}$$

$$\mathsf{P}(L(t) < \infty) = 1 \qquad \forall\, t \geq 0, \tag{1.3}$$

$$\lim_{t\to\infty} L(t) = +\infty \qquad \text{a.s.}, \tag{1.4}$$

where "a.s." stands for "almost surely".

Many of the classical limit theorems have been carried over from $\{S_n\}$ to $\{L(t)\}$ (cf., e.g., Gut [159]). In this chapter it will be proved that the sufficient moment conditions used there are, in fact, also necessary. This will be achieved by showing that many strong limit theorems, e.g. the strong law of large numbers (SLLN), the Marcinkiewicz–Zygmund strong law, or the law of the iterated logarithm (LIL), hold simultaneously for the sequence $\{S_n\}$ and the process $\{L(t)\}$. So, necessary and sufficient moment conditions, which are well-known for $\{S_n\}$, carry over to the renewal process $\{L(t)\}$, too. Therefore the behavior of the sequence $\{S_n\}$ determines that of its generalized inverse function $S^{\leftarrow}$ and vice versa.

Indeed, the results follow from the fact that the renewal sequence and process are inverses of each other. Hence it is natural to believe that strong (or weak) laws should hold simultaneously, and that even their normalizations should be inverses of each other. An inspection of the proofs below actually shows that the results, in fact, hold simultaneously for arbitrary positive summands and for (almost) arbitrary nonnegative summands. We state the results for one such case and leave the obvious analogs to the reader.

We also consider the case of renewal processes with infinite expectation. In this case, $L(t)/t \to 0$ a.s. as $t \to \infty$, i.e. the renewal process has "no linear drift". It turns out that there is an essential difference in the strong limiting behavior of $\{L(t)\}$ in the cases $\mathsf{E}X_1 < \infty$ and $\mathsf{E}X_1 = \infty$, respectively.

This chapter is organized as follows. In Sect. 1.2, we first collect some well-known facts about the strong limiting behavior of partial sums, which are essential preliminaries for the proofs of the equivalence statements given in Sects. 1.3 and 1.4.

In Sect. 1.3, we discuss renewal processes having "a linear drift". Along with known results we state several assertions on the convergence to infinity of normalized sums of random variables.

Section 1.4 treats renewal processes with infinite expectation ("no linear drift").

1.2 Strong Limit Theorems for Partial Sums

The proofs of our results in Sects. 1.3 and 1.4 make use of some classical results for partial sums. For the sake of completeness, these will be given next as the essential preliminaries.

Most of the results of this section are valid for general summands $\{X_n\}_{n\geq 1}$ and we do not assume that they are nonnegative. Nevertheless, some assertions need the nonnegativity which is explicitly mentioned in these cases. The proofs may be found in Petrov [291] if they are not given below. We assume throughout that $\{X_n\}_{n\geq 1}$ are i.i.d. (not necessarily nonnegative) random variables.

Proposition 1.1 (Kolmogorov's SLLN) *The strong law of large numbers*

$$\lim_{n\to\infty} \frac{S_n}{n} = \mu \qquad a.s. \quad \textit{for some} \quad \mu \in \mathbf{R}$$

holds if and only if

$$\mathsf{E}X_1 = \mu \quad \textit{for some} \quad \mu \in \mathbf{R}.$$

Proposition 1.1 can be completed by a result on the infinite limit in the case of nonnegative terms.

Proposition 1.2 *Let $X_1 \ge 0$ a.s. Then the relation*

$$\lim_{n\to\infty} \frac{S_n}{n} = +\infty \qquad a.s. \tag{1.5}$$

holds if and only if

$$\mathsf{E}X_1 = \infty. \tag{1.6}$$

Proof Condition (1.6) follows from (1.5) in view of the nonnegativity and Proposition 1.1. To prove the converse implication let $c > 0$ and $X_n^c = X_n \mathbb{I}_{\{X_1 < c\}}$. Put $\mu_c = \mathsf{E}X_1^c$. We obtain from Proposition 1.1 that

$$\liminf_{n\to\infty} \frac{S_n}{n} \ge \lim_{n\to\infty} \frac{X_1^c + \cdots + X_n^c}{n} = \mu_c \qquad \text{a.s.} \tag{1.7}$$

If (1.6) holds, then $\mu_c \to \infty$ as $c \to \infty$. Since $c > 0$ in (1.7) is arbitrary, (1.5) follows immediately. □

Proposition 1.3 (Marcinkiewicz–Zygmund SLLN) *Let $0 < r < 1$. The Marcinkiewicz–Zygmund strong law of large numbers*

$$\lim_{n\to\infty} \frac{S_n}{n^{1/r}} = 0 \qquad a.s.$$

holds if and only if

$$\mathsf{E}|X_1|^r < \infty.$$

Let $1 \le r < 2$. The Marcinkiewicz–Zygmund strong law of large numbers

$$\lim_{n\to\infty} \frac{S_n - n\mu}{n^{1/r}} = 0 \qquad a.s. \quad \textit{for some} \quad \mu \in \mathbf{R}$$

holds if and only if

$$\mathsf{E}X_1 = \mu, \qquad \mathsf{E}|X_1|^r < \infty \quad \textit{for some} \quad \exists\, \mu \in \mathbf{R}.$$

The following result is similar to the Marcinkiewicz–Zygmund SLLN but for infinite limits.

Proposition 1.4 *Assume that $X_1 \ge 0$ a.s.*

a) Let $0 < r \le 1$. In order that

$$\lim_{n\to\infty} \frac{S_n}{n^{1/r}} = +\infty \qquad a.s. \tag{1.8}$$

it is necessary and sufficient that

$$\sum_{n=1}^{\infty} \frac{1}{n} \mathsf{P}(S_n \le c n^{1/r}) < \infty \qquad \textit{for all} \qquad 0 < c < \infty. \tag{1.9}$$

If (1.8) holds, then

$$\mathsf{E} X_1^r = \infty. \tag{1.10}$$

b) *Let* $r > 1$. *If the distribution of the random variable* X_1 *is not concentrated in* 0, *that is, if* $\mathsf{P}(X_1 > 0) > 0$, *then (1.8) holds.*

Remark 1.5 In contrast to the case when $r \le 1$, assertion (1.8) holds in the case when $r > 1$ even if $\mathsf{E} X_1^r < \infty$.

Remark 1.6 If $r = 1$, then (1.8) and (1.10) are equivalent (see Proposition 1.2). This is not the case, however, for $0 < r < 1$. A counterexample for the converse can be found in Erickson [123], Theorem 5. Let the random variables X_i be nonnegative and the tail of the distribution be such that

$$\lim_{t \to \infty} \mathsf{P}(X_i > t) / t^{-r} (\log \log t)^{1-r} = c$$

for some constants $0 < c < \infty$ and $0 < r < 1$. Then

$$0 < \liminf_{n \to \infty} \frac{S_n}{n^r} < +\infty \qquad \text{a.s.},$$

which contradicts (1.8). Therefore (1.10) does not imply (1.8) if $r < 1$.

Proof of Proposition 1.4 The implication (1.9) $\Longrightarrow$ (1.8) of part a) follows from Lemma 1.7 below.

Now we prove the implication (1.8) $\Longrightarrow$ (1.9) of part a). If (1.9) is not satisfied for all $0 < c < \infty$ then there exists $0 < c_0 < \infty$ such that

$$\sum_{n=1}^{\infty} \frac{1}{n} \mathsf{P}(S_n \le c_0 n^r) = \infty.$$

Hence

$$\sum_{k=0}^{\infty} \mathsf{P}(n^{-r} S_n \in (a, b) \text{ for some } n \in [2^k, 2^{k+1})) = \infty$$

with $a < 0$ and $b > c_0 2^r$, since $X_i \geq 0$ a.s. and

$$\sum_{n=1}^{\infty} \frac{1}{n} \mathsf{P}(S_n \leq c_0 n^r) \leq \sum_{k=0}^{\infty} \mathsf{P}(S_{2^k} \leq c_0 2^{(k+1)r})$$

$$\leq \sum_{k=0}^{\infty} \mathsf{P}(n^{-r} S_n \leq c_0 2^r \text{ for some } n \in [2^k, 2^{k+1})).$$

Therefore, the Binmore–Katz lemma (see Lemma 1.8 below) implies that, for all $\varepsilon > 0$,

$$\mathsf{P}(n^{-r} S_n \in (a - \varepsilon, b + \varepsilon) \text{ i.o.}) = 1,$$

which contradicts (1.8). This proves the necessity of (1.9).

Condition (1.10) can be proved by the following reasoning. Let (1.8) hold. Then (1.9) also holds in view of part a). If $\mathsf{E} X_1^{1/r} < \infty$ were satisfied, then we would have

$$\lim_{n\to\infty} \frac{S_n}{n^r} = 0 \qquad \text{a.s.}$$

for any of the cases $r < 1$ or $r \geq 1$ according to the Marcinkiewicz–Zygmund SLLN (Proposition 1.3). Then

$$\mathsf{P}(S_n \leq cn^r) \to 1, \qquad n \to \infty,$$

for all positive c, which contradicts (1.9).

To prove part b) consider $c > 0$, define random variables $X_n^c = X_n \mathbb{I}_{\{X_n < c\}}$, and put $\mu_c = \mathsf{E} X_1^c$. Then $\mu_c > 0$ for sufficiently large $c > 0$. Proposition 1.3 implies that

$$\lim_{n\to\infty} \left[\frac{X_1^c + \cdots + X_n^c}{n^{1/r}} - \mu_c n^{1-\frac{1}{r}} \right] = 0 \qquad \text{a.s.},$$

whence $(X_1^c + \cdots + X_n^c)/n^{1/r} \to \infty$ a.s. This proves (1.8), since $X_n \geq X_n^c$ for all $n \geq 1$. □

Lemma 1.7 *Let $\{S_n, n \geq 1\}$ be a nondecreasing sequence of nonnegative random variables and let $\{b_n, n \geq 1\}$ be a nondecreasing sequence of positive constants such that*

$$b_{2n} = O(b_n). \tag{1.11}$$

If

$$\sum_{n=1}^{\infty} \frac{1}{n} \mathsf{P}(S_n \leq cb_n) < \infty \tag{1.12}$$

for all $0 < c < \infty$, *then*

$$\lim_{n\to\infty} \frac{S_n}{b_n} = \infty \qquad a.s. \tag{1.13}$$

Proof of Lemma 1.7 By (1.11), there exists a constant $M < \infty$ such that $b_{2n} \leq M b_n$ for all $n \geq 1$. Now, for $0 < c < \infty$,

$$\begin{aligned}
\infty &> \sum_{k=0}^{\infty} \sum_{n=2^k+1}^{2^{k+1}} \frac{1}{n} \mathsf{P}\left(S_n \leq c b_n\right) \\
&\geq \sum_{k=0}^{\infty} \frac{1}{2^{k+1}} \mathsf{P}\left(S_{2^{k+1}} \leq c b_{2^k+1}\right) (2^{k+1} - 2^k) \\
&\geq \frac{1}{2} \sum_{k=0}^{\infty} \mathsf{P}\left(S_{2^{k+1}} \leq \frac{c}{M} b_{2^{k+1}}\right),
\end{aligned}$$

so, since c is arbitrary, we have

$$\sum_{k=0}^{\infty} \mathsf{P}\left(S_{2^k} \leq c b_{2^k}\right) < \infty$$

for all $0 < c < \infty$. Then, by the Borel–Cantelli lemma,

$$\mathsf{P}\left(S_{2^k} \leq c b_{2^k} \text{ i.o.}\right) = 0$$

for all $0 < c < \infty$. Hence

$$\liminf_{k\to\infty} \frac{S_{2^k}}{b_{2^k}} \geq c \qquad \text{a.s.}$$

for all $0 < c < \infty$. Since c is arbitrary,

$$\lim_{k\to\infty} \frac{S_{2^k}}{b_{2^k}} = \infty \qquad \text{a.s.}$$

Then, for $2^{k-1} < n \leq 2^k$, recalling that $0 \leq S_n \uparrow$,

$$\frac{S_n}{b_n} \geq \frac{S_{2^{k-1}}}{b_{2^k}} \geq \frac{S_{2^{k-1}}}{M b_{2^{k-1}}} \to \infty \qquad \text{a.s.}$$

as $n \to \infty$. □

Lemma 1.8 (Binmore–Katz Type Result) *Let a sequence of real numbers* $\{\gamma(n)\}$ *be such that* $\gamma(n) > 0$, $n \ge 1$, *and* $\gamma(n) \to \infty$ *as* $n \to \infty$. *Assume that*

$$\lim_{\eta \downarrow 0} \sup_{1 \le n_1 \le \eta n_2} \frac{\gamma(n_1)}{\gamma(n_2)} = 0, \qquad \lim_{\eta \downarrow 0} \sup_{1 \le n_1 \le n_2 \le (1+\eta) n_1} \left| \frac{\gamma(n_2)}{\gamma(n_1)} - 1 \right| = 0.$$

If $D > 1$, $\varepsilon > 0$, *and* $-\infty < a < b < \infty$ *are fixed, then (1.14) implies (1.15) and (1.15) implies (1.16), where*

$$\mathsf{P}(\gamma(n)^{-1} S_n \in (a, b) \text{ i.o. }) = 1, \tag{1.14}$$

$$\sum_{k=0}^{\infty} \mathsf{P}(\gamma(n)^{-1} S_n \in (a, b) \text{ for some } n \in [D^k, D^{k+1})) = \infty, \tag{1.15}$$

$$\mathsf{P}(\gamma(n)^{-1} S_n \in (a - \varepsilon, b + \varepsilon) \text{ i.o. }) = 1. \tag{1.16}$$

Here "i.o." stands for "infinitely often".

Note also that the conditions imposed on the sequence $\{\gamma(n)\}$ are satisfied for both $\gamma(n) = n^r$, $r > 0$, and $\gamma(n) = n^r (\log \log n)^v$, $r > 0$, $v \in \mathbf{R}$.

Remark 1.9 The second assumption of Lemma 1.8 means in fact that the sequence $\{\gamma(n)\}$ possesses the PRV property (cf. Theorem 3.47 on the uniform convergence of PRV functions).

The following result is useful in many asymptotic problems concerning sums of independent, identically distributed random variables.

Lemma 1.10 (Spitzer's Theorem) *Let* $\{X_n\}_{n \ge 1}$ *be independent, identically distributed random variables.*

a) If for some $\mu \in \mathbf{R}$ *and all* $\varepsilon > 0$

$$\sum_{n=1}^{\infty} \frac{1}{n} \mathsf{P}(|S_n - n\mu| \ge n\varepsilon) < \infty, \tag{1.17}$$

then

$$\mathsf{E}X_1 \quad \textit{exists} \quad \textit{and} \quad \mathsf{E}X_1 = \mu. \tag{1.18}$$

b) If condition (1.18) holds, then condition (1.17) holds for all $\varepsilon > 0$.

Proposition 1.11 (LIL) *The law of the iterated logarithm*

$$\limsup_{n \to \infty} \frac{S_n - n\mu}{\sqrt{2n \log \log n}} = \sigma \qquad \text{a.s.} \quad \text{for some} \quad \mu \in \mathbf{R},\ \sigma > 0$$

holds if and only if

$$\mathsf{E}X_1 = \mu, \qquad \mathsf{E}(X_1 - \mu)^2 = \sigma^2 < \infty \quad \textit{for some} \quad \mu \in \mathbf{R},\ \sigma^2 > 0. \tag{1.19}$$

By symmetry, the latter conditions are also equivalent to

$$\liminf_{n\to\infty} \frac{S_n - n\mu}{\sqrt{2n \log\log n}} = -\sigma \qquad \text{a.s.} \quad \text{for some} \quad \mu \in \mathbf{R},\ \sigma > 0.$$

Proposition 1.12 *Assume that $X_1 \geq 0$ almost surely. In order that*

$$\liminf_{n\to\infty} \frac{S_n \log\log S_n}{n^2} = c_0 \qquad \textit{a.s.} \quad \textit{for some} \quad 0 < c_0 \leq \infty \tag{1.20}$$

it is necessary and sufficient that there exists a constant $0 < c < \infty$ such that

$$\sum_{n=27}^{\infty} \frac{1}{n} \mathsf{P}\left(S_n \leq c \frac{n^2}{\log\log n} \right) < \infty. \tag{1.21}$$

Remark 1.13 It can easily be seen from the proof of Proposition 1.12 that the constant c_0 in (1.20) is determined as follows:

$$c_0 = \sup\{0 < c < \infty\colon\ (1.21) \text{ holds}\},$$

i.e. (1.20) holds with $c_0 = \infty$ if and only if (1.21) holds for all $0 < c < \infty$ (cf. also Remark 1.31b below).

Proof of Proposition 1.12 Set $g(x) = x \log\log x$ for $x > e^e$ and $g(x) = x$ for $x \leq e^e$. Let $f(x) = g^{-1}(x)$ be the inverse function to g. It is clear that

$$f(x) \sim \frac{x}{\log\log x}, \qquad x \to \infty.$$

We first prove that (1.20) implies (1.21). If, on the contrary, for any $0 < c < \infty$,

$$\sum_{n=3}^{\infty} \frac{1}{n} \mathsf{P}\left(S_n \leq c \frac{n^2}{\log\log n} \right) = \infty,$$

then, for any $0 < c < \infty$, also

$$\sum_{k=0}^{\infty} \mathsf{P}\left(S_{2^k} \leq c f(2^{2k}) \right) = \infty, \tag{1.22}$$

since

$$\sum_{n=32}^{\infty} \frac{1}{n} \mathsf{P}\left(S_n \le c\frac{n^2}{\log\log n}\right) = \sum_{k=5}^{\infty} \sum_{2^k \le n < 2^{k+1}} \frac{1}{n} \mathsf{P}\left(S_n \le c\frac{n^2}{\log\log n}\right)$$
$$\le \sum_{k=5}^{\infty} \mathsf{P}\left(S_{2^k} \le c\frac{2^{2(k+1)}}{\log\log 2^k}\right).$$

Condition (1.22), however, implies that, for any $0 < c < \infty$,

$$\sum_{k=0}^{\infty} \mathsf{P}\left(\frac{S_n}{f(n^2)} \in (0, c) \text{ for some } n \in [2^k, 2^{k+1})\right) = \infty.$$

Using the above Binmore–Katz result (namely, the implication (1.15) $\Longrightarrow$ (1.14) of Lemma 1.8 for $a = 0, b = c$, and $D = 2$) we obtain

$$\mathsf{P}\left(\frac{S_n}{f(n^2)} \in (0, c) \text{ i.o.}\right) = 1,$$

whence $S_n/c < f(n^2)$ i.o. with probability 1 or

$$\mathsf{P}\left(n^{-2} g\left(\frac{S_n}{c}\right) < 1 \text{ i.o.}\right) = 1.$$

Since $c > 0$ is arbitrary, this implies, in particular, that

$$\liminf_{n\to\infty} \frac{g(S_n)}{n^2} = 0 \qquad \text{a.s.}$$

This contradicts (1.20) and proves (1.21).

Now we prove that (1.21) implies (1.20). From (1.21) it follows that

$$\sum_{n=32}^{\infty} \frac{1}{n} \mathsf{P}\left(S_n \le c\frac{n^2}{\log\log n}\right) \ge \sum_{k=5}^{\infty} \sum_{n=2^k}^{2^{n+1}-1} \frac{1}{n} \mathsf{P}\left(S_n \le c\frac{n^2}{\log\log n}\right)$$
$$\ge \sum_{k=5}^{\infty} \frac{2^{k+1} - 2^k}{2^{k+1}} \mathsf{P}\left(S_{2^{k+1}} \le c\frac{2^{2k}}{\log\log 2^{k+1}}\right)$$
$$\ge \frac{1}{2} \sum_{k=5}^{\infty} \mathsf{P}\left(S_{2^k} \le c_1 f(2^{2k})\right) < \infty$$

for some $0 < c_1 < \infty$. Since $g(cx)/g(x) \to c$ as $x \to \infty$ for all $c > 0$,

$$\sum_{k=5}^{\infty} \mathsf{P}\left(g(S_{2^k}) \le c_2 2^{2k}\right) < \infty \quad \text{for some} \quad 0 < c_2 < \infty.$$

On observing that $2^{-2k} g(S_{2^k}) \le 4n^{-2} g(S_n)$ for $2^k \le n < 2^{k+1}$, the Borel–Cantelli lemma implies

$$\liminf_{n\to\infty} \frac{S_n \log\log S_n}{n^2} \ge \frac{c_2}{4} \qquad \text{a.s.},$$

which completes the proof. □

1.3 Renewal Processes with Linear Drift

1.3.1 Results

Under the assumptions of Sect. 1.1, the following equivalence statements hold true. Note that in the theorems of this section, we only consider the case $0 < \mathsf{E}X_1 < \infty$. The case $\mathsf{E}X_1 = +\infty$ is dealt with in Sect. 1.4. The case $\mathsf{E}X_1 = 0$ is degenerate, since, by the nonnegativity of the random variable X_1, it means that $\mathsf{P}(X_1 = 0) = 1$.

In this section, we assume that the random variables $\{X_n\}$ are independent, identically distributed, and *nonnegative*. We denote their partial sums by S_n and let $\{L(t)\}$ be the renewal process defined in (1.1).

Theorem 1.14 (SLLN) *The following statements are equivalent:*

$$\lim_{t\to\infty} \frac{L(t)}{t} = a \qquad \textit{a.s.} \quad \textit{for some} \quad 0 < a < \infty; \tag{1.23}$$

$$\lim_{n\to\infty} \frac{S_n}{n} = \frac{1}{a} \qquad \textit{a.s.} \quad \textit{for some} \quad 0 < a < \infty; \tag{1.24}$$

$$\mathsf{E}X_1 = \frac{1}{a} \quad \textit{for some} \quad 0 < a < \infty. \tag{1.25}$$

The rate of convergence in (1.23)–(1.24) is given by the following result.

Theorem 1.15 (Marcinkiewicz–Zygmund SLLN) *Let $1 \le r < 2$. The following statements are equivalent:*

$$\lim_{t\to\infty} \frac{L(t) - ta}{t^{1/r}} = 0 \qquad \textit{a.s.} \quad \textit{for some} \quad 0 < a < \infty; \tag{1.26}$$

$$\lim_{n\to\infty} \frac{S_n - n\frac{1}{a}}{n^{1/r}} = 0 \qquad a.s. \quad \textit{for some} \quad 0 < a < \infty; \tag{1.27}$$

$$\mathsf{E}X_1 = \frac{1}{a}, \qquad \mathsf{E}X_1^r < \infty \quad \textit{for some} \quad 0 < a < \infty. \tag{1.28}$$

The rate of convergence in Theorem 1.14 can be described in a more precise way.

Theorem 1.16 (One-Sided LIL) *The following statements are equivalent:*

$$\limsup_{t\to\infty} \frac{L(t) - ta}{\sqrt{2t \log\log t}} = b \qquad a.s. \quad \textit{for some} \quad 0 < a, b < \infty; \tag{1.29}$$

$$\liminf_{n\to\infty} \frac{S_n - n\frac{1}{a}}{\sqrt{2n \log\log n}} = -\frac{b}{a^{3/2}} \qquad a.s. \quad \textit{for some} \quad 0 < a, b < \infty; \tag{1.30}$$

$$\mathsf{E}X_1 = \frac{1}{a}, \qquad \operatorname{var} X_1 = \frac{b^2}{a^3} \quad \textit{for some} \quad 0 < a, b < \infty; \tag{1.31}$$

$$\liminf_{t\to\infty} \frac{L(t) - ta}{\sqrt{2t \log\log t}} = -b \qquad a.s. \quad \textit{for some} \quad 0 < a, b < \infty; \tag{1.32}$$

$$\limsup_{n\to\infty} \frac{S_n - n\frac{1}{a}}{\sqrt{2n \log\log n}} = \frac{b}{a^{3/2}} \qquad a.s. \quad \textit{for some} \quad 0 < a, b < \infty. \tag{1.33}$$

Remark 1.17 By combining the statements of Theorem 1.16, it is obvious that the following "two-sided" versions are also equivalent to (1.29)–(1.33):

$$\limsup_{t\to\infty} \frac{|L(t) - ta|}{\sqrt{2t \log\log t}} = b \qquad \text{a.s.} \quad \text{for some} \quad 0 < a, b < \infty; \tag{1.34}$$

$$\limsup_{n\to\infty} \frac{|S_n - n\frac{1}{a}|}{\sqrt{2n \log\log n}} = \frac{b}{a^{3/2}} \qquad \text{a.s.} \quad \text{for some} \quad 0 < a, b < \infty. \tag{1.35}$$

A direct proof could be given by making use of Strassen [349] instead of Martikainen [262], Rosalsky [306], or Pruitt [297] (cf. the Proofs in Sects. 1.3 and 1.4).

1.3.2 *Proofs*

For the proofs below it will be necessary to consider the renewal counting process $\{L(t)\}$ at its "renewal points" $\{S_n\}$. From the definition of $\{L(t)\}$ it is obvious that, if $\mathsf{P}(X_1 > 0) = 1$,

$$L(S_n) = n \qquad \text{a.s.} \quad \forall\, n \geq 1.$$

In the case $0 < \mathsf{P}(X_1 = 0) < 1$ we set

$$R_n = \begin{cases} \max\{k : X_{n+1} = \cdots = X_{n+k}\}, & \text{if} \quad X_{n+1} = 0, \\ 0, & \text{if} \quad X_{n+1} > 0. \end{cases}$$

Then $S_n = S_{n+m}$ for $0 \le m \le R_n$, but $S_n < S_{n+R_n+1}$ and thus

$$L(S_n) = \cdots = L(S_{n+R_n}) = n + R_n. \tag{1.36}$$

Obviously, R_n has a geometric distribution, i.e.

$$\mathsf{P}(R_n = r) = qp^r, \qquad r = 0, 1, \ldots,$$

where $p = \mathsf{P}(X_1 = 0)$, $q = 1 - p$.

Lemma 1.18 *If* $0 < p = \mathsf{P}(X_1 = 0) < 1$, *then*

$$\limsup_{n\to\infty} \frac{R_n}{\log n} \le \frac{1}{\log(1/p)} \qquad \text{a.s.} \tag{1.37}$$

Proof of Lemma 1.18 For an arbitrary $\varepsilon > 0$ let $a = (1+\varepsilon)/\log(1/p)$. Then

$$\mathsf{P}(R_n \ge a \log n) \le \mathsf{P}(R_n \ge [a \log n]) \le \frac{1}{p} p^{a \log n} = \frac{1}{p} n^{-a \log(1/p)} = \frac{1}{p} n^{-(1+\varepsilon)}.$$

So, $\sum_{n=1}^{\infty} \mathsf{P}(R_n \ge a \log n) < \infty$, and the Borel–Cantelli lemma implies

$$\limsup_{n\to\infty} \frac{R_n}{\log n} \le \frac{1+\varepsilon}{\log(1/p)} \qquad \text{a.s.}$$

By letting $\varepsilon \to 0$, (1.37) is proved. □

Remark 1.19 It is easily verified that "=" holds in (1.37). We leave the details to the reader, since we only make later use of the order of R_n. Moreover, setting $R_n = 0$ if $p = 0$, Lemma 1.18 retains by defining $1/0 := \infty$, $1/\infty := 0$. So, in general,

$$0 \le R_n = O(\log n) \qquad \text{a.s.} \quad (n \to \infty). \tag{1.38}$$

Note that, if $p = 1$, then $R_n = \infty$ and (1.38) does not hold. We remark in passing that for strictly positive summands we obviously may replace (1.36) by

$$L(S_n) = n. \tag{1.39}$$

The proofs of our results make use of the following fundamental relations

$$L(t) \geq n \iff S_n \leq t, \qquad t \geq 0,\ n \geq 0, \tag{1.40}$$

$$L(t) < n \iff S_n > t, \qquad t \geq 0,\ n \geq 0, \tag{1.41}$$

$$S_{L(t)} \leq t < S_{L(t)+1}, \qquad t \geq 0. \tag{1.42}$$

Proof of Theorem 1.14 Assume (1.23). Since $L(t) < \infty$ a.s., $S_n \to \infty$ a.s. and we have (recall (1.36)),

$$\frac{L(S_n)}{S_n} = \frac{n + R_n}{S_n} = \frac{n}{S_n}\left(1 + \frac{R_n}{n}\right) \to a \qquad \text{a.s.}$$

Hence, in view of (1.38), $S_n/n \to 1/a$ a.s. $(n \to \infty)$.

Assume (1.24). From (1.42), we have (cf. Gut [159], (II.5.3))

$$\frac{S_{L(t)}}{L(t)} \leq \frac{t}{L(t)} < \frac{S_{L(t)+1}}{L(t)+1}\frac{L(t)+1}{L(t)}.$$

So, since $S_n \to \infty$ a.s., $L(t) \to \infty$ a.s. and we conclude that $t/L(t) \to 1/a$ a.s., i.e. that $L(t)/t \to a$ a.s. $(t \to \infty)$.

The equivalence of (1.24) and (1.25) is well-known and may be found, for example, in Petrov [291], Chapter 6.4, Theorem 6.11. □

Remark 1.20 Note also that, in order to prove the equivalence of (1.23) and (1.24), we only make use of the fact that

$$\lim_{n\to\infty} \frac{R_n}{n} = 0 \qquad \text{a.s.}, \tag{1.43}$$

together with $L(t) \to \infty$ a.s. $(t \to \infty)$ and $S_n \to +\infty$ a.s. $(n \to \infty)$. No independence and identical distribution of the X_n's is needed here. Moreover, if, in particular, the X_n's are strictly positive, then (recall (1.39)) this implies that we actually have the following, more general results.

Theorem 1.21 (SLLN) *Let $\{X_n\}_{n\geq 1}$ be arbitrary strictly positive random variables. The following statements are equivalent:*

$$\lim_{t\to\infty} \frac{L(t)}{t} = a \qquad \text{a.s.} \quad \text{for some} \quad 0 < a < \infty; \tag{1.44}$$

$$\lim_{n\to\infty} \frac{S_n}{n} = \frac{1}{a} \qquad \text{a.s.} \quad \text{for some} \quad 0 < a < \infty. \tag{1.45}$$

Theorem 1.22 (SLLN) *If* $\{X_n\}_{n\ge1}$ *are arbitrary nonnegative random variables such that* (1.43) *holds, then the following statements are equivalent:*

$$\lim_{t\to\infty}\frac{L(t)}{t}=a \qquad a.s. \quad for\ some \quad 0<a<\infty;$$

$$\lim_{n\to\infty}\frac{S_n}{n}=\frac{1}{a} \qquad a.s. \quad for\ some \quad 0<a<\infty.$$

Analogous remarks (precise formulations of which we omit) also apply to the strong limiting equivalences in Sects. 1.3 and 1.4, where it is sufficient to know that, for a suitably chosen $\alpha>0$,

$$\lim_{n\to\infty}\frac{R_n}{n^\alpha}=0 \qquad \text{a.s.} \tag{1.46}$$

In the i.i.d. case, relations (1.43) and (1.46) are, of course, immediate from Lemma 1.18.

Remark 1.23 General results like Theorems 1.21 or 1.22 are studied in Chap. 7 for the ratio of two arbitrary functions and that of their asymptotically quasi-inverse functions (see, for example, Theorem 7.57).

Proof of Theorem 1.15 Assume (1.26). Since $r\ge1$, this also implies (1.23) (and (1.24)), i.e. $\lim_{t\to\infty}L(t)/t=a$ a.s. and $\lim_{n\to\infty}S_n/n=1/a$ a.s. for some nonrandom number $0<a<\infty$. Now,

$$\frac{L(S_n)-S_na}{S_n^{1/r}}=\frac{n+R_n-S_na}{S_n^{1/r}}=\left(\frac{n}{S_n}\right)^{1/r}\left(\frac{R_n}{n^{1/r}}-a\frac{S_n-n\frac{1}{a}}{n^{1/r}}\right).$$

Since $(n/S_n)^{1/r}\to a^{1/r}$ a.s. and $R_n/n^{1/r}\to0$ a.s. ($n\to\infty$), (1.26) implies (1.27).

Assume (1.27). Using again the duality (1.42), we obtain that

$$\frac{L(t)-ta}{t^{1/r}}\le\frac{L(t)-S_{L(t)}a}{t^{1/r}}=-\left(\frac{L(t)}{t}\right)^{1/r}a\frac{S_{L(t)}-L(t)\frac{1}{a}}{L(t)^{1/r}}.$$

So,

$$\limsup_{t\to\infty}\frac{L(t)-ta}{t^{1/r}}\le0 \qquad \text{a.s.}$$

Similarly, using the upper estimate in (1.42),

$$\liminf_{t\to\infty}\frac{L(t)-ta}{t^{1/r}}\ge0 \qquad \text{a.s.,}$$

which completes the proof.

The equivalence of (1.27) and (1.28) is well-known from the Marcinkiewicz–Zygmund strong law of large numbers. □

Proof of Theorem 1.16 Since the X_n's are nonnegative, we know that $\mathsf{E}X_1$ exists and $0 \le \mathsf{E}X_1 \le \infty$. However, under (1.29) or (1.30), $\mathsf{E}X_1 \ne 1/a$ would contradict either (1.29) or (1.30) in view of Theorem 1.14, that is, we must have $\mathsf{E}X_1 = 1/a$. So, by Theorem 1.14,

$$\lim_{n\to\infty} \frac{S_n}{n} = \frac{1}{a} \text{ a.s.} \quad \text{and} \quad \lim_{t\to\infty} \frac{L(t)}{t} = a \text{ a.s.,}$$

which also implies that

$$\lim_{n\to\infty} \frac{\log\log S_n}{\log\log n} = 1 \text{ a.s.} \quad \text{and} \quad \lim_{t\to\infty} \frac{\log\log L(t)}{\log\log t} = 1 \text{ a.s.} \tag{1.47}$$

Now assume (1.29). Then, in view of (1.38),

$$\begin{aligned} b &\ge \limsup_{n\to\infty} \frac{L(S_n) - aS_n}{\sqrt{2S_n \log\log S_n}} = \limsup_{n\to\infty}\, (-a^{3/2}) \frac{S_n - (n + R_n)\frac{1}{a}}{\sqrt{2n\log\log n}} \\ &\ge (-a^{3/2}) \liminf_{n\to\infty} \frac{S_n - n\frac{1}{a}}{\sqrt{2n\log\log n}} \qquad \text{a.s.} \end{aligned}$$

This proves the lower part of (1.30), i.e. that

$$\liminf_{n\to\infty} \frac{S_n - n\frac{1}{a}}{\sqrt{2n\log\log n}} \ge -\frac{b}{a^{3/2}} \qquad \text{a.s.} \tag{1.48}$$

To obtain equality, observe that, from (1.29), for some event Ω_0 with $\mathsf{P}(\Omega_0) = 1$ and every $\omega \in \Omega_0$, there exists an increasing sequence $\{t_n\}$, $t_n = t_n(\omega) \to +\infty$, such that

$$\lim_{n\to\infty} \frac{L(t_n) - t_n a}{\sqrt{2t_n \log\log t_n}} = b.$$

Moreover, without loss of generality, $L(t_n)/t_n \to a$, $n \to \infty$, for $\omega \in \Omega_0$. So, invoking (1.42) once more,

$$\begin{aligned} \liminf_{n\to\infty} \frac{S_n - n\frac{1}{a}}{\sqrt{2n\log\log n}} &\le \liminf_{n\to\infty} \frac{S_{L(t_n)} - L(t_n)\frac{1}{a}}{\sqrt{2L(t_n)\log\log L(t_n)}} \\ &\le \liminf_{n\to\infty} \frac{t_n - L(t_n)\frac{1}{a}}{\sqrt{2L(t_n)\log\log L(t_n)}} \\ &= -\frac{1}{a^{3/2}} \limsup_{n\to\infty} \frac{L(t_n) - t_n a}{\sqrt{2t_n\log\log t_n}} = -\frac{b}{a^{3/2}}. \end{aligned}$$

By combining the latter result and (1.48), we finally get (1.30).

The proof of the converse, i.e. that (1.30) implies (1.29), is similar and will be omitted here.

For the equivalence of (1.30) and (1.31) we refer to the law of the iterated logarithm given in Sect. 1.2 (cf. Proposition 1.11).

The equivalences of (1.31)–(1.33) follow similarly. □

Remark 1.24 Note that, since (1.30) and (1.31) are equivalent by Proposition 1.11, the implication (1.30) $\Longrightarrow$ (1.29) could also be deduced via (1.31) and the fact that (1.31) implies (1.29).

Remark 1.25 We use a simple property of the slowly varying function $f(t) \stackrel{\text{def}}{=} \log\log t$ in the proof of (1.47). One can prove the same result as a corollary of the general theory we develop in Chap. 3. Namely, since $f \in \mathcal{SV}$, we have $f \in \mathcal{PRV}$ (see Definition 3.16). This implies that f preserves the equivalence of functions (see Theorem 3.42). The latter property is used in (1.47).

1.4 Renewal Processes Without Linear Drift

1.4.1 Results

The equivalence statements concerning the SLLN, in fact, carry over to the case of infinite expectation $\mathsf{E}X_1 = +\infty$. Marcinkiewicz–Zygmund and LIL type strong limit theorems, however, turn out to be entirely different here.

We assume throughout this section that $\{X_n\}$ are *nonnegative*, independent and identically distributed random variables.

Theorem 1.26 (SLLN) *The following statements are equivalent:*

$$\lim_{t\to\infty} \frac{L(t)}{t} = 0 \qquad a.s.; \tag{1.49}$$

$$\lim_{n\to\infty} \frac{S_n}{n} = +\infty \qquad a.s.; \tag{1.50}$$

$$\mathsf{E}X_1 = +\infty; \tag{1.51}$$

$$\int_1^\infty \frac{1}{t}\mathsf{P}(L(t) \ge ct)\,dt < \infty \quad \textit{for all} \quad c > 0; \tag{1.52}$$

$$\int_1^\infty \frac{1}{t}\mathsf{P}(L(t) < ct)\,dt = \infty \quad \textit{for all} \quad c > 0; \tag{1.53}$$

$$\sum_{n=1}^{\infty} \frac{1}{n} \mathsf{P}(S_n \le cn) < \infty \quad \textit{for all} \quad c > 0; \tag{1.54}$$

$$\sum_{n=1}^{\infty} \frac{1}{n} \mathsf{P}(S_n > cn) = \infty \quad \textit{for all} \quad c > 0. \tag{1.55}$$

Remark 1.27 For the case of general summands X_n, if $\mathsf{E}X_1^+ = \infty$, it is proved in Theorem 6 of Kesten [215] that condition (1.53) is equivalent to

$$\limsup_{n\to\infty} \frac{S_n}{n} = \infty \qquad \text{a.s.}$$

As we see from Theorem 1.26, condition (1.53) is, in fact, equivalent to (1.50) if the terms are nonnegative.

Theorem 1.28 (Marcinkiewicz–Zygmund-Type SLLN)

a) Let $0 < r \le 1$. *The following statements are equivalent:*

$$\lim_{t\to\infty} \frac{L(t)}{t^r} = 0 \qquad \textit{a.s.}; \tag{1.56}$$

$$\lim_{n\to\infty} \frac{S_n}{n^{1/r}} = +\infty \qquad \textit{a.s.}; \tag{1.57}$$

$$\int_1^{\infty} \frac{1}{t} \mathsf{P}(L(t) \ge ct^r)\, dt < \infty \quad \textit{for all} \quad 0 < c < \infty; \tag{1.58}$$

$$\sum_{n=1}^{\infty} \frac{1}{n} \mathsf{P}(S_n \le cn^{1/r}) < \infty \quad \textit{for all} \quad 0 < c < \infty. \tag{1.59}$$

b) Let $r > 1$. *If the distribution of the random variable* X_1 *is nondegenerate at* 0, *that is, if* $\mathsf{P}(X_1 = 0) < 1$, *then both conditions* (1.56) *and* (1.57) *hold.*

Remark 1.29 If one of the conditions (1.56)–(1.59) is satisfied (and, hence, all of them) and $0 < r \le 1$, then $\mathsf{E}X_1^{1/r} = +\infty$ (see Proposition 1.4), but, in general, the converse holds only for $r = 1$ (see Remark 1.6).

Theorem 1.30 (One-Sided LIL; Upper Part) *The following statements are equivalent:*

$$\limsup_{t\to\infty} \frac{L(t)}{\sqrt{t \log\log t}} = b_1 \qquad \textit{a.s.} \quad \textit{for some} \quad 0 \le b_1 < \infty; \tag{1.60}$$

$$\liminf_{n\to\infty} \frac{S_n \log\log S_n}{n^2} = \frac{1}{b_1^2} \quad \textit{a.s. for some} \quad 0 \le b_1 < \infty; \tag{1.61}$$

$$\int_e^\infty \frac{1}{t} \mathsf{P}(L(t) \ge c\sqrt{t \log\log t})\, dt < \infty \quad \textit{for some} \quad 0 < c < \infty; \tag{1.62}$$

$$\sum_{n=3}^\infty \frac{1}{n} \mathsf{P}\left(S_n \le c\frac{n^2}{\log\log n}\right) < \infty \quad \textit{for some} \quad 0 < c < \infty. \tag{1.63}$$

Remark 1.31

a) If one of the conditions (1.60)–(1.63) is satisfied (and, hence, all of them), then $\mathsf{E}\sqrt{X_1 \log^+ \log^+ X_1} = +\infty$ but, in general, the converse does not hold. Here $\log^+ z = \log(1 + |z|)$ for real numbers z.
b) The constant b_1 is determined as follows: $b_1 = c_1^{-1/2}$, where

$$c_1 = \sup\{0 < c < \infty\colon\ (1.63) \text{ holds}\}.$$

c) Note that, in case $b_1 = 0$, the equivalence of (1.60) and (1.61) means that

$$\lim_{t\to\infty} \frac{L(t)}{\sqrt{t \log\log t}} = 0 \quad \text{a.s.} \quad \text{if and only if} \quad \lim_{n\to\infty} \frac{S_n \log\log S_n}{n^2} = \infty \quad \text{a.s.}$$

Theorem 1.32 (One-Sided LIL; Lower Part) *The following statements are equivalent:*

$$\liminf_{t\to\infty} \frac{L(t)}{\sqrt{t \log\log t}} = b_0 \quad \textit{a.s. for some} \quad 0 < b_0 \le \infty; \tag{1.64}$$

$$\limsup_{n\to\infty} \frac{S_n \log\log S_n}{n^2} = \frac{1}{b_0^2} \quad \textit{a.s. for some} \quad 0 < b_0 \le \infty; \tag{1.65}$$

$$\mathsf{E}\sqrt{X_1 \log^+ \log^+ X_1} < \infty. \tag{1.66}$$

Remark 1.33 In fact, if one of the conditions (1.64)–(1.66) is satisfied (and, hence, all of them), then necessarily $b_0 = +\infty$ in the statements (1.64) and (1.65), where $1/\infty = 0$. This is an immediate consequence of the strong law of large numbers in Feller [126]. Moreover, liminf and limsup in (1.64) and (1.65), respectively, turn into limits again.

1.4.2 Proofs

Proof of Theorem 1.26 Assume (1.49). Since $L(t) < \infty$ a.s., $S_n \to \infty$ a.s. and we have (recall (1.36)),

$$\frac{L(S_n)}{S_n} = \frac{n + R_n}{S_n} = \frac{n}{S_n}\left(1 + \frac{R_n}{n}\right) \to 0 \qquad \text{a.s.}$$

Hence, in view of (1.38), $S_n/n \to \infty$ a.s. $(n \to \infty)$.

Assume (1.50). From (1.42), we have

$$\frac{S_{L(t)}}{L(t)} \le \frac{t}{L(t)} < \frac{S_{L(t)+1}}{L(t)+1}\,\frac{L(t)+1}{L(t)}.$$

So, since $S_n \to \infty$ a.s., $L(t) \to \infty$ a.s. and we conclude that $t/L(t) \to \infty$ a.s., i.e. that $L(t)/t \to 0$ a.s. $(t \to \infty)$.

The equivalence of (1.24) and (1.25) follows from Proposition 1.2, while that of relations (1.50) and (1.52) is proved in Proposition 1.4.

The implication (1.52) $\Longrightarrow$ (1.53) follows from the divergence of the harmonic series. To prove the implication (1.53) $\Longrightarrow$ (1.52) let condition (1.53) hold. Then

$$\sum_{n=1}^{\infty} \frac{1}{n} \mathsf{P}(S_n - n\mu \ge \varepsilon n) = \infty \quad \text{for all} \quad \mu \ge 0,\ \varepsilon > 0.$$

Thus Lemma 1.10 for nonnegative summands yields $\mathsf{E}X_1 = \infty$, whence we obtain (1.50). This proves (1.52) via Proposition 1.4.

The equivalence of conditions (1.54) and (1.52) can be derived as follows. Condition (1.54) holds if and only if

$$\int_1^{\infty} \frac{1}{s} \mathsf{P}(L(s/c) \ge s)\, ds < \infty \quad \text{for all} \quad c > 0.$$

If $c > 0$ is fixed, then the latter integral can be estimated from below by the series

$$\sum_{n=1}^{\infty} \frac{1}{n+1} \mathsf{P}\left(L\left(\frac{n}{c}\right) \ge n+1\right) = \sum_{n=1}^{\infty} \frac{1}{n+1} \mathsf{P}\left(S_{n+1} \le \frac{n}{c}\right).$$

An upper estimate is given by

$$\sum_{n=1}^{\infty} \frac{1}{n} \mathsf{P}\left(L\left(\frac{n+1}{c}\right) \ge n\right) = \sum_{n=1}^{\infty} \frac{1}{n} \mathsf{P}\left(S_n \le \frac{n+1}{c}\right).$$

Since $c > 0$ is arbitrary, the convergence of these series is equivalent to condition (1.52).

The proof of the equivalence of conditions (1.55) and (1.53) follows the lines of the latter reasoning concerning the equivalence of conditions (1.54) and (1.52). The only difference is that one uses relation (1.41) here instead of (1.40). □

Proof of Theorem 1.28 a) Assume (1.56). Since $r \le 1$, this also implies (1.49) (and (1.50)), i.e. $\lim_{t\to\infty} L(t)/t = 0$ a.s. and $\lim_{n\to\infty} S_n/n = \infty$ a.s. Now,

$$\frac{L(S_n)}{S_n^r} = \frac{n+R_n}{S_n^r} = \frac{n}{S_n^r}\left(1+\frac{R_n}{n}\right).$$

Since $L(S_n)/S_n^r \to 0$ a.s. and $R_n/n \to 0$ a.s. ($n \to \infty$), (1.56) implies (1.57).

Assume (1.57). Using again the duality (1.42), we obtain that

$$\frac{S_{L(t)}}{L^{1/r}(t)} \le \frac{t}{L^{1/r}(t)} \le \frac{S_{L(t)+1}}{L^{1/r}(t)}. \tag{1.67}$$

This implies condition (1.56).

The equivalence of (1.57) and (1.59) has been proved in Proposition 1.4 above.

Now we prove that (1.58) and (1.59) are equivalent. Put $v = 1/r$. First we prove the implication (1.58) $\Longrightarrow$ (1.59). Via the transformation $t = s^v/c^v$, (1.58) is equivalent to

$$\int_1^\infty \frac{1}{s}\mathsf{P}(L(s^v/c^v) \ge s)\,ds < \infty \quad \text{for all} \quad 0 < c < \infty. \tag{1.68}$$

Upon observing that $\mathsf{P}(L(x) \ge n) = \mathsf{P}(S_n \le x)$, $x > 0$, by (1.40) we obtain

$$\begin{aligned}\int_1^\infty \frac{1}{s}\mathsf{P}(L(s^v/c^v) \ge s)\,ds &\ge \sum_{n=1}^\infty \frac{1}{n+1}\mathsf{P}\left(L\left(\frac{n^v}{c^v}\right) \ge n+1\right)\\ &= \sum_{n=1}^\infty \frac{1}{n+1}\mathsf{P}\left(S_{n+1} \le \frac{n^v}{c^v}\right).\end{aligned}$$

Since $n^v \sim (n+1)^v$ and (1.68) holds for any $0 < c < \infty$, (1.59) follows from the above estimation.

The converse implication (1.59) $\Longrightarrow$ (1.68) follows from the inequality

$$\begin{aligned}\int_1^\infty \frac{1}{s}\mathsf{P}(L(s^v/c^v) \ge s)\,ds &\le \sum_{n=1}^\infty \frac{1}{n}\mathsf{P}\left(L\left(n^v/c^v\right) \ge n\right)\\ &= \sum_{n=1}^\infty \frac{1}{n}\mathsf{P}\left(S_n \le (n+1)^v/c^v\right).\end{aligned}$$

b) According to Proposition 1.4, condition (1.57) holds. Applying (1.67) we prove (1.57). □

Proof of Theorem 1.30 Assume (1.60). Since $L(S_n) = n + R_n \ge n$,

$$b_1^2 \ge \limsup_{n\to\infty} \frac{L^2(S_n)}{S_n \log\log S_n} = \limsup_{n\to\infty} \frac{(n+R_n)^2}{S_n \log\log S_n}$$
$$\ge \limsup_{n\to\infty} \frac{n^2}{S_n \log\log S_n} \qquad \text{a.s.}$$

So,

$$\liminf_{n\to\infty} \frac{S_n \log\log S_n}{n^2} \ge \frac{1}{b_1^2} \qquad \text{a.s.}$$

Equality is obtained as in the proof of Theorem 1.16, i.e., there is an event Ω_0 with $\mathsf{P}(\Omega_0) = 1$ such that, for every $\omega \in \Omega_0$, there exists a sequence $\{t_n\}$, $t_n = t_n(\omega) \to +\infty$ $(n \to \infty)$ satisfying

$$\lim_{n\to\infty} \frac{L^2(t_n)}{t_n \log\log t_n} = b_1^2.$$

Now,

$$\liminf_{n\to\infty} \frac{S_n \log\log S_n}{n^2} \le \liminf_{n\to\infty} \frac{S_{L(t_n)} \log\log S_{L(t_n)}}{L^2(t_n)} \le \liminf_{n\to\infty} \frac{t_n \log\log t_n}{L^2(t_n)} = \frac{1}{b_1^2},$$

which completes the proof of (1.61).

Next, assume (1.61). Then

$$\frac{1}{b_1^2} = \liminf_{n\to\infty} \frac{S_n \log\log S_n}{n^2} \le \liminf_{t\to\infty} \frac{S_{L(t)} \log\log S_{L(t)}}{L^2(t)}$$
$$\le \liminf_{t\to\infty} \frac{t \log\log t}{L^2(t)} \qquad \text{a.s.,}$$

since $S_{L(t)} \log\log S_{L(t)} \le r \log\log t$ by (1.42). This implies

$$\limsup_{t\to\infty} \frac{L(t)}{\sqrt{t \log\log t}} \le b_1 \qquad \text{a.s.}$$

To obtain equality, note that, for $\omega \in \Omega_0$ with $\mathsf{P}(\Omega_0) = 1$, there exists a sequence of integers $\{n_k\}$, $n_k = n_k(\omega) \to +\infty$ $(k \to \infty)$ such that

$$\lim_{k\to\infty} \frac{S_{n_k} \log\log S_{n_k}}{n_k^2} = \frac{1}{b_1^2} > 0.$$

Now,

$$\limsup_{t\to\infty} \frac{L(t)}{\sqrt{t\log\log t}} \geq \limsup_{k\to\infty} \frac{L(S_{n_k})}{\sqrt{S_{n_k}\log\log S_{n_k}}} = \limsup_{k\to\infty} \frac{n_k + R_{n_k}}{\sqrt{S_{n_k}\log\log S_{n_k}}}$$

$$\geq \lim_{k\to\infty} \frac{n_k}{\sqrt{S_{n_k}\log\log S_{n_k}}} = b_1,$$

which proves the equivalence of (1.60) and (1.61).

The equivalence of relations (1.61) and (1.63) has been proved in Proposition 1.12.

The equivalence of (1.62) and (1.63) follows similarly to the proof of Theorem 1.28. Note that, via setting $s = \sqrt{t\log\log t}$, (1.62) is equivalent to

$$\int_e^\infty \frac{1}{s}\mathsf{P}\left(L\left(\frac{s^2}{\log\log s}\right) \geq cs\right) ds < \infty \quad \text{for some} \quad 0 < c < \infty. \tag{1.69}$$

The rest of the proof follows again from the equality $\mathsf{P}(L(s) \geq n) = \mathsf{P}(S_n \leq s)$ as in the proof of Theorem 1.28. □

Proof of Remark 1.31

a) The conclusion follows similarly to the proof of Remark 1.29, but instead of the Marcinkiewicz–Zygmund SLLN one could, e.g., use the Feller [126] limit theorem for random variables with infinite moments. Note that assertion (1.20) in Proposition 1.12 could be replaced by

$$\liminf_{n\to\infty} \frac{S_n \log\log n}{n^2} = c \qquad \text{a.s.} \quad \text{for some} \quad 0 < c \leq \infty, \tag{1.70}$$

which can easily be seen from the arguments in the proof.

b) This statement also follows from the proof of Proposition 1.12 if, instead of the subsequence $\{2^k\}$ used there, one chooses $\{[a^k]\}$ with a greater than, but arbitrarily close to, 1. □

Proof of Theorem 1.32 The equivalence of (1.64) and (1.65) follows exactly as in the corresponding part in the proof of Theorem 1.30 (symmetry).

Now assume (1.65). Then, by nonnegativity,

$$\limsup_{n\to\infty} \frac{X_n \log\log X_n}{n^2} \leq \frac{1}{b_0^2} \qquad \text{a.s.},$$

which, together with the i.i.d. assumption and the Borel–Cantelli lemma, yields

$$\sum_{n=1}^\infty \mathsf{P}(X_1 \log^+\log^+ X_1 > cn^2) < \infty,$$

e.g., for $c > 2/b_0^2$. This proves (1.66).

Conversely, if (1.66) holds, then by the SLLN in Feller [126],

$$\lim_{n\to\infty} \frac{S_n \log\log n}{n^2} = 0 \qquad \text{a.s.,}$$

which, in view of $\log\log S_n = O(\log\log n)$ a.s., also implies

$$\lim_{n\to\infty} \frac{S_n \log\log S_n}{n^2} = 0 \qquad \text{a.s.}$$

This proves (1.65) with $b_0 = +\infty$. □

1.5 Comments

This chapter is based on Gut et al. [160]. The main idea of the proof is to use the duality between cumulative sums and renewal processes which, in a certain sense, means that they are inverses to each other. This idea is used in the monograph of Gut [159] as well. Vervaat [360] follows a similar idea to prove that the invariance principle for the law of the iterated logarithm for partial sums with positive drift is equivalent to that for their inverse process. Via a similar approach, Iglehart and Whitt [188] study a functional form of the central limit theorem for renewal processes.

Section 1.2 Proposition 1.1 is proved in Kolmogorov [234]. Proposition 1.2 is obtained in Kesten [215] (see Theorem 2 therein) for general random variables. Above we provide a different proof for nonnegative random variables.

Proposition 1.3 is proved in Marcinkiewicz and Zygmund [260]. The proof of an extension of Proposition 1.4 for general random variables is more involved. For example, if $\mathsf{E}|X_1| = \infty$, then either $\limsup |S_n/a_n| = \infty$ or $\lim S_n/a_n = 0$ almost surely for any sequence of positive numbers $\{a_n\}$ (see Chow and Robbins [88]). In the case when $a_n = n^{1/r}$, this result was obtained earlier by Feller [126].

Lemma 1.7 is proved in Klesov et al. [226] and generalizes the corresponding result in Gut et al. [160]. Lemma 1.8 is given in Kesten [215] (see also Theorem 6.3.3 in Stout [348]). Lemma 1.10 is proved in Spitzer [343].

The fact that (1.19) implies (1.35) is shown in Hartman and Wintner [175], while the inverse implication is proved in Strassen [349]. The necessity has independently been proved by several authors, see, e.g., Martikainen [262], Rosalsky [306], and Pruitt [297].

Other proofs of the two-sided law of the iterated logarithm have also been given (see, e.g., Heyde [179], Csörgő and Révész [97], Acosta [3]. A corresponding proof of Martikainen is included in the book by Petrov [291]). Various proofs of the one-sided law of the iterated logarithm (Proposition 1.11) are presented, for example, in Martikainen [262], Rosalsky [306], and Pruitt [297].

Section 1.3 The equivalence of (1.34) and (1.35) can be proved directly by using the two-sided criterion of Strassen [349] instead of the one-sided criterion of Martikainen [262], Rosalsky [306], or Pruitt [297] (see the proof of Theorem 1.16).

Section 1.4 Part b) of Theorem 1.28 is valid only for nonnegative random variables. A survey of possible generalizations for general summands is given by Stout [348] (Chapter 6).

Some related results for stopped functionals of Markov renewal processes are obtained in Alsmeyer and Gut [14].

Chapter 2
Almost Sure Convergence of Renewal Processes

2.1 Introduction

Consider some *renewal sequence*, that is, a sequence of partial sums $\{S_n\}_{n\ge 0}$ of independent identically distributed random variables $\{X_n\}_{n\ge 1}$:

$$S_0 = 0, \qquad S_n = \sum_{k=1}^{n} X_k, \quad n \ge 1.$$

The corresponding *renewal counting process* $\{L(t)\}_{t\ge 0}$ is defined in Chap. 1 as follows:

$$L(t) = \max\{n \ge 0\colon S_n \le t\}, \qquad t \ge 0,$$

(see Definition (1.1)). Recall that the random variables $\{X_n\}_{n\ge 1}$ are nonnegative in Chap. 1, however Definition (1.1) can obviously be extended to the case of general random variables.

The nonnegativity is an appropriate assumption if the underlying random variables $\{X_n\}_{n\ge 1}$ correspond to an expendable resource like the life span of a bulb. Other important applications, for example, risk models, may require models in which the random variables may have both signs. The main difference between these two cases is that $L(t)$ is not monotonic if the random variables $\{X_n\}_{n\ge 1}$ attain both positive and negative values.

In the case of nonnegativity, the meaning of $L(t)$ is that $L(t) + 1$ is the *first moment* when the sequence $\{S_n\}_{n\ge 1}$ leaves the strip $(-\infty, t]$. If the random variables $\{X_n\}_{n\ge 1}$ attain both positive and negative values, but nevertheless $S_n \to \infty$, $n \to \infty$, one might be interested in the *last moment* when $\{S_n\}_{n\ge 1}$ leaves the strip $(-\infty, t]$. Obviously the first moment coincides with the last one if the $\{X_n\}_{n\ge 1}$ are

V. V. Buldygin et al., *Pseudo-Regularly Varying Functions and Generalized Renewal Processes*, Probability Theory and Stochastic Modelling 91, https://doi.org/10.1007/978-3-319-99537-3_2

nonnegative, but if not, then many other functionals of interest can be introduced, for example, the *second moment* when $\{S_n\}_{n\ge1}$ leaves the strip $(-\infty, t]$ and so forth.

It seems obvious that the behavior of each of these functionals requires a separate study and, perhaps, different methods. To an even greater extent this applies to another important functional constructed from $\{S_n\}_{n\ge1}$, namely the *total time* spent by the sequence $\{S_n\}_{n\ge1}$ in the strip $(-\infty, t]$.

Our aim in this chapter is to show that all these functionals are asymptotically equivalent if one considers them from the point of view of *generalized renewal processes*.

We have seen in Chap. 1 that if the random variables $\{X_n\}_{n\ge1}$ are nonnegative, then many limit theorems for the renewal counting process $\{L(t)\}_{t\ge0}$ follow from their counterparts for the underlying renewal sequence by using the duality relation

$$L(t) = n \quad \Longleftrightarrow \quad S_n \le t,\ S_{n+1} > t \tag{2.1}$$

for all $t \ge 0$ and $n \ge 0$. For example, the strong law of large numbers for $\{L(t)\}$, i.e.

$$\lim_{t\to\infty} \frac{L(t)}{t} = \frac{1}{a} \qquad \text{a.s.,} \tag{2.2}$$

follows immediately from the strong law of large numbers for $\{S_n\}$, i.e.

$$\lim_{n\to\infty} \frac{S_n}{n} = a \qquad \text{a.s.,}$$

(see Theorem 1.14).

Some other limit theorems, however, such as the law of the iterated logarithm

$$\limsup_{t\to\infty} \frac{L(t) - t/a}{\sqrt{2t \ln\ln t}} = \sqrt{\frac{\sigma^2}{a^3}} \qquad \text{a.s.} \tag{2.3}$$

(see Theorem 1.16) or the central limit theorem

$$\frac{L(t) - t/a}{\sqrt{t\sigma^2/a^3}} \xrightarrow{d} Z, \qquad t \to \infty, \tag{2.4}$$

require a more sophisticated technique. Here Z is a standard normal random variable, $0 < \sigma^2 = \operatorname{var} X_1 < \infty$, and $\xrightarrow{d}$ denotes the weak convergence of random variables.

Note that the renewal process in the case of nonnegative random variables can be defined in an equivalent way by

$$L(t) = \sum_{n=1}^{\infty} \mathbb{I}\{S_n \le t\}, \qquad t \ge 0. \tag{2.5}$$

In Chap. 1, we have shown that there are some equivalences between limit theorems for sequences and corresponding counting processes in the case of independent, identically distributed random variables $\{X_n\}_{n\geq 1}$ (see also Gut et al. [160]). The general case of sums S_n is studied in Gut [159]. The terms of S_n may attain negative values but their expectation should be positive. Some results on the equivalence in the general case for independent, identically distributed random variables are obtained in Frolov et al. [138].

Much less is known about renewal processes for which either identical distribution, or independence, or nonnegativity, or all of these assumptions are dropped. The aim of this chapter is to develop a general approach to deriving limit theorems for "renewal processes" from their corresponding counterparts for the underlying "partial sum sequence". Indeed, it is not necessary to assume any structure of the underlying renewal sequence $\{S_n\}_{n\geq 0}$; one can just start from an appropriate limit theorem there.

Note, however, that certain regularity assumptions are sometimes crucial for the applicability of a duality argument. There are situations in which a limit theorem for the renewal process is almost immediate from its partial sum counterpart. But there are other cases where the desired inversion requires more sophisticated techniques. Finally, there are also examples in which a duality argument does not work at all, because either the partial sum sequence satisfies a certain limit theorem, but not so its corresponding renewal process, or vice versa. We restrict our attention to *strong limit theorems* for renewal processes. A similar approach applies to *weak limit theorems* and to *renewal functions*, i.e. the expected number of renewals.

In order to avoid confusion with the case of independent, identically distributed random variables, we change the notation from now on and let $\{Z_n\}_{n\geq 0}$ be a *general* sequence of real-valued random variables. In what follows we assume that

$$Z_n \to \infty \qquad \text{almost surely} \quad \text{as } n \to \infty. \tag{2.6}$$

The *generalized renewal process* $\{L(t)\}_{t\geq 0}$ is defined pointwise (for every elementary $\omega \in \Omega$) by

$$L(t) = \sup\{n \geq 0 : Z_n \leq t\} = \sum_{n=1}^{\infty} \mathbb{I}\{\inf(Z_n, Z_{n+1}, \dots) \leq t\}, \qquad t \geq 0, \tag{2.7}$$

where $\sup \varnothing = 0$, i.e. $L(t) + 1$ is the *last-passage time* of the sequence $\{Z_n\}_{n\geq 0}$ from the set $(-\infty, t]$.

Along with $\{L(t)\}_{t\geq 0}$ we study two other *generalized renewal processes*. The first of them is defined by

$$M(t) = \sup\{n \geq 0 : \max(Z_0, Z_1, \dots, Z_n) \leq t\} \tag{2.8}$$

$$= \sum_{n=1}^{\infty} \mathbb{I}\{\max(Z_0, Z_1, \dots, Z_n) \leq t\}, \qquad t \geq 0,$$

that is $M(t) + 1$ is the *first-passage time* of the sequence $\{Z_n\}_{n\ge 0}$ from the interval $(-\infty, t]$.

The other process is defined similarly to (2.5):

$$T(t) = \sum_{n=1}^{\infty} \mathbb{I}\{Z_n \le t\}, \qquad t \ge 0, \tag{2.9}$$

that is $T(t)$ is the *sojourn time* spent by the sequence $\{Z_n\}$ in the set $(-\infty, t]$.

If condition (2.6) holds, then the processes $T(t)$, $M(t)$, and $L(t)$ are finite almost surely for an arbitrary $t \ge 0$, since only a finite number of terms in the sum (2.9) are non-zero. For those $\omega \in \Omega$ for which (2.6) does not hold, we assume that $M(t) = L(t) = T(t) = 0$.

Formally speaking, both processes $\{L(t)\}$ and $\{M(t)\}$ are partial cases of the process $\{T(t)\}$, where $\max(Z_0, Z_1, \dots, Z_n)$ or $\inf\{Z_n, Z_{n+1}, \dots)$ is used instead of Z_n, respectively. Nevertheless we use their own notation for the processes $L(t)$ and $M(t)$, since these processes will later play a special role in our proofs because of the *dualities*:

$$L(t) \ge n \qquad \Longleftrightarrow \qquad \max(Z_0, Z_1, \dots, Z_n) \le t, \tag{2.10}$$

$$L(t) = n \qquad \Longleftrightarrow \qquad \max(Z_0, Z_1, \dots, Z_n) \le t,\ Z_{n+1} > t; \tag{2.11}$$

and

$$M(t) \ge n \qquad \Longleftrightarrow \qquad \inf(Z_n, Z_{n+1}, \dots) \le t, \tag{2.12}$$

$$M(t) = n \qquad \Longleftrightarrow \qquad Z_n \le t,\ \inf(Z_{n+1}, Z_{n+2}, \dots) > t. \tag{2.13}$$

Note that we have no such nice properties for $T(t)$, however it is obvious that, for any $t \ge 0$,

$$\mathbb{I}\{\max(Z_0, Z_1, \dots, Z_n) \le t\} \le \mathbb{I}\{Z_n \le t\} \le \mathbb{I}\{\inf(Z_n, Z_{n+1}, \dots) \le t\},$$

for an arbitrary $t \ge 0$, and thus

$$M(t) \le T(t) \le L(t). \tag{2.14}$$

Moreover, the following inequalities hold true for finite $L(t)$ and $M(t)$, respectively:

$$Z_{L(t)} \le t < Z_{L(t)+1}, \tag{2.15}$$

$$Z_{M(t)} \le t < Z_{M(t)+1}. \tag{2.16}$$

Since $Z_n < +\infty$ almost surely for all n,

$$M(t) = \min\{M(t), L(t), T(t)\} \to +\infty \qquad \text{a.s.}$$

Remark 2.1 If $0 = Z_0 \le Z_1 \le Z_2 \le \cdots$, then

$$L(t) = T(t) = M(t).$$

But if (say) $Z_n > Z_{n+1}$ for some n, then for $Z_{n+1} \le t < Z_n$,

$$\mathbb{I}\{\max(Z_0, Z_1, \dots, Z_{n+1}) \le t\} = 0, \qquad \mathbb{I}\{Z_{n+1} \le t\} = 1,$$

that is $M(t) < T(t)$, and also

$$\mathbb{I}\{Z_n \le t\} = 0, \qquad \mathbb{I}\{\inf(Z_n, Z_{n+1}, \dots) \le t\} = 1,$$

whence $T(t) < L(t)$ follows.

This chapter is organized as follows. In Sect. 2.2, strong laws of large numbers are presented for general renewal processes including rates of convergence statements such as Marcinkiewicz–Zygmund-type results.

The usefulness of this general approach is demonstrated via a series of examples in Sect. 2.3 including renewal sequences of independent, but non-identically distributed summands, martingales and mixing sequences, weighted sums and non-linear renewal processes, among others.

2.2 Almost Sure Limit Theorems

2.2.1 The Strong Law of Large Numbers

If the strong law of large numbers holds for the sequence of random variables $\{Z_n\}_{n\ge 0}$ and for some normalizing sequence of positive numbers $\{a_n\}$, then corresponding results hold for generalized renewal processes if the sequence $\{a_n\}$ has a certain regularity property. For example, if $a_n = a(n)$ and the function $a(\cdot)$ is continuous and increasing, then the normalization in the strong law of large numbers for the generalized renewal processes is the (generalized) inverse function $a^{-1}(\cdot)$.

The proof of the result mentioned above consists of two steps. The first step (Lemma 2.2) is to use inequalities (2.15) and (2.16) and prove that the superposition of $\{a_n\}$ and a generalized renewal process grow linearly (see (2.19)–(2.21)). The second step is to deduce the strong laws of large numbers (2.24)–(2.26) from (2.19)–(2.21) if the normalizing sequence $\{a_n\}$ satisfies condition (2.23).

Such an approach is quite common in similar problems and will be used again later on (see Chaps. 8 and 9) for a more general situation.

Lemma 2.2 *Assume that*

$$\frac{Z_n}{a_n} \to 1 \qquad a.s., \tag{2.17}$$

where $\{a_n\}_{n\ge1}$ *is a nonrandom sequence such that* $a_n \to \infty$ *as* $n \to \infty$ *and*

$$\frac{a_{n+1}}{a_n} \to 1. \tag{2.18}$$

Then

$$\frac{a_{L(t)}}{t} \to 1 \qquad a.s., \tag{2.19}$$

$$\frac{a_{M(t)}}{t} \to 1 \qquad a.s. \tag{2.20}$$

as $t \to \infty$. *Moreover, if* $\{a_n\}_{n\ge1}$ *is nondecreasing, then*

$$\frac{a_{T(t)}}{t} \to 1 \qquad a.s. \tag{2.21}$$

Proof of Lemma 2.2 In view of (2.15),

$$\frac{Z_{L(t)}}{a_{L(t)}} \le \frac{t}{a_{L(t)}} < \frac{Z_{L(t)+1}}{a_{L(t)+1}} \frac{a_{L(t)+1}}{a_{L(t)}}.$$

Since $L(t) \to \infty$ almost surely as $t \to \infty$, both sides of the latter relation tend to 1 almost surely by (2.17) and (2.18). This proves (2.19).

Similarly, (2.20) follows from (2.16), (2.17), and (2.18).

If $\{a_n\}_{n\ge1}$ is nondecreasing, (2.14) implies $a_{M(t)} \le a_{T(t)} \le a_{L(t)}$, so that (2.21) is immediate from (2.19) and (2.20). □

If the sequence $\{a_n\}_{n\ge0}$ increases, then one can introduce the function $\{a(t)\}_{t\ge0}$ by putting $a(n) = a_n$ for all $n = 0, 1, 2, \dots$ such that

$$\begin{gathered} a(\cdot) \text{ is continuous, increasing, and such that} \\ a(t) \to \infty, \quad t \to \infty. \end{gathered} \tag{2.22}$$

Put

$$a^{-1}(u) = \inf\{t\colon a(t) = u\}, \qquad u > u_0 = a_0.$$

It is clear that the function $a^{-1}(\cdot)$ is also continuous, increasing, and such that $a^{-1}(u) \to \infty$ as $u \to \infty$.

Suppose that

$$\lim_{\varepsilon \downarrow 0} \limsup_{t \to \infty} \left| \frac{a^{-1}((1 \pm \varepsilon)t)}{a^{-1}(t)} - 1 \right| = 0. \tag{2.23}$$

Then not only relations (2.19)–(2.21) are valid, but also the following strong law of large numbers holds for the generalized renewal processes $L(t)$, $M(t)$, and $T(t)$. Theorem 2.3 below extends the implication (1.24)$\Longrightarrow$(1.23) in Theorem 1.14 to the functionals $M(t)$ and $T(t)$.

Theorem 2.3 *Let the strong law of large numbers (2.17) hold. If conditions (2.18), (2.22) and (2.23) are satisfied, then*

$$\frac{L(t)}{a^{-1}(t)} \to 1 \qquad a.s., \tag{2.24}$$

$$\frac{M(t)}{a^{-1}(t)} \to 1 \qquad a.s., \tag{2.25}$$

$$\frac{T(t)}{a^{-1}(t)} \to 1 \qquad a.s. \tag{2.26}$$

as $t \to \infty$.

Proof of Theorem 2.3 Lemma 2.2 implies condition (2.19), whence

$$\frac{a\,(L(t))}{t} = \frac{a_{L(t)}}{t} \to 1 \qquad \text{a.s.}$$

as $t \to \infty$. Thus

$$t(1-\varepsilon) < a\,(L(t)) < t(1+\varepsilon)$$

for an arbitrary $0 < \varepsilon < 1$ and $t \ge t_0 = t_0(\varepsilon, \omega)$. Using the monotonicity of the function $a^{-1}(t)$, we get

$$a^{-1}\,(t(1-\varepsilon)) < L(t) < a^{-1}\,(t(1+\varepsilon))$$

for $t \ge t_0$. This proves (2.24) in view of (2.23).

The proof of (2.25) and (2.26) follows by the same arguments. □

Remark 2.4 There are situations in which the strong laws of Theorem 2.3 cannot be derived from their corresponding counterparts of Lemma 2.2, since e.g. condition (2.18) is violated. Consider, for example, partial sums $S_n = X_1 + \cdots + X_n$ of a sequence $\{X_n\}_{n \ge 1}$ of independent, identically distributed positive random variables with $\mathsf{E}X_1 = 1$. Choose $Z_n = S_{2^n}$, $n \geq 1$. Then, with $a_n = 2^n$, of course

$$\frac{Z_n}{a_n} \to 1 \qquad \text{a.s.}$$

as $n \to \infty$, but neither $a_{n+1}/a_n \to 1$, nor

$$\frac{Z_{n+1} - Z_n}{a_n} = \frac{X_{2^n+1} + \cdots + X_{2^{n+1}}}{2^n} \to 0 \qquad \text{a.s.},$$

so that the arguments used in the proof of Theorem 2.3 do not apply here.

So, the growth condition (2.18) on the normalizing sequence $\{a_n\}_{n \geq 1}$ is crucial for deriving the strong laws of Theorem 2.3 from their counterparts in Lemma 2.2. Yet, this condition can be avoided, and thus the regularity assumptions can be weakened, by applying a totally different technique of proof. Such a method was introduced in Klesov and Steinebach [229] for the case of renewal processes constructed from random walks with multidimensional time.

Theorem 2.5 *If conditions (2.17), (2.22), and (2.23) hold, then relations (2.24)–(2.26) follow as $t \to \infty$.*

Proof of Theorem 2.5 First consider

$$T(t) = \sum_{k=1}^{\infty} \mathbb{I}\{Z_k \leq t\}.$$

Let $0 < \varepsilon < 1$. Putting $n = \left[a^{-1}(t)\right]$ and $m^{\pm} = \left[a^{-1}(t(1 \pm \varepsilon))\right]$, we obtain

$$\begin{aligned} T(t) - n &= -\sum_{k=1}^{n} \mathbb{I}\{Z_k > t\} + \sum_{k=n+1}^{\infty} \mathbb{I}\{Z_k \leq t\} \\ &= -\sum_{k=1}^{m^-} \mathbb{I}\{Z_k > t\} - \sum_{k=m^-+1}^{n} \mathbb{I}\{Z_k > t\} \\ &\quad + \sum_{k=n+1}^{m^+} \mathbb{I}\{Z_k \leq t\} + \sum_{k=m^++1}^{\infty} \mathbb{I}\{Z_k \leq t\} \\ &= -T_1(t) - T_2(t) + T_3(t) + T_4(t). \end{aligned} \tag{2.27}$$

Since

$$\frac{t}{a_k} \geq \frac{t}{a\left(a^{-1}((1-\varepsilon)t)\right)} = \frac{1}{1-\varepsilon} > 1$$

for $k \leq m^- \leq a^{-1}((1-\varepsilon)t)$, the inequality $Z_k/a_k > 1/(1-\varepsilon)$ holds almost surely only finitely often by (2.17). Hence

$$\frac{T_1(t)}{a^{-1}(t)} \to 0 \qquad \text{a.s.} \tag{2.28}$$

as $t \to \infty$. A similar argument shows that also

$$\frac{T_4(t)}{a^{-1}(t)} \to 0 \qquad \text{a.s.} \tag{2.29}$$

as $t \to \infty$. Next, in view of (2.23), we have

$$0 \leq \frac{T_2(t)}{a^{-1}(t)} \leq \frac{a^{-1}(t) - a^{-1}((1-\varepsilon)t) + 1}{a^{-1}(t)} \to 0 \tag{2.30}$$

as $t \to \infty$ and $\varepsilon \downarrow 0$, and similarly

$$0 \leq \frac{T_3(t)}{a^{-1}(t)} \leq \frac{a^{-1}((1+\varepsilon)t) - a^{-1}(t) + 1}{a^{-1}(t)} \to 0 \tag{2.31}$$

as $t \to \infty$ and $\varepsilon \downarrow 0$.

A combination of (2.27)–(2.31) proves (2.26), since

$$\frac{T(t) - \left[a^{-1}(t)\right]}{a^{-1}(t)} = \frac{T(t)}{a^{-1}(t)} - 1 + o(1)$$

as $t \to \infty$.

For the proof of (2.24) and (2.25) we note that, from $Z_n/a_n \to 1$ and $a_n \uparrow \infty$, $n \to \infty$, it also follows that

$$\frac{\max(Z_0, Z_1, \ldots, Z_n)}{a_n} \to 1 \quad \text{and} \quad \frac{\inf(Z_n, Z_{n+1}, \ldots)}{a_n} \to 1$$

as $n \to \infty$. This allows us to apply the same arguments to $L(t)$ and $M(t)$ as to $T(t)$. □

2.2.2 *Non-equivalent Strong Laws*

Unfortunately, there are also situations in which the inversion techniques applied in Lemma 2.2 and Theorems 2.3–2.5 cannot work at all. Consider, for instance, a max-scheme of independent identically distributed random variables $\{X_n\}_{n\geq 1}$ with distribution function $F(t) = \mathsf{P}(X_1 \leq t), t \in \mathbf{R}$. For $Z_n = \max(X_1, \ldots, X_n), n \geq 1$,

$Z_0 = 0$, the corresponding renewal processes $\{L(t)\}_{t\geq 0}$, $\{T(t)\}_{t\geq 0}$, and $\{M(t)\}_{t\geq 0}$ coincide. Moreover, for any $t \geq 0$, $T(t)$ has a geometric distribution, i.e.

$$\mathsf{P}(T(t) = n) = \mathsf{P}(\max(X_1, \ldots, X_n) \leq t, X_{n+1} > t) = F^n(t)(1 - F(t)),$$
$$\mathsf{P}(T(t) \geq n) = F^n(t), \qquad n = 0, 1, \ldots .$$

If $F(t) < 1$ for all $t \geq 0$, then for all fixed $x \geq 0$

$$\begin{aligned}\mathsf{P}\left(T(t) > x/\big(1 - F(t)\big)\right) &= \mathsf{P}\left(T(t) \geq \left[x/\big(1 - F(t)\big)\right] + 1\right)\\ &= \exp\left\{x(\ln F(t))/(1 - F(t)) + O(1)\ln F(t)\right\}.\end{aligned}$$

Since $\ln(1-x)/x \to -1$ as $x \to 0$, the right-hand side tends to $\exp\{-x\}$ as $t \to \infty$. Thus

$$T(t)(1 - F(t)) \xrightarrow{d} E, \qquad t \to \infty,$$

where the random variable E has an exponential Exp(1)-distribution. The weak convergence means that the strong law of large numbers

$$\frac{T(t)}{b(t)} \to 1 \qquad \text{a.s.}, \qquad t \to \infty,$$

does not hold for any (nonrandom) normalizing family $\{b(t)\}_{t\geq 0}$. Moreover, even the law of large numbers does not hold in this case for the process $\{T(t)\}_{t\geq 0}$. Indeed, if the law of large numbers were satisfied, then, for all $\varepsilon > 0$,

$$\mathsf{P}(T(t) > (1 + \varepsilon)b(t)) \to 0, \qquad t \to \infty,$$

whence $b(t)(1 - F(t)) \to \infty$, $t \to \infty$, according to the above weak convergence. On the other hand,

$$\mathsf{P}(T(t) \leq (1 - \varepsilon)b(t)) \to 1, \qquad t \to \infty.$$

Nevertheless, the strong law of large numbers may hold for the sequence $\{Z_n\}_{n\geq 1}$. This is indeed the case, for example, under an exponential Exp(1)-distribution, that is for $F(t) = 1 - e^{-t}$ if $t \geq 0$ and $F(t) = 0$ otherwise, since

$$\frac{Z_n}{\ln n} = \frac{\max(X_1, \ldots, X_n)}{\ln n} \to 1 \qquad \text{a.s.}, \qquad n \to \infty,$$

(see Galambos [141]).

In this example, all the assumptions of Lemma 2.2 and Theorems 2.3 and 2.5 hold for $a_n = \ln n$, $a(t) = \ln t$, and $a^{-1}(t) = e^t$ with the exception of (2.23). So, the latter condition cannot be dropped in general.

Another example would be $F(t) = \Phi(t)$, $t \in \mathbf{R}$, a standard normal distribution function, in which case

$$\frac{Z_n}{\sqrt{2\ln n}} = \frac{\max(X_1, \dots, X_n)}{\sqrt{2\ln n}} \to 1 \qquad \text{a.s.}$$

as $n \to \infty$ (see Galambos [141]). Here $a^{-1}(t) = \exp\{t^2/2\}$ also violates assumption (2.23).

So, there are (renewal) sequences $\{Z_n\}_{n\geq 1}$ satisfying a SLLN for which their corresponding renewal processes $\{T(t)\}_{t\geq 0}$, $\{L(t)\}_{t\geq 0}$, and $\{M(t)\}_{t\geq 0}$ do not possess any (nondegenerate) strong limiting behavior.

Remark 2.6 Just for the sake of completeness we should mention that there are also cases in which the renewal process satisfies a strong law of large numbers, but not so its sequence of renewal times. Consider, for example, a nonhomogeneous Poisson process $\{\pi(t)\}_{t\geq 0}$ with cumulative intensity function $\{\lambda(t)\}_{t\geq 0}$, i.e. $\lambda(t) = \mathsf{E}\pi(t)$, $t \geq 0$. If e.g. $\lambda(t)$ is continuous and strictly increasing to infinity, it is well-known that

$$\{\pi(t)\}_{t\geq 0} \stackrel{d}{=} \{\widetilde{\pi}(\lambda(t))\}_{t\geq 0},$$

where $\{\widetilde{\pi}(t)\}_{t\geq 0}$ is a homogeneous Poisson process with renewal times $\widetilde{S}_0 = 0$, $\widetilde{S}_n = X_1 + \cdots + X_n$, $n \geq 1$, based on a sequence $\{X_n\}_{n\geq 1}$ of independent Exp(1)-random variables. In other words, $\pi(t)$ is a renewal process constructed from exponential random variables, so that $\pi(t) = M(t) = L(t) = T(t)$ in this case.

Let

$$\lambda(t) = \begin{cases} \ln t, & t \geq e, \\ t/e, & 0 \leq t \leq e. \end{cases}$$

Then the strong law of large numbers for $\{\widetilde{\pi}(t)\}_{t\geq 0}$ implies that

$$\frac{\pi(t)}{\ln t} = \frac{\widetilde{\pi}(\ln t)}{\ln t} \to 1 \qquad \text{a.s.}$$

as $t \to \infty$. On the other hand, since $Z_n = \exp(\widetilde{S}_n)$ are renewal times for the process $\{\pi(t)\}_{t\geq 0}$, the law of the iterated logarithm for sums $\{\widetilde{S}_n\}_{n\geq 1}$ yields that

$$\frac{Z_n}{e^n} = \exp\left\{\widetilde{S}_n - n\right\}$$

oscillates almost surely between 0 and $+\infty$ as $n \to \infty$.

2.2.3 Rate of Convergence

It may also be interesting to collect general conditions under which convergence rate statements apply to the laws of large numbers in Lemma 2.2 and Theorems 2.3–2.5.

Theorem 2.7 *Assume that*

$$\frac{Z_n - a_n}{b_n} \to 0 \qquad a.s. \tag{2.32}$$

as $n \to \infty$, where

$$a_n \to \infty \quad \text{and} \quad a_n - a_{n-1} = o(b_n), \tag{2.33}$$

$$0 < b_n \to \infty \quad \text{and} \quad b_n = o(a_n), \tag{2.34}$$

$$\frac{b_{n+1}}{b_n} = O(1). \tag{2.35}$$

Then

$$\frac{a_{L(t)} - t}{b_{L(t)}} \to 0 \qquad a.s., \tag{2.36}$$

$$\frac{a_{M(t)} - t}{b_{M(t)}} \to 0 \qquad a.s. \tag{2.37}$$

as $t \to \infty$. Moreover, if the sequence $\{a_n\}$ is nondecreasing, then

$$\frac{a_{T(t)} - t}{\max(b_{L(t)}, b_{M(t)})} \to 0 \qquad a.s. \tag{2.38}$$

Proof of Theorem 2.7 First note that, with $X_n = Z_n - Z_{n-1}$, $n \geq 1$, Assumptions (2.32), (2.33), and (2.35) yield

$$\frac{X_n}{b_n} = \frac{Z_n - a_n}{b_n} - \frac{Z_{n-1} - a_{n-1}}{b_{n-1}}\frac{b_{n-1}}{b_n} + \frac{a_n - a_{n-1}}{b_n} \to 0, \qquad n \to \infty. \tag{2.39}$$

Now, from (2.15),

$$\frac{a_{L(t)} - Z_{L(t)+1}}{b_{L(t)}} < \frac{a_{L(t)} - t}{b_{L(t)}} \leq \frac{a_{L(t)} - Z_{L(t)}}{b_{L(t)}}. \tag{2.40}$$

Since $L(t) \to \infty$ almost surely as $t \to \infty$, relation (2.32) implies that the right-hand side of (2.40) tends to 0 almost surely. On the other hand, by (2.39) and (2.35),

$$\frac{a_{L(t)} - Z_{L(t)}}{b_{L(t)}} - \frac{a_{L(t)} - Z_{L(t)+1}}{b_{L(t)}} = \frac{X_{L(t)+1}}{b_{L(t)+1}} \frac{b_{L(t)+1}}{b_{L(t)}} \to 0 \qquad \text{a.s.,}$$

and thus the left-hand side of (2.40) tends to 0 almost surely. This completes the proof of (2.36).

The proof of (2.37) can be given by using the same arguments.

Finally, (2.38) follows from (2.36) and (2.37), since

$$|a_{T(t)} - t| \le \max(|a_{L(t)} - t|, |a_{M(t)} - t|)$$

because of $\{a_n\}_{n\ge1}$ being monotone. □

Now assume that $a(\cdot)$ has a continuous derivative $a'(\cdot)$ on $(t_0, +\infty)$ satisfying

$$a'(t) \asymp a'(s) \quad \text{as} \quad t \asymp s, \tag{2.41}$$

i.e. $|a'(t)/a'(s)|$ is bounded away from 0 and ∞, if $|t/s|$ is bounded away from 0 and ∞, as $t, s \to \infty$. Moreover, let $\{b(t), t \ge 0\}$ be an extension of $\{b_n\}$ such that

$$b(t) \asymp b(s) \quad \text{as} \quad t \asymp s. \tag{2.42}$$

Corollary 2.8 *Assume the strong law of large numbers (2.32) together with (2.22), (2.23), (2.33), (2.34), (2.41) and (2.42). Then, as $t \to \infty$,*

$$\frac{a'(a^{-1}(t))}{b(a^{-1}(t))}\big(L(t) - a^{-1}(t)\big) \to 0 \qquad a.s., \tag{2.43}$$

$$\frac{a'(a^{-1}(t))}{b(a^{-1}(t))}\big(T(t) - a^{-1}(t)\big) \to 0 \qquad a.s., \tag{2.44}$$

$$\frac{a'(a^{-1}(t))}{b(a^{-1}(t))}\big(M(t) - a^{-1}(t)\big) \to 0 \qquad a.s. \tag{2.45}$$

Proof of Corollary 2.8 We only prove relation (2.43). The proofs of (2.44) and (2.45) are similar.

Observe that, in view of (2.32) and (2.34), as $n \to \infty$,

$$\frac{Z_n}{a_n} - 1 = \frac{Z_n - a_n}{b_n}\frac{b_n}{a_n} = o(1) \qquad \text{a.s.} \tag{2.46}$$

Since conditions (2.33) and (2.34) also yield (2.18), assertion (2.46) implies (2.19) and, moreover, (2.24) under the given assumptions, i.e. $L(t) \sim a^{-1}(t)$ almost surely, $t \to \infty$.

Now, by the mean value theorem,

$$a_{L(t)} - t = a(L(t)) - a(a^{-1}(t)) = a'(\xi(t))(L(t) - a^{-1}(t)), \tag{2.47}$$

where $\xi(t) \sim a^{-1}(t)$ almost surely as $t \to \infty$. On the other hand, relations (2.36) and (2.42) imply

$$\frac{a_{L(t)} - t}{b(a^{-1}(t))} = \frac{a_{L(t)} - t}{b_{L(t)}} \frac{b(L(t))}{b(a^{-1}(t))} = o(1) \qquad \text{a.s.} \tag{2.48}$$

as $t \to \infty$, so that a combination of (2.41), (2.47), and (2.48) completes the proof of (2.43). □

The following result extends the sufficiency part of Theorem 1.15, i.e. the implication (1.27)$\Longrightarrow$(1.26), to the functionals $T(t)$ and $M(t)$. It is also worth mentioning that Corollary 2.9 below holds in the general case, while Theorem 1.15 is valid only for independent, identically distributed random variables.

Corollary 2.9 *Assume that, for some $a > 0$ and $r \geq 1$, as $n \to \infty$,*

$$\frac{Z_n - na}{n^{1/r}} \to 0 \qquad a.s. \tag{2.49}$$

Then, as $t \to \infty$,

$$\frac{L(t) - t/a}{t^{1/r}} \to 0 \qquad a.s., \tag{2.50}$$

$$\frac{T(t) - t/a}{t^{1/r}} \to 0 \qquad a.s., \tag{2.51}$$

$$\frac{M(t) - t/a}{t^{1/r}} \to 0 \qquad a.s. \tag{2.52}$$

Proof of Corollary 2.9 If $r > 1$, the functions $a(t) = ta$, $b(t) = t^{1/r}$ satisfy the assumptions of Corollary 2.8 with $a^{-1}(t) = t/a$, $a'(t) \equiv a$. If $r = 1$, the result follows from Theorem 2.3 with $a^{-1}(t) = t/a$. □

Remark 2.10 By the same technique as used to prove Corollary 2.8, one can also derive general LIL type results for renewal processes. The details will be omitted.

Similar to Theorem 2.5 the regularity assumptions of Corollary 2.8 can be considerably weakened if a different technique of proof is applied.

Theorem 2.11 *Assume (2.32) together with*

$$b(t) \uparrow \infty \qquad as \quad t \to \infty, \tag{2.53}$$

$$\frac{a(t)}{b(t)} \uparrow \qquad as \quad t \to \infty, \tag{2.54}$$

$$a'\left(a^{-1}(t)\right) \asymp a'\left(a^{-1}(s)\right) \qquad as \quad t \asymp s, \tag{2.55}$$

$$b\left(a^{-1}(t)\right) \asymp b\left(a^{-1}(s)\right) \qquad as \quad t \asymp s, \tag{2.56}$$

where $a(t)$ is continuously differentiable on (t_0, ∞) with

$$a'(t) = o(b(t)) \qquad as \quad t \to \infty. \tag{2.57}$$

Then, as $t \to \infty$, assertions (2.43)–(2.45) retain.

Proof of Theorem 2.11 We prove only (2.44). The arguments for (2.43) and (2.45) are similar.

First, for any $\varepsilon > 0$, set $n = \left[a^{-1}(t)\right]$ and $m^{\pm} = \left[a^{-1}(t \pm \varepsilon(t))\right]$, where $\varepsilon(t) = \varepsilon b(a^{-1}(t))$. Note that $\varepsilon(t)/t \asymp \varepsilon b(a^{-1}(t))/a(a^{-1}(t)) \asymp \varepsilon$ as $t \to \infty$.

Then, as in the proof of (2.27),

$$T(t) - n = -T_1(t) - T_2(t) + T_3(t) + T_4(t)$$

and we have to show that the four terms are of order $o\left(b\left(a^{-1}(t)\right)/a'\left(a^{-1}(t)\right)\right)$ as $t \to \infty$.

For example, if $k \le m^- \le a^{-1}(t - \varepsilon(t))$,

$$\frac{t - a_k}{b_k} \ge \frac{t - a\left(a^{-1}(t - \varepsilon(t))\right)}{b\left(a^{-1}(t)\right)} = \frac{\varepsilon(t)}{b\left(a^{-1}(t)\right)} = \varepsilon.$$

Thus (2.57) implies that

$$\frac{a'\left(a^{-1}(t)\right)}{b\left(a^{-1}(t)\right)} T_1(t) \to 0 \qquad \text{a.s.}$$

as $t \to \infty$, since only a finite number of terms in $T_1(t)$ are nonvanishing.

Similarly, for $k \ge m^+ + 1 > a^{-1}(t + \varepsilon(t))$,

$$\frac{t - a_k}{b_k} \le \frac{t - a\left(a^{-1}(t + \varepsilon(t))\right)}{b\left(a^{-1}(t + \varepsilon(t))\right)} = -\frac{\varepsilon(t)}{b(a^{-1}(t + \varepsilon(t)))} \asymp -\varepsilon, \qquad t \to \infty,$$

which, by the same reasoning, implies

$$\frac{a'\left(a^{-1}(t)\right)}{b\left(a^{-1}(t)\right)} T_4(t) \to 0 \qquad \text{a.s.,} \qquad t \to \infty.$$

It remains to prove

$$\frac{a'\left(a^{-1}(t)\right)}{b\left(a^{-1}(t)\right)} (T_2(t) + T_3(t)) \to 0 \qquad \text{a.s.,} \qquad t \to \infty.$$

The latter relation (similarly (2.30) and (2.31)) follows from the decomposition

$$a^{-1}(t \pm \varepsilon(t)) - a^{-1}(t) = \pm \frac{\varepsilon(t)}{a'\left(a^{-1}(\tau^{\pm})\right)} \asymp \pm \varepsilon \frac{b\left(a^{-1}(t)\right)}{a'\left(a^{-1}(t)\right)}.$$

Since

$$\frac{a'\left(a^{-1}(t)\right)}{b\left(a^{-1}(t)\right)} \left(T(t) - \left[a^{-1}(t)\right]\right) = \frac{a'\left(a^{-1}(t)\right)}{b\left(a^{-1}(t)\right)} \left(T(t) - a^{-1}(t)\right) + o(1)$$

in view of (2.56), the proof of (2.44) is complete. □

Example 2.12 There are still situations in which the assumptions of Theorems 2.5 and 2.11 do not hold true, but yet a strong law of large numbers may be available. Consider, for instance, a sequence $\{Z_n\}_{n \geq 1}$ satisfying

$$\frac{Z_n}{\ln n} \to 1 \qquad \text{a.s.}$$

as $n \to \infty$, and assume a rate of convergence therein, e.g.

$$\limsup_{n \to \infty} \left| \frac{Z_n - \ln n}{b(n)} \right| \leq B$$

for some nonrandom constant $B > 0$. Let the function $\{b(t)\}_{t \geq 0}$ be such that, for all $A > B$,

$$a_{\pm}(t) = \ln t \pm A b(t)$$

have inverse functions (say) $a_{\pm}^{-1}(t)$ satisfying

$$a_{\pm}^{-1}(t) = e^t \mp o\left(e^t\right)$$

as $t \to \infty$. Then the SLLN for $\{T(t), t \geq 0\}$ retains, i.e.

$$\lim_{t\to\infty} \frac{T(t)}{e^t} = 1 \qquad \text{a.s.}$$

The proof is similar to that of Theorem 2.11 by dividing the series

$$T(t) = \sum_{n=1}^{\infty} \mathbb{I}\{Z_n \leq t\}$$

into four subseries according to the conditions $n \leq a_{+}^{-1}(t)$, $a_{+}^{-1}(t) < n \leq e^t$, $e^t < n \leq a_{-}^{-1}(t)$, $a_{+}^{-1}(t) < n$, respectively. The details are omitted.

Note that the function $a(t) = \log t$ with $a^{-1}(t) = e^t$ violates conditions (2.23) and (2.55), so that neither Theorem 2.5 nor Theorem 2.11 is applicable in this situation. Nevertheless, a strong law of large numbers for the renewal process holds true.

2.3 Examples

In this section, we demonstrate the applicability of our general results above with a series of examples. Various situations of renewal processes are discussed related to a scheme of independent, identically distributed random variables, to independent, but nonidentically distributed renewal times, and also to certain dependent sequences.

2.3.1 Renewal Processes Constructed from Independent, Identically Distributed Random Variables

For Examples 2.13–2.16 we assume that $\{X_n\}_{n\geq 1}$ are independent, identically distributed random variables, and set $S_0 = 0$, $S_n = X_1 + \cdots + X_n$, $n \geq 1$.

Example 2.13 (Linear Renewal Process) Let $\mathsf{E}X_1 = a > 0$ and $a(t) = ta$. Then $a^{-1}(t)$ equals t/a. The Kolmogorov strong law of large numbers for $Z_n = S_n$ means that

$$\frac{Z_n}{na} \to 1 \qquad \text{a.s.}$$

as $n \to \infty$. If $L(t)$, $M(t)$ and $T(t)$ are defined by (2.7)–(2.9), respectively, Theorem 2.3 implies

$$\frac{M(t)}{t/a} \to 1, \qquad \frac{L(t)}{t/a} \to 1, \qquad \frac{T(t)}{t/a} \to 1 \quad \text{a.s.} \tag{2.58}$$

as $t \to \infty$.

For a convergence rate statement we refer to an extension of a classical result of Feller [126] obtained in Martikainen and Petrov [264]: *If the function $b(t)$ is positive, increasing and such that*

$$\sum_{k=n}^{\infty} \frac{1}{b^2(k)} = O\left(\frac{n}{b^2(n)}\right), \tag{2.59}$$

the following statements are equivalent:

$$\frac{S_n - na}{b(n)} \to 0 \quad \text{a.s.}, \qquad n \to \infty, \tag{2.60}$$

$$\sum_{n=1}^{\infty} \mathsf{P}(|Y| \ge b(n)) < \infty, \qquad \lim_{n\to\infty} \frac{n}{b(n)} \int_{\{|x|<b(n)\}} x\, dF(x) = 0, \tag{2.61}$$

where $Y = X_1 - a$ and where F is the distribution function of the random variable Y.

Combining the latter result with Theorem 2.11, we deduce from (2.61) that

$$\lim_{t\to\infty} \frac{M(t) - t/a}{b(t)} = 0, \qquad \lim_{t\to\infty} \frac{L(t) - t/a}{b(t)} = 0, \qquad \lim_{t\to\infty} \frac{T(t) - t/a}{b(t)} = 0$$

almost surely.

One of the possible choices for b is given by the (so-called) Marcinkiewicz–Zygmund normalization $b(t) = t^{1/r}$, where $t > 0$ and $1 < r < 2$: *if $\mathsf{E}|X_1|^r < \infty$ for some $1 < r < 2$, then*

$$\lim_{t\to\infty} \frac{M(t) - t/a}{t^{1/r}} = 0, \qquad \lim_{t\to\infty} \frac{L(t) - t/a}{t^{1/r}} = 0, \qquad \lim_{t\to\infty} \frac{T(t) - t/a}{t^{1/r}} = 0$$

almost surely.

Example 2.14 (Nonlinear Renewal Process) As before, consider a sequence of independent, identically distributed random variables $\{X_n\}_{n\ge1}$ with $\mathsf{E}X_1 = a > 0$. Now we put $Z_n = S_n/\alpha(n)$, where α is a positive continuous function such that

$$\frac{t}{\alpha(t)} \uparrow \infty, \qquad t \uparrow \infty.$$

For example, the first-exit time

$$L(t)+1=\inf\{n : Z_n > t\}=\inf\{n : S_n > t\alpha(n)\},$$

where $\inf\varnothing = +\infty$, is of some statistical importance in sequential analysis and plays a key role in what is called nonlinear renewal theory (cf., e.g., Woodroofe [366] and Siegmund [335]). By Theorem 2.5, if the inverse function $a^{-1}(t)$ for $a(t)=ta/\alpha(t)$ satisfies condition (2.23), then

$$\lim_{t\to\infty}\frac{L(t)}{a^{-1}(t)}=1,\qquad \lim_{t\to\infty}\frac{M(t)}{a^{-1}(t)}=1,\qquad \lim_{t\to\infty}\frac{T(t)}{a^{-1}(t)}=1\qquad \text{a.s.} \tag{2.62}$$

as $t\to\infty$, where $L(t)$, $M(t)$, and $T(t)$ are the processes defined by (2.7)–(2.9), respectively.

If, additionally, $\mathsf{E}|X_1|^r<\infty$ for some $1<r<2$, put $b(n)=n^{1/r}/\alpha(n)$. Then

$$\frac{Z_n-a(n)}{b(n)}=\frac{S_n-na}{n^{1/r}}\to 0\qquad \text{a.s.}$$

as $n\to\infty$. Theorem 2.11 implies

$$\lim_{t\to\infty}\frac{L(t)-a^{-1}(t)}{(a^{-1}(t))^{1/r}}=\lim_{t\to\infty}\frac{T(t)-a^{-1}(t)}{(a^{-1}(t))^{1/r}}=\lim_{t\to\infty}\frac{M(t)-a^{-1}(t)}{(a^{-1}(t))^{1/r}}=0\qquad \text{a.s.,} \tag{2.63}$$

as $t\to\infty$ (see Gut [159], Theorem 5.5 of Chapter IV).

Example 2.15 (Renewal Processes Constructed from Subsequences) Strong laws of large numbers and other convergence properties have also been extensively studied for subsequences of partial sums of a sequence of independent, identically distributed random variables. Corresponding properties of their renewal processes, which may be viewed as being related to certain nonlinear inspection schemes, can also be derived from the general results of Sect. 2.2. Consider e.g. a subsequence $\{a_n\}_{n\ge1}$ of integers with an extension $\{a(t)\}_{t\ge0}$ and inverse $\{a^{-1}(t)\}_{t\ge0}$. Let $Z_n = S_{a_n}$, $n\ge 1$, and the renewal processes $L(t)$, $M(t)$, and $T(t)$ be defined as in (2.7)–(2.9).

Suppose that the sequence a_n/n increases, that is the function $a(t)/t$ is also increasing. This assumption is sufficient for (2.23). Indeed, for all $\varepsilon>0$,

$$\frac{a\left((1+\varepsilon)a^{-1}(t)\right)}{(1+\varepsilon)a^{-1}(t)}\ge\frac{a\left(a^{-1}(t)\right)}{a^{-1}(t)}=\frac{t}{a^{-1}(t)},$$

whence $a\left((1+\varepsilon)a^{-1}(t)\right)\ge(1+\varepsilon)t$ or $(1+\varepsilon)a^{-1}(t)\ge a^{-1}((1+\varepsilon)t)$. A similar approach allows us to obtain corresponding estimates with $1-\varepsilon$ instead of $1+\varepsilon$. This, in turn, implies (2.23), so that (2.24)–(2.26) are also satisfied.

In order to get a rate of convergence, one has to assume conditions (2.54)–(2.57) in Theorem 2.11. Let us consider the case of a Marcinkiewicz–Zygmund normalization. On assuming $\mathsf{E}X_1 > 0$ and $\mathsf{E}|X_1|^r < \infty$ for some $1 < r < 2$, we obtain

$$\frac{S_{a_n} - a_n \mathsf{E}X_1}{a_n^{1/r}} \to 0 \qquad \text{a.s.}$$

as $n \to \infty$. Choose $b(n) = a_n^{1/r}$. One would now expect a related result for the corresponding renewal processes, but this is not always true. For the sake of simplicity, consider $a_n = [n^\nu]$ with $\nu > 1$. Then $a(t) \asymp t^\nu$, $a'(t) \asymp t^{\nu-1}$, $a^{-1}(t) \asymp t^{1/\nu}$, and $b(t) \asymp t^{\nu/r}$. Conditions (2.54)–(2.56) are obvious in this case, but condition (2.57) requires a restriction on r, i.e. $r < (\nu - 1)/\nu$.

Example 2.16 (Renewal Processes Constructed from Sums When Extremes Are Excluded) It is well-known that, under certain assumptions, partial sums of independent, identically distributed random variables with infinite expectation can still satisfy a strong law of large numbers if the extremal terms are removed from the sums. Consider, for example, S_n as above, and put

$$Z_n = S_n - \max_{1 \le k \le n} |X_k|,$$

i.e. Z_n is the n-th partial sum with the maximal term being excluded. It is proved in Mori [278] that

$$\lim_{n\to\infty} \frac{Z_n - nc_n}{n} = 0 \qquad \text{a.s.}$$

for some nonrandom sequence $\{c_n\}_{n\ge 1}$ if and only if

$$\int_0^\infty x\overline{F}^2(x)\,dx < \infty, \tag{2.64}$$

where $\overline{F}(x) = \mathsf{P}(|X_1| > x)$. Without loss of generality, one can assume that $c_n = \mathsf{E}X_1 \mathbb{I}\{|X_1| < n\tau\}$ for some positive number τ, for example for $\tau = 1$. Put $c(t) = \mathsf{E}X_1 \mathbb{I}\{|X_1| < t\}$ and $a(t) = tc(t)$. Then, under the assumptions of Theorem 2.5,

$$\frac{M(t)}{a^{-1}(t)} \to 1, \qquad \frac{L(t)}{a^{-1}(t)} \to 1, \qquad \frac{T(t)}{a^{-1}(t)} \to 1, \qquad \text{a.s.}$$

as $t \to \infty$. Moreover, under the assumptions of Theorem 2.11 with $b(t) = t$, we have

$$\lim_{t\to\infty} \frac{a'(a^{-1}(t))}{a^{-1}(t)}\left(M(t) - a^{-1}(t)\right) = 0 \qquad \text{a.s.},$$

$$\lim_{t\to\infty} \frac{a'(a^{-1}(t)}{a^{-1}(t)}\left(L(t) - a^{-1}(t)\right) = 0 \qquad \text{a.s.},$$

$$\lim_{t\to\infty} \frac{a'(a^{-1}(t)}{a^{-1}(t)}\left(T(t) - a^{-1}(t)\right) = 0 \qquad \text{a.s.}$$

as $t \to \infty$ provided condition (2.64) holds.

For example, choose $F(x) = \mathsf{P}(X_1 \le x)$ as follows:

$$F(x) = \begin{cases} 1 - \frac{1}{x \ln x}, & x \ge e, \\ 0, & x < e. \end{cases}$$

The expectation does not exist for this distribution and therefore the renewal process based on sums has no linear asymptotic. On the other hand, the renewal process based on sums with excluded maximal term has a (non-linear) asymptotic. Indeed, in this case $c(t) = \ln\ln t + O(1)$, $a(t) = t \ln\ln t + O(t)$, $a'(t) = \ln\ln t + o(1)$, $a^{-1}(t) = t/\ln\ln t + O\left(t/(\ln\ln t)^2\right)$. Thus

$$\frac{(\ln\ln t)^2}{t}\left(T(t) - \frac{t}{\ln\ln t}\right) \to 0 \qquad \text{a.s.}$$

as $t \to \infty$. Similar results can be obtained for sums with $r \ge 1$ maximal terms excluded, if condition (2.64) is replaced by

$$\int_0^\infty x^r \overline{F}^{r+1}(x)\, dx < \infty. \tag{2.65}$$

Example 2.17 (Generalized Renewal Process) The so-called strong approximation is studied in Horváth [184] for generalized renewal processes of the following type. Let $\{\mathbf{X}_n\}_{n\ge 1}$ be a sequence of d-dimensional independent identically distributed random vectors with $\mathsf{E}\mathbf{X}_1 = \mathbf{a}$. Put $\mathbf{S}_0 = \mathbf{0}$, $\mathbf{S}_n = \mathbf{X}_1 + \cdots + \mathbf{X}_n$, $n \ge 1$. Let $h : \mathbf{R}^d \to \mathbf{R}^1$ be a first-order homogeneous and continuously differentiable function such that $h(\mathbf{a}) > 0$. For a fixed $0 \le p < 1$, put $Z_n = h(\mathbf{S}_n)/n^p$ and

$$L(t) + 1 = \inf\{n : Z_n > t\} = \inf\{n : h(\mathbf{S}_n) > tn^p\},$$

$\inf \varnothing = +\infty$. If $\mathsf{E}|\mathbf{X}|^r < \infty$ for some $r > 2$, the strong approximation of

$$L(t) - (t/h(\mathbf{a}))^{1/q}, \qquad q = 1 - p, \tag{2.66}$$

by a Wiener process is obtained in Horváth [184]. In turn, this result has various corollaries for the almost sure convergence of $\{L(t)\}_{t\geq 0}$.

Under weaker assumptions, we are still able to retain some SLLN or convergence rate type results by the methods of Sect. 2.2. For example, if only $\mathsf{E}|\mathbf{X}_1| < \infty$, then similar to Example 2.14,

$$\frac{M(t)}{(t/h(\mathbf{a}))^{1/q}} \to 1, \qquad \frac{L(t)}{(t/h(\mathbf{a}))^{1/q}} \to 1, \qquad \frac{T(t)}{(t/h(\mathbf{a}))^{1/q}} \to 1 \qquad \text{a.s.}$$

as $t \to \infty$. If $\mathsf{E}|\mathbf{X}_1|^r < \infty$ for some $1 < r < 2$, then, as $t \to \infty$,

$$\frac{M(t) - (t/h(\mathbf{a}))^{1/q}}{t^{1/rq}} \to 0 \qquad \text{a.s.,}$$

$$\frac{L(t) - (t/h(\mathbf{a}))^{1/q}}{t^{1/rq}} \to 0 \qquad \text{a.s.,}$$

$$\frac{T(t) - (t/h(\mathbf{a}))^{1/q}}{t^{1/rq}} \to 0 \qquad \text{a.s.}$$

2.3.2 *Nonidentically Distributed or Dependent Interarrival Times*

Example 2.18 (Weighted Independent, Identically Distributed Interarrival Times) Consider a sequence $\{X_n\}_{n\geq 1}$ of independent, identically distributed random variables with $\mathsf{E}X_1 = 1$ and some positive function w. Put $Z_n = \sum_{k=1}^n w(k)X_k$ and introduce the renewal processes $\{L(t)\}_{t\geq 0}$, $\{M(t)\}_{t\geq 0}$, and $\{T(t)\}_{t\geq 0}$ by (2.7), (2.8) and (2.9), respectively.

If $a(n) = w(1) + \cdots + w(n)$, $n \geq 1$, then $\mathsf{E}Z_n = a(n)$. By Kolmogorov's strong law of large numbers for nonidentically distributed random variables

$$\lim_{n\to\infty} \frac{S_n - a(n)}{b(n)} = 0 \qquad \text{a.s.}$$

if $\mathsf{E}X_1^2 < \infty$, where the function b is positive, nondecreasing, unbounded and such that

$$\sum_{n=1}^{\infty} \frac{w^2(n)}{b^2(n)} < \infty.$$

The moment condition can be weakened by using a truncation procedure. Nevertheless we retain the assumption that the second moment exists to make the presentation simpler. Moreover, we choose $w(t) = t^\theta - (t-1)^\theta \sim \theta t^{\theta-1}$ for some $\theta > 0$.

Then $a(n) = n^\theta$. Put $a(t) = t^\theta$ and note that $a^{-1}(t) = t^{1/\theta}$, $a'(t) = \theta t^{\theta-1}$. The assumptions of Theorem 2.11 hold if one assumes that

$$\lim_{n\to\infty} \frac{b(n)}{n^\theta} = 0, \qquad \lim_{n\to\infty} \frac{b(n)}{n^{\theta-1}} = \infty.$$

Then

$$\sum_{n=1}^{\infty} \frac{n^{2(\theta-1)}}{b^2(n)} < \infty,$$

implies that

$$\lim_{t\to\infty} \left(M(t) - t^{1/\theta}\right) \frac{t^{(\theta-1)/\theta}}{b(t^{1/\theta})} = 0 \qquad \text{a.s.,}$$

$$\lim_{t\to\infty} \left(L(t) - t^{1/\theta}\right) \frac{t^{(\theta-1)/\theta}}{b(t^{1/\theta})} = 0 \qquad \text{a.s.,}$$

$$\lim_{t\to\infty} \left(T(t) - t^{1/\theta}\right) \frac{t^{(\theta-1)/\theta}}{b(t^{1/\theta})} = 0 \qquad \text{a.s.}$$

As a normalization $b(\cdot)$ one can take, for example, $b(t) = t^\delta$, $t > 0$, with $\theta - 1 < \delta < \theta$. Then

$$\lim_{t\to\infty} \frac{M(t) - t^{1/\theta}}{t^{(\delta-\theta+1)/\theta}} = 0 \qquad \text{a.s.,}$$

$$\lim_{t\to\infty} \frac{L(t) - t^{1/\theta}}{t^{(\delta-\theta+1)/\theta}} = 0 \qquad \text{a.s.,}$$

$$\lim_{t\to\infty} \frac{T(t) - t^{1/\theta}}{t^{(\delta-\theta+1)/\theta}} = 0 \qquad \text{a.s.}$$

A general approach to the strong law of large numbers is developed in Fazekas and Klesov [124]. Their key idea was to show that Hajek–Rényi type inequalities can be obtained from appropriate maximal inequalities for cumulative sums, and that the latter, in turn, imply the strong law of large numbers. By this method, no assumptions on the dependency structure of the summands are required, and a number of examples can be covered including sums of independent, but nonidentically distributed summands, martingale difference schemes, mixing sequences, mixingales, orthogonal sequences, sequences with superadditive moment structure and many others.

All these examples may, under appropriate conditions, be converted into strong laws for their corresponding renewal processes. Just for the sake of demonstration, we consider some further examples.

Example 2.19 (Renewal Processes Under ρ-Mixing) Let $\{X_n\}_{n\geq 1}$ be a sequence of identically distributed, ρ-mixing random variables. Suppose that $\mathsf{E}X_1 = a > 0$ and $\mathsf{E}|X_1|^r < \infty$ for some $1 \leq r < 2$. Putting $a(t) = ta$ and $b(t) = t^{1/r}$, we get

$$\frac{M(t) - t/a}{t^{1/r}} \to 0, \qquad \frac{L(t) - t/a}{t^{1/r}} \to 0, \qquad \frac{T(t) - t/a}{t^{1/r}} \to 0 \qquad \text{a.s.}$$

as $t \to \infty$, if $\sum \rho(2^n) < \infty$, where ρ is the Kolmogorov–Rozanov mixing coefficient for the sequence $\{X_n\}_{n\geq 1}$.

The above mixing condition can even be weakened (cf. Shao [331] or Fazekas and Klesov [124], Theorem 5.1). Extensions to mixing sequences of nonidentically distributed random variables are also available (see [124], Theorem 5.2). Naturally, the case of m-dependent renewal times is included (see also Janson [199] for further asymptotics in the latter case).

Example 2.20 (Martingale-Difference Scheme) Let $\{X_n\}_{n\geq 1}$ be a martingale-difference scheme with respect to the filtration of σ-algebras $\{\mathfrak{F}_n\}_{n\geq 1}$, where $\mathfrak{F}_n$ is generated by the random variables $X_1, \dots, X_n$. Suppose that $\mathsf{E}X_1 = a > 0$, $q > 1/2$. Let $\{b_n\}_{n\geq 1}$ be a nondecreasing, unbounded sequence such that

$$\sum_{n=1}^{\infty} \frac{\mathsf{E}|Z_n|^{2q} - \mathsf{E}|Z_{n-1}|^{2q}}{b_n^{2q}} < \infty.$$

If $a(t) = ta$ and $b(t)$ satisfy the conditions of Theorems 2.11, then

$$\frac{M(t) - t/a}{b(t/a)} \to 0, \qquad \frac{L(t) - t/a}{b(t/a)} \to 0, \qquad \frac{T(t) - t/a}{b(t/a)} \to 0 \tag{2.67}$$

almost surely as $t \to \infty$ (see Fazekas and Klesov [124], Theorem 3.1).

As a consequence, we obtain a Brunk–Prokhorov type strong law for renewal processes based on martingale difference schemes: Let $\{X_n\}_{n\geq 1}$ and $\{b_n\}_{n\geq 1}$ be as above, but assume either $q = 1$, or $q > 1$ and $n^{-\delta} b_n$ be nondecreasing for some $\delta > (q-1)/2q$. If

$$\sum_{n=1}^{\infty} \frac{\mathsf{E}|X_n|^{2q}}{b_n^{2q}} n^{q-1} < \infty,$$

then (2.67) holds (see Fazekas and Klesov [124], Corollary 3.1).

Example 2.21 (Banach Space Schemes) Let $\{\mathbf{X}_n\}_{n\geq 1}$ be a sequence of independent, identically distributed random variables assuming values in a separable Banach space equipped with the norm $\|\cdot\|$. Put $\mathbf{S}_n = \mathbf{X}_1 + \cdots + \mathbf{X}_n$ and $Z_n = \|\mathbf{S}_n\|$. If $\mathsf{E}\|\mathbf{X}_1\| < \infty$, then (2.17) holds for $a_n = n\mu$, where $\mu = \|\mathsf{E}\mathbf{X}_1\| > 0$ (see Mourier [279]). Since conditions (2.22) and (2.23) are satisfied for such a sequence $\{a_n\}_{n\geq 1}$, we can deduce the asymptotics of the renewal process constructed from a random walk in a Banach space via Theorem 2.5, namely

$$\frac{1}{t}\sum_{n=1}^{\infty}\mathbb{I}\{\|\mathbf{S}_n\| \le t\} \to \frac{1}{\mu} \qquad \text{a.s.}$$

as $t \to \infty$.

2.4 Comments

This chapter is based on Klesov et al. [227].

The state of the art in renewal theory up to the middle of the 20th century is described in Smith [337] and Cox [93]. For a more recent monograph see, e.g., Alsmeyer [11]. A rich collection of works up to 1974 devoted to classical renewal processes and various generalizations are also included in Sevastyanov [328]. Numerous applications of renewal functions and processes from reliability theory are discussed in Kovalenko [238] and corresponding ones from queueing theory in Lipsky [251]. Markov renewal processes are studied in Pyke [298] (see also Alsmeyer [13]). A short introduction to renewal theory can be found in Feller [129], whereas discrete renewal processes are investigated in Barbu and Limnios [24]. Ergodic properties of trajectories in renewal theory are studied in Shurenkov [333, 334]. Some statistical problems arising in renewal theory are discussed in Cox and Lewis [94].

Section 2.2 The strong law of large numbers for renewal processes constructed from sums of independent, identically distributed random variables is proved in Doob [111]. Doob [111] noticed that the result "*follows from the law of large numbers for mutually independent chance variables with a common distribution function*" and that the existence of the first moment is just needed to verify the strong law of large numbers for the partial sums which is then used to obtain the strong law of large numbers for the renewal process. The method to prove Theorem 2.3 is developed in Klesov and Steinebach [229] for renewal processes constructed from multiindexed sums of independent random variables.

Theorem 2.3 states that strong laws of large numbers (2.24)–(2.26) follow from (2.17). The latter relation is the strong law of large numbers for the sequence $\{Z_n\}_{n\ge1}$. If $\{Z_n\}_{n\ge1}$ is a sequence of partial sums of independent, identically distributed random variables, then (2.17) coincides with the Kolmogorov strong law of large numbers for sums [234], while (2.24) is the Doob strong law of large numbers for renewal processes [111]. Theorem 2.3 is wider than the implication (1.24)$\Longrightarrow$(1.23) in Theorem 1.14 because it is valid for general random variables $\{X_n\}_{n\ge1}$ and for the functionals $M(t)$ and $T(t)$, as well.

Corollary 2.9 derives strong laws of large numbers (2.50)–(2.52) for generalized renewal processes from (2.49). The latter is called the Marcinkiewicz–Zygmund strong law of large numbers in the case where $\{Z_n\}_{n\geq 1}$ is a sequence of partial sums of independent, identically distributed random variables (see [260]).

Section 2.3 The strong law of large numbers for partial sums mentioned in Example 2.13 is due to Martikainen and Petrov [264].

The nonlinear renewal processes considered in Example 2.14 are studied by Woodroofe [366] and Siegmund [335] (see also Gut [159], pp. 133–138). We obtain (2.62) and (2.63) as simple corollaries of our general results.

The generalized renewal processes considered in Example 2.17 are studied in Horváth [184].

Some other examples of strong laws of large numbers for generalized renewal processes can be provided, e.g., for alternating renewal processes. Further examples are reward processes (see Khorshidian [219]), cyclic renewal processes (see Serfozo [327]), sojourn times for networks (see Serfozo [326], renewal processes constructed from non-identically distributed terms (see Borovkov [47]) and others.

Chapter 3
Generalizations of Regularly Varying Functions

3.1 Introduction

Karamata [205] introduced the (so-called) *regularly varying functions* (RV-functions) in 1930 and proved a number of fundamental results for them (see also Karamata [206]). These results and their further generalizations formed the basis of a developed theory with a wide range of applications (cf., e.g., Seneta [324], Bingham et al. [41], Alsmeyer [11], Bingham [38], Marić [261], Korevaar [237], Resnick [300, 302], Yakymiv [373]).

Following Karamata's [205] work, there appeared a number of various generalizations of regularly varying functions. The first of such generalizations is due to Avakumović [19] and has since been studied by a number of mathematicians (see, for example, Karamata [207], Feller [128], Aljančić and Arandelović [8], Bari and Stechkin [26], Arandelović [18]). The corresponding functions are called *O-regularly varying functions* (ORV-functions).

This chapter is devoted to the discussion of certain classes of functions which generalize, in one way or another, the notion of regular variation. Additionally to the classes $\mathcal{RV}$ and $\mathcal{ORV}$ of functions, an important place in the theory is occupied by the classes $\mathcal{PRV}$, $\mathcal{PI}$, $\mathcal{SQI}$, and $\mathcal{POV}$. The main attention in this chapter is paid to these four classes. Keeping the notation $\mathcal{RV}$, $\mathcal{ORV}$, $\mathcal{PI}$, etc. for the classes of measurable functions, we also consider their extensions, $\mathcal{WRV}$, $\mathcal{WORV}$, $\mathcal{WPI}$ etc., containing nonmeasurable functions, as well. Some properties of measurable functions of a specific class can be transferred into their counterparts for nonmeasurable functions.

The class $\mathcal{PRV}$ ($\mathcal{WPRV}$) of *pseudo-regularly varying functions* (PRV-functions (WPRV-functions)) is considered first. The characterizing property of WPRV-functions is that they preserve the asymptotic equivalence of functions and sequences. PRV- and WPRV-functions are studied, under different names, in Korenblyum [235], Matuszewska [267, 270], Gihman and Skorohod [149], Stadtmüller and Trautner

V. V. Buldygin et al., *Pseudo-Regularly Varying Functions and Generalized Renewal Processes*, Probability Theory and Stochastic Modelling 91, https://doi.org/10.1007/978-3-319-99537-3_3

[344, 345], Berman [30, 32], Yakymiv [372], Cline [90], Djurčić [105], Djurčić and Torgašev [108], Klesov et al. [227], and Buldygin et al. [59, 61–67].

The class $\mathcal{PI}$ ($\mathcal{WPI}$) of *positively increasing functions* (PI- and WPI-functions) is studied next. This class is introduced independently in [39, 40, 168, 305] and [59, 62, 66, 69].

Then the class $\mathcal{SQI}$ ($\mathcal{WSQI}$) of *sufficiently quickly increasing* functions (SQI- and WSQI-functions) is considered. This class is studied in Yakymiv [371], Djurčić and Torgašev [108], and Buldygin et al. [59, 61–67]. The WSQI-property is "inverse" to the WPRV-property in the sense that, under certain assumptions, the inverse or quasi-inverse functions to WSQI-functions preserve the asymptotic equivalence of functions and sequences, that is they belong to $\mathcal{WPRV}$.

The class $\mathcal{POV}$ ($\mathcal{WPOV}$) of functions with *positive order of variation* closes the list. These functions are introduced in Buldygin et al. [59] as a generalization of RV-functions with positive index. The characterizing property of POV-functions is that they together with their quasi-inverses preserve the asymptotic equivalence of functions and sequences.

The theory and applications of all the classes of functions mentioned above are developed not only in this chapter; some results are also obtained in other appropriate places of the book.

Chapter 3 is organized as follows. In Sect. 3.2, necessary general definitions and notations are given. Some simple results of this section are motivated by the corresponding results for RV-, SV-, ORV-functions. In particular, we consider the *upper* and *lower limit functions*.

Some preliminary results concerning PRV-, PI-, SQI-, and POV-functions are given in Sect. 3.3.

The main result of Sect. 3.4 states that only WPRV-functions preserve the asymptotic equivalence of functions and sequences.

Section 3.5 is devoted to the generalizations of known results for RV- and ORV-functions.

A generalization of Potter's bounds for PRV-functions is obtained in Sect. 3.6.

Conditions for the convergence to infinity of WPI- and PI-functions are given in Sect. 3.7.

Some characterizations of POV-functions are discussed in Sect. 3.8.

The final Sect. 3.9 is devoted to the proof of existence of equivalent continuous and increasing versions of POV-functions.

3.2 RV- and ORV-Functions

3.2.1 Some Notation

Let $\mathbf{R}$ be the set of real numbers, $\mathbf{R}_0$ the set of nonnegative numbers, $\mathbf{R}_+$ the set of positive numbers, $\mathbf{Q}$ the set of rational numbers, $\mathbf{Z}$ the set of integers, $\mathbf{N}$ the set of positive integers, and $\mathbf{N}_0 = \mathbf{N} \cup \{0\}$ the set of nonnegative integers.

We say that a real-valued function $\phi = (\phi(t), t \in D)$, $D \subset \mathbf{R}$, *increases* (*decreases*) in the set M ($M \subset D$) if $\phi(t_1) < \phi(t_2)$ ($\phi(t_1) > \phi(t_2)$) for all $t_1, t_2 \in M$ such that $t_1 < t_2$. Similarly, we say that a function ϕ is *non-decreasing* (*non-increasing*) in the set M ($M \subset D$) if $\phi(t_1) \le \phi(t_2)$ ($\phi(t_1) \ge \phi(t_2)$) for all $t_1, t_2 \in M$ such that $t_1 < t_2$. We also say that a function ϕ increases (decreases) or is non-decreasing (non-increasing) *at infinity* if there exists an $a > 0$ such that the corresponding property holds in the set $M = [a, \infty)$. We assume that M coincides with the domain of a function ϕ if M is not specified when talking about the increase or decrease of ϕ.

Let $\mathbb{F}$ denote the set of real-valued functions $f = (f(t), t \ge 0)$,

$$\mathbb{F}_+(A) = \{f \in \mathbb{F} : f(t) > 0, t \in [A, \infty)\}$$

and let

$$\mathbb{F}_+ = \bigcup_{A>0} \mathbb{F}_+(A).$$

Therefore, $f \in \mathbb{F}_+$ if and only if $f(t) > 0$ for all sufficiently large t.

Let $\mathbb{F}^{(\infty)}$ be the set of functions $f \in \mathbb{F}_+$ such that

$$\limsup_{t\to\infty} f(t) = \infty,$$

and let $\mathbb{F}^{\infty}$ be the subset of functions of $\mathbb{F}^{(\infty)}$ such that

$$\lim_{t\to\infty} f(t) = \infty.$$

Further, let $\mathbb{F}_{\text{ndec}}$ ($\mathbb{F}_{\text{inc}}$) be the set of functions $f \in \mathbb{F}_+$ such that f is non-decreasing (increasing) for large t. Put

$$\mathbb{F}^{\infty}_{\text{ndec}} = \mathbb{F}_{\text{ndec}} \cap \mathbb{F}^{\infty},$$
$$\mathbb{F}^{\infty}_{\text{inc}} = \mathbb{F}_{\text{inc}} \cap \mathbb{F}^{\infty}.$$

By $\mathbb{C}^{(\infty)}$, $\mathbb{C}^{\infty}$, $\mathbb{C}_{\text{ndec}}$, $\mathbb{C}_{\text{inc}}$, $\mathbb{C}^{\infty}_{\text{ndec}}$, and $\mathbb{C}^{\infty}_{\text{inc}}$ we denote the subsets comprising the continuous functions of the spaces $\mathbb{F}^{(\infty)}$, $\mathbb{F}^{\infty}$, $\mathbb{F}_{\text{ndec}}$, $\mathbb{F}_{\text{inc}}$, $\mathbb{F}^{\infty}_{\text{ndec}}$, and $\mathbb{F}^{\infty}_{\text{inc}}$, respectively.

In what follows "measurability" means "Lebesgue measurability" and "meas" stands for the Lebesgue measure in $\mathbf{R}$.

3.2.2 Upper and Lower Limit Functions

For $f \in \mathbb{F}_+$, we consider the *upper* and *lower limit functions* denoted by f^* and f_*, respectively:

$$f^*(c) = \limsup_{t\to\infty} \frac{f(ct)}{f(t)} \quad \text{and} \quad f_*(c) = \liminf_{t\to\infty} \frac{f(ct)}{f(t)}, \qquad c > 0. \tag{3.1}$$

The values of these two functions belong to $[0, \infty]$.

The limit functions are the main tool of the Karamata and similar approaches. The following properties of limit functions are direct consequences of the definitions.

Lemma 3.1 *Let $f \in \mathbb{F}_+$. Then:*

(i) for all $c > 0$,

$$0 \le f_*(c) \le f^*(c);$$

(ii) for all $c > 0$,

$$f_*(c) = \frac{1}{f^*(1/c)},$$

where $1/\infty = 0$ and $1/0 = \infty$;

(iii) for all $c_1 > 0$ and $c_2 > 0$, the following inequalities hold provided they do not contain expressions like $0 \cdot \infty$ and $\infty \cdot 0$:

$$\begin{aligned} f_*(c_1)f_*(c_2) &\le f_*(c_1c_2) \le \min\{f_*(c_1)f^*(c_2),\ f_*(c_2)f^*(c_1)\} \\ &\le \max\{f_*(c_1)f^*(c_2),\ f_*(c_2)f^*(c_1)\} \le f^*(c_1c_2) \\ &\le f^*(c_1)f^*(c_2); \end{aligned}$$

(iv) $f_*(1) = f^*(1) = 1$;

(v) if f is a non-decreasing (non-increasing) function at infinity, then f_* and f^* are non-decreasing (non-increasing) functions in the set $\mathbf{R}_+$;

(vi) for any constant $a > 0$ and for all $c > 0$,

$$(af)_*(c) = f_*(c) \quad \text{and} \quad (af)^*(c) = f^*(c);$$

(vii) if $f_1 \in \mathbb{F}_+$, $f_2 \in \mathbb{F}$, and

$$\lim_{t\to\infty} \frac{|f_2(t)|}{f_1(t)} = 0, \tag{3.2}$$

then, for all $c > 0$,

$$(f_1 + f_2)_*(c) = (f_1)_*(c) \quad \text{and} \quad (f_1 + f_2)^*(c) = (f_1)^*(c);$$

(viii) *if* $f_k \in \mathbb{F}_+$, $a_k > 0$, $k = 1, \dots, n$, *and*

$$f = \sum_{k=1}^{n} a_k f_k,$$

then, for all $c > 0$,

$$\min_{k=1,\dots,n} \{(f_k)^*(c)\} \le f^*(c) \le \max_{k=1,\dots,n} \{(f_k)^*(c)\}$$

and

$$\min_{k=1,\dots,n} \{(f_k)_*(c)\} \le f_*(c) \le \max_{k=1,\dots,n} \{(f_k)_*(c)\};$$

(ix) *if* $f_1 \in \mathbb{F}_+$, $f_2 \in \mathbb{F}_+$, *and*

$$f = f_1 \cdot f_2,$$

then, for all $c > 0$, *the following inequalities hold provided they do not contain expressions like* $0 \cdot \infty$ *and* $\infty \cdot 0$*:*

$$\begin{aligned}(f_1)_*(c) \cdot (f_2)_*(c) \le f_*(c) &\le \min\{(f_1)_*(c) \cdot (f_2)^*(c), (f_2)_*(c) \cdot (f_1)^*(c)\} \\ &\le \max\{(f_1)_*(c) \cdot (f_2)^*(c), (f_2)_*(c) \cdot (f_1)^*(c)\} \\ &\le f^*(c) \le (f_1)^*(c) \cdot (f_2)^*(c);\end{aligned}$$

(x) *for all* $c > 0$,

$$\left(\frac{1}{f}\right)_*(c) = f_*\left(\frac{1}{c}\right) \quad \text{and} \quad \left(\frac{1}{f}\right)^*(c) = f^*\left(\frac{1}{c}\right).$$

3.2.3 RV- and SV-Functions

First we provide two definitions.

Definition 3.2 We say that a measurable function $f \in \mathbb{F}_+$ is *regularly varying* (or, an RV-function) if

$$f_*(c) = f^*(c) = \kappa_f(c) \in (0, \infty) \quad \text{for all} \quad c > 0. \tag{3.3}$$

In other words, f is regularly varying if the limit

$$\kappa_f(c) = \lim_{t \to \infty} \frac{f(ct)}{f(t)}$$

exists and is positive and finite for all $c > 0$.

The class of RV-functions is denoted by $\mathcal{RV}$.

Definition 3.3 We say that a measurable function $f \in \mathbb{F}_+$ is *slowly varying* (or, an SV-function) if f is regularly varying and

$$\kappa_f(c) = 1 \quad \text{for all} \quad c > 0. \tag{3.4}$$

The class of all SV-functions is denoted by $\mathcal{SV}$.

The above definitions were introduced in Karamata [205]; the properties of regularly varying functions are studied in Karamata [206].

If $f \in \mathcal{RV}$, then

$$\kappa_f(c) = \kappa(c) = c^\rho, \qquad c > 0, \tag{3.5}$$

for some real number ρ, called the *index* of the function f. The index $\rho = 0$ characterizes the SV-functions.

RV-functions f admit the following representation:

$$f(t) = t^\rho \ell(t), \quad t > 0, \tag{3.6}$$

where $(\ell(t), t > 0)$ is a slowly varying function. Moreover, relations (3.4), (3.5), and (3.6) are equivalent.

Remark 3.4 It is known (see, for example, Seneta [324], Bingham et al. [41]) that

$$\lim_{t\to\infty} \frac{\ln r(t)}{\ln t} = \rho$$

for every RV-function $(r(t), t > 0)$ of index ρ.

In the class $\mathcal{RV}$ we consider subclasses $\mathcal{RV}_\rho$ of all RV-functions with index ρ; similarly, $\mathcal{RV}_+$ denotes the subclass of all RV-functions with positive indices.

One may consider the special case $\rho = \infty$ that corresponds to those functions f for which

$$\lim_{t\to\infty} \frac{f(ct)}{f(t)} = \infty \quad \text{for all} \quad c > 1.$$

The functions $f \in \mathcal{RV}_\infty$ are called *quickly varying* functions. An example of $f \in \mathcal{RV}_\infty$ is presented by the exponential function a^t, $t \geq 0$, if $a > 1$.

Definition 3.5 A function $f \in \mathbb{F}_+$ is called *weakly regularly varying* (WRV-function) if relation (3.3) holds.

Similarly, a function $f \in \mathbb{F}_+$ is called *weakly slowly varying* (WSV-function) if relation (3.4) holds.

The class of all WRV- (WSV-) functions is denoted by $\mathcal{WRV}$ ($\mathcal{WSV}$).

Remark 3.6 The function f in Definition 3.5 is not assumed to be measurable. The word "*wide sense*" (W) here and in what follows indicates that the function is not assumed to be measurable.

We will turn to RV- and WRV-functions several times throughout the book; in what follows we will study their properties in more detail.

3.2.4 ORV- and OSV-Functions

Among the various generalizations of the notion of *regular variation*, the first place is occupied by the notion of *O-regular variation.*

Definition 3.7 A function $f \in \mathbb{F}_+$ is called *weakly O-regularly varying* (WORV-function) if

$$f^*(c) < \infty \quad \text{for all} \quad c > 0.$$

Correspondingly, every measurable WORV-function is called *O-regularly varying* (ORV-function).

The class of all ORV- (WORV-) functions is denoted by $\mathcal{ORV}$ ($\mathcal{WORV}$).

ORV functions were introduced in Avakumović [19] and were further studied in Karamata [207] and in many other papers.

Example 3.8 It is clear that every RV- (WRV-) function belongs to $\mathcal{ORV}$ ($\mathcal{WORV}$). Note also that measurable quickly varying functions, say exponential functions $a^t, t \geq 0$, for some $a > 0, a \neq 1$, do not belong to $\mathcal{ORV}$.

The function $f(t) = t(\sin(t) + 2)$ belongs to $\mathcal{ORV}$, but does not belong to $\mathcal{RV}$.

Remark 3.9 Lemma 3.1 applied to $f \in \mathbb{F}_+$ implies that the following conditions are equivalent:

(i) $f \in \mathcal{WORV}$;
(ii) $f_*(c) > 0$ for all $c > 0$;
(iii) there exists an interval $[a, b]$, $0 < a < 1 < b < \infty$, such that $f^*(c) < \infty$ or equivalently that $f_*(c) > 0$ for all $c \in [a, b]$;
(iv) for all $c > 0$,

$$0 < f_*(c) \leq f^*(c) < \infty; \tag{3.7}$$

(v) relation (3.7) holds for all $c \geq 1$.

Note also that, for non-decreasing functions $f \in \mathbb{F}_+$, $f \in \mathcal{ORV}$ if and only if

$$f^*(c_0) < \infty \quad \text{for some} \quad c_0 > 1.$$

Similarly, for non-increasing functions $f \in \mathbb{F}_+$, $f \in \mathcal{ORV}$ if and only if

$$f_*(c_0) > 0 \quad \text{for some} \quad c_0 \in (0, 1).$$

The so-called OSV-functions were studied in Drasin and Seneta [114].

Definition 3.10 A function $f \in \mathbb{F}_+$ is called *weakly O-slowly varying* (WOSV-function) if

$$\sup_{c>0} f^*(c) < \infty.$$

Correspondingly, a measurable WOSV-function is called *O-slowly varying* (OSV-function).

The class of all OSV- (WOSV-) functions is denoted by $\mathcal{OSV}$ ($\mathcal{WOSV}$).

Example 3.11 Note that every slowly varying function f belongs to $\mathcal{OSV}$, since $f^*(c) = 1$, $c > 0$.

The function $f(t) = \log(t)(\sin(t) + 2)$ belongs to $\mathcal{OSV}$, but does not belong to $\mathcal{SV}$.

3.2.5 OURV-Functions

Now we consider the so-called OURV-functions.

Definition 3.12 A function $f \in \mathbb{F}_+$ is called *weakly O-uniformly regularly varying* (WOURV-function) if, for all intervals $[a, b] \subset (0, \infty)$,

$$\sup_{c \in [a,b]} f^*(c) < \infty. \tag{3.8}$$

Correspondingly, a measurable WOURV-function is called *O-uniformly regularly varying* (OURV-function).

The class of all OURV- (WUORV-) functions is denoted by $\mathcal{OURV}$ ($\mathcal{WOURV}$).

Remark 3.13 Lemma 3.1 applied to $f \in \mathbb{F}_+$ implies that the following conditions are equivalent:

(i) $f \in \mathcal{WOURV}$;
(ii) for all intervals $[a, b] \subset (0, \infty)$,

$$\inf_{c \in [a,b]} f_*(c) > 0;$$

(iii) there exists an interval $[a, b]$, $0 < a < 1 < b < \infty$, such that (3.8) holds;

(iv) there exist an interval $[a, b]$, $1 \le a < b < \infty$, and a number $c_0 \in [a/b, 1)$ such that

$$\sup_{c\in[a,b]} f^*(c) < \infty \qquad \text{and} \qquad f^*(c_0) < \infty\,;$$

(v) there exist an interval $[a, b]$, $0 < a < b \le 1$, and a number $c_0 \in (1, b/a]$ such that

$$\sup_{c\in[a,b]} f^*(c) < \infty \qquad \text{and} \qquad f^*(c_0) < \infty.$$

Example 3.14 Let $f \in \mathbb{F}_+$ and $\lim_{\lambda\to 1} f^*(\lambda) = 1$. Then condition (iv) holds (see Remark 3.13) and $f \in \mathcal{WOURV}$.

Remark 3.15 Definitions 3.7 and 3.12 imply that every WOURV- (OURV-) function belongs to $\mathcal{WORV}$ ($\mathcal{ORV}$). It is known that $\mathcal{ORV} \subset \mathcal{OURV}$ (see Aljančić and Arandelović [8], Theorem 1, or Bingham et al. [41], Theorems 2.0.1 and 2.0.4). Thus

$$\mathcal{ORV} = \mathcal{OURV}. \tag{3.9}$$

Moreover, if $f \in \mathcal{ORV}$, then

$$0 < \liminf_{t\to\infty} \inf_{\lambda\in[a,b]} \frac{f(\lambda t)}{f(t)} \le \limsup_{t\to\infty} \sup_{\lambda\in[a,b]} \frac{f(\lambda t)}{f(t)} < \infty$$

for all $[a, b] \subset (0, \infty)$.

3.3 Four Important Classes of Functions

3.3.1 PRV-Functions

For every RV-function f (see Definition 3.2),

$$f^*(c) \to 1 \qquad \text{as} \qquad c \to 1.$$

The latter relation, called "*pseudo-regularity*" leads to a wider class of functions.

Definition 3.16 A function $f \in \mathbb{F}_+$ is said to be *weakly pseudo-regularly varying* (WPRV-function) if

$$\limsup_{c\to 1} f^*(c) = 1. \tag{3.10}$$

Correspondingly, a measurable WPRV-function is called *pseudo-regularly varying* (PRV-function).

The class of all PRV- (WPRV-) functions is denoted by $\mathcal{PRV}$ ($\mathcal{WPRV}$).

Remark 3.17 If $f \in \mathbb{F}_+$ and

$$\limsup_{c\to 1} f^*(c) < \infty,$$

then $f \in \mathcal{WORV}$. Indeed, otherwise there exists a number $c > 0$ such that $f^*(c) = \infty$, whence we conclude

$$f^*\left(\sqrt{c}\right) = \infty,$$

since

$$f^*(c) \le \left(f^*\left(\sqrt{c}\right)\right)^2$$

by Lemma 3.1. Repeating this procedure provides a sequence $c_n = c^{1/2^n}, n \ge 1$, tending to 1 and such that $f^*(c_n) = \infty, n \ge 1$. This is a contradiction.

Therefore, Definition 3.16 implies that every WPRV- (PRV-) function f belongs to $\mathcal{WORV}$ ($\mathcal{ORV}$).

Note that every quickly varying function, say $f(t) = e^t, t \ge 0$, cannot belong to $\mathcal{PRV}$.

As shown above, every PRV-function belongs to $\mathcal{ORV}$, but the converse does not hold true.

Example 3.18 The measurable function $f(t) = 2 + (-1)^{[t]}, t \ge 0$, belongs to $\mathcal{ORV}$, but $f \notin \mathcal{PRV}$.

Further, every RV-function belongs to $\mathcal{PRV}$, but not vice versa.

Example 3.19 Let α be a given real number. The function

$$f(t) = \begin{cases} 0, & \text{if } t = 0, \\ t^\alpha \exp\{\sin(\ln t)\}, & \text{if } t > 0, \end{cases}$$

belongs to $\mathcal{PRV}$, but $f \notin \mathcal{RV}$.

Example 3.20 The function

$$f(t) = \begin{cases} 1, & \text{if } t \in [0, 1), \\ 2^k, & \text{if } t \in \left[2^{2k}, 2^{2k+1}\right), \quad k = 0, 1, 2, \ldots, \\ t/2^{k+1}, & \text{if } t \in \left[2^{2k+1}, 2^{2k+2)}\right), \quad k = 0, 1, 2, \ldots, \end{cases}$$

belongs to $\mathcal{PRV}$, but $f \notin \mathcal{RV}$.

It is clear that

$$\liminf_{c\to 1} f_*(c) = 1 \tag{3.11}$$

is equivalent to the WPRV-condition (3.10). Some other equivalent conditions are listed in the following proposition.

Proposition 3.21 *Let $f \in \mathbb{F}_+$. Then:*

1. *condition (3.10) holds if and only if the upper limit function f^* (lower limit function f_*) is continuous at the point $c = 1$, that is if*

$$\lim_{c\to 1} f^*(c) = 1 \quad (\lim_{c\to 1} f_*(c) = 1),$$

   or equivalently if the function f^* (f_*) is continuous on $(0, \infty)$;
2. *condition (3.10) is equivalent to every of the following four conditions:*
   (i) *relation (3.11) hold,*
   (ii) $\lim_{c\to 1} \limsup_{t\to\infty} \left|\frac{f(ct)}{f(t)} - 1\right| = 0,$
   (iii) $\lim_{c\downarrow 1} f^*(c) = \lim_{c\downarrow 1} f_*(c) = 1,$
   (iv) $\lim_{c\uparrow 1} f^*(c) = \lim_{c\uparrow 1} f_*(c) = 1;$
3. *if the function f^* is nondecreasing (nonincreasing), then (3.10) holds if and only if*

$$\lim_{c\downarrow 1} f^*(c) = 1 \quad \left(\lim_{c\uparrow 1} f_*(c) = 1\right).$$

Proof of Proposition 3.21 By Lemma 3.1,

$$\lim_{c\to 1} f^*(c) = 1 \quad \Longleftrightarrow \quad \lim_{c\to 1} f_*(c) = 1.$$

Since $f^*(1) = f_*(1) = 1$, the condition that $\lim_{c\to 1} f^*(c) = 1$ is equivalent to the continuity of the functions f^* and f_* at the point $c = 1$. Moreover,

$$\lim_{c\to 1} f^*(c) = 1 \quad \Longrightarrow \quad (3.10).$$

Now assume that (3.10) holds. Then (3.11) holds as well, and hence

$$1 = \liminf_{c\to 1} f_*(c) \le \liminf_{c\to 1} f^*(c) \le \limsup_{c\to 1} f^*(c) = 1.$$

Thus $\lim_{c\to 1} f^*(c) = 1$ and (3.10) is equivalent to the continuity of the function $f^*(f_*)$ at the point $c = 1$.

Now we show that (3.10) (or, the continuity of the functions f^* and f_* at $c = 1$) implies the continuity of these functions on $(0, \infty)$. Indeed, since $f \in \mathcal{ORV}$ (see Remark 3.17), Lemma 3.1 implies

$$f_*(q) \le \frac{f^*(qc)}{f^*(c)} \le f^*(q)$$

for all $c, q \in (0, \infty)$. In view of

$$\lim_{q \to 1} f_*(q) = \lim_{q \to 1} f^*(q) = 1,$$

we have for all $c \in (0, \infty)$

$$\lim_{q \to 1} \frac{f^*(qc)}{f^*(c)} = 1.$$

Thus the function f^* and, as a result, the function f_*, too, are continuous on $(0, \infty)$. So, Statement 1 is proved.

Using Lemma 3.1, we show that

$$\begin{aligned}
\limsup_{c \to 1} \limsup_{t \to \infty} \left| \frac{f(ct)}{f(t)} - 1 \right| &= \limsup_{c \to 1} \limsup_{t \to \infty} \max \left\{ \frac{f(ct)}{f(t)} - 1, 1 - \frac{f(ct)}{f(t)} \right\} \\
&= \limsup_{c \to 1} \max \left\{ f^*(c) - 1, 1 - f_*(c) \right\} \\
&= \max \left\{ \limsup_{c \to 1} f^*(c) - 1, 1 - \liminf_{c \to 1} f_*(c) \right\} \\
&= \max \left\{ \limsup_{c \to 1} f^*(c) - 1, 1 - \liminf_{c \to 1} f_* \left(\frac{1}{c} \right) \right\} \\
&= \max \left\{ \limsup_{c \to 1} f^*(c) - 1, 1 - \left[\limsup_{c \to 1} f^*(c) \right]^{-1} \right\}.
\end{aligned}$$

This implies that conditions (ii), (3.10), and (i) are equivalent. By Lemma 3.1, (iii) and (iv) are equivalent. Thus they are also equivalent to (3.10) by Statement 1. This proves Statement 2.

By Lemma 3.1,

$$\lim_{c \downarrow 1} f^*(c) = 1 \quad \Longleftrightarrow \quad \lim_{c \uparrow 1} f_*(c) = 1.$$

Assume that the function f^* is nondecreasing and that

$$\lim_{c \downarrow 1} f^*(c) = 1.$$

Since

$$1 = \lim_{c\uparrow 1} f_*(c) \le \lim_{c\uparrow 1} f^*(c) \le \lim_{c\downarrow 1} f^*(c) = 1,$$

we have $\lim_{c\to 1} f^*(c) = 1$. This together with Statement 1 implies

$$\lim_{c\downarrow 1} f^*(c) = 1 \quad\Longleftrightarrow\quad \lim_{c\to 1} f^*(c) = 1 \quad\Longleftrightarrow\quad (3.10).$$

Therefore, Statement 3 is proved. □

Remark 3.22 Statement 2 of Proposition 3.21 implies that a function f does not belong to $\mathcal{PRV}$ if and only if there exist two subsequences $\{c_n\}$ and $\{t_n\}$ such that $c_n \to 1$ and $t_n \to \infty$ as $n \to \infty$, but

$$\limsup_{n\to\infty} \frac{f(c_n t_n)}{f(t_n)} > 1 \quad \text{or} \quad \liminf_{n\to\infty} \frac{f(c_n t_n)}{f(t_n)} < 1.$$

The two latter relations can be exchanged with the following ones:

$$\limsup_{n\to\infty} \frac{f(c_n t_n)}{f(t_n)} \neq 1 \quad \text{or} \quad \liminf_{n\to\infty} \frac{f(c_n t_n)}{f(t_n)} \neq 1.$$

The converse statement says that (3.10) holds *if and only if*

$$\lim_{\substack{t\to\infty \\ c\to 1}} \frac{f(ct)}{f(t)} = 1 \tag{3.12}$$

or, equivalently, *if and only if*

$$\lim_{\substack{t\to\infty \\ \frac{s}{t}\to 1}} \frac{f(s)}{f(t)} = 1.$$

3.3.1.1 Some Simple Properties of WPRV- and PRV-Functions

Lemma 3.1 together with Proposition 3.21 implies that the classes $\mathcal{WPRV}$ and $\mathcal{PRV}$ are closed with respect to the usual arithmetic operations.

Proposition 3.23 *Let $f_1 \in \mathbb{F}_+$ and $f_2 \in \mathbb{F}$. Then:*

1) if $f_1 \in \mathcal{WPRV}$($f_1 \in \mathcal{PRV}$, f_2 is a measurable function) and (3.2) holds, then

$$f_1 + f_2 \in \mathcal{WPRV}\,(\mathcal{PRV});$$

2) if $f_1 \in \mathcal{WPRV}\,(\mathcal{PRV})$, $f_2 \in \mathcal{WPRV}\,(\mathcal{PRV})$ *and* $a_1 > 0, a_2 > 0$, *then*

$$\left\{a_1 f_1 + a_2 f_2,\ f_1 \cdot f_2,\ \frac{1}{f_2},\ \frac{f_1}{f_2}\right\} \subset \mathcal{WPRV}\,(\mathcal{PRV}).$$

The operation of subtraction is missing in Proposition 3.23. The first reason is that the corresponding difference of functions has to be nonnegative, which is not a nice assumption. But even if this assumption is made, the difference of functions may not belong to the corresponding class as Statement (vii) of Lemma 3.1 shows together with the following example.

Example 3.24 Let $f_1 = f_2 + f_3$, where $f_2(t) = t, t \geq 0$, and f_3 is an arbitrary positive, uniformly bounded function on $[0, \infty)$ belonging to $\mathcal{WPRV}$. For example, the function $(2 + (-1)^{[t]}, t \geq 0)$ introduced in Example 3.18 meets all these assumptions. Since

$$\lim_{t\to\infty} \frac{f_1(ct)}{f_1(t)} = \lim_{t\to\infty} \frac{ct + f_3(ct)}{t + f_3(t)} = c,$$

we conclude that f_1, f_2 are RV-functions and thus they belong to $\mathcal{PRV}$. At the same time, their difference $f_1 - f_2 = f_3$ is not a WPRV-function, but it is positive and measurable.

Among other things, Example 3.24 shows that $\mathcal{RV}$ is also not closed with respect to the operation of subtraction of functions.

Example 3.25 Since the classes $\mathcal{WPRV}$ $(\mathcal{PRV})$ are closed with respect to the operation of division of functions, the multiplication of a WPRV- (PRV-) function f_1 by an arbitrary $f_2 \notin \mathcal{WPRV}$ cannot, in general, result in a WPRV- (PRV-) function. For example, the measurable function $(t(2 + (-1)^{[t]}), t \geq 0)$ does not belong to $\mathcal{WPRV}$, since $(2 + (-1)^{[t]}, t \geq 0)$ is not a WPRV-function, see Examples 3.18 and 3.24.

3.3.2 *PI- and SQI-Functions*

Let us consider two more classes of functions.

Definition 3.26 A function $f \in \mathbb{F}_+$ is called *weakly positively increasing* (WPI-function) if

$$f_*(c_0) > 1 \quad \text{for some} \quad c_0 > 1. \tag{3.13}$$

Correspondingly, every measurable WPI-function is called *positively increasing* (PI-function).

The class of all PI- (WPI-) functions is denoted by $\mathcal{PI}$ ($\mathcal{WPI}$). It is clear that every RV-function with positive index belongs to $\mathcal{PI}$. On the other hand, every nonincreasing function $f \in \mathbb{F}_+$ does not belong to $\mathcal{PI}$, since $\sup_{c\ge 1} f_*(c) \le 1$ in this case.

Proposition 3.27 *For $f \in \mathbb{F}_+$, the following three properties are equivalent:*

(a) $f \in \mathcal{WPI}$;
(b) $\limsup_{c\to\infty} f_*(c) = \infty$;
(c) $\limsup_{c\to\infty} f_*(c) > 1$.

Proof of Proposition 3.27 If $f \in \mathcal{WPI}$, then (3.13) implies

$$\limsup_{c\to\infty} f_*(c) \ge \limsup_{m\to\infty} f_*(c_0{}^m) \ge \limsup_{m\to\infty} (f_*(c_0))^m = \infty.$$

Therefore, the implication (a)$\Longrightarrow$(b) is proved. On the other hand, the implications (b)$\Longrightarrow$(c) and (c)$\Longrightarrow$(a) are obvious. □

We will prove in Sect. 3.7 that every WPI-function f belongs to $\mathbb{F}^{(\infty)}$ and, under some extra assumptions, every WPI- (PI-) function belongs to $\mathbb{F}^{\infty}$.

Now we consider an important subclass of WPI- (PI-) functions.

Definition 3.28 A function $f \in \mathbb{F}_+$ is called *weakly sufficiently quickly increasing* (WSQI-function) if

$$f_*(c) > 1 \quad \text{for all} \quad c > 1. \tag{3.14}$$

Correspondingly, a measurable WSQI-function is called *sufficiently quickly increasing* (SQI-function).

The class of all SQI- (WSQI-) functions is denoted by $\mathcal{SQI}$ ($\mathcal{WSQI}$).

Note that every slowly varying function does not belong to $\mathcal{SQI}$. On the other hand, every RV-function with positive index as well as every quickly increasing function, for example $f(t) = e^t$, $t \ge 0$, belongs to $\mathcal{SQI}$.

For every WSQI-function f, its limit functions f^* and f_* are non-decreasing in the set $[1, \infty)$. Indeed, if $c_2 > c_1 \ge 1$, then relation (3.14) and Lemma 3.1 imply that

$$f^*(c_2) \ge f_*(c_2/c_1) f^*(c_1) \ge f^*(c_1)$$

and

$$f_*(c_2) \ge f_*(c_2/c_1) f_*(c_1) \ge f_*(c_1).$$

The latter inequalities become equalities if the functions f_* and f^* equal ∞. For example, for $f(t) = e^t$, $t \ge 0$, we have $f_*(c) = f^*(c) = \infty$, $c > 1$.

If $f_*(c_1), f^*(c_1) \in (0, \infty)$, then relation (3.14) and Lemma 3.1 imply that, for all $c_2 > c_1 > 0$,

$$f^*(c_2) \geq f_*(c_2/c_1) f^*(c_1) > f^*(c_1)$$

and

$$f_*(c_2) \geq f_*(c_2/c_1) f_*(c_1) > f_*(c_1). \tag{3.15}$$

These inequalities show that the limit functions f^* and f_* increase on $(0, \infty)$ if $f \in \mathcal{WSQI}$. Moreover, the monotonicity of f_* characterizes the WSQI-property in the class of WORV-functions.

Proposition 3.29 *For $f \in \mathcal{WORV}$, the following three conditions are equivalent:*

(a) $f \in \mathcal{WSQI}$;
(b) f_* *increases to* ∞ *on* $(0, \infty)$;
(c) f_* *increases on* $[1, \infty)$.

Proof of Proposition 3.29 The implications (b)$\Longrightarrow$(c) and (c)$\Longrightarrow$(a) are obvious. The implication (a)$\Longrightarrow$(b) follows from (3.15). □

3.3.2.1 Some Simple Properties of WSQI- and SQI-Functions

Lemma 3.1 implies that the classes $\mathcal{WSQI}$ and $\mathcal{SQI}$ are closed with respect to addition and multiplication.

Proposition 3.30 *Let $f_1 \in \mathbb{F}_+$ and $f_2 \in \mathbb{F}$.*

1) If $f_1 \in \mathcal{WSQI}(f_1 \in \mathcal{SQI})$ satisfies condition (3.2), then $f_1 + f_2 \in \mathcal{WSQI}$ $(\mathcal{SQI})$.
2) If $f_1 \in \mathcal{WSQI}$ $(f_1 \in \mathcal{SQI})$ and

$$(f_2)_*(c) \geq 1 \quad \textit{for all} \quad c > 1, \tag{3.16}$$

then $f_1 \cdot f_2 \in \mathcal{WSQI}$ $(\mathcal{SQI})$.
3) If $f_1 \in \mathcal{WSQI}$ $(\mathcal{SQI})$, $f_2 \in \mathcal{WSQI}$ $(\mathcal{SQI})$ and $a_1 > 0, a_2 > 0$, then

$$\{a_1 f_1 + a_2 f_2,\ f_1 \cdot f_2\} \subset \mathcal{WSQI}\ (\mathcal{SQI}).$$

Remark 3.31 Proposition 3.29 and statement (x) of Lemma 3.1 show that $f \in \mathcal{WSQI}$ implies $(1/f) \notin \mathcal{WSQI}$. Thus if $f_1 \in \mathcal{WSQI}$ and $f_2 \in \mathcal{WSQI}$, then in general $(f_1/f_2) \notin \mathcal{WSQI}$. This means that the classes $\mathcal{SQI}$ and $\mathcal{SQI}$ are not closed with respect to division. For example, the functions $f_1 = (t,\ t > 0)$ and $f_2 = (t^2,\ t > 0)$ belong to $\mathcal{RV}_+$ and thus they are SQI-functions. But $f_1/f_2 = ((1/t), t > 0)$ is not an SQI-function.

Moreover, statement (vii) of Lemma 3.1 implies that the classes $\mathcal{WSQI}$ and $\mathcal{SQI}$ are not closed with respect to subtraction. For example, $(t + \ln t,\ t > 1)$ and $(t,\ t > 1)$ belong to $\mathcal{RV}_1$ and thus are SQI-functions. On the other hand, the difference of these functions, $(\ln t,\ t > 1)$, does not belong to $\mathcal{SQI}$.

3.3.3 POV-Functions

Condition (3.14) is used to characterize a subclass of $\mathcal{PRV}$ whose members possess properties similar to those of regularly varying functions with positive indices.

Definition 3.32 A function $f \in \mathbb{F}_+$ is called *weakly positively varying* (WPOV-function) if both conditions (3.10) and (3.14) hold.

Correspondingly, any measurable WPOV-function is called *positively varying* (POV-function).

The class of POV- (WPOV-) functions is denoted by $\mathcal{POV}$ ($\mathcal{WPOV}$). It is clear that

$$\mathcal{WPOV} = \mathcal{WPRV} \cap \mathcal{WSQI}, \qquad \mathcal{POV} = \mathcal{PRV} \cap \mathcal{SQI}.$$

Remark 3.33 Every slowly varying function as well as every sufficiently quickly increasing function does not belong to $\mathcal{POV}$. On the other hand, every regularly varying function with positive index belongs to $\mathcal{POV}$. Moreover, the PRV-function of Example 3.20 does not belong to $\mathcal{RV}$ nor to $\mathcal{POV}$. Further, the PRV-function of Example 3.19 with $\alpha \geq 1$ does not belong to $\mathcal{RV}$, but does belong to $\mathcal{POV}$.

The following proposition contains some characterizing properties of WPOV-functions. The proof of Proposition 3.34 follows directly from Proposition 3.21 and 3.29.

Proposition 3.34 *For $f \in \mathbb{F}_+$, the following four properties are equivalent:*

(a) $f \in \mathcal{WPOV}$;
(b) f_* increases to ∞ and is continuous on $(0, \infty)$;
(c) f_* increases on $[1, \infty)$ and is continuous at $c = 1$;
(d) f_* increases on $[1, \infty)$ and condition (3.11) holds.

3.3.3.1 Some Simple Properties of POV-Functions

Propositions 3.23 and 3.30 imply that the classes $\mathcal{WPOV}$ and $\mathcal{POV}$ are closed with respect to addition and multiplication.

Proposition 3.35 *Let $f_1 \in \mathbb{F}_+$ and $f_2 \in \mathbb{F}$. Then:*

1) if $f_1 \in \mathcal{WPOV}$ ($f_1 \in \mathcal{POV}$, f_2 is measurable) and condition (3.2) holds, then

$$f_1 + f_2 \in \mathcal{WPOV}\ (\mathcal{POV});$$

2) if $f_1 \in \mathcal{WPOV}\,(\mathcal{POV})$, $f_2 \in \mathcal{WPRV}\,(\mathcal{PRV})$ *and condition (3.16) holds, then*

$$f_1 \cdot f_2 \in \mathcal{WPOV}\,(\mathcal{POV});$$

3) if $f_1 \in \mathcal{WPOV}\,(\mathcal{POV})$, $f_2 \in \mathcal{WPOV}\,(\mathcal{POV})$ *and* $a_1 > 0, a_2 > 0$, *then*

$$\{a_1 f_1 + a_2 f_2,\ f_1 \cdot f_2\} \subset \mathcal{WPOV}\,(\mathcal{POV}).$$

Remark 3.36 Proposition 3.29 and statement (x) of Lemma 3.1 (see also Remark 3.31) imply that $f \in \mathcal{WPOV}$ results in $(1/f) \notin \mathcal{WPOV}$. Thus if $f_1 \in \mathcal{WPOV}$ and $f_2 \in \mathcal{WPOV}$, then in general $(f_1/f_2) \notin \mathcal{WPOV}$. This implies that the classes $\mathcal{WPOV}$ and $\mathcal{POV}$ are not closed with respect to division. For example, the functions $f_1 = (t,\ t > 0)$ and $f_2 = (t^2,\ t > 0)$ belong to $\mathcal{RV}_+$ and thus they are POV-functions. But $f_1/f_2 = ((1/t), t > 0)$ does not belong to $\mathcal{POV}$.

Moreover, statement (vii) of Lemma 3.1 implies that the classes $\mathcal{WSQI}$ and $\mathcal{SQI}$ are not closed with respect to subtraction (see also Remark 3.31). For example, $(t + \ln t,\ t > 1)$ and $(t,\ t > 1)$ belong to $\mathcal{RV}_1$. Thus they are POV-functions. But their difference, $(\ln t,\ t > 1)$, is not a POV-function.

3.4 Functions Preserving the Asymptotic Equivalence

In the space $\mathbb{F}_+$, we consider the following equivalence relation.

Definition 3.37 Two functions $u, v \in \mathbb{F}_+$ are called *(asymptotically) equivalent* if

$$\lim_{t\to\infty} \frac{u(t)}{v(t)} = 1. \tag{3.17}$$

The asymptotic equivalence of two functions u and v is denoted by $u \sim v$, or $u(t) \sim v(t)$ as $t \to \infty$, or $u(t) \underset{t\to\infty}{\sim} v(t)$.

Remark 3.38 If $u, v \in \mathbb{F}_+$ and $u \sim v$, then $u^* = v^*$ ($u_* = v_*$). Indeed, for all $c > 0$,

$$\lim_{t\to\infty} \frac{u(ct)}{v(ct)} = 1,$$

and thus

$$u^*(c) = \limsup_{t\to\infty} \frac{u(ct)}{u(t)} = \limsup_{t\to\infty} \frac{u(ct)}{v(ct)} \cdot \frac{v(ct)}{v(t)} \cdot \frac{v(t)}{u(t)} = \limsup_{t\to\infty} \frac{v(ct)}{v(t)} = v^*(c).$$

If an equivalence relation is given, then the natural problem arises how to characterize the functions that preserve this relation. In other words, the problem is to describe the functions f such that

$$\lim_{t\to\infty} \frac{f(u(t))}{f(v(t))} = 1 \tag{3.18}$$

for all equivalent functions u and v.

If the equivalence relation is considered for functions tending to a positive finite limit as $t \to \infty$, then this equivalence is preserved by continuous functions in $\mathbf{R}_+$.

Another case is if $u \sim v$ and $u, v \in \mathbb{F}^\infty$, that is, if

$$\lim_{t\to\infty} u(t) = \infty \quad \text{and} \quad \lim_{t\to\infty} v(t) = \infty, \tag{3.19}$$

or if u and v do not have limits at infinity. In what follows we restrict our consideration to the case where condition (3.19) holds.

Definition 3.39 We say that a function $f \in \mathbb{F}_+$ *preserves (asymptotic) equivalence of functions* in the space $\mathbb{F}^\infty$ if (3.18) holds for all equivalent functions u and v for which condition (3.19) holds.

In other words, a function f preserves the equivalence of functions if

$$f \circ u \sim f \circ v$$

for all u and v such that $u \sim v$ and $u, v \in \mathbb{F}^\infty$.

In a similar way, one can introduce functions preserving the equivalence of sequences. Throughout this section, we assume that the sequences $\{u_n\} = \{u_n\}_{n\ge1}$ and $\{v_n\} = \{v_n\}_{n\ge1}$ are such that $u_n > 0$ and $v_n > 0$ for sufficiently large n. The space of such sequences is denoted by $\mathbb{S}_+$; the subspace of sequences tending to ∞ is denoted by $\mathbb{S}^\infty$, and the subspaces of sequences of $\mathbb{S}^\infty$ that are eventually nondecreasing (increasing) by $\mathbb{S}^\infty_{\text{ndec}}$ ($\mathbb{S}^\infty_{\text{inc}}$).

Definition 3.40 Two sequences $\{u_n\}$ and $\{v_n\}$ are called *(asymptotically) equivalent* if

$$\lim_{n\to\infty} \frac{u_n}{v_n} = 1. \tag{3.20}$$

The equivalence of sequences $\{u_n\}$ and $\{v_n\}$ is denoted by $\{u_n\} \sim \{v_n\}$ (or $u_n \sim v_n$ as $n \to \infty$).

Definition 3.41 We say that a function $f \in \mathbb{F}_+$ *preserves the equivalence of sequences* (in the space $\mathbb{S}^\infty$) if

$$\lim_{n\to\infty} \frac{f(u_n)}{f(v_n)} = 1 \tag{3.21}$$

for all equivalent sequences $\{u_n\}$ and $\{v_n\}$ such that

$$\lim_{n\to\infty} u_n = \infty \quad \text{and} \quad \lim_{n\to\infty} v_n = \infty. \tag{3.22}$$

The key property of WPRV-functions (see Definition 3.16) is that they preserve the equivalence of functions and the equivalence of sequences.

Theorem 3.42 *For $f \in \mathbb{F}_+$, the following six properties are equivalent:*

(P1) *f preserves the equivalence of functions;*
(P2) *f preserves the equivalence of continuous functions;*
(P3) *f preserves the equivalence of continuous increasing functions;*
(P4) *f preserves the equivalence of increasing sequences;*
(P5) *f preserves the equivalence of sequences;*
(P6) *$f \in \mathcal{WPRV}$.*

The proof of Theorem 3.42 is based on some auxiliary results. In what follows the symbol $\lfloor t \rfloor$ denotes the integer part of a real number t.

For a real sequence $\{x_n\} = \{x_n\}_{n\ge 0}$ we define its *piecewise-linear interpolation*, i.e.

$$\hat{x}(t) = (\lfloor t \rfloor + 1 - t)x_{\lfloor t \rfloor} + (t - \lfloor t \rfloor)x_{\lfloor t \rfloor+1}, \quad t \ge 0. \tag{3.23}$$

Note that the piecewise-linear interpolation is a continuous function.

Some simple properties of piecewise-linear interpolations are listed in the following result.

Lemma 3.43 *For all real sequences $\{x_n\}$ and $\{y_n\}$,*

(i) $\hat{x} \in \mathbb{F}_+ \iff \{x_n\} \in \mathbb{S}_+$;
(ii) $\hat{x} \in \mathbb{F}^\infty \iff \{x_n\} \in \mathbb{S}^\infty$;
(iii) $\hat{x} \in \mathbb{F}^\infty_{\text{ndec}} \iff \{x_n\} \in \mathbb{S}^\infty_{\text{ndec}}$;
(iv) $\hat{x} \in \mathbb{F}^\infty_{\text{inc}} \iff \{x_n\} \in \mathbb{S}^\infty_{\text{inc}}$;
(v) $\hat{x} \sim \hat{y} \iff \{x_n\} \sim \{y_n\}$.

Proof of Lemma 3.43 Statements (i) and (ii) follow directly from (3.23). Moreover, since

$$\hat{x}(t) - \hat{x}(\lfloor t \rfloor) = (t - \lfloor t \rfloor)(x_{\lfloor t \rfloor+1} - x_{\lfloor t \rfloor}), \quad t \ge 0,$$

statements (iii) and (iv) also hold. Finally, statement (v) follows from the inequalities

$$\min\left\{\frac{x_{\lfloor t \rfloor}}{y_{\lfloor t \rfloor}}, \frac{x_{\lfloor t \rfloor+1}}{y_{\lfloor t \rfloor+1}}\right\} \le \frac{\widehat{x}(t)}{\widehat{y}(t)} \le \max\left\{\frac{x_{\lfloor t \rfloor}}{y_{\lfloor t \rfloor}}, \frac{x_{\lfloor t \rfloor+1}}{y_{\lfloor t \rfloor+1}}\right\},$$

which, in turn, follow from (3.23) and from

$$\min\left\{\frac{a}{c}, \frac{b}{d}\right\} \le \frac{a+b}{c+d} \le \max\left\{\frac{a}{c}, \frac{b}{d}\right\}$$

for all positive numbers a, b, c, and d. □

Proof of Theorem 3.42 The implications (P1)⟹(P2)⟹(P3) are obvious.

The implication (P3)⟹(P4) follows from Lemma 3.43.

Assume that (P4) holds, but not (P5). Then there exists a number $\varepsilon > 0$ and two asymptotically equivalent sequences $\{x_n\}$ and $\{y_n\}$, both tending to infinity, such that

$$\text{either} \quad \frac{f(x_n)}{f(y_n)} \ge 1+\varepsilon \quad \text{or} \quad \frac{f(x_n)}{f(y_n)} \le 1-\varepsilon \tag{3.24}$$

for all $n \ge 1$. Now choose a monotonic subsequence $\{x_{n_k}\}$ of the sequence $\{x_n\}$ and then a monotonic subsequence $\{y_{n_{k_i}}\}$ of the sequence $\{y_{n_k}\}$. Define $x_i' = x_{n_{k_i}}$ and $y_i' = y_{n_{k_i}}$ for $i \ge 1$. Then both sequences $\{x_i'\}$ and $\{y_i'\}$ are monotonic and tend to infinity. Moreover, $x_i' \sim y_i'$, $i \to \infty$. Thus, by (P4),

$$\frac{f(x_i')}{f(y_i')} \to 1, \qquad i \to \infty,$$

contradicting (3.24). This proves the implication (P4)⟹(P5).

If (P5) holds, but not (P6), then (see Remark 3.22) there are two sequences $\{c_n\}$ and $\{t_n\}$ such that $c_n \to 1$ and $t_n \to \infty$ as $n \to \infty$, but

$$\limsup_{n\to\infty} \frac{f(c_n t_n)}{f(t_n)} \ne 1, \quad \text{or} \quad \liminf_{n\to\infty} \frac{f(c_n t_n)}{f(t_n)} \ne 1.$$

On the other hand,

$$\lim_{n\to\infty} (c_n t_n)/t_n = 1$$

and (P5) implies

$$\lim_{n\to\infty} f(c_n t_n)/f(t_n) = 1.$$

This contradiction proves the implication (P5)⟹(P6).

To complete the proof of Theorem 3.42, assume that (P1) does not hold. In this case, there are two functions $u \in \mathbb{F}^\infty$ and $v \in \mathbb{F}^\infty$ such that $u \sim v$, but

$$\limsup_{t\to\infty} \frac{f(u(t))}{f(v(t))} \ne 1,$$

or

$$\liminf_{t\to\infty} \frac{f(u(t))}{f(v(t))} \neq 1.$$

This implies that there exists a sequence $\{t_n\}$ such that $u(t_n) \to \infty$, $v(t_n) \to \infty$ as $n \to \infty$, and

$$\{u(t_n)\} \sim \{v(t_n)\},$$

but

$$\limsup_{n\to\infty} \frac{f(u(t_n))}{f(v(t_n))} \neq 1,$$

or

$$\liminf_{n\to\infty} \frac{f(u(t_n))}{f(v(t_n))} \neq 1.$$

This means (see Remark 3.22) that the function f does not belong to $\mathcal{WPRV}$, that is, (P6) does not hold, which yields the implication (P6)$\Longrightarrow$(P1). □

Remark 3.44 Theorem 3.42 and Remark 3.22 imply that a function $f \in \mathbb{F}_+$ preserves the equivalence of functions (sequences) if and only if condition (3.12) holds. This complement of Theorem 3.42 has been used in Korenblyum [235], Gihman and Skorohod [149], Stadtmüller and Trautner [344], and Djurčić [104].

Consider the following simple corollary to Theorem 3.42, which will be useful in the sequel (see, e.g., Theorem 3.51).

Corollary 3.45 *If* $f \in \mathcal{WPRV}$ $(\mathcal{PRV})$, *then* $f \circ \ln \in \mathcal{WSV}$ $(\mathcal{SV})$.

Proof of Corollary 3.45 Since

$$\lim_{t\to\infty} \frac{\ln ct}{\ln t} = 1$$

for all $c > 0$, Theorem 3.42 implies that

$$\lim_{t\to\infty} \frac{f(\ln ct)}{f(\ln t)} = 1$$

for all $c > 0$. □

Remark 3.46 The function $f(t) = \exp\{\sqrt{t}\}$, $t \geq 0$, does not belong to $\mathcal{PRV}$, while $f \circ \ln$ belongs to $\mathcal{SV}$. These functions serve as a counterexample to the converse of Corollary 3.45.

Moreover, the function $f \circ \ln$ may not belong to $\mathcal{SV}$ if $f \in \mathcal{ORV}$ (see Example 3.18). On the other hand, one can prove that the function $f \circ \ln$ belongs to $\mathcal{OSV}$ in this case. At the same time, the converse statement does not hold. A corresponding example is given by the function $f(t) = \exp\{\sqrt{t}\}$, $t \geq 0$.

3.4.1 A Uniform Convergence Theorem

Theorem 3.42 implies the following uniform convergence theorem for WPRV-functions (see also Yakymiv [372] and Buldygin et al. [59]).

Theorem 3.47 *Let $f \in \mathcal{WPRV}$. Then*

1)

$$\lim_{a \downarrow 1} \limsup_{t \to \infty} \sup_{a^{-1} \leq c \leq a} \left| \frac{f(ct)}{f(t)} - 1 \right| = 0;$$

2)

$$\lim_{a \downarrow 1} \limsup_{t \to \infty} \sup_{a^{-1} \leq c \leq a} \frac{f(ct)}{f(t)} = 1.$$

Proof of Theorem 3.47 Indeed, let

$$\lim_{a \downarrow 1} \limsup_{t \to \infty} \sup_{a^{-1} \leq c \leq a} \left| \frac{f(ct)}{f(t)} - 1 \right| > 0,$$

whence there are two sequences $\{c_n\}$ and $\{t_n\}$ such that $c_n \to 1$, $t_n \to \infty$ as $n \to \infty$, but

$$\lim_{n \to 1} \frac{f(c_n t_n)}{f(t_n)} \neq 1.$$

This inequality contradicts Theorem 3.42, since $\{c_n t_n\} \sim \{t_n\}$. So, Statement 1 of Theorem 3.47 is proved. Statement 2 can be proved in the same way. □

3.4.2 A Uniform Convergence Theorem for SV- and RV-Functions

The classical Karamata uniform convergence theorem for SV- and RV-functions (see, for example, Karamata [205, 206] and Bingham et al. [41]) is a stronger result than Theorem 3.47. We compare this result with Theorem 3.47 in the following remark.

Remark 3.48 The uniform convergence for SV functions (see Bingham et al. [41], p. 22) asserts that *for SV-functions* ℓ,

$$\ell(ct)/\ell(t) \to 1, \quad t \to \infty,$$

uniformly with respect to c in every interval $[a, b] \subset \mathbf{R}_+$, that is

$$\lim_{t\to\infty} \sup_{c\in[a,b]} |\ell(ct)/\ell(t) - 1| = 0,$$

and

$$\lim_{t\to\infty} \sup_{c\in[a,b]} \left|r(ct)/r(t) - c^\rho\right| = 0, \tag{3.25}$$

where r is an RV-function with index ρ.

3.5 Integral Representations

An integral representation for a regularly varying function is the second key result of the Karamata [205] theory. Various proofs of this result are known (see, for example, Seneta [324], Bingham et al. [41]). A generalization of this representation for ORV-functions has been obtained in Karamata [207], Aljančić and Aranđelović [8].

In this section, we prove two integral representations for PRV-functions. The corresponding result is obtained for POV-functions, too.

3.5.1 Integral Representations for RV- and ORV-Functions

The integrals in the representations below are to be understood in the Lebesgue sense.

Recall that $f \in \mathcal{RV}$ *if and only if*

$$f(t) = \exp\left\{\alpha(t) + \int_{t_0}^{t} \beta(s)\frac{ds}{s}\right\} \tag{3.26}$$

for some $t_0 > 0$ *and all* $t \geq t_0$, *where* α *and* β *are measurable bounded functions for which the limits*

$$\lim_{t\to\infty} \alpha(t) \quad \text{and} \quad \lim_{t\to\infty} \beta(t)$$

exist. Moreover, $f \in \mathcal{SV}$ *if and only if*

$$\lim_{t\to\infty} \beta(t) = 0.$$

In turn, $f \in \mathcal{ORV}$ *if and only if relation* (3.26) *holds, where* α *and* β *are measurable bounded functions* (see Aljančić and Arandelović [8]).

Similarly, $f \in \mathcal{OSV}$ *if and only if relation* (3.26) *holds, where* α *and* β *are measurable bounded functions*, and $\lim_{t\to\infty} \beta(t) = 0$ (cf. Drasin and Seneta [114]).

Note that the above integral representation for a fixed function f is not unique. For example, one can start with a representation involving a discontinuous function β and then pass to a similar representation with other functions $\tilde{\alpha}$ and $\tilde{\beta}$, where $\tilde{\beta}$ is continuous or even infinitely often differentiable (see, for example, Seneta [324], Bingham et al. [41]).

3.5.2 Integral Representations for PRV-Functions

Consider two versions of the integral representation for PRV-functions. The proof of the first one is based on the integral representation for ORV-functions.

Theorem 3.49 *A function* f *belongs to* $\mathcal{PRV}$ *if and only if representation (3.26) holds, where* α *and* β *are measurable bounded functions such that*

$$\lim_{c\to 1} \limsup_{t\to\infty} |\alpha(ct) - \alpha(t)| = 0. \tag{3.27}$$

Remark 3.50 Condition (3.27) characterizes the so-called *slowly oscillating* functions (see Bingham et al. [41]). This condition is equivalent to

$$\limsup_{c\to 1} \limsup_{t\to\infty} (\alpha(ct) - \alpha(t)) = 0.$$

Proof of Theorem 3.49 Let $f \in \mathcal{ORV}$. Then (3.26) holds by the integral representation theorem for ORV-functions, where α and β are measurable bounded functions.

Since

$$\left| \int_t^{ct} \beta(s) \frac{ds}{s} \right| \le \big| \ln(c) \big| \operatorname*{ess\,sup}_{t\ge t_0} |\beta(t)|,$$

and since β is bounded,

$$\lim_{c\to 1} \limsup_{t\to\infty} \left| \int_t^{ct} \beta(s) \frac{ds}{s} \right| = 0. \tag{3.28}$$

In view of

$$\frac{f(ct)}{f(t)} = \exp\{\alpha(ct) - \alpha(t)\} \exp\left\{\int_t^{ct} \beta(s)\frac{ds}{s}\right\},$$

and by (3.28) we obtain

$$f^*(c) = \limsup_{t\to\infty} \frac{f(ct)}{f(t)} = \exp\{\limsup_{t\to\infty} (\alpha(ct) - \alpha(t))\},$$

whence

$$\limsup_{c\to 1} f^*(c) = \exp\{\limsup_{c\to 1} \limsup_{t\to\infty} (\alpha(ct) - \alpha(t))\}.$$

So,

$$\limsup_{c\to 1} f^*(c) = 1 \iff \limsup_{c\to 1} \limsup_{t\to\infty} (\alpha(ct) - \alpha(t)) = 0 \tag{3.29}$$

$$\iff \limsup_{c\to 1} \limsup_{t\to\infty} |\alpha(ct) - \alpha(t)| = 0.$$

Relation (3.29) shows that an ORV-function f, that is, a function for which (3.26) holds with measurable bounded α and β, belongs to $\mathcal{PRV}$ if and only if condition (3.27) holds.

To complete the proof of Theorem 3.49 just recall that every PRV-function is an ORV-function (see Remark 3.17). □

Along with (3.26), we consider another integral representation for PRV-functions that follows from that for SV-functions by using the superposition $f \circ \ln$, being an SV-function (see Corollary 3.45).

Theorem 3.51 *A function f belongs to $\mathcal{PRV}$ if and only if*

$$f(t) = \exp\left\{a(t) + \int_{t_0}^{t} b(u)\,du\right\} \tag{3.30}$$

for some $t_0 > 0$ and all $t \geq t_0$, where the measurable functions a and b are such that the limit $\lim_{t\to\infty} a(t)$ exists, and

$$\limsup_{c\to 1} \limsup_{t\to\infty} \int_t^{ct} b(u)\,du = 0. \tag{3.31}$$

Remark 3.52 Condition (3.31) is equivalent to

$$\lim_{c\to 1} \limsup_{t\to\infty} \left|\int_t^{ct} b(u)\,du\right| = 0.$$

Proof of Theorem 3.51 Let $f \in \mathcal{PRV}$. Then $f \circ \ln \in \mathcal{SV}$ (see Corollary 3.45) and one can use the representation (3.26) for SV-functions. Thus

$$f(\ln(u)) = \exp\left\{\alpha(u) + \int_{u_0}^{u} \beta(s)\frac{ds}{s}\right\}$$

for some $t_0 > 0$ and all $t \geq t_0$, where α and β are measurable functions such that the limit $\lim_{u\to\infty} \alpha(u)$ exists and $\lim_{u\to\infty} \beta(u) = 0$. This implies relation (3.30) with $a(t) = \alpha(e^t)$ and $b(t) = \beta(e^t)$.

It is clear that

$$\lim_{t\to\infty} a(t) = \lim_{t\to\infty} \alpha(t),$$

$$\lim_{t\to\infty} b(t) = \lim_{t\to\infty} \beta(t) = 0.$$

Moreover, if (3.30) holds and $\lim_{t\to\infty} a(t)$ exists, then

$$f^*(c) = \limsup_{t\to\infty} \frac{f(ct)}{f(t)} = \limsup_{t\to\infty} \int_t^{ct} b(s)\,ds,$$

whence

$$\limsup_{c\to 1} f^*(c) = 1 \iff \limsup_{c\to 1} \limsup_{t\to\infty} \int_t^{ct} b(u)\,du = 0.$$

Therefore, if relation (3.30) holds for f and $\lim_{t\to\infty} a(t)$ exists, then $f \in \mathcal{PRV}$ if and only if condition (3.31) holds. This completes the proof of the "necessity" part of Theorem 3.51 and proves its "sufficiency" part as well. □

Remark 3.53 One can see from the proof of Theorem 3.51 that

$$\lim_{t\to\infty} b(t) = 0$$

for PRV-functions f.

3.5.3 Integral Representations for POV-Functions

Integral representations for POV-functions follow from the integral representations for PRV-functions.

Theorem 3.54 *A function f belongs to $\mathcal{POV}$ if and only if representation (3.26) holds, where α and β are measurable bounded functions for which condition (3.27)* *is satisfied and*

$$\liminf_{t\to\infty} \int_t^{ct} \beta(s)\frac{ds}{s} > 0 \quad \textit{for all} \quad c > 1. \tag{3.32}$$

Proof of Theorem 3.54 Theorem 3.54 follows from Theorem 3.49 since, under assumption (3.27), the function f belongs to $\mathcal{POV}$ (see Definition 3.32) and condition (3.14) holds if and only if condition (3.32) is satisfied. □

Example 3.55 If

$$\liminf_{t\to\infty} \beta(t) > 0,$$

then condition (3.32) retains.

Another integral representation for POV-functions follows directly from Theorem 3.51.

Theorem 3.56 *A function f belongs to $\mathcal{POV}$ if and only if the representation (3.30) holds, where a and b are measurable functions for which the limit* $\lim_{t\to\infty} a(t)$ *exists, condition (3.31) holds, and*

$$\liminf_{t\to\infty} \int_t^{ct} b(u)\, du > 0 \quad \textit{for all} \quad c > 1. \tag{3.33}$$

Remark 3.57 Remarks 3.52 and 3.53 also apply to Theorem 3.56.

Example 3.58 If

$$\liminf_{t\to\infty} tb(t) > 0,$$

then condition (3.33) retains.

3.6 Potter's Bounds for PRV-Functions

Potter's theorem and its applications occupy a prominent place in the theory of regularly varying functions (see, e.g., Potter [293] and Bingham et al. [41]). In this section, we prove a generalization of Potter's theorem for pseudo-regularly varying functions (Theorem 3.59). The proof of this generalization is based on the integral representation for PRV-functions (Theorem 3.49) and on the uniform convergence theorem for PRV-functions (Theorem 3.47). Theorem 3.59 is later applied to study the properties of asymptotically quasi-inverse functions in Sect. 7.2.

Theorem 3.59 *Let $f \in \mathcal{PRV}$. Then there exists a number $p \ge 0$ and, for all $A > 1$, there exist numbers $\lambda_A > 1$ and $t_A > 0$ such that*

$$A^{-1}\lambda^{-p} f(s) \le f(t) \le A\lambda^p f(s) \tag{3.34}$$

for $\lambda \in (1, \lambda_A]$, $t \ge t_A$, and $s \in [\lambda^{-1}t, \lambda t]$.

Proof of Theorem 3.59 By Theorem 3.49, there exist a number $t_0 > 0$ and measurable bounded functions α and β such that condition (3.27) holds, and, for all $t \ge t_0$,

$$f(t) = \Phi(t) \exp\left\{\int_{t_0}^{t} \beta(u)\frac{du}{u}\right\}, \tag{3.35}$$

where $\Phi(t) = \exp\{\alpha(t)\}$.

In view of (3.27), Φ is a measurable bounded PRV-function. Thus by Theorem 3.47, for all $A > 1$, there exist numbers $\lambda_A > 1$ and $\tau_A > 0$ such that

$$\frac{\Phi(t)}{\Phi(s)} < A \tag{3.36}$$

for all $t \ge \tau_A$ and $s \in [\lambda_A^{-1}t, \lambda_A t]$. Relations (3.35) and (3.36) imply that

$$\frac{f(t)}{f(s)} \le A \exp\left\{\int_{s}^{t} \beta(u)\frac{du}{u}\right\} \tag{3.37}$$

for $t \ge t_A = \lambda_A \max\{t_0, \tau_A\}$, $\lambda \in (1, \lambda_A]$, and $s \in [\lambda^{-1}t, \lambda t]$.

Since $\lambda > 1$, we have

$$\exp\left\{\int_{s}^{t} \beta(u)\frac{du}{u}\right\} \le \exp\left\{p\int_{s}^{t} \frac{du}{u}\right\} = \left(\frac{t}{s}\right)^p \le \lambda^p$$

for all $s \in [\lambda^{-1}t, t]$, where

$$p = \operatorname*{ess\,sup}_{t \ge t_0} |\beta(t)|.$$

Moreover, for all $s \in [t, \lambda t]$,

$$\exp\left\{\int_{s}^{t} \beta(u)\frac{du}{u}\right\} = \exp\left\{-\int_{t}^{s} \beta(u)\frac{du}{u}\right\} \le \exp\left\{p\int_{t}^{s} \frac{du}{u}\right\} = \left(\frac{s}{t}\right)^p \le \lambda^p.$$

The latter two inequalities together with (3.37) imply that

$$\frac{f(t)}{f(s)} \le A\lambda^p \quad \text{and} \quad \frac{f(s)}{f(t)} \le A\lambda^p$$

for $t \geq t_A$, $\lambda \in (1, \lambda_A]$ and $s \in [\lambda^{-1}t, \lambda t]$. This results in inequality (3.34). □

Remark 3.60 Let $f \in \mathbb{F}_+$. If relation (3.34) holds, then $f \in \mathcal{WPRV}$.

Indeed, if (3.34) holds, then, for all $A > 1$, there exists a $\lambda_A > 1$ such that

$$A^{-1}\lambda^{-p} \leq f^*(c) \leq A\lambda^p$$

for all $\lambda \in (1, \lambda_A]$ and $c \in [\lambda^{-1}, \lambda]$. This yields

$$A^{-1}\lambda^{-p} \leq \liminf_{c\to 1} f^*(c) \leq \limsup_{c\to 1} f^*(c) \leq A\lambda^p.$$

Passing to the limit as $\lambda \downarrow 1$ and then as $A \downarrow 1$, we see that $\lim_{c\to 1} f^*(c) = 1$, that is $f \in \mathcal{WPRV}$ (see Definition 3.16).

Theorem 3.59 and Remark 3.60 provide a characterization of PRV-functions in terms of Potter's bounds.

Corollary 3.61 *Let $f \in \mathbb{F}_+$. If f is measurable, then $f \in \mathcal{PRV}$ if and only if (3.34) holds.*

Remark 3.62 In Theorem 3.59,

$$p = \operatorname*{ess\,sup}_{t\geq t_0} |\beta(t)| = 0$$

if and only if

$$f(t) = \exp\{\alpha(t)\}, \quad t \geq t_0,$$

where α is a measurable bounded function for which condition (3.27) holds.

3.7 Convergence to Infinity of PI-Functions

In this section, we study the limits of WPI- (PI-) functions at infinity (see Definition 3.26).

Proposition 3.27 characterizes WPI-functions in terms of their lower limit functions. In particular, $f \in \mathcal{WPI}$ if and only if the lower limit function f_* belongs to the class $\mathbb{F}^{(\infty)}$, that is

$$\limsup_{c\to\infty} f_*(c) = \infty.$$

This implies that the upper limit function f^* belongs to the class $\mathbb{F}^{(\infty)}$ as well. Now we show that WPI-functions themselves belong to the class $\mathbb{F}^{(\infty)}$.

Lemma 3.63 *If* $f \in \mathcal{WPI}$, *then* $f \in \mathbb{F}^{(\infty)}$.

Proof of Lemma 3.63 Let $f \in \mathcal{WPI}$, that is, condition (3.13) holds. Then there are constants $c_0 > 1$, $t_0 > 0$, and $r > 1$ such that $f(t) > 0$ and

$$\frac{f(c_0 t)}{f(t)} \geq r \quad \text{for all} \quad t \geq t_0.$$

Thus

$$\frac{f(c_0^m t)}{f(t)} = \frac{f(c_0^m t)}{f(c_0^{m-1} t)} \cdots \frac{f(c_0 t)}{f(t)} \geq r^m$$

and

$$f(c_0^m t) \geq r^m f(t) \tag{3.38}$$

for all $t \geq t_0$ and all $m \in \mathbf{N}_0$. This implies

$$\limsup_{t\to\infty} f(t) \geq \limsup_{m\to\infty} f(c_0^m t_0) \geq f(t_0) \lim_{m\to\infty} r^m = \infty,$$

completing the proof. □

The next result is a direct corollary of Lemma 3.63 for WPI-functions being asymptotically equivalent to nondecreasing functions.

Corollary 3.64 *Let* $f \in \mathcal{WPI}$. *If there is a nondecreasing function* h *such that* $f \sim h$, *then* $f \in \mathbb{F}^\infty$ *and* $h \in \mathbb{F}^\infty_{\text{ndec}}$. *In particular, if* f *is a nondecreasing WPI-function, then* $f \in \mathbb{F}^\infty$.

The following two examples complete Lemma 3.63 and show that the asymptotic behavior of f may be arbitrary if assumption (3.13) is omitted.

Example 3.65 Let $\theta_1 \in (0, 1)$ and $\theta_2 \in (1, \infty)$. Consider two sequences of positive numbers $\{v_n\}$ and $\{t_n\}$ defined by the following recurrence relations:

$$v_1 = 1, \qquad v_{n+1} = \theta_1 \theta_2 v_n, \quad n \geq 1;$$
$$t_1 = 1, \qquad t_{n+1} = n^2 t_n, \quad n \geq 1.$$

Via these sequences, we introduce the function

$$f(t) = \begin{cases} 1, & 0 \leq t < t_2, \\ v_n, & t \in L_n, \ n \geq 2, \\ \theta_1 v_n, & t \in R_n, \ n \geq 2, \end{cases}$$

where $L_n = [t_n, nt_n)$ and $R_n = [nt_n, t_{n+1})$.

Fix an arbitrary $c > 1$. If t is sufficiently large, then

$$t \in L_n \cup R_n \Longrightarrow ct \in L_n \cup R_n \cup L_{n+1}.$$

Thus, for t sufficiently large,

$$\frac{f(ct)}{f(t)} = \begin{cases} 1, & \text{if} \quad ct,\ t \in L_n \quad \text{or} \quad ct,\ t \in R_n, \\ \theta_1, & \text{if} \quad t \in L_n \quad \text{and} \quad ct \in R_n, \\ \theta_2, & \text{if} \quad t \in R_n \quad \text{and} \quad ct \in L_{n+1}. \end{cases}$$

This implies that

$$f_*(c) = \liminf_{t\to\infty} \frac{f(ct)}{f(t)} = \theta_1 \quad \text{and} \quad f^*(c) = \limsup_{t\to\infty} \frac{f(ct)}{f(t)} = \theta_2$$

for all $c > 1$. Therefore the function f is measurable, but $f \notin \mathcal{PI}$, since $\theta_1 < 1$. Moreover,

$$\begin{cases} \lim_{t\to\infty} f(t) = \infty, & \text{if} \quad \theta_1\theta_2 > 1, \\ \liminf_{t\to\infty} f(t) = \theta_1,\ \limsup_{t\to\infty} f(t) = 1, & \text{if} \quad \theta_1\theta_2 = 1, \\ \lim_{t\to\infty} f(t) = 0, & \text{if} \quad \theta_1\theta_2 < 1. \end{cases}$$

Example 3.66 Let $B = \{1!, 2!, 3!, \dots\}$ and $A = \mathbf{R}_+ \setminus B$. Put

$$f(t) = \mathbb{I}_A(t) + \sum_{n=1}^{\infty} \frac{1}{n} \mathbb{I}_{\{n!\}}(t), \qquad t > 0.$$

Thus

$$f(t) = \begin{cases} 1, & t \in A, \\ \frac{1}{n}, & t = n!, \quad n \ge 1. \end{cases}$$

The definition of the function f implies that

$$\frac{f(c(n!))}{f(n!)} = n \quad \text{and} \quad \frac{f(n!)}{f(n!/c)} = \frac{1}{n}$$

for all $c > 0$ and $n > \max\{c, \frac{1}{c}\}$. Hence

$$f^*(c) = \limsup_{t\to\infty} \frac{f(ct)}{f(t)} \ge \limsup_{n\to\infty} \frac{f(c(n!))}{f(n!)} = \infty$$

and

$$f_*(c) = \liminf_{t\to\infty} \frac{f(ct)}{f(t)} \le \liminf_{n\to\infty} \frac{f(c(n!/c))}{f(n!/c)} = 0.$$

Therefore, $f^*(c) = \infty$ and $f_*(c) = 0$ for all $c > 0$. It remains to note that the function f is measurable, $0 < f(t) \le 1$, and $f \notin \mathcal{PI}$.

The upper limit function of f introduced in Example 3.66 is the same as that of $g(x) = e^x$, $x > 0$. It is worth mentioning that, in contrast to a quickly increasing function g, f is bounded.

Looking at Lemma 3.63, the natural question arises on the conditions under which WPI-functions tend to infinity. The corresponding answers are given in the following two propositions. The first of these propositions does not require measurability of the function f.

Proposition 3.67 *Let $f \in \mathcal{WPI}$ and $c_0 > 1$, $t_0 > 0$, and $r > 1$ be the constants involved in inequality (3.38). If there is a $T_0 \ge t_0$ such that*

$$\varepsilon_0 = \inf_{T_0 \le \theta \le c_0 T_0} f(\theta) > 0, \tag{3.39}$$

then $f \in \mathbb{F}^\infty$.

Proof of Proposition 3.67 Inequality (3.38) implicitly implies that

$$f(t) \ge f\left(\frac{t}{c_0^{\mathfrak{m}(t)}}\right) r^{\mathfrak{m}(t)}$$

for all $t \ge T_0$, where

$$\mathfrak{m}(t) = \max\left\{n \in \mathbf{N}_0 : c_0^n T_0 \le t\right\}.$$

It is clear that $\mathfrak{m}(t)$ is the integer part of $\ln_{c_0}(t/T_0)$ and

$$\frac{t}{c_0^{\mathfrak{m}(t)}} \in [T_0, c_0 T_0]$$

for all $t \ge T_0$. Thus

$$f(t) \ge \varepsilon_0 r^{\mathfrak{m}(t)}, \quad t \ge T_0.$$

Since $\lim_{t\to\infty} \mathfrak{m}(t) = \infty$ and $r > 1$, we get

$$\lim_{t\to\infty} r^{\mathfrak{m}(t)} = \infty.$$

Taking condition (3.39) into account, we get

$$\liminf_{t\to\infty} f(t) \geq \varepsilon_0 \lim_{t\to\infty} r^{\mathrm{m}(t)} = \infty,$$

which had to be proved. □

Remark 3.68 Condition (3.39) can be exchanged by any of the following two conditions: *for all sufficiently large $s > 0$ there exists a $T = T(s) \geq s$ such that*

$$\inf_{T\leq t\leq c_0 T} f(t) > 0$$

or

$$\liminf_{t\to\infty} f(t) > 0.$$

Each of the latter two conditions implies (3.39) and thus is sufficient for the convergence of a WPI-function f to ∞.

Moreover, condition (3.39) holds, for example, if *for all sufficiently large $s > 0$, there exists a $T \geq s$ such that the function f is continuous in the interval $[T, c_0 T]$.*

Remark 3.69 We have shown in the proof of Proposition 3.67 that

$$f(t) \geq \varepsilon_0 r^{\mathrm{m}(t)}, \quad t \geq T_0,$$

that is, a WPI-*function f increases no slower than a power function if condition (3.39) holds.*

Proposition 3.67 immediately yields the following result.

Corollary 3.70 *Let $f \in \mathcal{WPI}$. If there is a number $T > 0$ such that f is continuous on $[T, \infty)$, then $f \in \mathbb{F}^\infty$.*

Note that the assumptions of Corollary 3.70 are only a bit weaker than $f \in \mathcal{PI}$. The latter assumption is used in the following result.

Proposition 3.71 *If $f \in \mathcal{ORV} \cap \mathcal{PI}$, then $f \in \mathbb{F}^\infty$.*

Remark 3.72 Since $\mathcal{ORV} \cap \mathcal{PI} = \mathcal{ORV} \cap \mathcal{WPI}$, Proposition 3.71 implies that the ORV-property, similar to condition (3.39), is an additional condition ensuring that a WPI-function tends to ∞. Recall that the ORV-property requires both the WORV-property (3.7) and the measurability of the corresponding function.

Remark 3.73 Since $\mathcal{RV}_+ \subset \mathcal{ORV} \cap \mathcal{PI}$, Proposition 3.71 implies the well-known fact that *every RV-function with positive index tends to infinity* (see Seneta [324] and Bingham et al. [41]).

To prove Proposition 3.71 we need an auxiliary result.

Lemma 3.74 *Let $f \in \mathcal{ORV} \cap \mathcal{PI}$. Then*

$$\lim_{n\to\infty} \frac{f(a_n t_n)}{f(t_n)} = \infty \tag{3.40}$$

for all sequences of positive numbers $\{a_n\}$ and $\{t_n\}$ such that $\lim_{n\to\infty} a_n = \infty$ and $\lim_{n\to\infty} t_n = \infty$.

Proof of Lemma 3.74 Let $\{a_n\}$ and $\{t_n\}$ be two sequences of positive numbers such that $\lim_{n\to\infty} a_n = \infty$ and $\lim_{n\to\infty} t_n = \infty$.

According to condition (3.13), there exist two numbers $t_0 > 0$ and $r > 1$ such that $f(t) > 0$ and

$$\frac{f(c_0 t)}{f(t)} \geq r \quad \text{for all} \quad t \geq t_0,$$

where $c_0 > 1$ is the constant involved in relation (3.13).

Via the sequence $\{a_n\}$ and the number c_0, we define a sequence $\{m_n\} \subset \mathbf{N}$ such that, for some $n_0 \in \mathbf{N}$,

$$c_0^{m_n} \leq a_n < c_0^{m_n+1}, \quad n \geq n_0.$$

Since $\lim_{n\to\infty} m_n = \infty$, we obtain $\lim_{n\to\infty} a_n = \infty$.

It is clear that

$$\frac{f(a_n t_n)}{f(t_n)} = \frac{f(a_n t_n)}{f(c_0^{m_n} t_n)} \prod_{k=1}^{m_n} \frac{f(c_0^k t_n)}{f(c_0^{k-1} t_n)}$$

for all $n \geq 1$, whence

$$\frac{f(a_n t_n)}{f(t_n)} \geq r^{m_n} \frac{f(a_n t_n)}{f(c_0^{m_n} t_n)}$$

for all sufficiently large n.

By the assumptions of the lemma, $f \in \mathcal{ORV}$. Thus (see Theorem 3.49) there are measurable bounded functions α and β such that the integral representation

$$f(t) = \Phi(t) \exp\left\{\int_{t_0}^{t} \beta(u) \frac{du}{u}\right\}$$

holds for sufficiently large t, where $\Phi(t) = \exp\{\alpha(t)\}$. This implies that

$$\frac{f(a_n t_n)}{f(c_0^{m_n} t_n)} = \frac{\Phi(a_n t_n)}{\Phi(c_0^{m_n} t_n)} \exp\left\{\int_{c_0^{m_n} t_n}^{a_n t_n} \beta(u) \frac{du}{u}\right\}$$
$$\geq \frac{\Phi(a_n t_n)}{\Phi(c_0^{m_n} t_n)} \exp\left\{-B \ln\left(\frac{a_n}{c_0^{m_n}}\right)\right\} \geq K\,\lambda^{-B}$$

for sufficiently large n, where

$$K = \frac{\liminf_{t\to\infty} \Phi(t)}{2 \limsup_{t\to\infty} \Phi(t)} > 0 \quad \text{and} \quad B = |\inf_{t\in[t_0,\infty)} \beta(t)| < \infty.$$

Therefore,

$$\liminf_{n\to\infty} \frac{f(a_n t_n)}{f(t_n)} \geq \left(K c_0^{-B}\right) \liminf_{n\to\infty} r^{m_n} = \infty,$$

whence relation (3.40) follows. □

Proof of Proposition 3.71 Assume the converse, that is, let $f(t)$ not tend to infinity as $t \to \infty$. Then there exist a sequence of positive numbers $\{u_n\}$ and a number $q \in [0, \infty)$ such that $u_n \to \infty$ and $\lim_{n\to\infty} f(u_n) = q$.

Let $q \in (0, \infty)$. In this case, there is a sequence of positive integers $\{n_k\}$ for which

$$\lim_{k\to\infty} s_k = \infty, \qquad \lim_{k\to\infty} s_{k+1}/s_k = \infty \quad \text{and} \quad \lim_{k\to\infty} f(s_{k+1})/f(s_k) = 1,$$

where $s_k = u_{n_k}$, $k \geq 1$.

Now let $q = 0$. Then there is a sequence of positive integer numbers $\{n_k\}$ for which

$$\lim_{k\to\infty} s_k = \infty, \qquad \lim_{k\to\infty} s_{k+1}/s_k = \infty \quad \text{and} \quad \lim_{k\to\infty} f(s_{k+1})/f(s_k) = 0,$$

where $s_k = u_{n_k}$, $k \geq 1$.

In both cases, we obtain a contradiction to Lemma 3.74, which proves $f \in \mathbb{F}^{\infty}$. □

The following result follows immediately from Proposition 3.71, since $\mathcal{POV} = \mathcal{PRV} \cap \mathcal{SQI} \subset \mathcal{ORV} \cap \mathcal{PI}$.

Corollary 3.75 *If $f \in \mathcal{POV}$, then $f \in \mathbb{F}^{\infty}$.*

3.7.1 A Counterexample

Analyzing the proof of Proposition 3.71, one can see that assumption (3.7) can be weakened. Indeed, Proposition 3.71 remains true for measurable functions f for which the integral representation (3.26) holds for sufficiently large t, where the measurable function α is bounded and β is bounded from below and locally bounded from above. Related to this observation is the following question: is it possible to omit the assumption on the measurability in Proposition 3.71 and add to the WPI-property (3.13) the following "one-sided" WORV-property

$$f_*(c) > 0 \quad \text{for all} \quad c > 1$$

or the property

$$f_*(c) > 1 \quad \text{for all} \quad c > 1,$$

being stronger than the WPI-property (3.13)?

The following assertion gives a negative answer to this question.

Proposition 3.76 *There exists a nonmeasurable function $f \in \mathbb{F}_+$ such that*

$$\lim_{t\to\infty} \frac{f(ct)}{f(t)} = \infty \quad \textit{for all} \quad c > 1, \tag{3.41}$$

but

$$\liminf_{t\to\infty} f(t) = 0. \tag{3.42}$$

Proof of Proposition 3.76 Let $\mathbf{H}$ be a Hamel basis, that is, a family of real numbers such that every real number $x \neq 0$ can be represented uniquely as a finite linear combination of elements of $\mathbf{H}$ with rational coefficients:

$$x = \sum_{i=1}^{n(x)} r_i(x) b_i(x),$$

where $n(x) \in \mathbf{N}$, $r_i(x) \in \mathbf{Q} \setminus \{0\}$, and $b_i(x) \in \mathbf{H}$ (see, for example, Greub [154] and Bingham et al. [41]).

Note that $(n(x), x \in \mathbf{R})$ is a nonmeasurable semiadditive function, that is,

$$n(x + y) \leq n(x) + n(y) \quad \text{for all} \quad x, y \in \mathbf{R} \tag{3.43}$$

(see, e.g., Korevaar [236] and Bingham et al. [41]). Moreover, for all fixed $n \geq 1$ and for all fixed and different $b_1, \ldots, b_n \in \mathbf{H}$,

$$\text{the set } M_n = \left\{ \sum_{i=1}^{n} r_i b_i;\ r_1, \ldots, r_n \in \mathbf{Q} \setminus \{0\} \right\} \text{ is dense in } \mathbf{R}. \tag{3.44}$$

Let

$$h(x) = x^2 - n(x), \quad x > 0,$$

and

$$f(t) = \exp\{h(\ln t)\}, \quad t > 0.$$

It is clear that $f \in \mathbb{F}_+$. Moreover, inequality (3.43) implies that

$$h(x+u) - h(x) = 2xu + u^2 - (n(x+u) - n(x)) \geq 2xu - n(u)$$

for all $x > 0$ and $u > 0$. Thus

$$\lim_{x\to\infty} \big(h(x+u) - h(x)\big) = \infty \quad \text{for all} \quad u > 0.$$

This results in

$$\lim_{t\to\infty} \frac{f(ct)}{f(t)} = \lim_{t\to\infty} \exp\{h(\ln t + \ln c) - h(\ln t)\}$$

$$= \exp\{\lim_{t\to\infty} (h(\ln t + \ln c) - h(\ln t))\} = \infty$$

for all $c > 1$. Therefore, relation (3.41) is proved.

On the other hand, in view of (3.44) there exists a sequence $\{x_k\}_{k\geq 1}$ such that

$$x_k \in (k-1, k) \cap M_{k^3}, \quad k \geq 1.$$

Obviously,

$$h(x_k) < k^2 - n(x_k) = k^2 - k^3 \quad \text{for all} \quad k \geq 1,$$

and hence

$$\liminf_{x\to\infty} h(x) = -\infty.$$

This yields relation (3.42), whence Proposition 3.76 follows. □

3.8 Characterizations of POV-Functions

In this section, we obtain conditions which are equivalent to (3.14) in the case of PRV-functions. According to Definition 3.32, these conditions are necessary and sufficient for a PRV-function to be a POV-function. As a result, we obtain some equivalent characterizations of the class of POV-functions.

Proposition 3.77 *Let $f \in \mathcal{PRV}$. Then condition (3.14) is equivalent to each of the following two conditions:*

1) *for all sequences of positive numbers $\{c_n\}$ and $\{t_n\}$ such that* $\limsup_{n\to\infty} c_n > 1$ *and* $\lim_{n\to\infty} t_n = \infty$,

$$\limsup_{n\to\infty} \frac{f(c_n t_n)}{f(t_n)} > 1; \tag{3.45}$$

2) *relation (3.45) holds for all sequences of positive numbers $\{c_n\}$ and $\{t_n\}$ such that* $1 < \limsup_{n\to\infty} c_n < \infty$ *and* $\lim_{n\to\infty} t_n = \infty$.

Proof of Proposition 3.77 The implications 1) $\Longrightarrow$ 2) and 2) $\Longrightarrow$ (3.14) are obvious. Note that both implications hold even if $f \notin \mathcal{PRV}$.

Now, let condition (3.14) hold. Consider an arbitrary sequence of positive numbers $\{c_n\}$ such that $1 < c = \limsup_{n\to\infty} c_n < \infty$. Then there exists a sequence $\{c_{n'}\} \subset \{c_n\}$ such that $\lim_{n'\to\infty} c_{n'} = c$. Since f belongs to $\mathcal{PRV}$, Theorem 3.42 implies that

$$\limsup_{n\to\infty} \frac{f(c_n t_n)}{f(t_n)} \ge \limsup_{n'\to\infty} \frac{f(c_{n'} t_{n'})}{f(t_{n'})} \ge \liminf_{n'\to\infty} \frac{f(c_{n'} t_{n'})}{f(c t_{n'})} \cdot \liminf_{n'\to\infty} \frac{f(c t_{n'})}{f(t_{n'})}$$

$$= \liminf_{n'\to\infty} \frac{f(c t_{n'})}{f(t_{n'})} \ge \liminf_{t\to\infty} \frac{f(ct)}{f(t)} > 1,$$

for an arbitrary sequence of positive numbers $\{t_n\}$ such that $\lim_{n\to\infty} t_n = \infty$. This, in particular, means that inequality (3.45) holds. Therefore, the implication (3.14) $\Longrightarrow$ 2) is proved.

Assume condition (3.14) again and consider two arbitrary sequences of positive numbers $\{c_n\}$ and $\{t_n\}$ such that $\limsup_{n\to\infty} c_n = \infty$ and $\lim_{n\to\infty} t_n = \infty$. Then there exists a sequence $\{c_{n'}\} \subset \{c_n\}$ such that $\lim_{n'\to\infty} c_{n'} = \infty$. The function f belongs to $\mathcal{PRV}$ and condition (3.14) holds. Hence $f \in \mathcal{ORV} \cap \mathcal{PI}$. This together with Lemma 3.74 implies that

$$\limsup_{n\to\infty} \frac{f(c_n t_n)}{f(t_n)} \ge \limsup_{n'\to\infty} \frac{f(c_{n'} t_{n'})}{f(t_{n'})} = \infty,$$

that is, inequality (3.45) holds. This together with the implication (3.14) $\Longrightarrow$ 2) proved above yields the implication (3.14) $\Longrightarrow$ 1). This completes the proof of Proposition 3.77. □

Corollary 3.78 *If $f \in \mathbb{F}_+$, then the following three conditions are equivalent:*

(a) $f \in \mathcal{POV}$;
(b) $f \in \mathcal{PRV}$ *and assumption 1) of Proposition 3.77 holds;*
(c) $f \in \mathcal{PRV}$ *and assumption 2) of Proposition 3.77 holds.*

3.9 Asymptotically Equivalent Monotone Versions of POV-Functions

Many problems concerning various classes of functions simplify essentially if the functions increase or, at least, have (asymptotically) equivalent increasing versions. It is known that every RV-function with positive index possesses an equivalent increasing version. In this section, we prove that every POV-function possesses an equivalent continuous version which increases to infinity and is a POV-function. Since every RV-function with positive index belongs to $\mathcal{POV}$ (see Remark 3.33), Theorem 3.79 generalizes the corresponding property of regularly varying functions.

Theorem 3.79 *Let $f \in \mathcal{POV}$. Then there exists a function f_1 such that*

$$f_1 \in \mathbb{C}^{\infty}_{\mathrm{inc}} \cap \mathcal{POV} \quad \textit{and} \quad f_1 \sim f.$$

Proof of Theorem 3.79 We use the integral representation for POV-functions, i.e.

$$f(t) = \exp\left\{a(t) + \int_{t_0}^{t} b(u)\,du\right\},$$

proved in Theorem 3.56. According to (3.33),

$$\liminf_{t\to\infty} \int_t^{ct} b(s)\,ds > 0 \quad \text{for all} \quad c > 1.$$

This implies that, given $c > 1$ there exists a $t_c > t_0$ such that

$$\int_t^{ct} b(s)\,ds > 0 \quad \text{for all} \quad t \geq t_c. \tag{3.46}$$

Consider a sequence of positive numbers $\{c_n, n \geq 1\}$ such that:

(i) $c_n \downarrow 1, n \to \infty$,
(ii) $c_1 c_2 \cdots c_n \to \infty, n \to \infty$.

Let $\tau_n = t_{c_n}$, where t_c is defined in (3.46) and $\{m_n, n \ge 1\}$ is a sequence of positive integer numbers such that

$$\begin{aligned}
m_1 \ge 1: &\quad c_1^{m_1} \ge \tau_2, \\
m_2 \ge 1: &\quad c_1^{m_1} c_2^{m_2} \ge \tau_3, \\
&\ldots\ldots\ldots\ldots\ldots\ldots\ldots\ldots \\
m_n \ge 1: &\quad c_1^{m_1} c_2^{m_2} \ldots c_n^{m_n} \ge \tau_{n+1}, \\
&\ldots\ldots\ldots\ldots\ldots\ldots\ldots\ldots
\end{aligned}$$

Put

$$\begin{aligned}
T_k &= c_1^{m_1} c_2^{m_2} \ldots c_k^{m_k}, \qquad k \in \mathbf{N}, \\
T_{k+1,l} &= T_k c_{k+1}^l, \quad 0 \le l \le m_{k+1}.
\end{aligned}$$

It is clear that $T_k = T_{k+1,0} < \cdots < T_{k+1,m_{k+1}} = T_{k+1}$. Moreover, $T_{k+1,l+1} = c_{k+1} T_{k+1,l}$ and assumption (ii) implies that $T_k \to \infty$ as $k \to \infty$. Now, consider the following functions b_1 and g_1:

$$b_1(t) = \frac{1}{T_{k+1,l+1} - T_{k+1,l}} \int_{T_{k+1,l}}^{T_{k+1,l+1}} b(s)\, ds, \qquad T_{k+1,l} \le t < T_{k+1,l+1},$$

$$g_1(t) = \begin{cases} 1, & 0 \le t < T_1, \\ \exp\left\{\int_{T_1}^t b_1(s)\, ds\right\}, & t \ge T_1. \end{cases}$$

For every $t \ge T_1$, there exist k and l such that

$$T_{k+1,l} \le t < T_{k+1,l+1}.$$

Thus

$$b_1(t) = \frac{1}{T_{k+1,l+1} - T_{k+1,l}} \int_{T_{k+1,l}}^{c_{k+1} T_{k+1,l}} b(s)\, ds > 0$$

in view of (3.46). The latter equality means that the function g_1 increases for $t \ge T_1$. Moreover, it is clear that this function is continuous.

Put

$$g(t) = \begin{cases} 1, & 0 \le t < T_1, \\ \exp\left\{\int_{T_1}^t b(s)\, ds\right\}, & t \ge T_1. \end{cases}$$

We show that $g \sim g_1$, which is equivalent to

$$\int_{T_1}^{t} (b_1(s) - b(s))\, ds \to 0 \qquad \text{as} \quad t \to \infty. \tag{3.47}$$

Indeed, the definition of the function b_1 implies that

$$\int_{T_{k+1,l}}^{T_{k+1,l+1}} (b_1(s) - b(s))\, ds = 0 \qquad \text{for all } k \text{ and } l.$$

Given an arbitrary fixed $t > 0$, there exist k and l such that $T_{k+1,l} \le t < T_{k+1,l+1}$. Thus

$$\begin{aligned}\int_{T_1}^{t} (b_1(s) - b(s))\, ds &= \int_{T_{k+1,l}}^{t} (b_1(s) - b(s))\, ds \\ &= \frac{t - T_{k+1,l}}{T_{k+1,l+1} - T_{k+1,l}} \int_{T_{k+1,l}}^{T_{k+1,l+1}} b(s)\, ds - \int_{T_{k+1,l}}^{t} b(s)\, ds,\end{aligned}$$

whence

$$\left| \int_{T_1}^{t} (b_1(s) - b(s))\, ds \right| \le 2 \sup_{T_{k+1,l} \le t < T_{k+1,l+1}} \left| \int_{T_{k+1,l}}^{t} b(s)\, ds \right|. \tag{3.48}$$

Since $f \in \mathcal{POV}$, Theorem 3.56 with the function b implies condition (3.31). This together with Theorem 3.51 yields $g \in \mathcal{PRV}$. Now, Theorem 3.47 implies

$$\lim_{k\to\infty} \sup_{l} \sup_{T_{k+1,l} \le t < T_{k+1,l+1}} \left| \frac{g(t)}{g(T_{k+1,l})} - 1 \right| = 0,$$

whence (3.47) follows in view of (3.48). Therefore, relation $g \sim g_1$ is proved.

Moreover, let

$$h(t) = \exp\left\{ a(t) + \int_{t_0}^{T_1} b(s)\, ds \right\}.$$

Obviously,

$$f(t) = h(t) g(t),$$

if $t > \max\{t_0, T_1\}$. Now, let

$$h_1(t) = \exp\left\{ a_\infty + \int_{t_0}^{T_1} b(s)\, ds \right\} \cdot \left(1 - e^{-t}\right), \quad t \ge 0,$$

where a_∞ denotes the limit $\lim_{t\to\infty} a(t)$ which exists in view of Theorem 3.56. It is clear that $h \sim h_1$ and that the function h_1 is continuous and increasing.

Finally, consider the function

$$f_1(t) = h_1(t)g_1(t), \quad t \geq 0.$$

Clearly $f \sim f_1$, since $g \sim g_1$ and $h \sim h_1$. The function f_1 is continuous and increasing, since both functions g_1 and h_1 are so. It remains to note that $f_1 \in \mathcal{POV}$, in view of $f \in \mathcal{POV}$ and $f \sim f_1$, and that Corollary 3.75 implies that f_1 increases to ∞ as $t \to \infty$. Therefore, the proof of Theorem 3.79 is complete. □

3.10 Comments

This chapter is based on the works of Buldygin et al. [59, 61, 62, 65, 69]. Some complementary results can be found in the monographs of Bingham et al. [41] and Seneta [324].

Section 3.2 The limit functions in (3.1) are also called *index functions* (see, e.g., Djurčić [104]).

In 1930, Karamata [205] introduced the notion of *regularly varying* functions (RV-functions) and proved a number of fundamental results for them (see also Karamata [206]). These results and further generalizations are discussed in the monographs of Bingham et al. [41] and Seneta [324]. *WRV-functions* have been studied in Matuszewska [266], Seneta [323], Bingham et al. [41], and Seneta [324].

Karamata's fruitful notion of the regular variation of functions has been generalized in various ways. The first generalization is due to Avakumović [19] and has been studied in many papers (see, e.g., Karamata [207], Feller [128], Aljančić and Arandelović [8], Bari and Stechkin [26], and Arandelović [18]; further references can be found in Bingham et al. [41] and Seneta [324]). The corresponding functions are said to possess the property of *O-regular variation* (ORV).

Condition (R) in Avakumović [19] defining ORV functions is stated in the form of a uniform convergence theorem. In a short note Avakumović [20] used another name (R–O) for the same notion. Karamata [207] follows the Avakumović [20] definition for this class of functions and uses the same notation. Seneta [324, p. 92] uses a slightly different name, i.e. RO, for the definition of such functions, while Aljančić and Arandelović [8] change the name to O–RV and formulate the condition in the form of Definition 3.7. Bingham et al. [41] switch to the name OR. Perhaps the name *Avakumović–Karamata functions* would also be appropriate here, however we prefer to shorten the Aljančić and Arandelović [8] notation to ORV in the spirit of Seneta [324]. We also mention de Haan [165] for another generalization of RV functions.

Every increasing ORV-function is also said to have *dominated variation*. This notion appeared in Krasnoselskiĭ and Rutickiĭ [242] and has been further developed by Feller [127, 128]. Many results on functions of dominated variation are collected in Seneta [324, Section A.3].

OSV-functions, being a subclass of ORV-functions, have been introduced in Drasin and Seneta [114]. Another subclass, the so-called OURV-functions, is studied in Aljančić and Arandelović [8].

Section 3.3 *Pseudo regularly varying* (PRV) functions and their applications have been studied in Korenblyum [235], Matuszewska [267, 270], Gihman and Skorohod [149], Stadtmüller and Trautner [344, 345], Berman [30, 32], Yakymiv [372], Cline [90], Djurčić [105], Djurčić and Torgašev [108], Klesov et al. [227], and Buldygin et al. [56–59, 61–70]. Note that PRV-functions are also called *regularly oscillating* in Berman [30], *weakly oscillating* in Yakymiv [372], *intermediate regular functions* in Cline [90] (see Cline and Samorodnitsy [92] for a closely related discussion), or *CRV-functions* in Djurčić [105]. Throughout this book we continue to use the name *PRV-functions* introduced in Buldygin et al. [59].

Nondecreasing PRV-functions are considered in Korenblyum [235] and Stadtmüller and Trautner [344] in the study of generalizations of Tauberian theorems for Laplace transforms. In particular, it is shown in Stadtmüller and Trautner [344] that the Tauberian theorem for the Laplace transform of a nondecreasing positive function f holds if and only if f is a PRV-function. A multidimensional PRV-property is studied in Yakymiv [372], which is, of course, important for the one-dimensional case, too.

Some applications of $\mathcal{PRV}$ functions in queueing theory and ruin theory are exhibited in Schlegel [329] and Jelenković and Lazar [201]. Ng et al. [284], Tang [355], and Yang et al. [374] discuss the class of $\mathcal{PRV}$ functions in the large deviations problem for sums of random variables and for ruin probabilities in renewal models (the distribution functions with $\mathcal{PRV}$ tails are called *consistently varying* therein). $\mathcal{PRV}$ functions appear quite naturally in studies of quasilinear differential equations, (see, for example, Takaŝi et al. [353, 354]).

Monotone functions satisfying condition (3.13) are studied by Bari and Stechkin [26] (see condition (L) therein). *Positive increasing* (PI) functions and their applications are studied in Bingham and Goldie [39, 40], de Haan and Stadtmüller [168], Rogozin [305], and Buldygin et al. [59, 62, 66, 69]. *Sufficiently quickly increasing* (SQI) functions and their applications are discussed in Yakymiv [371], Djurčić and Torgašev [108], and Buldygin et al. [59, 61–67]. The notion of a function possessing *positive order of variation* (POV) is introduced in Buldygin et al. [59] as a generalization of the class of RV-functions with positive index. Properties and applications of POV-functions are studied in Buldygin et al. [59, 61–70].

Section 3.4 The class of functions preserving the asymptotic equivalence of functions is denoted by CRV in Djurčić [104], where the asymptotic equivalence of functions is called *strong asymptotic equivalence*. It is proved in [104] that only CRV-functions satisfy condition (3.12). A version of Theorem 3.42 can also be found (without proof) in Korenblyum [235], Gihman and Skorohod [149], and Stadtmüller and Trautner [344].

The uniform convergence theorem, Theorem 3.47, appears in Cline [90] in a slightly different form. The uniform convergence theorems for the class $\mathcal{ORV}$ is proved in Aljancic and Arandelovic [8] (a slightly more general result, not specifically attributed to $\mathcal{ORV}$, can be found in Bingham and Goldie [39, 40], see also Bingham et al. [41, Section 2.0.1]). Earlier versions of this result are obtained by Delange [99] and Csiszár and Erdös [96].

Section 3.5 The integral representation theorem for RV-functions is proved in Karamata [205] (see also Bingham et al. [41] or Seneta [324]). The case of RV-functions of several arguments is considered by Yakymiv [370]. An extended integral representation theorem is studied in Karamata [207] and Aljančić and Arandelović [8] for ORV-functions and in Drasin and Seneta [114] for OSV-functions. A more general result for the class $\mathcal{ORV}$ can be found in Bojanić and Seneta [45]. An application of the class $\mathcal{ORV}$ to Hörmander spaces and elliptic operators is discussed by Mikhaĭlets and Murach [271]. ORV functions in approximation theory are studied by Jansche [198]. Another form of the integral representation given in Theorem 3.49 is obtained in Cline [90].

Section 3.6 The classical *Potter theorem* is proved in Potter [293] and plays an important role in the theory of regularly varying functions (see Bingham et al. [41]). It was Karamata [207] who proved such a result for the class $\mathcal{ORV}$ (see also Seneta [324, Appendix]). Theorem 3.59 improves Corollary 4 in Yakymiv [372]. Keldysh [210] used conditions like the Potter bounds in several problems concerning the spectral theory of operators.

Section 3.7 A special case of Corollary 3.75 for regularly varying functions is Proposition 1.5.1 in Bingham et al. [41].

Section 3.9 Analogs of Theorem 3.79 for regularly varying functions are proved by Karamata [205] and Matuszewska [266] (see also Bingham et al. [41, Section 1.5.2] or Seneta [324, Section 1.5, Property 4°]).

Chapter 4
Properties of Absolutely Continuous Functions

4.1 Introduction

Absolutely continuous functions are an important class of functions for both applications and theory. Every polynomial of a finite order as well as every differentiable function is absolutely continuous. Moreover, any solution of an ordinary differential equation is absolutely continuous, since the latter is at least one times differentiable. These examples are not exhaustive.

More applied examples can be found in probability theory where stochastic processes whose paths are absolutely continuous occur in many models. For example, a useful tool in the theory of stationary stochastic processes is given by so-called *filters* or, more generally, by *integral transformations*. From the point of view of calculus, the input stochastic process is a weighted density of the output process and thus the latter is absolutely continuous.

Another example comes from the theory of stochastic differential equations. If the paths of a solution of a stochastic differential equation are differentiable, then the paths are absolutely continuous functions.

Marić [261] studies the question of whether or not the solution of a second-order ordinary differential equation is regularly varying (we do not go deeper into this actively developing subject; see, however, Trajković and Manojlović [357] and the references therein, where the same question is studied for fourth-order nonlinear differential equations). A more general question is as follows: *Which properties of an absolutely continuous function guarantee that it belongs to the class* $\mathcal{RV}$? This question, of course, can be extended from $\mathcal{RV}$ to either $\mathcal{PRV}$, $\mathcal{PI}$, $\mathcal{SQI}$, or $\mathcal{POV}$.

In this chapter, we consider conditions under which *absolutely continuous functions* belong to the classes $\mathcal{PRV}$, $\mathcal{PI}$, $\mathcal{SQI}$, or $\mathcal{POV}$ (see Definitions 3.16–3.32). The results obtained in this chapter will be used throughout the book, for example, in Chaps. 5–9.

V. V. Buldygin et al., *Pseudo-Regularly Varying Functions and Generalized Renewal Processes*, Probability Theory and Stochastic Modelling 91, https://doi.org/10.1007/978-3-319-99537-3_4

Chapter 4 is organized as follows. Section 4.2 contains necessary definitions and auxiliary results for the class of absolutely continuous functions and their densities.

In Sect. 4.3, two-sided inequalities are proved between limit functions of an absolutely continuous function with a positive density and those of the density itself.

In Sect. 4.4, an integral representation is given for absolutely continuous functions and for the corresponding upper and lower limit functions. Using these representations, we find characterizing conditions under which an absolutely continuous function possesses some properties of a "regular variation" type. Some transformations of densities that preserve their regularity properties are also studied in this section.

The integral representations given in Sect. 4.4 contain logarithmic densities, which are used in Sect. 4.5 to introduce the "elasticity" of a function with respect its density and to define corresponding indices of "asymptotic elasticity".

In Sect. 4.6, two-sided inequalities are obtained for limit functions of absolutely continuous functions.

The classes of functions and densities with bounded indices of asymptotic elasticity are considered in Sect. 4.7. We also study properties of these classes and their relationships to the class $\mathbb{ER}$ introduced there.

In Sects. 4.8–4.10, some conditions are collected under which absolutely continuous functions possess certain regularity properties.

In Sect. 4.11, we study properties of the "regularity" of piecewise interpolations of sequences of real numbers.

4.2 Some Classes of Absolutely Continuous Functions

We start with the definition of the classes of positive absolutely continuous functions $\mathbb{DL}$ and $\mathbb{DL}^\infty$.

Definition 4.1 We say that a *continuous positive function*

$$f = (f(t), t \geq t_0), \qquad t_0 \geq 0,$$

belongs to the *class of absolutely continuous functions* $\mathbb{DL}$ if

$$f(t) = f(t_0) + \int_{t_0}^{t} \theta(u)\, du, \qquad t \geq t_0, \tag{4.1}$$

where the integrals on the right-hand side are to be understood in the Lebesgue sense and where $\theta = (\theta(t), t \geq t_0)$ is a measurable and locally Lebesgue integrable real-valued function, that is

$$\int_{t_0}^{t} |\theta(u)|\, du < \infty \tag{4.2}$$

for all $t > t_0$.

If, in addition to conditions (4.1) and (4.2),

$$\lim_{t\to\infty}\int_{t_0}^{t}\theta(u)\,du=\infty, \tag{4.3}$$

that is, if $\lim_{t\to\infty} f(t)=\infty$, then we say that f is an *absolutely continuous function of the class* $\mathbb{DL}^\infty$.

The function θ in relation (4.1) is called a *density* (*Lebesgue density*) of the function f. A function f with a density θ is also denoted by f_θ. A density of a function is defined up to a null set with respect to the Lebesgue measure. More precisely, if θ_1 and θ_2 coincide almost everywhere (with respect to the Lebesgue measure), then $f_{\theta_1}=f_{\theta_2}$.

If a function f is continuously differentiable for all $t\ge t_0$, then its derivative $f'(t)$, $t\ge t_0$, may serve as its density. Correspondingly, if a density θ is continuous, then the corresponding function f is continuously differentiable and its derivative coincides with the density.

For $f\in\mathbb{DL}$, the symbol $\mathfrak{D}(f)$ denotes the set of densities of f. It is clear that any two members of $\mathfrak{D}(f)$ coincide almost everywhere. Moreover, if $\theta\in\mathfrak{D}(f)$, then the set $\mathfrak{D}(f)$ contains all real-valued measurable functions $\tilde{\theta}=(\tilde{\theta}(t), t\ge t_0)$ that coincide with θ almost everywhere.

Members of the class $\mathfrak{D}(f)$ may attain negative values. Thus, along with the class $\mathfrak{D}(f)$, we consider the classes $\mathfrak{D}_{sp}(f)$ and $\mathfrak{D}_p(f)$ of all nonnegative and positive densities, respectively.

Below is a further list of some important subclasses of $\mathbb{DL}$:

(1) $f\in\mathbb{DLP}$ if $f\in\mathbb{DL}$ and there exists a $\theta\in\mathfrak{D}(f)$ such that $\theta\in\mathfrak{D}_p(f)$;
(2) $f\in\mathbb{DLSP}$ if $f\in\mathbb{DLP}$ and there exists a $\theta\in\mathfrak{D}(f)$ such that $\theta\in\mathfrak{D}_{\mathrm{sp}}(f)$;
(3) $\mathbb{DLP}^\infty=\mathbb{DLP}\cap\mathbb{DL}^\infty$;
(4) $\mathbb{DLSP}^\infty=\mathbb{DLSP}\cap\mathbb{DL}^\infty$.

The families of all densities of the classes $\mathbb{DL}$ and $\mathbb{DL}^\infty$ are denoted by $\mathfrak{D}(\mathbb{DL})$ and $\mathfrak{D}(\mathbb{DL}^\infty)$, respectively. The families of densities of functions of the above subclasses are denoted by $\mathfrak{D}(\mathbb{DLP})$, $\mathfrak{D}(\mathbb{DLP}^\infty)$, $\mathfrak{D}(\mathbb{DLSP})$, and $\mathfrak{D}(\mathbb{DLSP}^\infty)$.

The main aim of this chapter is to study conditions under which functions represented in the form of (4.1) belong to the classes $\mathcal{PRV}$, $\mathcal{PI}$, $\mathcal{SQI}$, $\mathcal{POV}$, or $\mathcal{RV}$. In Chap. 6, we continue the study of those functions in order to establish a generalization of Karamata's theorem on the asymptotic behavior of integrals of RV-functions to the class of log-periodic functions.

4.3 Relationships Between Limit Functions

In this section, we establish two-sided inequalities between the limit functions of an absolutely continuous function $f\in\mathbb{DLSP}^\infty$ and those of its density θ

(see Proposition 4.2). First, we introduce the so-called *essential limit functions* of a density θ.

4.3.1 Essential Limit Functions

As noted above, every absolutely continuous function $f \in \mathbb{DL}$ possesses a family of densities $\mathfrak{D}(f)$ coinciding almost everywhere.

Let $f = f_\theta \in \mathbb{DLSP}$. By analogy with the essential supremum and essential infimum of a function, consider the *essential upper limit function* $\theta^{\text{es}*}$ and *essential lower limit function* $\theta_{\text{es}*}$ of a density $\theta \in \mathfrak{D}(f)$ defined by

$$\theta^{\text{es}*}(c) = \inf_{\tilde{\theta} \in \mathfrak{D}_{\text{sp}}(f_\theta)} \limsup_{t\to\infty} \frac{\tilde{\theta}(ct)}{\tilde{\theta}(t)}$$

and

$$\theta_{\text{es}*}(c) = \sup_{\tilde{\theta} \in \mathfrak{D}_{\text{sp}}(f_\theta)} \liminf_{t\to\infty} \frac{\tilde{\theta}(ct)}{\tilde{\theta}(t)}.$$

Note that, for all $c > 0$,

$$\begin{aligned}\theta^{\text{es}*}(c) &= \inf_{\tilde{\theta} \in \mathfrak{D}_{\text{sp}}(f_\theta)} \inf_{a>t_0} \sup_{t\ge a} \frac{\tilde{\theta}(ct)}{\tilde{\theta}(t)} = \inf_{a>t_0} \inf_{\tilde{\theta} \in \mathfrak{D}_{\text{sp}}(f_\theta)} \sup_{t\ge a} \frac{\tilde{\theta}(ct)}{\tilde{\theta}(t)} \\ &= \inf_{a>t_0} \operatorname*{ess\,sup}_{t\ge a} \frac{\theta(ct)}{\theta(t)} = \lim_{a\to\infty} \operatorname*{ess\,sup}_{t\ge a} \frac{\theta(ct)}{\theta(t)} = \operatorname*{ess\,lim\,sup}_{t\to\infty} \frac{\theta(ct)}{\theta(t)}\end{aligned}$$

and

$$\begin{aligned}\theta_{\text{es}*}(c) &= \sup_{\tilde{\theta} \in \mathfrak{D}_{\text{sp}}(f_\theta)} \sup_{a>t_0} \inf_{t\ge a} \frac{\tilde{\theta}(ct)}{\tilde{\theta}(t)} = \sup_{a>t_0} \sup_{\tilde{\theta} \in \mathfrak{D}_{\text{sp}}(f_\theta)} \inf_{t\ge a} \frac{\tilde{\theta}(ct)}{\tilde{\theta}(t)} \\ &= \sup_{a>t_0} \operatorname*{ess\,inf}_{t\ge a} \frac{\theta(ct)}{\theta(t)} = \lim_{a\to\infty} \operatorname*{ess\,inf}_{t\ge a} \frac{\theta(ct)}{\theta(t)} = \operatorname*{ess\,lim\,inf}_{t\to\infty} \frac{\theta(ct)}{\theta(t)},\end{aligned}$$

that is,

$$\theta^{\text{es}*}(c) = \inf_{\tilde{\theta} \in \mathfrak{D}_{\text{sp}}(f_\theta)} \limsup_{t\to\infty} \frac{\tilde{\theta}(ct)}{\tilde{\theta}(t)} = \operatorname*{ess\,lim\,sup}_{t\to\infty} \frac{\theta(ct)}{\theta(t)} \tag{4.4}$$

and

$$\theta_{\text{es}*}(c) = \sup_{\tilde{\theta} \in \mathfrak{D}_{\text{sp}}(f_\theta)} \liminf_{t\to\infty} \frac{\tilde{\theta}(ct)}{\tilde{\theta}(t)} = \operatorname*{ess\,lim\,inf}_{t\to\infty} \frac{\theta(ct)}{\theta(t)}. \tag{4.5}$$

Moreover, for all $c > 0$,

$$\theta^{\mathrm{es}*}(c) = \frac{1}{\theta_{\mathrm{es}*}(1/c)} \tag{4.6}$$

and if the density θ is positive, then

$$\theta_*(c) \le \theta_{\mathrm{es}*}(c) \le \theta^{\mathrm{es}*}(c) \le \theta^*(c). \tag{4.7}$$

Essential upper and essential lower limit functions are characteristics of the whole class $\mathfrak{D}_{\mathrm{sp}}(f)$, that is, $\theta^{\mathrm{es}*} = \tilde{\theta}^{\mathrm{es}*}$ and $\theta_{\mathrm{es}*} = \tilde{\theta}_{\mathrm{es}*}$ for all $\theta \in \mathfrak{D}_{\mathrm{sp}}(f)$ and $\tilde{\theta} \in \mathfrak{D}_{\mathrm{sp}}(f)$. Thus relation (4.7) implies that

$$\tilde{\theta}_*(c) \le \theta_{\mathrm{es}*}(c) \le \theta^{\mathrm{es}*}(c) \le \tilde{\theta}^*(c) \tag{4.8}$$

for any density $\tilde{\theta} \in \mathfrak{D}_{\mathrm{sp}}(f_\theta)$.

4.3.2 Inequalities Between the Limit Functions of *f* and Its Density *θ*

Proposition 4.2 *Let $f \in \mathbb{DLSP}^\infty$ and $\theta \in \mathfrak{D}_{\mathrm{sp}}(f)$. Then*

$$c\,\tilde{\theta}_*(c) \le c\,\theta_{\mathrm{es}*}(c) \le f_*(c) \le f^*(c) \le c\,\theta^{\mathrm{es}*}(c) \le c\,\tilde{\theta}^*(c) \tag{4.9}$$

for all $\tilde{\theta} \in \mathfrak{D}_{\mathrm{sp}}(f)$ and $c > 0$.

Proof of Proposition 4.2 Since $f = f_\theta \in \mathbb{DLSP}^\infty$ and taking into account relations (4.2) and (4.3), we get

$$\begin{aligned}
\liminf_{t\to\infty}\left(\int_{t_0}^{ct}\theta(u)\,du \Big/ \int_{t_0}^{t}\theta(u)\,du\right) &= \liminf_{t\to\infty}\left(\int_{ca}^{ct}\theta(u)\,du \Big/ \int_{a}^{t}\theta(u)\,du\right)\\
&= \liminf_{t\to\infty}\left(c\int_{a}^{t}\theta(cu)\,du \Big/ \int_{a}^{t}\theta(u)\,du\right)\\
&\ge \liminf_{t\to\infty}\left(c\operatorname*{ess\,inf}_{a\le u\le t}\frac{\theta(cu)}{\theta(u)}\right)\\
&\ge c\operatorname*{ess\,inf}_{u\ge a}\frac{\theta(cu)}{\theta(u)}.
\end{aligned}$$

This implies

$$\liminf_{t\to\infty} \frac{f(ct)}{f(t)} = \liminf_{t\to\infty} \left(\int_{t_0}^{ct} \theta(u)\,du \Big/ \int_{t_0}^{t} \theta(u)\,du \right)$$
$$\geq \lim_{a\to\infty} \left(c \operatorname*{ess\,inf}_{u\geq a} \frac{\theta(cu)}{\theta(u)} \right) = c\,\theta_{\mathrm{es}*}(c)$$

for all $c > 0$ and for all $a > t_0$, that is,

$$c\,\theta_{\mathrm{es}*}(c) \leq f_*(c)$$

for all $c > 0$.

The latter inequality together with (4.6) implies that, for all $c > 0$,

$$f^*(c) = \frac{1}{f_*(1/c)} \leq \frac{c}{\theta_{\mathrm{es}*}(1/c)} = c\,\theta^{\mathrm{es}*}(c).$$

Thus, for all $c > 0$,

$$c\,\theta_{\mathrm{es}*}(c) \leq f_*(c) \leq f^*(c) \leq c\,\theta^{\mathrm{es}*}(c),$$

whence (4.9) follows in view of (4.8). □

In order to clarify how a "property of regularity" of an absolutely continuous function depends on that of its density, we introduce the corresponding notion for the whole class of densities $\mathfrak{D}_{\mathrm{sp}}(f_\theta)$.

Definition 4.3 Let $f = f_\theta \in \mathbb{DLSP}$. We say that

(1) $\mathfrak{D}_{\mathrm{sp}}(f)$ is the ORV-class if $\theta^{\mathrm{es}*}(c) < \infty$ for all $c > 0$;
(2) $\mathfrak{D}_{\mathrm{sp}}(f)$ is the PRV-class if $\limsup_{c\to 1} \theta^{\mathrm{es}*}(c) = 1$;
(3) $\mathfrak{D}_{\mathrm{sp}}(f)$ is the PI-class if $c\theta_{\mathrm{es}*}(c) > 1$ for some $c > 1$;
(4) $\mathfrak{D}_{\mathrm{sp}}(f)$ is the SQI-class if $c\theta_{\mathrm{es}*}(c) > 1$ for all $c > 1$;
(5) $\mathfrak{D}_{\mathrm{sp}}(f)$ is the POV-class if $\mathfrak{D}_{\mathrm{sp}}(f)$ is both the PRV-class and SQI-class;
(6) $\mathfrak{D}_{\mathrm{sp}}(f)$ is the RV-class if $\theta_{\mathrm{es}*}(c) = \theta^{\mathrm{es}*}(c) \in (0, \infty)$ for all $c > 0$.

Remark 4.4 Let $\mathfrak{K} \in \{\mathcal{ORV}, \mathcal{PRV}, \mathcal{PI}, \mathcal{SQI}, \mathcal{POV}, \mathcal{RV}\}$ and $f \in \mathbb{DLSP}$. In view of relation (4.9) and Definition 4.3, we conclude that $\mathfrak{D}_{\mathrm{sp}}(f)$ is the $\mathfrak{K}$-class if there exists a density $\tilde{\theta} \in \mathfrak{D}_{\mathrm{sp}}(f)$ such that $\tilde{\theta} \in \mathfrak{K}$.

A density may lose or may acquire a certain property if the density is changed on a null set with respect to Lebesgue measure. The following example shows that a density θ may not have a certain property, but the class $\mathfrak{D}_{\mathrm{sp}}(f_\theta)$ may have that property.

Example 4.5 Let a density θ of a function f be such that $\theta(t) = 2^t$ for $t \in \mathbf{N}$, and $\theta(t) = 1$ for $t \geq 0$ and $t \notin \mathbf{N}$. Further, let $\theta_1(t) = 1$ for $t \geq 0$. It is clear that $\theta_1 \in \mathfrak{D}_{\mathrm{sp}}(f)$ and θ_1 belongs to $\mathcal{SV}$, that is, θ_1 is an RV-function with index $\rho = 0$. This together with Remark 4.4 implies that $\mathfrak{D}_{\mathrm{sp}}(f)$ is the RV-class. At the same time,

$$\limsup_{t\to\infty} \frac{\theta(ct)}{\theta(t)} \geq \limsup_{n\to\infty} \frac{\theta(cn)}{\theta(n)} = \limsup_{n\to\infty} \frac{2^{cn}}{2^n} = \limsup_{n\to\infty} 2^{(c-1)n} = \infty$$

for any integer $c \geq 2$.

Thus, $\theta^*(c) = \infty$ for any integer $c \geq 2$, that is, θ does not belong to $\mathcal{ORV}$.

The following Corollary 4.6 shows that a function f inherits the "properties of regularity" of the class $\mathfrak{D}_{\mathrm{sp}}(f)$.

Corollary 4.6 *Let $f \in \mathbb{DLSP}^\infty$. Then:*

(1) if $\mathfrak{D}_{\mathrm{sp}}(f)$ is the ORV-class, then $f \in \mathcal{ORV}$;
(2) if $\mathfrak{D}_{\mathrm{sp}}(f)$ is the PRV-class, then $f \in \mathcal{PRV}$;
(3) if $\mathfrak{D}_{\mathrm{sp}}(f)$ is the PI-class, then $f \in \mathcal{PI}$;
(4) if $\mathfrak{D}_{\mathrm{sp}}(f)$ is the SQI-class, then $f \in \mathcal{SQI}$;
(5) if $\mathfrak{D}_{\mathrm{sp}}(f)$ is the POV-class, then $f \in \mathcal{POV}$;
(6) if $\mathfrak{D}_{\mathrm{sp}}(f)$ is the RV-class, then $f \in \mathcal{RV}$.

Remark 4.7 The relation

$$\theta \in \mathcal{RV} \Longrightarrow f_\theta \in \mathcal{RV},$$

which follows from statement (6) of Corollary 4.6, can be improved. Indeed, let a function f_θ belong to the class $\mathbb{DLSP}^\infty$ and let its density θ be an RV-function with index ρ (this is only possible if $\rho \geq -1$). Then, since

$$\theta_*(c) = \theta^*(c) = c^\rho, \qquad c > 0,$$

inequalities (4.9) imply that

$$c^{\rho+1} \leq f_*(c) \leq f^*(c) \leq c^{\rho+1}, \qquad c > 0,$$

that is, f is an RV-function with index $\rho + 1$. Thus, if $\rho \geq -1$,

$$\theta \in \mathcal{RV}_\rho \Longrightarrow f \in \mathcal{RV}_{\rho+1}.$$

Note that the condition $\theta \in \mathcal{RV}_\rho$ implies condition (4.3) if $\rho > -1$.

Let $\mathbb{G}_r(\theta)$ be the set of all numbers $c > 0$ such that

$$\theta_*(c) = \theta^*(c) \in (0, \infty).$$

In view of inequalities (4.9), we conclude that

$$f_*(c) = f^*(c) = c\,\theta_*(c) \tag{4.10}$$

for all $c \in \mathbb{G}_r(\theta)$.

In Chap. 5, we show that the set $\mathbb{G}_r(\theta)$, called the *set of regular points* of the function θ, is a multiplicative subgroup of positive numbers. Thus, (4.10) implies

$$\mathbb{G}_r(\theta) \subset \mathbb{G}_r(f),$$

that is, the group of regular points of a function cannot shrink after passing to the integral of this function.

Since the set of regular points of a function coincides with the set of positive numbers if and only if the function is of regular variation, (4.10) contains the above relations for RV-functions.

The above consideration may be viewed as an "Introduction" to the Karamata theorem on the asymptotic behavior of integrals of RV-functions and its generalization to the class of log-periodic functions (see Chap. 6).

4.4 Logarithmic Density and Integral Representations

The dependence of the "properties of regularity" of absolutely continuous functions in the class $\mathbb{DLSP}^\infty$ on those of their densities has been studied in the preceding section where we found some sufficient conditions imposed on the limit functions of densities (or, on the essential limit functions of a class of densities) under which the corresponding function possesses a certain property (see Proposition 4.2 and Corollary 4.6).

In this section, we use integral representations and logarithmic densities to obtain characterizing conditions for the corresponding "properties of regularity" of absolutely continuous functions in the class $\mathbb{DL}$.

4.4.1 Integral Representations for Functions in the Class $\mathbb{DL}$

Proposition 4.8 *Assume $f \in \mathbb{DL}$ and let θ be a density of f. Then*

$$f(t) = f(t_0)\exp\left\{\int_{t_0}^{t} \frac{\theta(u)}{f(u)}\,du\right\}, \qquad t \ge t_0, \tag{4.11}$$

$$f^*(c) = \limsup_{t\to\infty} \frac{f(ct)}{f(t)} = \exp\left\{\limsup_{t\to\infty} \int_t^{ct} \frac{\theta(u)}{f(u)}\,du\right\}, \qquad c > 0, \tag{4.12}$$

and

$$f_*(c) = \liminf_{t\to\infty} \frac{f(ct)}{f(t)} = \exp\left\{\liminf_{t\to\infty} \int_t^{ct} \frac{\theta(u)}{f(u)}\,du\right\}, \qquad c > 0. \tag{4.13}$$

Proof of Proposition 4.8 Relation (4.11) holds if the density θ is continuous, since $\theta(t) = f'(t)$, $t \ge t_0$, in this case, and thus

$$\int_{t_0}^{t} \frac{\theta(u)}{f(u)}\,du = \int_{t_0}^{t} \frac{f'(u)}{f(u)}\,du = \ln\left(f(t)/f(t_0)\right).$$

Since relation (4.11) holds for continuous densities θ, the continuity of the function f implies that it holds for piecewise constant densities θ as well.

To complete the proof of relation (4.11) we note that the function θ is locally integrable by definition of the class $\mathbb{DL}$. Hence it can be approximated in the mean by a sequence of piecewise constant functions on each fixed interval $[t_0, T]$.

Now, relation (4.11) implies

$$\frac{f(ct)}{f(t)} = \exp\left\{\int_t^{ct} \frac{\theta(u)}{f(u)}\,du\right\}$$

for $c > 0$ and $t \ge \max\{t_0, t_0/c\}$, whence (4.12) and (4.13) follow. □

4.4.2 Logarithmic Densities

Assume $f \in \mathbb{DL}$ and let θ be a density of f. The function

$$f'_{\ln}(\theta; t) = \frac{\theta(t)}{f(t)}, \qquad t \ge t_0, \tag{4.14}$$

is called the *logarithmic density* of the function f with respect to the density θ. If f is continuously differentiable, its logarithmic density coincides with the *logarithmic derivative* $(\ln f)' = f'/f$.

The logarithmic density $f'_{\ln}$ depends on f and θ. Thus, if f is fixed, the function $f'_{\ln}$ is defined (like a density θ) up to a null set with respect to Lebesgue measure.

4.4.3 A Characterization of Properties of Absolutely Continuous Functions

The integral representations (4.11)–(4.13) imply characterizing conditions under which some "properties of regularity" hold for absolutely continuous functions in the class $\mathbb{DL}$.

Theorem 4.9 *Assume $f \in \mathbb{DL}$ and let θ be a density of f. Then*

1) $f \in \mathcal{ORV}$ if and only if

$$-\infty < \liminf_{t\to\infty} \int_t^{ct} f'_{\ln}(\theta; u)du \quad and \quad \limsup_{t\to\infty} \int_t^{ct} f'_{\ln}(\theta; u)\, du < \infty$$

for all $c > 1$;

2) $f \in \mathcal{PRV}$ if and only if

$$\limsup_{c\to 1} \limsup_{t\to\infty} \int_t^{ct} f'_{\ln}(\theta; u)\, du = 0; \tag{4.15}$$

3) if $f \in \mathbb{DLP}$, then $f \in \mathcal{PRV}$ if and only if

$$\lim_{c\downarrow 1} \limsup_{t\to\infty} \int_t^{ct} f'_{\ln}(\theta; u)\, du = 0;$$

4) $f \in \mathcal{PI}$ if and only if there exists a $c_0 > 1$ such that

$$\liminf_{t\to\infty} \int_t^{c_0 t} f'_{\ln}(\theta; u)\, du > 0;$$

5) $f \in \mathcal{SQI}$ if and only if, for all $c > 1$,

$$\liminf_{t\to\infty} \int_t^{ct} f'_{\ln}(\theta; u)\, du > 0; \tag{4.16}$$

6) $f \in \mathcal{POV}$ if and only if conditions (4.15) and (4.16) hold;

7) f is an RV-function with index ρ if and only if the limit

$$\lim_{t\to\infty} \int_t^{ct} f'_{\ln}(\theta; u)\, du = \rho \in (-\infty, \infty)$$

exists for all $c > 1$.

4.4.4 Admissible Transformations of Densities

Theorem 4.9 allows us to solve the problem of admissible transformations of densities that leave the density in the same class.

Definition 4.10 Let $\theta \in \mathfrak{D}(\mathbb{DL})$ and $\mathfrak{K} \in \{\mathcal{PRV}, \mathcal{PI}, \mathcal{SQI}, \mathcal{POV}\}$. We call θ a $\mathfrak{K}$-*density* if there exists a function $f \in \mathbb{DL}$ such that $\theta \in \mathfrak{D}(f)$ and $f \in \mathfrak{K}$. The class of all $\mathfrak{K}$-densities is denoted by $\mathfrak{D}(\mathfrak{K})$ and the class of all nonnegative $\mathfrak{K}$-densities is denoted by $\mathfrak{D}_p(\mathfrak{K})$.

According to Definition 4.10, a density θ belongs to the class $\mathfrak{D}(\mathfrak{K})$ if the function

$$f(t) = f(t_0) + \int_{t_0}^{t} \theta(u)\,du, \qquad t \geq t_0,$$

belongs to the class $\mathfrak{K}$. If, in addition, the density θ is nonnegative, then $\theta \in \mathfrak{D}_p(\mathfrak{K})$.

We call an arithmetical transformation of densities *$\mathfrak{K}$-admissible* if the class $\mathfrak{D}(\mathfrak{K})$ is closed with respect to this transformation.

The classes $\mathfrak{D}(\mathcal{PRV})$, $\mathfrak{D}(\mathcal{SQI})$ and $\mathfrak{D}(\mathcal{POV})$ are closed under summation and multiplication by a positive constant (see Propositions 3.23, 3.30, and 3.35). These results are generalized below to the classes of nonnegative densities and also extended to the class $\mathfrak{D}_p(\mathcal{PI})$.

Definition 4.11 We say that a real-valued measurable bounded positive function $d = (d(t), t \geq t_0)$ is *almost constant* if it is uniformly bounded away from zero and from infinity for sufficiently large t, that is, if there exists $t_1 \geq t_0$ such that

$$0 < d_1 \leq d(t) \leq d_2 < \infty \qquad \text{for all} \quad t \geq t_1.$$

The numbers $d_1 \leq d_2$ are called *boundaries of the almost constant function d*.

Now we prove that all four classes of $\mathfrak{K}$-densities above are closed with respect to multiplication by an almost constant function, that is, multiplication by an almost constant function is admissible.

Lemma 4.12 *Let $\mathfrak{K} \in \{\mathcal{PRV}, \mathcal{PI}, \mathcal{SQI}, \mathcal{POV}\}$, $\theta \in \mathfrak{D}_p(\mathfrak{K})$ and let $d = (d(t), t \geq t_0)$ be an almost constant function. Then $d \cdot \theta \in \mathfrak{D}_p(\mathfrak{K})$, where $d \cdot \theta = (d(t)\,\theta(t), t \geq t_0)$.*

Proof of Lemma 4.12 Since $\theta \in \mathfrak{D}_p(\mathfrak{K})$, there exists a function $f \in \mathbb{DLP}$ such that $\theta \in \mathfrak{D}_p(f)$ and $f \in \mathfrak{K}$. Moreover, since the function d is almost constant,

$$\theta_1(t) = d(t)\,\theta(t) \in [d_1\,\theta(t), d_2\,\theta(t)]$$

for all $t \geq t_1 \geq t_0$. This implies that the function θ_1 is nonnegative and locally integrable.

Put

$$f_1(t) = f_1(t_0) + \int_{t_0}^{t} \theta_1(u)\,du, \qquad t \geq t_0,$$

where the number $f_1(t_0)$ is chosen such that

$$f_1(t_0) \in [d_1\,f(t_0), d_2\,f(t_0)]. \tag{4.17}$$

Hence, for all $t \geq t_1$,

$$\theta_1(t) \in [d_1\,\theta(t), d_2\,\theta(t)] \qquad \text{if} \quad f_1(t) \in [d_1 f(t), d_2 f(t)],$$

whence

$$\left(\frac{d_1}{d_2}\right) f'_{ln}(\theta; t) \leq (f_1)'_{ln}(\theta_1; t) \leq \left(\frac{d_2}{d_1}\right) f'_{ln}(\theta; t)$$

for all $t \geq t_1$. Thus

$$\begin{aligned}\left(\frac{d_1}{d_2}\right) \liminf_{t\to\infty} \int_t^{ct} f'_{\ln}(\theta; u)\,du &\leq \liminf_{t\to\infty} \int_t^{ct} (f_1)'_{ln}(\theta_1; u)\,du \\ &\leq \left(\frac{d_2}{d_1}\right) \liminf_{t\to\infty} \int_t^{ct} f'_{\ln}(\theta; u)\,du\end{aligned}$$

and

$$\begin{aligned}\left(\frac{d_1}{d_2}\right) \limsup_{t\to\infty} \int_t^{ct} f'_{\ln}(\theta; u)\,du &\leq \limsup_{t\to\infty} \int_t^{ct} (f_1)'_{ln}(\theta_1; u)\,du \\ &\leq \left(\frac{d_2}{d_1}\right) \limsup_{t\to\infty} \int_t^{ct} f'_{\ln}(\theta; u)\,du\end{aligned}$$

for all $c > 0$.

These inequalities together with Theorem 4.9 imply that $f_1 \in \mathfrak{K}$ for every class of $\mathfrak{K}$-densities mentioned in Lemma 4.12. □

Remark 4.13 Since Lemma 4.12 only deals with nonnegative densities, one can assume without loss of generality that $t_0 > 0$ and that the densities θ are defined on the interval $[A, \infty)$, where $A < t_0$. Moreover, if

$$f(t) = \int_A^t \theta(u)\,du, \qquad t \geq A,$$

then

$$f(t_0) = \int_A^{t_0} \theta(u)\,du > 0.$$

Thus

$$d_1 f_{d\cdot\theta}(t_0) \leq f_{d\cdot\theta}(t_0) = \int_A^{t_0} d(t)\,\theta(u)\,du \leq d_2 f_{d\cdot\theta}(t_0),$$

that is, condition (4.17) holds automatically.

Theorem 4.9 and Lemma 4.12 imply that a linear combination of densities in one of the classes $\mathfrak{D}_p(\mathcal{PRV})$, $\mathfrak{D}_p(\mathcal{PI})$, $\mathfrak{D}_p(\mathcal{SQI})$, and $\mathfrak{D}_p(\mathcal{POV})$ belongs to the same class (see also Remark 4.13).

Proposition 4.14 *Let* $\mathfrak{K} \in \{\mathcal{PRV}, \mathcal{PI}, \mathcal{SQI}, \mathcal{POV}\}$, $\theta_k \in \mathfrak{D}_p(\mathfrak{K})$, $k = 1, \dots, n$, *and let* $(d_k, k = 1, \dots, n)$ *be almost constant functions. Then*

$$d_1 \cdot \theta_1 + \dots + d_n \cdot \theta_n \in \mathfrak{D}_p(\mathfrak{K}).$$

Another corollary of Lemma 4.12 claims that, if densities are positive and their ratios are almost constant, then the corresponding integrals belong to the same class.

Proposition 4.15 *Assume* $f_1 \in \mathbb{DLSP}$, $f_2 \in \mathbb{DLSP}$ *and let* θ_1, θ_2 *be positive densities of* f_1 *and* f_2*, respectively. If* θ_1/θ_2 *is almost constant, then*

$$f_1 \in \mathfrak{K} \Longleftrightarrow f_2 \in \mathfrak{K}$$

for each of the classes $\mathfrak{K} \in \{\mathcal{PRV}, \mathcal{PI}, \mathcal{SQI}, \mathcal{POV}\}$.

4.5 Elasticity and Indices of Asymptotic Elasticity

It can sometimes be complicated to check the characteristic assumptions for the "properties of regularity" of absolutely continuous functions given in terms of logarithmic densities in Theorem 4.9. Some improvement of the corresponding sufficient conditions can be achieved with the help of indices of *asymptotic elasticity*.

4.5.1 Elasticity of Functions

Let $f \in \mathbb{DL}$, θ be a density of f and let $f'_{\ln}$ be its logarithmic density (see (4.14)).

Definition 4.16 The function

$$\mathscr{E}(t) = \mathscr{E}(f, \theta; t) = t f'_{\ln}(\theta; t) = \frac{t\theta(t)}{f(t)}, \qquad t \geq t_0, \tag{4.18}$$

is called the *elasticity* of the function f *with respect to the density* θ.

The elasticity $\mathscr{E}$ depends on both f and θ. Thus, given a function f, its elasticity $\mathscr{E}$ is defined, like its density θ, up to a null set with respect to Lebesgue measure.

Theorem 4.9 contains the characteristic conditions for some properties of absolutely continuous functions given in terms of integrals of logarithmic densities or, equivalently, in terms of integrals of the elasticity. Some sufficient conditions, which sometimes are easier to check, can also be provided.

Proposition 4.17 *Let $f \in \mathbb{DL}$, θ be a density of f and let $\mathcal{E}(f,\theta;\cdot)$ be the elasticity of the function f with respect to the density θ. Then:*

1) if

$$-\infty < \liminf_{t\to\infty} \mathcal{E}(f,\theta;t) \quad \textit{and} \quad \limsup_{t\to\infty} \mathcal{E}(f,\theta;t) < \infty \tag{4.19}$$

or, equivalently, if

$$\limsup_{t\to\infty} |\mathcal{E}(f,\theta;t)| = \limsup_{t\to\infty} \frac{t|\theta(t)|}{f(t)} < \infty, \tag{4.20}$$

then $f \in \mathcal{PRV}$;

2) if

$$\liminf_{t\to\infty} \mathcal{E}(f,\theta;t) > 0,$$

then $f \in \mathcal{SQI}$;

3) if

$$0 < \liminf_{t\to\infty} \mathcal{E}(f,\theta;t) \quad \textit{and} \quad \limsup_{t\to\infty} \mathcal{E}(f,\theta;t) < \infty,$$

then $f \in \mathcal{POV}$;

4) if the limit

$$\lim_{t\to\infty} \mathcal{E}(f,\theta;t) = \rho \in (-\infty,\infty)$$

exists, then f is an RV-function of index ρ.

Proof of Proposition 4.17 Let condition (4.20) hold. Since

$$\left|\int_t^{ct} f'_{\ln}(\theta;u)\,du\right| = \left|\int_t^{ct} \mathcal{E}(f,\theta;u)\,\frac{du}{u}\right| \le \left(\sup_{u\ge t}|\mathcal{E}(f,\theta;u)|\right)\cdot\left|\int_t^{ct}\frac{du}{u}\right|$$
$$\le \left(\sup_{u\ge t}|\mathcal{E}(f,\theta;u)|\right)|\ln c|$$

for all $c > 0$, we conclude that

$$\limsup_{c\to1}\limsup_{t\to\infty}\left|\int_t^{ct} f'_{\ln}(\theta;u)\,du\right| \le \limsup_{c\to1}|\ln c|\cdot\limsup_{t\to\infty}|\mathcal{E}(f,\theta;u)| = 0.$$

Hence condition (4.15) holds and the first statement of Proposition 4.17 follows from statement 2 of Theorem 4.9. All other statements of Proposition 4.17 can be proved similarly. □

Remark 4.18 The first and fourth statement of Proposition 4.17 can also be proved by comparing the integral representation (4.11) with the corresponding integral representations collected in Sect. 3.5.

4.5.2 Indices of Asymptotic Elasticity

Proposition 4.17 highlights the importance of the characteristics

$$\liminf_{t\to\infty} \mathscr{E}(f,\theta;t) \quad \text{and} \quad \limsup_{t\to\infty} \mathscr{E}(f,\theta;t).$$

Thus it is worthwhile to give a special name to these indices.

Definition 4.19 If $f = f_\theta \in \mathbb{DL}$, then the *upper bound of elasticity*

$$\epsilon^*(f_\theta) = \epsilon^*(f,\theta) = \limsup_{t\to\infty} \mathscr{E}(f,\theta;t) = \limsup_{t\to\infty} \frac{t\theta(t)}{f(t)}$$

and the *lower bound of elasticity*

$$\epsilon_*(f_\theta) = \epsilon_*(f,\theta) = \liminf_{t\to\infty} \mathscr{E}(f,\theta;t) = \liminf_{t\to\infty} \frac{t\theta(t)}{f(t)}$$

are called *upper* and *lower indices of asymptotic elasticity* of the function f.

Note that, in general, $\epsilon^*(f_\theta), \epsilon_*(f_\theta) \in [-\infty, \infty]$.

Proposition 4.17 reads as follows in terms of the indices of asymptotic elasticity.

Proposition 4.20 *Assume $f \in \mathbb{DL}$ and let θ be a density of f. Then:*

1) if

$$-\infty < \epsilon_*(f,\theta) \quad \textit{and} \quad \epsilon^*(f,\theta) < \infty,$$

then $f \in \mathcal{PRV}$;

2) if

$$\epsilon_*(f,\theta) > 0,$$

then $f \in \mathcal{SQI}$;

3) if

$$0 < \epsilon_*(f,\theta) \quad \textit{and} \quad \epsilon^*(f,\theta) < \infty,$$

then $f \in \mathcal{POV}$;

4) if

$$\epsilon(f,\theta) = \epsilon^*(f,\theta) = \epsilon_*(f,\theta) \in (-\infty,\infty),$$

then f *is an RV-function with index* $\rho = \epsilon(f,\theta)$.

4.5.3 Essential Indices of Asymptotic Elasticity

Along with the upper and lower indices of asymptotic elasticity of a function f with respect to a density θ, we consider the *essential upper index of asymptotic elasticity*

$$\epsilon^{\mathrm{es}*}(f) = \inf_{\theta \in \mathfrak{D}(f)} \epsilon^*(f,\theta)$$

and the *essential lower index of asymptotic elasticity*

$$\epsilon_{\mathrm{es}*}(f) = \sup_{\theta \in \mathfrak{D}(f)} \epsilon_*(f,\theta)$$

of a function f.

Note that

$$\begin{aligned}
\epsilon^{\mathrm{es}*}(f) &= \inf_{\theta \in \mathfrak{D}(f)} \inf_{a>t_0} \sup_{t\ge a} \frac{t\theta(t)}{f(t)} = \inf_{a>t_0} \inf_{\theta \in \mathfrak{D}(f)} \sup_{t\ge a} \frac{t\theta(t)}{f(t)} \\
&= \inf_{a>t_0} \operatorname*{ess\,sup}_{t\ge a} \frac{t\theta(t)}{f(t)} = \lim_{a\to\infty} \operatorname*{ess\,sup}_{t\ge a} \frac{t\theta(t)}{f(t)} \\
&= \lim_{a\to\infty} \operatorname*{ess\,sup}_{t\ge a} \mathscr{E}(f,\theta;t) = \operatorname*{ess\,lim\,sup}_{t\to\infty} \mathscr{E}(f,\theta;t)
\end{aligned}$$

and

$$\begin{aligned}
\epsilon_{\mathrm{es}*}(c) &= \sup_{\theta \in \mathfrak{D}(f)} \sup_{a>t_0} \inf_{t\ge a} \frac{t\theta(t)}{f(t)} = \sup_{a>t_0} \sup_{\theta \in \mathfrak{D}(f)} \inf_{t\ge a} \frac{t\theta(t)}{f(t)} \\
&= \sup_{a>t_0} \operatorname*{ess\,inf}_{t\ge a} \frac{t\theta(t)}{f(t)} = \lim_{a\to\infty} \operatorname*{ess\,inf}_{t\ge a} \frac{t\theta(t)}{f(t)} \\
&= \lim_{a\to\infty} \operatorname*{ess\,inf}_{t\ge a} \mathscr{E}(f,\theta;t) = \operatorname*{ess\,lim\,inf}_{t\to\infty} \mathscr{E}(f,\theta;t).
\end{aligned}$$

Thus

$$-\infty \le \epsilon_*(f,\theta) \le \epsilon_{es*}(f) \le \epsilon^{es*}(f) \le \epsilon^*(f,\theta) \le \infty, \tag{4.21}$$

where $\theta \in \mathfrak{D}(f)$ is an arbitrary density. If $\theta \in \mathfrak{D}_p$, then $\epsilon_*(f,\theta) \ge 0$, hence $\epsilon_{es*}(f) \ge 0$ for $f \in \mathbb{DLP}$.

Remark 4.21 If a density θ is continuous, then a change of this density on a null set with respect to Lebesgue measure cannot decrease its upper asymptotic elasticity and cannot increase its lower asymptotic elasticity. Thus

$$\epsilon_*(f,\theta) = \epsilon_{es*}(f) \quad \text{and} \qquad \epsilon^*(f,\theta) = \epsilon^{es*}(f).$$

Similar equalities are also satisfied for densities possessing right or left limits at every point under the condition that the density is continuous at every point at least from one side.

4.5.4 Examples

Consider some simple examples of the logarithmic density, elasticity, and indices of asymptotic elasticity.

Example 4.22 Let $t_0 = 1$; $f_\alpha(1) = 1$; $\theta_\alpha(t) = |\alpha+1|t^\alpha$, $t \ge 1$; $\alpha \in (-\infty,\infty)$, $\alpha \ne -1$, and $\theta_{-1}(t) = t^{-1}$, $t \ge 1$. Then:

(1) if $\alpha > -1$, then

$$f_\alpha(t) = f_{\theta_\alpha}(t) = 1 + \int_1^t \theta_\alpha(u)du = t^{\alpha+1}, \qquad t \ge 1,$$

$$(f_\alpha)'_{ln}(t) = \frac{\theta_\alpha(t)}{f_\alpha(t)} = \frac{\alpha+1}{t}, \qquad t \ge 1,$$

$$\mathscr{E}(f_\alpha,\theta_\alpha;t) = \frac{t\theta_\alpha(t)}{f_\alpha(t)} = \alpha+1, \qquad t \ge 1,$$

$$\epsilon_*(f_\alpha,\theta_\alpha) = \epsilon_{es*}(f_\alpha) = \epsilon^{es*}(f_\alpha) = \epsilon^*(f_\alpha,\theta_\alpha) = \alpha+1,$$

whence we conclude that f_α is an RV-function of index $\rho = \alpha + 1$ (see Proposition 4.20);

(2) if $\alpha = -1$, then

$$f_{-1}(t) = 1 + \ln t, \qquad t \geq 1,$$

$$(f_{-1})'_{ln}(t) = \frac{1}{t(1+\ln t)}, \qquad \mathscr{E}(f_{-1}, \theta_{-1}; t) = \frac{1}{1+\ln t}, \qquad t \geq 1,$$

$$\epsilon_*(f_{-1}, \theta_{-1}) = \epsilon_{\mathrm{es}*}(f_{-1}) = \epsilon^{\mathrm{es}*}(f_{-1}) = \epsilon^*(f_{-1}, \theta_{-1}) = 0,$$

whence we obtain that f_{-1} is an SV-function (see Proposition 4.20);

(3) if $\alpha < -1$, then

$$f_\alpha(t) = 2 - t^{\alpha+1}, \qquad t \geq 1,$$

$$(f_\alpha)'_{ln}(t) = \frac{|\alpha+1|t^\alpha}{2 - t^{\alpha+1}}, \quad \mathscr{E}(f_\alpha, \theta_\alpha; t) = \frac{|\alpha+1|}{2t^{|\alpha+1|} - 1}, \qquad t \geq 1,$$

$$\epsilon_*(f_\alpha, \theta_\alpha) = \epsilon_{\mathrm{es}*}(f_\alpha) = \epsilon^{\mathrm{es}*}(f_\alpha) = \epsilon^*(f_\alpha, \theta_\alpha) = 0,$$

whence we have that f_α is an SV-function (see Proposition 4.20).

Example 4.23 Let $t_0 = 0$; $f_a(0) = 1$; $\theta_a(t) = |a|e^{at}$, $t \geq 0$; $a \in (-\infty, \infty)$. Then:

(1) if $a > 0$, then

$$f_a(t) = f_{\theta_a}(t) = 1 + \int_0^t \theta_a(u)du = e^{at}, \qquad t \geq 0,$$

$$(f_a)'_{ln}(t) = a, \quad \mathscr{E}(f_a, \theta_a; t) = at, \qquad t \geq 0,$$

$$\epsilon_*(f_a, \theta_a) = \epsilon_{\mathrm{es}*}(f_a) = \epsilon^{\mathrm{es}*}(f_a) = \epsilon^*(f_a, \theta_a) = \infty;$$

(2) if $a \leq 0$, then

$$f_a(t) = 2 - e^{at}, \qquad t \geq 0,$$

$$(f_a)'_{ln}(t) = \frac{|a|}{2e^{|a|t} - 1}, \quad \mathscr{E}(f_a, \theta_a; t) = \frac{|a|t}{2e^{|a|t} - 1}, \qquad t \geq 0,$$

$$\epsilon_*(f_a, \theta_a) = \epsilon_{\mathrm{es}*}(f_a) = \epsilon^{\mathrm{es}*}(f_a) = \epsilon^*(f_a, \theta_a) = 0,$$

whence we conclude that f_α is an SV-function (see Proposition 4.20).

Example 4.24 Assume $f = f_\theta \in \mathbb{DL}$ and let θ be an almost constant density (see Definition 4.11), that is, there exists $t_1 \geq t_0$ such that

$$0 < a \leq \theta(t) \leq b < \infty, \qquad t \geq t_1. \tag{4.22}$$

Then

$$f(t_0) + a(t - t_0) \leq f(t) \leq f(t_0) + b(t - t_0), \qquad t \geq t_1,$$

whence

$$\frac{at}{f(t_0) + b(t - t_0)} \leq \mathscr{E}(f, \theta; t) \leq \frac{bt}{f(t_0) + a(t - t_0)}, \qquad t \geq t_1,$$

and

$$0 < \frac{a}{b} \leq \epsilon_*(f, \theta) \leq \epsilon_{\text{es}*}(f) \leq \epsilon^{\text{es}*}(f) \leq \epsilon^*(f, \theta) \leq \frac{b}{a} < \infty.$$

This implies that f is a POV-function (see Proposition 4.20).

Example 4.24 suggests the following corollary.

Corollary 4.25 *Assume $f_\theta \in \mathbb{DL}$ and let a density θ be almost constant (see Definition 4.11). Then f_θ is a POV-function.*

The following example shows that the sufficient conditions of Proposition 4.20 are, in general, not necessary.

Example 4.26 Let $t_0 = 1$, $b \in (0, \infty)$, $f_b(1) = 1 + \sin 2$ and

$$\theta_b(t) = 1 + 2bt^{b-1} \cos(2t^b), \qquad t \geq 1.$$

Then

$$f_b(t) = t + \sin(2t^b), \qquad t \geq 1,$$

$$(f_b)'_{ln}(t) = \frac{1 + 2bt^{b-1} \cos(2t^b)}{t + \sin(2t^b)}, \qquad t \geq 1,$$

and

$$\mathscr{E}(f_b, \theta_b; t) = \frac{t(1 + 2bt^{b-1} \cos(2t^b))}{t + \sin(2t^b)}, \qquad t \geq 1.$$

Moreover,

(1) if $b \in (0, 1)$, then

$$\epsilon_*(f_b, \theta_b) = \epsilon_{\text{es}*}(f_b) = \epsilon^{\text{es}*}(f_b) = \epsilon^*(f_b, \theta_b) = 1,$$

whence we obtain that f_b is an RV-function of index $\rho = 1$ (see Proposition 4.20);

(2) if $b = 1$, then

$$\epsilon_{\text{es}*}(f_1) = \epsilon_*(f_1, \theta_1) = -1 \quad \text{and} \quad \epsilon^{\text{es}*}(f_1) = \epsilon^*(f_1, \theta_1) = 3,$$

whence we conclude that f_b is a PRV-function (see Proposition 4.20);

(3) if $b > 1$, then

$$\epsilon_{\text{es}*}(f_b) = \epsilon_*(f_b, \theta_b) = -\infty \quad \text{and} \quad \epsilon^{\text{es}*}(f_b) = \epsilon^*(f_b, \theta_b) = \infty.$$

Note that

$$\lim_{t\to\infty} \frac{f_b(ct)}{f_b(t)} = \lim_{t\to\infty} \frac{ct + \sin(2(ct)^b)}{t + \sin(2t^b)} = c$$

for all $b \in (0, \infty)$ and $c > 0$. Therefore, f_b is an RV-function of index $\rho = 1$ for all $b \in (0, \infty)$. Note that this conclusion, for $b \in (0, 1)$, follows from Proposition 4.20 as well.

4.5.5 Arithmetic Properties of Indices of Asymptotic Elasticity

The following result contains some statements that follow directly from the definition of indices of asymptotic elasticity in combination with Lemma 4.12. These statements show how the indices of asymptotic elasticity change after some arithmetical transformations.

Lemma 4.27 *Let $f = f_\theta \in \mathbb{DL}$, $f_1 = f_{\theta_1} \in \mathbb{DL}$ and $f_2 = f_{\theta_2} \in \mathbb{DL}$. Then:*

(i) for all $a > 0$,

$$\epsilon_*(af_\theta) = \epsilon_*(f_\theta) \quad \textit{and} \quad \epsilon^*(af_\theta) = \epsilon^*(f_\theta),$$

and

$$\epsilon_{\text{es}*}(af) = \epsilon_{\text{es}*}(f) \quad \textit{and} \quad \epsilon^{\text{es}*}(af) = \epsilon^{\text{es}*}(f);$$

(ii) if a density θ is positive,

$$d \cdot \theta = (d(t)\,\theta(t), t \geq t_0),$$

where $d = (d(t), t \geq t_0)$ *is an almost constant function with boundaries* $d_1 \leq d_2$ *(see Definition 4.11) and*

$$f_{d\cdot\theta}(t_0) \in [d_1 f(t_0), d_2 f(t_0)],$$

then

$$\frac{d_1}{d_2} \cdot \epsilon_*(f_\theta) \leq \epsilon_*(f_{d\cdot\theta}) \leq \frac{d_2}{d_1} \cdot \epsilon_*(f_\theta),$$

and

$$\frac{d_1}{d_2} \cdot \epsilon^*(f_\theta) \leq \epsilon^*(f_{d\cdot\theta}) \leq \frac{d_2}{d_1} \cdot \epsilon^*(f_\theta);$$

(iii) *for all* $c > 0$,

$$\min\{\epsilon_*(f_{\theta_1}), \epsilon_*(f_{\theta_2})\} \leq \epsilon_*(f_{\theta_1} + f_{\theta_2}) \leq \max\{\epsilon_*(f_{\theta_1}), \epsilon_*(f_{\theta_2})\},$$

$$\min\{\epsilon^*(f_{\theta_1}), \epsilon^*(f_{\theta_2})\} \leq \epsilon^*(f_{\theta_1} + f_{\theta_2}) \leq \max\{\epsilon^*(f_{\theta_1}), \epsilon^*(f_{\theta_2})\},$$

and

$$\min\{\epsilon_{\mathrm{es}^*}(f_1), \epsilon_{\mathrm{es}^*}(f_2)\} \leq \epsilon_{\mathrm{es}^*}(f_1 + f_2) \leq \max\{\epsilon_{\mathrm{es}^*}(f_1), \epsilon_{\mathrm{es}^*}(f_2)\},$$

$$\min\{\epsilon^{\mathrm{es}^*}(f_1), \epsilon^{\mathrm{es}^*}(f_2)\} \leq \epsilon^{\mathrm{es}^*}(f_1 + f_2) \leq \max\{\epsilon^{\mathrm{es}^*}(f_1), \epsilon^{\mathrm{es}^*}(f_2)\};$$

(iv) *if* f *is a continuously differentiable, positive function, then*

$$\epsilon_*\left(1/f, (1/f)'\right) = -\epsilon^*\left(f, f'\right) \quad \textit{and} \quad \epsilon^*\left(1/f, (1/f)'\right) = -\epsilon_*\left(f, f'\right);$$

(v) *if* f *is a continuously differentiable, positive and monotone function, then*

$$\epsilon_*\left(f^{-1}, (f^{-1})'\right) = \frac{1}{\epsilon^*(f, f')} \quad \textit{and} \quad \epsilon^*\left(f^{-1}, \left(f^{-1}\right)'\right) = \frac{1}{\epsilon_*(f, f')},$$

where f^{-1} *is the inverse function of* f.

4.6 Relationships in Terms of the Indices of Asymptotic Elasticity

The indices of asymptotic elasticity are helpful for obtaining two-sided inequalities for limit functions of $f \in \mathbb{DL}$. Recall that we set $c^{-\infty} = 0$ and $c^{\infty} = \infty$, if $c > 1$, and $c^{-\infty} = \infty$ and $c^{\infty} = 0$, if $c < 1$.

Theorem 4.28 *Let $f \in \mathbb{DL}$. Then, for all $\theta \in \mathfrak{D}(f)$,*

$$c^{\epsilon_*(f,\theta)} \le c^{\epsilon_{\mathrm{es}*}(f)} \le f_*(c) \le f^*(c) \le c^{\epsilon^{\mathrm{es}*}(f)} \le c^{\epsilon^*(f,\theta)}, \tag{4.23}$$

if $c \ge 1$, and

$$c^{\epsilon^*(f,\theta)} \le c^{\epsilon^{\mathrm{es}*}(f)} \le f_*(c) \le f^*(c) \le c^{\epsilon_{\mathrm{es}*}(f)} \le c^{\epsilon_*(f,\theta)}, \tag{4.24}$$

if $c \in (0, 1)$.

Proof of Theorem 4.28 Let $c \ge 1$. Then, according to the integral representation (4.12),

$$\begin{aligned} f^*(c) &= \exp\left\{\limsup_{t\to\infty} \int_t^{ct} \frac{\theta(u)}{f(u)}\,du\right\} = \exp\left\{\limsup_{t\to\infty} \int_t^{ct} \frac{\mathscr{E}(f,\theta;u)}{u}\,du\right\} \\ &\le \exp\left\{\limsup_{t\to\infty} \left(\left(\operatorname*{ess\,sup}_{u\ge t} \mathscr{E}(f,\theta;u)\right) \int_t^{ct} \frac{du}{u}\right)\right\} \\ &= \exp\left\{\limsup_{t\to\infty} \left(\operatorname*{ess\,sup}_{u\ge t} \mathscr{E}(f,\theta;u)\right) \ln c\right\} = \exp\left\{\epsilon^{\mathrm{es}*}(f) \ln c\right\} = c^{\epsilon^{\mathrm{es}*}(f)}. \end{aligned}$$

Taking (4.21) into account, we obtain the inequality on the right-hand side of (4.23).

Similarly, starting with the integral representation (4.13), we get

$$\begin{aligned} f_*(c) &= \exp\left\{\liminf_{t\to\infty} \int_t^{ct} \frac{\theta(u)}{f(u)}\,du\right\} = \exp\left\{\liminf_{t\to\infty} \int_t^{ct} \frac{\mathscr{E}(f,\theta;u)}{u}\,du\right\} \\ &\ge \exp\left\{\liminf_{t\to\infty} \left(\left(\operatorname*{ess\,inf}_{u\ge t} \mathscr{E}(f,\theta;u)\right) \int_t^{ct} \frac{du}{u}\right)\right\} \\ &= \exp\left\{\liminf_{t\to\infty} \left(\operatorname*{ess\,inf}_{u\ge t} \mathscr{E}(f,\theta;u)\right) \ln c\right\} = \exp\left\{\epsilon_{\mathrm{es}*}(f) \ln c\right\} = c^{\epsilon_{\mathrm{es}*}(f)}. \end{aligned}$$

In view of (4.21), the inequality on the left-hand side of (4.23) is proved, which renders the proof of (4.23) complete.

If $c \in (0, 1)$, then (4.23) and (4.21) imply that, for all $\theta \in \mathfrak{D}(f)$,

$$f^*(c) = \frac{1}{f_*(1/c)} \le \frac{1}{c^{-\epsilon_{\mathrm{es}*}(f)}} = c^{\epsilon_{\mathrm{es}*}(f)} \le c^{\epsilon_*(f,\theta)}$$

and

$$f_*(c) = \frac{1}{f^*(1/c)} \ge \frac{1}{c^{-\epsilon^{\mathrm{es}*}(f)}} = c^{\epsilon^{\mathrm{es}*}(f)} \ge c^{\epsilon^*(f,\theta)},$$

whence (4.24) follows. □

4.7 Functions with Bounded Asymptotic Elasticity

Proposition 4.20 and Theorem 4.28 are helpful in studying the "properties of regularity" of absolutely continuous functions. To go deeper into the analysis of the corresponding properties, we consider the classes of absolutely continuous functions as well as the classes of the corresponding densities, whose indices of asymptotic elasticity belong to some intervals.

Definition 4.29 Let $-\infty \le a \le b \le \infty$ and $f = f_\theta \in \mathbb{DL}$. We say that a function f belongs to the class $\mathbb{DL}_{\text{ind}}(a, b)$ and, respectively, that its density θ belongs to the class $\mathfrak{D}_{\text{ind}}(a, b)$ if

$$a < \epsilon_*(f, \theta) \le \epsilon^*(f, \theta) < b. \tag{4.25}$$

The classes of functions $\mathbb{DL}_{\text{ind}}(a, b]$, $\mathbb{DL}_{\text{ind}}[a, b)$, $\mathbb{DL}_{\text{ind}}[a, b]$ and classes of densities $\mathfrak{D}_{\text{ind}}(a, b]$, $\mathfrak{D}_{\text{ind}}[a, b)$, $\mathfrak{D}_{\text{ind}}[a, b]$ are defined analogously.

For the classes of densities defined above we consider the corresponding subclasses of nonnegative densities denoted by $\mathfrak{D}^{(p)}_{\text{ind}}(a, b)$, $\mathfrak{D}^{(p)}_{\text{ind}}(a, b]$, $\mathfrak{D}^{(p)}_{\text{ind}}[a, b)$, and $\mathfrak{D}^{(p)}_{\text{ind}}[a, b]$.

Moreover, we say that a function f_θ belongs to the class $\mathbb{DL}_{\text{esind}}(a, b)$, if

$$a < \epsilon_{\text{es}*}(f) \le \epsilon^{\text{es}*}(f) < b. \tag{4.26}$$

The classes $\mathbb{DL}_{\text{esind}}(a, b]$, $\mathbb{DL}_{\text{esind}}[a, b)$, and $\mathbb{DL}_{\text{esind}}[a, b]$ are defined in a similar manner.

The inequalities (4.21) show that

$$\mathbb{DL}_{\text{ind}} M \subset \mathbb{DL}_{\text{esind}} M \tag{4.27}$$

for a set M denoting any of the intervals (a, b), $(a, b]$, $[a, b)$, $[a, b]$.

We are more interested in the classes of *functions with finite indices of asymptotic elasticity* , i.e.

$$\mathbb{DL}_{\text{ind}}(-\infty, \infty) \quad \text{and} \quad \mathbb{DL}_{\text{esind}}(-\infty, \infty),$$

of *functions with positive, finite indices of asymptotic elasticity*,

$$\mathbb{DL}_{\text{ind}}(0, \infty) \quad \text{and} \quad \mathbb{DL}_{\text{esind}}(0, \infty),$$

and of *functions with positive indices of asymptotic elasticity*:

$$\mathbb{DL}_{\text{ind}}(0, \infty] \quad \text{and} \quad \mathbb{DL}_{\text{esind}}(0, \infty].$$

To understand better what the above classes mean, we refer to Proposition 4.20 and Theorem 4.28. According to these results

$$\mathbb{DL}_{\text{ind}}(-\infty, \infty) \subset \mathcal{PRV}, \quad \mathbb{DL}_{\text{ind}}(0, \infty] \subset \mathcal{SQI}, \quad \mathbb{DL}_{\text{ind}}(0, \infty) \subset \mathcal{POV} \tag{4.28}$$

for an arbitrary density $\theta \in \mathfrak{D}(f)$. Moreover, if $r \in (-\infty, \infty)$, then

$$\mathbb{DL}_{\text{ind}}[r, r] \subset \mathcal{RV}, \tag{4.29}$$

and r is the index of the RV-function f.

Remark 4.30 Example 4.26 shows that the inclusions in relation (4.28) and (4.29) are, in general, strict.

4.7.1 Admissible Transformations of Densities

Similar to Sect. 4.4, now with the classes of nonnegative densities,

$$\mathfrak{D}_{\text{ind}}^{(p)}[0, \infty), \quad \mathfrak{D}_{\text{ind}}^{(p)}(0, \infty], \qquad \mathfrak{D}_{\text{ind}}^{(p)}(0, \infty) \tag{4.30}$$

consider the problem of closeness of these classes with respect to arithmetical operations (see Lemma 4.12 and Proposition 4.14).

Since we are considering nonnegative densities, we assume without loss of generality that $t_0 > 0$ and that the densities θ are defined on the interval $[A, \infty)$, where $A < t_0$ and

$$f(t) = \int_A^t \theta(u)\, du, \qquad t \geq A,$$

(see Remark 4.13).

First we answer the question of how boundaries of the asymptotic elasticity change for functions with nonnegative boundaries when passing to linear combinations with almost constant coefficients (see Definition 4.11). The following result follows from statements (ii) and (iii) of Lemma 4.27.

Lemma 4.31 *Let $\theta_k \in \mathfrak{D}_p(\mathbb{DL})$, $k = 1, \ldots, n$, and $(d_k, k = 1, \ldots, n)$ be almost constant functions with boundaries $d_{1,k} \leq d_{2,k}$, $k = 1, \ldots, n$. Then*

$$\min_{k=1,\ldots,n} \left\{ \frac{d_{1,k}}{d_{2,k}} \cdot \epsilon_*(f_{\theta_k}) \right\} \leq \epsilon_*(f_{d_1 \cdot \theta_1 + \ldots + d_n \cdot \theta_n}) \leq \max_{k=1,\ldots,n} \left\{ \frac{d_{2,k}}{d_{1,k}} \cdot \epsilon_*(f_{\theta_k}) \right\}$$

and

$$\min_{k=1,\ldots,n}\left\{\frac{d_{1,k}}{d_{2,k}}\cdot\epsilon^*(f_{\theta_k})\right\}\le\epsilon^*(f_{d_1\cdot\theta_1+\ldots+d_n\cdot\theta_n})\le\max_{k=1,\ldots,n}\left\{\frac{d_{2,k}}{d_{1,k}}\cdot\epsilon^*(f_{\theta_k})\right\}.$$

Lemma 4.31 implies the closeness of the classes (4.30) under linear transformations of densities with almost constant coefficients (see also Remark 4.13).

Proposition 4.32 *Let* $\mathfrak{D}_p$ *denote any of the classes of nonnegative densities from (4.30). If* $\theta_k\in\mathfrak{D}_p$, $k=1,\ldots,n$, *and* $(d_k, k=1,\ldots,n)$ *are almost constant functions, then*

$$d_1\cdot\theta_1+\cdots+d_n\cdot\theta_n\in\mathfrak{D}_p.$$

Lemma 4.31 also implies an assertion similar to Proposition 4.15.

Proposition 4.33 *Let* $f_1\in\mathbb{DLSP}$, $f_2\in\mathbb{DLSP}$ *and* θ_1, θ_2 *be positive densities of the functions* f_1, f_2, *respectively. If the ratio* θ_1/θ_2 *is almost constant, then for each of the classes*

$$\mathbb{D}\in\{\mathbb{DL}_{\text{lind}}[0,\infty),\ \mathbb{DL}_{\text{lind}}(0,\infty],\ \mathbb{DL}_{\text{lind}}(0,\infty)\}$$

we have

$$f_1\in\mathbb{D}\Longleftrightarrow f_2\in\mathbb{D}.$$

4.7.2 The ER-Property of Absolutely Continuous Functions

As mentioned above, Proposition 4.20 and Theorem 4.28 are helpful in studying properties of absolutely continuous functions with the help of indices of asymptotic elasticity. At the same time, the inequalities of Theorem 4.28 provide better bounds as compared to Proposition 4.20 if the indices of asymptotic elasticity are known exactly or if, at least, one knows bounds for these indices. In the first place, this observation concerns the ER-*property* of absolutely continuous functions given as follows.

Definition 4.34 Let $-\infty\le a\le b\le\infty$ and $f\in\mathbb{F}_+$. We say that a function f belongs to the class $\mathbb{ER}(a,b)$ if, for all $c>1$,

$$c^a<f_*(c)\le f^*(c)<c^b. \tag{4.31}$$

Correspondingly, if, for all $c>1$,

$$c^a<f_*(c)\le f^*(c)\le c^b,$$

then we say that a function f belongs to the class $\mathbb{ER}(a,b]$.

The classes of functions $\mathbb{ER}[a, b)$ and $\mathbb{ER}[a, b]$ are defined similarly. We set $c^{\infty} = \infty$ and $c^{-\infty} = 0$. To shorten the notation, we write $\mathbb{ER} = \mathbb{ER}(-\infty, \infty)$ and $\mathbb{ER}_+ = \mathbb{ER}(0, \infty)$.

Definition 4.34 implies that

$$\mathcal{RV}_\rho = \mathbb{ER}[\rho, \rho], \qquad \mathcal{RV} \subset \mathbb{ER}, \qquad \mathcal{RV}_+ \subset \mathbb{ER}_+. \tag{4.32}$$

Moreover, Definition 4.34 also implies that every $\mathbb{ER}$-function belongs to $\mathcal{PRV}$, every $\mathbb{ER}(0, \infty]$-function belongs to $\mathcal{SQI}$, and every $\mathbb{ER}_+$-function belongs to $\mathcal{POV}$. Thus, in addition to (4.32),

$$\mathbb{ER} \subset \mathcal{PRV}, \qquad \mathbb{ER}(0, \infty] \subset \mathcal{SQI}, \qquad \mathbb{ER}_+ \subset \mathcal{POV}. \tag{4.33}$$

Theorem 4.28, Definitions 4.29 and 4.34, and relation (4.33) allow us to obtain some ER-properties with the help of indices of asymptotic elasticity.

Theorem 4.35 *Let $f = f_\theta \in \mathbb{DL}$. Then:*

1) $f \in \mathbb{ER}[\epsilon_{\mathrm{es}*}(f), \epsilon^{\mathrm{es}*}(f)]$ *and* $f \in \mathbb{ER}[\epsilon_*(f, \theta), \epsilon^*(f, \theta)]$;
2) $\mathbb{DL}_{\mathrm{esind}} M \subset \mathbb{ER} M$ *and* $\mathbb{DL}_{\mathrm{ind}} M \subset \mathbb{ER} M$ *for* $M \in \{(a, b), (a, b], [a, b), [a, b]\}$;
3) $\mathbb{DL}_{\mathrm{esind}}(-\infty, \infty) \subset \mathbb{ER} \subset \mathcal{PRV}$ *and* $\mathbb{DL}_{\mathrm{ind}}(-\infty, \infty) \subset \mathbb{ER} \subset \mathcal{PRV}$;
4) $\mathbb{DL}_{\mathrm{esind}}(0, \infty] \subset \mathbb{ER}(0, \infty] \subset \mathcal{SQI}$ *and* $\mathbb{DL}_{\mathrm{ind}}(0, \infty] \subset \mathbb{ER}(0, \infty] \subset \mathcal{SQI}$;
5) $\mathbb{DL}_{\mathrm{esind}}(0, \infty) \subset \mathbb{ER}_+ \subset \mathcal{POV}$ *and* $\mathbb{DL}_{\mathrm{ind}}(0, \infty) \subset \mathbb{ER}_+ \subset \mathcal{POV}$;
6) *if* $f \in \mathbb{DLP}$, *then* $f \in \mathbb{ER}[0, \epsilon^{\mathrm{es}*}(f)]$ *and* $f \in \mathbb{ER}[0, \epsilon^*(f, \theta)]$;
7) *if* $f \in \mathbb{DLP}$ *and* $\epsilon^{\mathrm{es}*}(f) < \infty$, *then*
$$f \in \mathbb{ER}[0, \epsilon^{\mathrm{es}*}(f)] \subset \mathbb{ER}[0, \infty) \subset \mathcal{PRV};$$
8) *if* $f \in \mathbb{DLP}$ *and* $\epsilon^*(f, \theta) < \infty$, *then*
$$f \in \mathbb{ER}[0, \epsilon^*(f, \theta)] \subset \mathbb{ER}[0, \infty) \subset \mathcal{PRV};$$
9) *if* $\epsilon_{\mathrm{es}*}(f) > 0$, *then* $f \in \mathbb{ER}[\epsilon_{\mathrm{es}*}(f), \infty] \subset \mathbb{ER}(0, \infty] \subset \mathcal{SQI}$;
10) *if* $\epsilon_*(f, \theta) > 0$, *then* $f \in \mathbb{ER}[\epsilon_*(f, \theta), \infty] \subset \mathbb{ER}(0, \infty] \subset \mathcal{SQI}$;
11) *if* $0 < \epsilon_{\mathrm{es}*}(f)$ *and* $\epsilon^{\mathrm{es}*}(f) < \infty$, *then*
$$f \in \mathbb{ER}[\epsilon_{\mathrm{es}*}(f), \epsilon^{\mathrm{es}*}(f)] \subset \mathbb{ER}(0, \infty) \subset \mathcal{POV};$$
12) *if* $0 < \epsilon_*(f, \theta)$ *and* $\epsilon^*(f, \theta) < \infty$, *then*
$$f \in \mathbb{ER}[\epsilon_*(f, \theta), \epsilon^*(f, \theta)] \subset \mathbb{ER}(0, \infty) \subset \mathcal{POV}.$$

Remark 4.36 Let $f \in \mathbb{DL}^\infty$ and $\theta \in \mathfrak{D}(f)$. If

$$\limsup_{t\to\infty} t|\theta(t)| < \infty,$$

then

$$\limsup_{t\to\infty} \frac{t\theta(t)}{f(t)} = 0.$$

This implies

$$\rho_*(f,\theta) = \rho^*(f,\theta) = 0,$$

and, by Theorem 4.28,

$$f^*(c) = f^*(c) = 1$$

for all $c \geq 1$, that is, $f \in \mathcal{SV}$. Thus, to prove the SV-property of a function f, one only needs to study the case when

$$\limsup_{t\to\infty} t|\theta(t)| = \infty.$$

4.7.3 Boundaries of the Indices of Asymptotic Elasticity for Almost Monotone Densities

We say that a nonnegative density θ is *almost decreasing* (or *μ-decreasing*), if there are $\mu \in (0, 1]$ and $t_1 \geq t_0$ such that

$$\theta(s) \geq \mu\theta(t) \qquad \text{for all} \qquad t \geq s \geq t_1. \tag{4.34}$$

Further, we say that a nonnegative density θ is said to be *almost increasing* (or *μ-increasing*) if there are $\mu \in (0, 1]$ and $t_1 \geq t_0$ such that

$$\theta(t) \geq \mu\theta(s) \qquad \text{for all} \qquad t \geq s \geq t_1. \tag{4.35}$$

Note that a positive density θ is almost decreasing if and only if

$$\lim_{s\to\infty} \inf_{t\geq s} \frac{\theta(s)}{\theta(t)} > 0. \tag{4.36}$$

Similarly, a positive density θ is almost increasing if and only if

$$\lim_{s\to\infty}\inf_{t\geq s}\frac{\theta(t)}{\theta(s)} > 0. \tag{4.37}$$

If a nonnegative density θ is either almost decreasing or almost increasing, then we say that the density is *almost monotone*.

Only densities θ that are nonincreasing and nonnegative, starting at some $t_1 \geq t_0$, are μ-decreasing with $\mu = 1$. Correspondingly, only densities θ that are nondecreasing and nonnegative, starting at some $t_1 \geq t_0$, are μ-increasing with $\mu = 1$.

Remark 4.37 The definition of almost monotone densities immediately implies that

1) every almost decreasing density θ is uniformly bounded from above for sufficiently large t, that is

$$\limsup_{t\to\infty}\theta(t) < \infty;$$

2) every almost increasing density θ is uniformly separated from zero for sufficiently large t, that is

$$\liminf_{t\to\infty}\theta(t) > 0;$$

3) a nonnegative density θ is both almost decreasing and almost increasing if and only if it is an almost constant function (see Definition 4.11), that is

$$\frac{\liminf_{t\to\infty}\theta(t)}{\limsup_{t\to\infty}\theta(t)} > 0.$$

Moreover, if $\liminf_{t\to\infty}\theta(t) > 0$, $\limsup_{t\to\infty}\theta(t) < \infty$, and

$$\mu \in \left(0, \frac{\liminf_{t\to\infty}\theta(t)}{\limsup_{t\to\infty}\theta(t)}\right), \tag{4.38}$$

then θ is both μ-decreasing and μ-increasing.

Below we list some properties of almost monotone densities that follow directly from the definitions.

Lemma 4.38 *Let $f_k \in \mathbb{DLSP}$ and $\theta_k \in \mathfrak{D}_p(f_k)$, $k = 1, 2$. Then:*

1) if the densities θ_1 and θ_2 are μ-decreasing (μ-increasing) with $\mu = \mu_1$ and $\mu = \mu_2$ respectively, then $\theta_1 + \theta_2$ is μ-decreasing (μ-increasing) with $\mu = \min\{\mu_1, \mu_2\}$;

2) *if the density* θ_1 *is* μ*-decreasing (*μ*-increasing) with* $\mu = \mu_1$ *and* $\theta_2 = d \cdot \phi$*, where* ϕ *is a* μ*-decreasing (*μ*-increasing) function with* $\mu = \mu_\phi$*, then* θ_2 *is* μ*-decreasing (*μ*-increasing) with* $\mu = \mu_\phi \mu_1$*;*
3) *if the density* θ_1 *is* μ*-decreasing (*μ*-increasing) with* $\mu = \mu_1$ *and* $\theta_2 = d \cdot \theta_1$*, where* d *is an almost constant function (see Definition 4.11), then* θ_2 *is* μ*-decreasing (*μ*-increasing) with* $\mu = \mu_d \mu_1$*, where*

$$\mu_d \in \left(0, \frac{\liminf_{t\to\infty} d(t)}{\limsup_{t\to\infty} d(t)}\right).$$

The following result provides estimates for the interval where the indices of asymptotic elasticity of positive, almost monotone densities assume their values.

Lemma 4.39 *Let* $f \in \mathbb{DLSP}$ *and* $\theta \in \mathfrak{D}_{\text{sp}}(f)$*. The following three statements hold true:*

1) *if the density* θ *is* μ*-decreasing, then*

$$0 \le \epsilon_*(f, \theta) \le \epsilon^*(f, \theta) \le \frac{1}{\mu} < \infty;$$

2) *if the density* θ *is* μ*-increasing, then*

$$0 < \mu \le \epsilon_*(f, \theta);$$

3) *if the density* θ *is almost constant (see Remark 4.37), then*

$$0 < \mu \le \epsilon_*(f, \theta) \le \epsilon^*(f, \theta) \le \frac{1}{\mu} < \infty,$$

where μ *is the same as in (4.38).*

Proof of Lemma 4.39 If the positive density θ is μ-decreasing, then

$$0 \le \epsilon_*(f, \theta) \le \epsilon^*(f, \theta) = \limsup_{t\to\infty} \frac{t\theta(t)}{f(t)}$$
$$\le \limsup_{t\to\infty} \frac{t\theta(t)}{\int_{t_0}^t \theta(u)du} \le \limsup_{t\to\infty} \frac{t\theta(t)}{\mu(t-t_0)\theta(t)} = \frac{1}{\mu} < \infty.$$

In turn, if the positive density θ is μ-increasing, then

$$\lim_{t\to\infty} \int_{t_0}^t \theta(u)du = \infty$$

and

$$\rho_*(f,\theta) = \liminf_{t\to\infty} \frac{t\theta(t)}{f(t)} = \liminf_{t\to\infty} \frac{t\theta(t)}{f(t_0) + \int_{t_0}^t \theta(u)du}$$
$$= \liminf_{t\to\infty} \frac{t\theta(t)}{\int_{t_0}^t \theta(u)du} \ge \limsup_{t\to\infty} \frac{\mu t\theta(t)}{(t-t_0)\theta(t)} = \mu > 0.$$

Thus the first two statements of Lemma 4.39 are proved, and the third one follows from them.

□

Lemma 4.39 together with Theorem 4.35 implies the following result.

Corollary 4.40 *Let $f \in \mathbb{DLSP}$ and $\theta \in \mathfrak{D}_{\text{sp}}(f)$. The following three statements hold true:*

1) if the density θ is μ-decreasing, then

$$\theta \in \mathfrak{D}_{\text{lind}}[0, 1/\mu] \quad \textit{and} \quad f \in \mathbb{DL}_{\text{lind}}[0, 1/\mu],$$

in particular, $f \in \mathbb{ER}[0, 1/\mu]$ and $f \in \mathcal{PRV}$;

2) if the density θ is μ-increasing, then

$$\theta \in \mathfrak{D}_{\text{lind}}[\mu, \infty] \quad \textit{and} \quad f \in \mathbb{DL}_{\text{lind}}[\mu, \infty],$$

in particular, $f \in \mathbb{ER}[\mu, \infty]$ and $f \in \mathcal{SQI}$;

3) if the density θ is almost constant (see Remark 4.37), then

$$\theta \in \mathfrak{D}_{\text{lind}}[\mu, 1/\mu] \quad \textit{and} \quad f \in \mathbb{DL}_{\text{lind}}[\mu, 1/\mu],$$

where μ is the same as in (4.38), in particular, $f \in \mathbb{ER}[\mu, 1/\mu]$ and $f \in \mathcal{POV}$.

Corollary 4.40 implies that, for all $c > 1$,

$$1 \le f_*(c) \le f^*(c) \le c^{1/\mu}, \tag{4.39}$$

if the density θ is μ-decreasing, and

$$c^\mu \le f_*(c) \tag{4.40}$$

if the density θ is μ-increasing.

The boundaries for the limit functions of almost monotone functions can be improved for sufficiently large c with the help of Proposition 4.2.

Corollary 4.41 *Let* $f \in \mathbb{DLSP}$ *and* $\theta \in \mathfrak{D}_{\mathrm{sp}}(f)$. *The following three statements hold true:*

1) if the density θ *is* μ*-decreasing, then, for all* $c > 1$,

$$1 \le f_*(c) \le f^*(c) \le \min\left\{c^{1/\mu}, c/\mu\right\};$$

2) if the density θ *is* μ*-increasing, then, for all* $c > 1$,

$$\max\left\{c^{\mu}, c\mu\right\} \le f_*(c);$$

3) if the density θ *is almost constant (see Remark 4.37), then, for all* $c > 1$,

$$\max\left\{c^{\mu}, c\mu\right\} \le f_*(c) \le f^*(c) \le \min\left\{c^{1/\mu}, c/\mu\right\},$$

where μ *is the same as in (4.38).*

Proof of Corollary 4.41 If the positive density θ is μ-decreasing, then, for all $c > 1$,

$$\theta^*(c) = \limsup_{t\to\infty} \frac{\theta(ct)}{\theta(t)} \le \limsup_{t\to\infty} \frac{\theta(ct)}{\mu\theta(ct)} \le \frac{1}{\mu}.$$

Using Proposition 4.2, we get

$$1 \le f_*(c) \le f^*(c) \le \frac{c}{\mu}$$

for all $c > 1$. This together with (4.39) implies the first statement.

Correspondingly, if the positive density θ is μ-increasing, then, for all $c > 1$,

$$\theta_*(c) = \liminf_{t\to\infty} \frac{\theta(ct)}{\theta(t)} \ge \liminf_{t\to\infty} \frac{\mu\theta(t)}{\theta(t)} \ge \mu.$$

Applying Proposition 4.2, we see that

$$c/\mu \le f_*(c)$$

for all $c > 1$. This together with (4.39) implies the second statement, whence we obtain the third statement as well. □

4.8 The PRV-Property of Absolutely Continuous Functions

In this section, some conditions are collected under which the PRV-property holds for functions in the class $\mathbb{DL}$.

Recall that, if $f = f_\theta \in \mathbb{DL}$, then $f \in \mathcal{PRV}$ if and only if

$$\limsup_{c\to 1}\limsup_{t\to\infty}\int_t^{ct}\frac{\theta(u)}{f(u)}du = 0$$

(see Theorem 4.9). In turn, if $f \in \mathbb{DLP}$, then $f \in \mathcal{PRV}$ if and only if

$$\lim_{c\downarrow 1}\limsup_{t\to\infty}\int_t^{ct}\frac{\theta(u)}{f(u)}du = 0.$$

4.8.1 Necessary PRV-Properties

First we discuss conditions for the PRV-property that follow from Theorems 4.9 and 4.28.

Proposition 4.42 *Let $f \in \mathbb{DL}$ and $f \in \mathcal{PRV}$. Then:*

1) the inequalities

$$\epsilon^{\mathrm{es}*}(f) > -\infty, \qquad \epsilon_{\mathrm{es}*}(f) < \infty \tag{4.41}$$

hold; in addition, for every density $\theta \in \mathfrak{D}(f)$,

$$\epsilon^*(f,\theta) > -\infty \quad \textit{and} \quad \epsilon_*(f,\theta) < \infty;$$

2) if $f \in \mathbb{DLP}$, then $\epsilon_{\mathrm{es}*}(f) \in [0,\infty)$; in addition, for every density $\theta \in \mathfrak{D}_p(f)$

$$\epsilon_*(f,\theta) \in [0,\infty);$$

3) if $f \in \mathbb{DLP}$, $\theta \in \mathfrak{D}_p(f)$ and there are numbers $t_1 \ge t_0$ and $\mu \in (0,1]$ such that

$$\frac{\theta(s)}{f(s)} \ge \mu \cdot \frac{\theta(t)}{f(t)} \qquad \textit{for all} \qquad t \ge s \ge t_1, \tag{4.42}$$

then

$$\epsilon^*(f,\theta) \in [0,\infty); \tag{4.43}$$

4) if $f \in \mathbb{DLP}$, $\theta \in \mathfrak{D}_p(f)$, and a density θ is μ-decreasing (almost decreasing) (see (4.34) and (4.36)), then (4.43) holds.

Proof of Proposition 4.42 Let

$$\epsilon^{\mathrm{es}*}(f) = -\infty \quad \text{or} \quad \epsilon_{\mathrm{es}*}(f) = \infty.$$

Then Theorem 4.28 implies that, for all $c > 1$,

$$f^*(c) = 0 \quad \text{or} \quad f^*(c) = \infty$$

which contradicts the PRV-property. This proves (4.41). The other relations of the first statement can be proved similarly.

Since the function f is nondecreasing in view of $f \in \mathbb{DLP}$, the second statement follows from the first one.

If (4.42) holds, then

$$\left(\frac{\mu(c-1)}{c}\right)\left(\frac{(ct)\theta(ct)}{f(ct)}\right) \le \int_t^{ct} \frac{\theta(u)}{f(u)}\,du$$

for all $c > 1$ and $t \ge t_1$, whence we conclude that

$$\limsup_{t\to\infty} \frac{t\theta(t)}{f(t)} = \limsup_{t\to\infty} \frac{(ct)\theta(ct)}{f(ct)} \le \left(\frac{c}{\mu(c-1)}\right) \limsup_{t\to\infty} \int_t^{ct} \frac{\theta(u)}{f(u)}\,du$$

for all $c > 1$.

Since $f \in \mathcal{PRV}$, the second statement of Theorem 4.9 implies that there exists a number $c_1 > 1$ such that

$$\limsup_{t\to\infty} \int_t^{ct} \frac{\theta(u)}{f(u)}\,du \le 1.$$

Thus

$$\limsup_{t\to\infty} \frac{t\theta(t)}{f(t)} \le \frac{c_1}{\mu(c_1-1)} < \infty.$$

Now the third statement follows, since the function f is nondecreasing; this, in turn, proves the fourth statement. □

Remark 4.43 For $f \in \mathbb{DLP}$ and $\theta \in \mathfrak{D}_p(f)$, condition (4.42) holds if the density θ is μ-decreasing (see (4.34)), in particular, if, for sufficiently large arguments, either θ/f or θ is a nonincreasing function.

Moreover, if $f \in \mathbb{DLSP}$ and $\theta \in \mathfrak{D}_{\mathrm{sp}}(f)$, then (4.42) is equivalent to

$$\lim_{s\to\infty} \inf_{t\ge s} \frac{f(t)\theta(s)}{\theta(t)f(s)} > 0. \tag{4.44}$$

In turn, this condition follows from (4.36), that is,

$$\lim_{s\to\infty} \inf_{t\geq s} \frac{\theta(s)}{\theta(t)} > 0.$$

Next we consider sufficient conditions for the PRV-property that follow from Theorems 4.9, 4.28, and 4.35.

Proposition 4.44 *Let $f \in \mathbb{DL}$. Then:*

1) if

$$-\infty < \epsilon_{\mathrm{es}*}(f) \quad \textit{and} \quad \epsilon^{\mathrm{es}*}(f) < \infty,$$

then $f \in \mathbb{ER}$ and $f \in \mathcal{PRV}$; in particular, this is the case if there exists a density $\theta \in \mathfrak{D}(f)$ such that

$$-\infty < \epsilon_*(f,\theta) \quad \textit{and} \quad \epsilon^*(f,\theta) < \infty;$$

2) if $f \in \mathbb{DLP}$ and

$$\epsilon^{\mathrm{es}*}(f) < \infty, \tag{4.45}$$

then $f \in \mathbb{ER}[0,\infty)$ and $f \in \mathcal{PRV}$; in particular, this is the case if there exists a density $\theta \in \mathfrak{D}_p(f)$ such that

$$\epsilon^*(f,\theta) < \infty;$$

3) if there exist a density $\theta \in \mathfrak{D}(f)$ and number $\alpha > -1$ such that

$$\liminf_{t\to\infty}(t^{-\alpha}\theta(t)) > 0 \quad \textit{and} \quad \limsup_{t\to\infty}(t^{-\alpha}\theta(t)) < \infty,$$

then $f \in \mathbb{ER}(0,\infty)$ and $f \in \mathcal{PRV}$;

4) if there exists a density $\theta \in \mathfrak{D}(f)$ such that

$$\liminf_{t\to\infty}\theta(t) > 0 \quad \textit{and} \quad \limsup_{t\to\infty}\theta(t) < \infty,$$

then $f \in \mathbb{ER}(0,\infty)$ and $f \in \mathcal{PRV}$;

5) if $f \in \mathbb{DLSP}$ and there exists a density $\theta \in \mathfrak{D}_{\mathrm{sp}}(f)$ which is μ-decreasing (almost decreasing) (see (4.34) and (4.36)), then $f \in \mathbb{ER}[0,\frac{1}{\mu}]$ and $f \in \mathcal{PRV}$.

Proof of Proposition 4.44 The first two statements are just copied from Theorem 4.35 (see also Proposition 4.20).

Under the assumption of the third statement, there exists a number $b_1 \geq t_0$ such that

$$B_1 = \inf_{t \geq b_1} (t^{-\alpha}\theta(t)) > 0.$$

Thus

$$f(t, b_1) = \int_{b_1}^{t} \theta(u)du \geq C_1 t^{1+\alpha}$$

for $t \geq b_1$, where $C_1 = B_1/(1+\alpha)$. This implies

$$\inf_{t \geq b_1} (t^{-1-\alpha} f(t, b_1)) \geq C_1 > 0.$$

Hence

$$\begin{aligned}\epsilon^{\mathrm{es}*}(f) \leq \epsilon^{*}(f, \theta) &= \limsup_{t\to\infty} \frac{t\theta(t)}{f(t)} = \limsup_{t\to\infty} \frac{t\theta(t)}{f(t, b_1)} \\ &= \limsup_{t\to\infty} \frac{t^{-\alpha}\theta(t)}{t^{-1-\alpha} f(t, b_1)} \leq \frac{1}{C_1} \limsup_{t\to\infty}(t^{-\alpha}\theta(t)) < \infty.\end{aligned}$$

Moreover, there exists a number $b_2 \geq t_0$ such that

$$B_2 = \sup_{t \geq b_2} (t^{-\alpha}\theta(t)) < \infty.$$

Thus

$$f(t, b_2) = \int_{b_2}^{t} \theta(u)du \leq C t^{1+\alpha},$$

for $t \geq b_2$, where $C = B/(1+\alpha)$. Therefore

$$\sup_{t \geq b_2} (t^{-1-\alpha} f(t, b_2)) \leq C < \infty.$$

This implies

$$\begin{aligned}\epsilon_{\mathrm{es}*}(f) \geq \epsilon_{*}(f, \theta) &= \liminf_{t\to\infty} \frac{t\theta(t)}{f(t)} = \liminf_{t\to\infty} \frac{t\theta(t)}{f(t, b_2)} \\ &= \liminf_{t\to\infty} \frac{t^{-\alpha}\theta(t)}{t^{-1-\alpha} f(t, b_2)} \geq \frac{1}{C} \liminf_{t\to\infty}(t^{-\alpha}\theta(t)) > 0.\end{aligned}$$

Now the third statement follows from the second one; in turn, the third statement with $\alpha = 0$ implies the fourth statement.

The fifth statement is a part of Corollary 4.40. □

The following result shows that condition (4.45) is a characterization of the PRV-property provided (4.42) holds.

Corollary 4.45 *Let $f \in \mathbb{DLP}$ and $\theta \in \mathfrak{D}_p(f)$. Then:*

1) *if condition (4.42) holds, then $f \in \mathcal{PRV}$ if and only if condition (4.45) holds;*
2) *if $f \in \mathbb{DLSP}$, $\theta \in \mathfrak{D}_{\rm sp}(f)$, and condition (4.44) holds, then $f \in \mathcal{PRV}$ if and only if condition (4.45) holds;*
3) *if $f \in \mathbb{DLSP}$, $\theta \in \mathfrak{D}_{\rm sp}(f)$, and condition (4.36) holds, then $f \in \mathcal{PRV}$ (without any additional assumption).*

Corollary 4.45 follows from Propositions 4.42 and 4.44 (see also Remark 4.43).

4.8.2 The Measurability of Limit Functions of Densities and the ER- (PRV)-Property of Absolutely Continuous Functions

In what follows we need the assumption that the function θ_* is measurable or, equivalently, that the function θ^* is measurable. This is the case if, for example, the function θ is nondecreasing, or nonincreasing, or continuous, or piecewise continuous etc.

Corollary 4.46 *Let $f \in \mathbb{DLSP}$ and $\theta \in \mathfrak{D}_{\rm sp}(f)$. If θ_* is measurable and*

$$\int_{0+}^{1} \theta_*(c)dc > 0 \tag{4.46}$$

or, equivalently, if

$$\int_{1}^{\infty} \frac{dc}{c^2\theta^*(c)} > 0,$$

then $f \in \mathbb{ER}[0,\infty)$ and $f \in \mathcal{PRV}$.

Proof of Corollary 4.46 According to Fatou's lemma,

$$\liminf_{t\to\infty} \frac{f(t)}{t\theta(t)} \ge \liminf_{t\to\infty} \int_{t_0}^{t} \frac{\theta(u)}{t\theta(t)} du = \liminf_{t\to\infty} \int_{\frac{t_0}{t}}^{1} \frac{\theta(ct)}{\theta(t)} dc \ge \int_{0+}^{1} \theta_*(c)dc > 0,$$

that is, $\epsilon^*(f,\theta) < \infty$ and $\epsilon^{\rm es^*}(f) < \infty$. This together with Proposition 4.44 implies that $f \in \mathbb{ER}$ and $f \in \mathcal{PRV}$. □

Corollary 4.46 implies the following result.

Corollary 4.47 *Let $f \in \mathbb{DLSP}$, $\theta \in \mathfrak{D}_{\mathrm{sp}}(f)$ and θ_* be measurable. Then:*

1) if the set

$$\{c \in (0, 1] : \theta_*(c) > 0\}$$

has positive Lebesgue measure or, equivalently, if the set

$$\{c \in [1, \infty) : \theta^*(c) < \infty\}$$

has positive Lebesgue measure, then $f \in \mathbb{ER}[0, \infty)$ and $f \in \mathcal{PRV}$;

2) if $\theta \in \mathcal{ORV}$, then $f \in \mathbb{ER}[0, \infty)$ and $f \in \mathcal{PRV}$.

Proof of Corollary 4.47 The first statement follows immediately from Corollary 4.46. Then the second statement follows as well, since

$$\{c > 0 : \theta_*(c) > 0\} = (0, \infty)$$

if $\theta \in \mathcal{ORV}$. □

4.9 The SQI-Property of Absolutely Continuous Functions

In this section sufficient conditions are collected under which the PI- or SQI-property holds for functions in the class $\mathbb{DL}$.

Recall that, according to Theorem 4.9, if $f = f_\theta \in \mathbb{DL}$, then $f \in \mathcal{PI}$ if and only if there exists a $c_0 > 1$ such that

$$\liminf_{t\to\infty} \int_t^{c_0 t} \frac{\theta(u)}{f(u)} du > 0;$$

analogously, $f \in \mathcal{SQI}$ if and only if

$$\liminf_{t\to\infty} \int_t^{ct} \frac{\theta(u)}{f(u)} du > 0 \quad \text{for all} \quad c > 1.$$

Remark 4.48 Any of the conditions that the density θ is positive or that its lower limit function θ_* is positive is not sufficient for f to be a PI-function. Moreover, each of the above conditions is not sufficient for f_* to have at least a single point of increase. Indeed, let $t_0 = 1$ and $\theta(t) = 1/t$, if $t \geq 1$. Then $\theta_*(c) = 1/c$ for all $c > 0$. This implies

$$f(t) = \ln t, \quad t \geq 1, \qquad \text{and} \qquad f_*(c) = 1, \quad c > 0.$$

Thus f does not belong to $\mathcal{PI}$ and its lower limit function f_* does not increase at any point.

This pathological case can be neglected, since according to the fourth statement of Theorem 4.9,

$$\liminf_{t\to\infty}\int_t^{ct}\frac{\theta(u)}{f(u)}\,du=\liminf_{t\to\infty}\int_t^{ct}\frac{du}{u\ln u}$$
$$=\liminf_{t\to\infty}\int_{\ln t}^{\ln ct}\frac{du}{u}=\liminf_{t\to\infty}\ln\left(\frac{\ln ct}{\ln t}\right)=0$$

for all $c>0$.

4.9.1 Necessary Conditions for the PI-Property

Below we discuss some necessary conditions for the PI-property that follow from Theorems 4.9 and 4.28 and Proposition 4.2. Since $\mathcal{SQI}\subset\mathcal{PI}$, these conditions are also necessary for $f\in\mathcal{SQI}$.

Proposition 4.49 *Let $f\in\mathbb{DL}$ and $f\in\mathcal{PI}$. Then:*

1) the inequality

$$\epsilon^{\mathrm{es}*}(f)>0 \tag{4.47}$$

holds and, for any density $\theta\in\mathfrak{D}(f)$,

$$\epsilon^*(f,\theta)>0; \tag{4.48}$$

2) if $f\in\mathbb{DLP}$, $\theta\in\mathfrak{D}_p(f)$ and condition (4.42) holds, that is, there are numbers $t_1\ge t_0$ and $\mu\in(0,1]$ such that

$$\frac{\theta(s)}{f(s)}\ge\mu\cdot\frac{\theta(t)}{f(t)}\qquad\textit{for all}\qquad t\ge s\ge t_1,$$

then

$$\epsilon_*(f,\theta)>0; \tag{4.49}$$

3) if $f\in\mathbb{DLP}$, $\theta\in\mathfrak{D}_p(f)$, and the density θ is μ-decreasing (almost decreasing) (see (4.34) and (4.36)), then relation (4.49) holds;

4) if $f\in\mathbb{DL}^\infty$, then, for any density $\theta\in\mathfrak{D}(f)$,

$$\limsup_{t\to\infty}(t\theta(t))=\infty;$$

5) if $f \in \mathbb{DLSP}^\infty$ *and* $\theta \in \mathfrak{D}_{\text{sp}}(f)$, *then*

$$\sup_{c>1} \left(c\theta^{\text{es}^*}(c) \right) > 1 \tag{4.50}$$

and

$$\sup_{c>1} \left(c\theta^*(c) \right) > 1. \tag{4.51}$$

Proof of Proposition 4.49 Assume that (4.47) does not hold, that is, $\epsilon^{\text{es}^*}(f) \le 0$. This together with (4.23) implies that, for all $c > 1$,

$$f_*(c) \le f^*(c) \le 1,$$

which contradicts the PI-property. This proves (4.47). In turn, (4.48) follows from (4.47). Thus the first statement is completely proved.

Property (4.42) implies that, for all $t \ge t_1$ and $c > 1$,

$$\left(\frac{\mu}{c-1} \right) \int_t^{ct} \frac{\theta(u)}{f(u)} du \le \frac{t\theta(t)}{f(t)}.$$

Since $f \in \mathcal{PI}$, there exists a number $c_0 > 1$ such that

$$\liminf_{t\to\infty} \int_t^{c_0 t} \frac{\theta(u)}{f(u)} du > 0.$$

Thus

$$0 < \frac{\mu}{c_0 - 1} \liminf_{t\to\infty} \int_t^{c_0 t} \frac{\theta(u)}{f(u)} du \le \liminf_{t\to\infty} \frac{t\theta(t)}{f(t)}.$$

This means that the second statement is also proved.

If $\theta \in \mathfrak{D}_p(f)$, then the function f is nondecreasing. Thus, if the density θ is μ-decreasing, then condition (4.42) holds. Therefore the second statement implies the third one.

The fourth statement follows from Remark 4.36, since any PI-function cannot be an SV-function.

To prove the fifth statement we assume that (4.50) does not hold. Then

$$\sup_{c>1} \left(c\theta^{\text{es}^*}(c) \right) \le 1,$$

whence we deduce that

$$f_*(c) \le f^*(c) \le \sup_{c>1} \left(c\theta^{\text{es}^*}(c) \right) \le 1,$$

for all $c > 1$ by inequality (4.9). This cannot happen for any PI-function. This proves (4.50). In turn, (4.51) follows from (4.50). Therefore the fifth statement is also proved. □

4.9.2 Sufficient SQI-Conditions

Now we pass to sufficient conditions for the SQI-property, which follow from Theorems 4.9, 4.28 and 4.35.

Proposition 4.50 *Assume $f \in \mathbb{DL}$. Then:*

1) if condition

$$\epsilon_{\text{es}*}(f) > 0$$

holds, then $f \in \mathbb{ER}(0, \infty]$ and $f \in \mathfrak{SQI}$; in particular, this is the case if there exists a density $\theta \in \mathfrak{D}(f)$ such that condition (4.49) holds, that is, if

$$\epsilon_*(f, \theta) > 0;$$

2) if $f \in \mathbb{DL}^\infty$, $\theta \in \mathfrak{D}(f)$, and there exists an $\alpha > -1$ such that

$$\liminf_{t\to\infty}(t^{-\alpha}\theta(t)) > 0 \quad \text{and} \quad \limsup_{t\to\infty}(t^{-\alpha}\theta(t)) < \infty,$$

then $f \in \mathbb{ER}(0, \infty)$ and $f \in \mathfrak{SQI}$;

3) if $f \in \mathbb{DL}^\infty$, $\theta \in \mathfrak{D}(f)$, and

$$\liminf_{t\to\infty}\theta(t) > 0 \quad \text{and} \quad \limsup_{t\to\infty}\theta(t) < \infty,$$

then $f \in \mathbb{ER}(0, \infty)$ and $f \in \mathfrak{SQI}$;

4) if $f \in \mathbb{DLSP}$, $\theta \in \mathfrak{D}_{\text{sp}}(f)$ and the density θ is μ-increasing (almost increasing) (see (4.35) and (4.37)), then $f \in \mathbb{ER}[\mu, \infty]$ and $f \in \mathfrak{SQI}$.

Proof of Proposition 4.50 The first statement is contained in Theorem 4.35, the second one is covered by Proposition 4.44 and, with $\alpha = 0$, implies the third one. In turn, the fourth statement is proved in Corollary 4.40. □

Condition (4.49) characterizes the SQI-property provided (4.42) holds; this result follows from Propositions 4.49 and 4.50 (see also Remark 4.43).

Corollary 4.51 *Let $f \in \mathbb{DLP}$ and $\theta \in \mathfrak{D}_p(f)$. We have,*

1) if condition (4.42) holds, then $f \in \mathfrak{SQI}$ if and only if condition (4.49) holds;

2) if $f \in \mathbb{DLSP}$, $\theta \in \mathfrak{D}_{\text{sp}}(f)$ and condition (4.44) holds, then $f \in \mathfrak{SQI}$ if and only if condition (4.49) holds;

3) if $f \in \mathbb{DLSP}$, $\theta \in \mathfrak{D}_{\mathrm{sp}}(f)$ and condition (4.37) holds, then $f \in \mathcal{SQI}$ (without any additional assumption).

The following result expresses the SQI-property in terms of the limit function of a density θ.

Corollary 4.52 *Let $f = f_\theta \in \mathbb{DLSP}$. If*

$$c\,\theta_{\mathrm{es}*}(c) > 1 \qquad \text{for all} \qquad c > 1,$$

then $f \in \mathcal{SQI}$.

Proof of Corollary 4.52 Corollary 4.52 follows directly from (4.9). □

Remark 4.53 Since every SQI-function is a PI-function, the example considered in Remark 4.48 shows that the condition of Corollary 4.52 cannot be exchanged with

$$c\,\theta_*(c) \geq 1 \qquad \text{for all} \qquad c > 1.$$

4.10 The POV-Property of Absolutely Continuous Functions

Since a function f belongs to $\mathcal{POV}$ if and only if it belongs to $\mathcal{PRV} \cap \mathcal{SQI}$, we may obtain sufficient conditions for $f \in \mathcal{POV}$ by intersecting the above conditions for the PRV-property with those for the SQI-property. Similarly, applying the union of necessary conditions for the PRV-property with those for the SQI-property we obtain necessary conditions for the POV-property.

Recall that, according to Theorem 4.9, $f \in \mathcal{POV}$ if and only if

$$\limsup_{c\to 1}\limsup_{t\to\infty}\int_t^{ct}\frac{\theta(u)}{f(u)}du = 0$$

and

$$\liminf_{t\to\infty}\int_t^{ct}\frac{\theta(u)}{f(u)}du > 0$$

for all $c > 1$ provided $f = f_\theta \in \mathbb{DL}$.

Propositions 4.44 and 4.50 imply the following sufficient conditions for the POV-property.

Proposition 4.54 *Let $f \in \mathbb{DL}$. We have,*

1) if

$$\epsilon_{\mathrm{es}*}(f) > 0 \quad \textit{and} \quad \epsilon^{\mathrm{es}*}(f) < \infty, \tag{4.52}$$

then $f \in \mathbb{ER}(0, \infty)$ *and* $f \in \mathcal{POV}$; *in particular, this is the case if there exists a density* $\theta \in \mathfrak{D}(f)$ *such that*

$$\epsilon_*(f, \theta) > 0 \quad \text{and} \quad \epsilon^*(f, \theta) < \infty;$$

2) *if* $f \in \mathbb{DLP}$, $\theta \in \mathfrak{D}_p(f)$ *and there exists an* $\alpha > -1$ *such that*

$$\liminf_{t \to \infty}(t^{-\alpha}\theta(t)) > 0 \quad \text{and} \quad \limsup_{t \to \infty}(t^{-\alpha}\theta(t)) < \infty,$$

then $f \in \mathbb{ER}(0, \infty)$ *and* $f \in \mathcal{POV}$; *in particular, if*

$$\liminf_{t \to \infty} \theta(t) > 0 \quad \text{and} \quad \limsup_{t \to \infty} \theta(t) < \infty,$$

then $f \in \mathbb{ER}(0, \infty)$ *and* $f \in \mathcal{POV}$;

3) *if* $f \in \mathbb{DLSP}$, $\theta \in \mathfrak{D}_{\text{sp}}(f)$, *the density* θ *is* μ*-decreasing (see (4.34)), and* $\epsilon_*(f, \theta) > 0$, *then* $f \in \mathbb{ER}(0, \frac{1}{\mu}]$ *and* $f \in \mathcal{POV}$;
4) *if* $f \in \mathbb{DLSP}$, $\theta \in \mathfrak{D}_{\text{sp}}(f)$, *the density* θ *is* μ*-increasing (see (4.35)), and* $\epsilon^*(f, \theta) < \infty$, *then* $f \in \mathbb{ER}[\mu, \infty)$ *and* $f \in \mathcal{POV}$;
5) *if* $f \in \mathbb{DLP}$, $\theta \in \mathfrak{D}_p(f)$ *and condition (4.42) holds, then* $f \in \mathcal{POV}$ *if and only if condition (4.52) holds;*
6) *if* $f \in \mathbb{DLSP}$, $\theta \in \mathfrak{D}_{\text{sp}}(f)$, *and condition (4.44) holds, then* $f \in \mathcal{POV}$ *if and only if condition (4.52) holds.*

The following result requires the measurability of the limit functions of the density θ. Corollary 4.55 follows from Corollaries 4.47 and 4.52.

Corollary 4.55 *Let* $f \in \mathbb{DLSP}$, $\theta \in \mathfrak{D}_{\text{sp}}(f)$ *and let* θ_* *be measurable. Then:*

1) *if the set* $\{c \in (0, 1] : \theta_*(c) > 0\}$ *(or, equivalently, if* $\{c \in [1, \infty) : \theta^*(c) < \infty\}$*) has positive Lebesgue measure and* $c\,\theta_*(c) > 1$ *for all* $c > 1$, *then* $f \in \mathcal{POV}$;
2) *if* $\theta \in \mathcal{ORV}$ *and* $c\,\theta_*(c) > 1$ *for all* $c > 1$, *then* $f \in \mathcal{POV}$.

The latter result has several corollaries for functions with monotone densities.

Corollary 4.56 *Let* $f \in \mathbb{DLSP}$, $\theta \in \mathfrak{D}_{\text{sp}}(f)$ *and let* θ_* *be measurable. Then* $f \in \mathcal{POV}$ *if at least one of the following three conditions holds:*

1) θ *is a nonincreasing function and* $c\,\theta_*(c) > 1$ *for all* $c > 1$;
2) θ *is a nondecreasing function and the set* $\{c \in [1, \infty) : \theta^*(c) < \infty\}$ *has positive Lebesgue measure;*
3) θ *is a nondecreasing ORV-function.*

4.11 Properties of Piecewise Linear Interpolations

The preceding sections dealt with conditions under which an absolutely continuous function f possesses a given "property of regularity". For a continuously differentiable function f, any "property of regularity" follows directly from the results obtained before in this chapter if the density θ in the statements is exchanged with the derivative f'. For other classes of functions the procedure is more complicated. In this section, we consider "properties of regularity" for piecewise linear interpolations as an important example of absolutely continuous functions with discontinuous densities.

4.11.1 Properties of Piecewise Linear Interpolations of Positive Sequences

Given a sequence $\{a_n\} = \{a_n\}_{n\geq 0}$, where $a_0 = 0$ and $a_n > 0$, $n \geq 1$, consider its *piecewise linear interpolation*, that is, the continuous function

$$\widehat{a}(t) = a_{\lfloor t \rfloor} + \left(a_{\lfloor t \rfloor+1} - a_{\lfloor t \rfloor}\right)(t - \lfloor t \rfloor), \qquad t \geq 0, \tag{4.53}$$

where the symbol $\lfloor t \rfloor$ stands for the integer part of a number t. Recall that such functions already appeared in Sect. 3.4 (see relation (3.23) and Lemma 3.43 therein).

The function $\widehat{a}$ is absolutely continuous and its density $\widehat{\theta} = (\widehat{\theta}(t),\ t \geq 0)$ is given by

$$\widehat{\theta}(t) = a_{\lfloor t \rfloor+1} - a_{\lfloor t \rfloor}, \qquad t \geq 0. \tag{4.54}$$

According to Definition 4.1, the function $\widehat{a}$ belongs to the class $\mathbb{DL}$ for all $t_0 > 0$. If $a_{n+1} - a_n \geq 0$, $n \geq 1$, that is, if the sequence $\{a_n\}$ is nondecreasing, then $\widehat{\theta}(t) \geq 0$, $t \geq 0$, and $\widehat{a} \in \mathbb{DLP}$. Analogously, if $a_{n+1} - a_n > 0$, $n \geq 1$, that is, if the sequence $\{a_n\}$ is increasing, then $\widehat{\theta}(t) > 0$, $t \geq 0$, and $\widehat{a} \in \mathbb{DLSP}$.

Using relations (4.53) and (4.54) we evaluate the logarithmic density of $\widehat{a}$ with respect to the density $\widehat{\theta}$, i.e.

$$\frac{\widehat{\theta}(t)}{\widehat{a}(t)} = \frac{1}{\delta_{\lfloor t \rfloor} + t - \lfloor t \rfloor}, \qquad t \geq 0. \tag{4.55}$$

Also we evaluate the elasticity of $\widehat{a}$ with respect to the density $\widehat{\theta}$, i.e.

$$\frac{t\widehat{\theta}(t)}{\widehat{a}(t)} = \frac{t}{\delta_{\lfloor t \rfloor} + t - \lfloor t \rfloor}, \qquad t \geq 0, \tag{4.56}$$

where

$$\delta_{\lfloor t\rfloor} = \frac{a_{\lfloor t\rfloor}}{a_{\lfloor t\rfloor+1} - a_{\lfloor t\rfloor}}.$$

Here we set

$$\delta_{\lfloor t\rfloor} = \infty \quad \text{and} \quad \frac{\widehat{\theta}(t)}{\widehat{a}(t)} = 0,$$

if $a_{\lfloor t\rfloor+1} = a_{\lfloor t\rfloor}$.

Using (4.56) we obtain the indices of asymptotic elasticity (see Definition 4.19) of $\widehat{a}$ with respect to the density $\widehat{\theta}$, i.e.

$$\limsup_{t\to\infty} \frac{t\widehat{\theta}(t)}{\widehat{a}(t)} = \limsup_{t\to\infty} \frac{t}{\delta_{\lfloor t\rfloor} + t - \lfloor t\rfloor} = \limsup_{t\to\infty} \lfloor t\rfloor \Big(\frac{a_{\lfloor t\rfloor+1}}{a_{\lfloor t\rfloor}} - 1 \Big),$$

and

$$\liminf_{t\to\infty} \frac{t\widehat{\theta}(t)}{\widehat{a}(t)} = \liminf_{t\to\infty} \frac{t}{\delta_{\lfloor t\rfloor} + t - \lfloor t\rfloor} = \liminf_{t\to\infty} \lfloor t\rfloor \Big(\frac{a_{\lfloor t\rfloor+1}}{a_{\lfloor t\rfloor}} - 1 \Big).$$

Hence

$$\epsilon^*(\widehat{a}, \widehat{\theta}) = \limsup_{n\to\infty} \frac{n(a_{n+1} - a_n)}{a_n} \tag{4.57}$$

and

$$\epsilon_*(\widehat{a}, \widehat{\theta}) = \liminf_{n\to\infty} \frac{n(a_{n+1} - a_n)}{a_n}. \tag{4.58}$$

Theorem 4.9 allows us to state general characterizing conditions for PRV- and SQI-functions together with other "properties of regularity" of such functions $\widehat{a}$.

Theorem 4.57 *Let $\widehat{a}$ be the piecewise linear interpolation of a sequence $\{a_n\}$. We have,*

1) $\widehat{a} \in \mathcal{ORV}$ if and only if, for all $c > 1$,

$$-\infty < \liminf_{t\to\infty} \int_t^{ct} \frac{du}{\delta_{\lfloor u\rfloor} + u - \lfloor u\rfloor} \quad \text{and} \quad \limsup_{t\to\infty} \int_t^{ct} \frac{du}{\delta_{\lfloor u\rfloor} + u - \lfloor u\rfloor} < \infty;$$

2) $\widehat{a} \in \mathcal{PRV}$ if and only if

$$\limsup_{c\to1} \limsup_{t\to\infty} \int_t^{ct} \frac{du}{\delta_{\lfloor u\rfloor} + u - \lfloor u\rfloor} = 0; \tag{4.59}$$

3) *if the sequence* $\{a_n\}$ *is nondecreasing, then* $f \in \mathcal{PRV}$ *if and only if*

$$\lim_{c\downarrow 1}\limsup_{t\to\infty}\int_t^{ct}\frac{du}{\delta_{\lfloor u\rfloor}+u-\lfloor u\rfloor}=0;$$

4) $\widehat{a} \in \mathcal{PI}$ *if and only if there exists a* $c_0 > 1$ *such that*

$$\liminf_{t\to\infty}\int_t^{c_0t}\frac{1}{\delta_{\lfloor u\rfloor}+u-\lfloor u\rfloor}\,du>0;$$

5) $\widehat{a} \in \mathcal{SQI}$ *if and only if, for all* $c > 1$,

$$\liminf_{t\to\infty}\int_t^{ct}\frac{du}{\delta_{\lfloor u\rfloor}+u-\lfloor u\rfloor}>0; \tag{4.60}$$

6) $\widehat{a} \in \mathcal{POV}$ *if and only if conditions (4.59) and (4.60) hold;*
7) f *is an RV-function of index* ρ *if and only if, for all* $c > 1$, *the limit*

$$\lim_{t\to\infty}\int_t^{ct}\frac{du}{\delta_{\lfloor u\rfloor}+u-\lfloor u\rfloor}\,du=\rho\in(-\infty,\infty)$$

exists.

Some simple sufficient conditions for "properties of regularity" for the functions $\widehat{a}$ can be given in terms of the sequence $\{a_n\}$ by using Proposition 4.20 and relations (4.57) and (4.58).

Theorem 4.58 *Let* $\widehat{a}$ *be the piecewise linear interpolation of a sequence* $\{a_n\}$. *We have,*

1) *if*

$$\liminf_{n\to\infty}\frac{n(a_{n+1}-a_n)}{a_n}>-\infty \quad \textit{and} \quad \limsup_{n\to\infty}\frac{n(a_{n+1}-a_n)}{a_n}<\infty, \tag{4.61}$$

then $\widehat{a} \in \mathcal{PRV}$;
2) *if the sequence* $\{a_n\}$ *is nondecreasing and*

$$\limsup_{n\to\infty}\frac{n(a_{n+1}-a_n)}{a_n}<\infty, \tag{4.62}$$

then $\widehat{a} \in \mathcal{PRV}$;
3) *if*

$$\liminf_{n\to\infty}\frac{n(a_{n+1}-a_n)}{a_n}>0, \tag{4.63}$$

then $\widehat{a} \in \mathcal{SQI}$;

4) if

$$\liminf_{n\to\infty} \frac{n(a_{n+1}-a_n)}{a_n} > 0 \quad \textit{and} \qquad \limsup_{n\to\infty} \frac{n(a_{n+1}-a_n)}{a_n} < \infty, \tag{4.64}$$

then $\widehat{a} \in \mathcal{POV}$;

5) if the limit

$$\lim_{n\to\infty} \frac{n(a_{n+1}-a_n)}{a_n} = \rho \in (-\infty, \infty) \tag{4.65}$$

exists, then $\widehat{a}$ *is an RV-function of index* ρ.

Remark 4.59 Note that

1) any of the conditions (4.61), (4.62), (4.64), or (4.65) implies that

$$\lim_{n\to\infty} \frac{a_{n+1}}{a_n} = 1, \tag{4.66}$$

whence we deduce that

$$\lim_{t\to\infty} \frac{a_{\lfloor t\pm\tau\rfloor}}{a_{\lfloor t\rfloor}} = \lim_{t\to\infty} \frac{a_{\lfloor t\pm\tau\rfloor}}{\widehat{a}(t)} = \lim_{t\to\infty} \frac{\widehat{a}(t\pm\tau)}{\widehat{a}(t)} = 1 \tag{4.67}$$

for all $\tau > 0$;

2) relation (4.66) holds if and only if

$$\lim_{t\to\infty} \frac{\widehat{a}(t)}{a_{\lfloor t\rfloor}} = 1; \tag{4.68}$$

3) if relation (4.66) holds, then

$$\liminf_{t\to\infty} \frac{\widehat{a}(ct)}{\widehat{a}(t)} = \liminf_{t\to\infty} \frac{a_{\lfloor ct\rfloor}}{a_{\lfloor t\rfloor}} \quad \text{and} \quad \limsup_{t\to\infty} \frac{\widehat{a}(ct)}{\widehat{a}(t)} = \limsup_{t\to\infty} \frac{a_{\lfloor ct\rfloor}}{a_{\lfloor t\rfloor}} \tag{4.69}$$

for all $c > 0$.

The following result means that the sequence $\{a_n\}$ is increasing, starting at some $N > 0$, and tends to infinity provided relation (4.63) holds.

Lemma 4.60 *If condition (4.63) holds for the sequence* $\{a_n\}$, *then there exists an* $N > 0$ *such that*

$$a_{n+1} > a_n, \qquad n \ge N, \tag{4.70}$$

and

$$\lim_{n\to\infty} a_n = \infty. \tag{4.71}$$

Proof of Lemma 4.60 Assume that condition (4.63) holds.

We show that the sequence $\{a_n\}$ is increasing for sufficiently large n, that is, there exists a number $N > 0$ such that (4.70) holds. Indeed, if relation (4.70) were not true, then there would exist an increasing sequence of natural numbers $\{n_k\}_{k\geq 1}$ such that

$$a_{n_k+1} - a_{n_k} \leq 0, \qquad k \geq 1.$$

Hence

$$\liminf_{n\to\infty} \frac{n(a_{n+1} - a_n)}{a_n} \leq \liminf_{k\to\infty} \frac{n_k(a_{n_k+1} - a_{n_k})}{a_{n_k}} \leq 0$$

which contradicts (4.63). Thus, relation (4.70) is proved.

Now we assume that (4.71) does not hold. Taking relation (4.70) into account, we see that

$$\lim_{n\to\infty} a_n = A \in (0, \infty). \tag{4.72}$$

This together with (4.63) implies that

$$\liminf_{n\to\infty} n(a_{n+1} - a_n) = A \liminf_{n\to\infty} \frac{n(a_{n+1} - a_n)}{a_n} > C,$$

for some $C > 0$, and thus there exists an integer M such that

$$A = \sum_{n=0}^{\infty}(a_{n+1} - a_n) \geq \sum_{n=0}^{M}(a_{n+1} - a_n) + C \sum_{n=M+1}^{\infty} \frac{1}{n} = \infty.$$

This contradicts (4.72) and proves that (4.63) implies condition (4.71). □

Remark 4.61 One can prove (4.71) by using Corollary 3.70 telling us that a continuous PI-function tends to infinity. Indeed, if condition (4.63) holds, then the continuous function $\widehat{a}$ belongs to $\mathcal{SQI}$ by Theorem 4.58 and thus $\widehat{a} \in \mathcal{PI}$. Now Corollary 3.70 yields

$$\lim_{t\to\infty} \widehat{a}(t) = \infty$$

which is equivalent to (4.71).

Remark 4.62 Applying Lemma 4.60 together with Remark 4.59 we show that condition (4.64) as well as condition (4.65) with $\rho > 0$ imply (4.66), (4.70), and (4.71).

Under some extra assumptions, the sufficient conditions for the "properties of regularity" of a function $\widehat{a}$ considered in Theorem 4.58 become necessary as well.

Theorem 4.63 *If the sequence* $\{a_n\}$ *is increasing,*

$$\limsup_{n\to\infty} \frac{a_{n+1}}{a_n} < \infty \tag{4.73}$$

and

$$\lim_{m\to\infty} \inf_{n\ge m} \left(\left(\frac{a_{m+1}}{a_m} - 1 \right) \Big/ \left(\frac{a_{n+1}}{a_n} - 1 \right) \right) > 0, \tag{4.74}$$

then

1) $\widehat{a} \in \mathcal{PRV}$ *if and only if condition (4.62) holds;*
2) $\widehat{a} \in \mathcal{SQI}$ *if and only if condition (4.63) holds;*
3) $\widehat{a} \in \mathcal{POV}$ *if and only if condition (4.64) holds.*

Remark 4.64 Conditions (4.73) and (4.74) are satisfied if

$$\frac{a_{n+1}}{a_n} \ge \frac{a_{n+2}}{a_{n+1}}, \qquad n \ge N, \tag{4.75}$$

for some $N > 0$, that is, if

$$a_{n+1}^2 \ge a_n a_{n+1}, \qquad n \ge N.$$

The sequences $a_n = \ln(n+1)$, $n \ge 1$; $a_n = n^b$, $n \ge 1$, where $b > 0$; $a_n = c^n$, $n \ge 1$, with $c > 0$, serve as examples of sequences for which condition (4.75) as well as (4.73) and (4.74) are valid.

Proof of Theorem 4.63 To prove the first statement we apply the second statement of Corollary 4.45. First we check assumption (4.44) for $f = \widehat{a}$ and $\theta = \widehat{\theta}$.

Since

$$\lim_{s\to\infty} \inf_{t\ge s} \left(\frac{\widehat{\theta}(s)}{\widehat{a}(s)} \right) \Big/ \left(\frac{\widehat{\theta}(t)}{\widehat{a}(t)} \right)$$
$$= \lim_{s\to\infty} \inf_{t\ge s} \left(\frac{a_{\lfloor s\rfloor+1} - a_{\lfloor s\rfloor}}{a_{\lfloor t\rfloor+1} - a_{\lfloor t\rfloor}} \cdot \frac{a_{\lfloor t\rfloor} + \{a_{\lfloor t\rfloor+1} - a_{\lfloor t\rfloor}\}(t - \lfloor t\rfloor)}{a_{\lfloor s\rfloor} + \{a_{\lfloor s\rfloor+1} - a_{\lfloor s\rfloor}\}(s - \lfloor s\rfloor)} \right), \tag{4.76}$$

we get

$$\begin{aligned}
&\lim_{s\to\infty}\inf_{t\ge s}\left(\frac{\widehat{\theta}(s)}{\widehat{a}(s)}\right)\Big/\left(\frac{\widehat{\theta}(t)}{\widehat{a}(t)}\right)\\
&\ge \lim_{s\to\infty}\inf_{t\ge s}\left(\frac{\min\{a_{\lfloor t\rfloor},\, a_{\lfloor t\rfloor+1}\}}{\max\{a_{\lfloor s\rfloor},\, a_{\lfloor s\rfloor+1}\}}\cdot\frac{a_{\lfloor s\rfloor+1}-a_{\lfloor s\rfloor}}{a_{\lfloor t\rfloor+1}-a_{\lfloor t\rfloor}}\right)\\
&\ge \liminf_{m\to\infty}\left(\frac{a_m}{a_{m+1}}\right)\lim_{m\to\infty}\inf_{n\ge m}\left(\frac{a_n(a_{m+1}-a_m)}{a_m(a_{n+1}-a_n)}\right).
\end{aligned}$$

Thus relations (4.73) and (4.74) imply that

$$\lim_{s\to\infty}\inf_{t\ge s}\left(\frac{\widehat{\theta}(s)}{\widehat{a}(s)}\right)\Big/\left(\frac{\widehat{\theta}(t)}{\widehat{a}(t)}\right) > 0. \tag{4.77}$$

This means that condition (4.44) holds. Now, applying the second statement of Corollary 4.45 and taking relations (4.57) and (4.58) into account, we complete the proof of the first statement of Theorem 4.63.

The second statement of Theorem 4.63 can be proved similarly, if the second statement of Corollary 4.51 is used instead of the second statement of Corollary 4.45. In turn, the third statement of Theorem 4.63 follows from the first two statements. □

Remark 4.65 Relations (4.76) and (4.66) imply that

$$\lim_{s\to\infty}\inf_{t\ge s}\left(\frac{\widehat{\theta}(s)}{\widehat{a}(s)}\right)\Big/\left(\frac{\widehat{\theta}(t)}{\widehat{a}(t)}\right) = \lim_{m\to\infty}\inf_{n\ge m}\left(\left(\frac{a_{m+1}}{a_m}-1\right)\Big/\left(\frac{a_{n+1}}{a_n}-1\right)\right).$$

Thus (4.77) and (4.74) are equivalent provided (4.66) holds.

The sequences $a_n = \ln(n+1)$, $n \ge 1$, and $a_n = n^b$, $n \ge 1$, with $b > 0$, serve as examples of increasing sequences for which conditions (4.73), (4.74), and (4.62) are satisfied. In these two cases, $\widehat{a} \in \mathcal{PRV}$. For a sequence $a_n = c^n$, $n \ge 1$, with $c > 0$, conditions (4.73) and (4.74) are satisfied, but (4.62) does not hold.

Similarly, the sequences $a_n = n^b$, $n \ge 1$, with $b > 0$, and $a_n = c^n$, $n \ge 1$, with $c > 0$, serve as examples of increasing sequences for which conditions (4.73), (4.74) and (4.63) are satisfied. For these sequences, $\widehat{a} \in \mathcal{SQI}$. In contrast, conditions (4.73) and (4.74) are satisfied, but (4.63) does not hold for the sequence $a_n = \ln(n+1)$, $n \ge 1$.

Moreover, conditions (4.73), (4.74), and (4.63) are satisfied for the sequence $a_n = n^b$, $n \ge 1$, with $b > 0$. For this sequence, $\widehat{a} \in \mathcal{POV}$. On the other hand, conditions (4.73) and (4.74) are satisfied, but (4.64) does not hold for the sequences $a_n = \ln(n+1)$, $n \ge 1$, and $a_n = c^n$, $n \ge 1$, with $c > 0$.

The following result contains simple conditions for the PRV- and SQI-property.

Proposition 4.66 *Let $\{a_n\}$ be an increasing sequence. We have,*

1) if

$$\lim_{m\to\infty} \inf_{n\geq m} \frac{a_{m+1}-a_m}{a_{n+1}-a_n} > 0,$$

then $\widehat{a} \in \mathcal{PRV}$; in particular, this is the case if there exists an $N > 0$ such that

$$a_{n+2} - 2a_{n+1} + a_n \leq 0, \qquad n \geq N;$$

2) if

$$\lim_{m\to\infty} \inf_{n\geq m} \frac{a_{n+1}-a_n}{a_{m+1}-a_m} > 0,$$

then $\widehat{a} \in \mathcal{SQI}$; in particular, this is the case if there exists an $N > 0$ such that

$$a_{n+2} - 2a_{n+1} + a_n \geq 0, \qquad n \geq N.$$

Proposition 4.66 follows directly from the third statement of Corollary 4.45 and the third statement of Corollary 4.51.

4.11.2 Properties of Positive Piecewise Linear Functions

Given a sequence $\{a_n\} = \{a_n\}_{n\geq 0}$, where $a_0 = 0$, $a_n > 0$, $n \geq 1$, and a sequence $\{t_n\} = \{t_n\}_{n\geq 0}$, where $t_0 = 0$, $t_{n+1} - t_n > 0$, $n \geq 0$, $t_n \to \infty$, consider the *piecewise linear interpolation* with respect to $\{t_n\}$, that is, the function

$$\mathscr{L}(t) = \mathscr{L}(\{a_n\}, \{t_n\}; t) = a_n + \frac{\Delta a_n}{\Delta t_n}(t - t_n), \qquad t \in [t_n, t_{n+1}), \qquad n \geq 0, \tag{4.78}$$

where $\Delta a_n = a_{n+1} - a_n$, $\Delta t_n = t_{n+1} - t_n$, $n \geq 0$.

The function $\mathscr{L}$ is a positive, piecewise linear function on $[0, \infty)$ such that $L(0) = 0$. $\mathscr{L}$ is absolutely continuous with density $\theta_{\mathscr{L}} = (\theta_{\mathscr{L}}(t),\ t \geq 0)$ given by

$$\theta_{\mathscr{L}}(t) = \frac{\Delta a_n}{\Delta t_n}, \qquad t \in [t_n, t_{n+1}), \qquad n \geq 0. \tag{4.79}$$

Thus $\mathscr{L}$ belongs to the class $\mathbb{DL}$ for all $t_0 > 0$. If

$$a_{n+1} - a_n \ge 0, \qquad n \ge 1,$$

that is, if the function $\mathscr{L}$ and the sequence $\{a_n\}$ are nondecreasing, then $\theta_{\mathscr{L}}(t) \ge 0$, $t \ge 0$, and $L \in \mathbb{DLP}$. Similarly, if

$$a_{n+1} - a_n > 0, \qquad n \ge 1,$$

that is, if the function $\mathscr{L}$ and the sequence $\{a_n\}$ are increasing, then $\theta_{\mathscr{L}}(t) > 0$, $t \ge 0$, and $\mathscr{L} \in \mathbb{DLSP}$.

If $t_n = n$, $n \ge 0$, then $\mathscr{L}$ is the piecewise linear interpolation of the sequence $\{a_n\}$ with respect to the sequence of natural numbers. The properties of such piecewise linear functions have been studied in the previous sections. In this subsection, we consider a more general case where the asymptotic behavior of both sequences $\{a_n\}$ and $\{t_n\}$ plays a crucial role.

For simplicity, we suppose that

$$\lim_{n\to\infty} \frac{t_{n+1}}{t_n} = 1. \tag{4.80}$$

The latter conditions holds, for example, if $t_n = n^\beta$, $n \ge 1$, where $\beta > 0$.

We are not going to prove a general result like Theorem 4.57, rather we show how indices of asymptotic elasticity may be applied here (see Theorem 4.58).

Using relations (4.78) and (4.79) we find the indices of asymptotic elasticity of the function $\mathscr{L}$ with respect to its density $\theta_{\mathscr{L}}$ (see Definition 4.19):

$$\begin{aligned}\epsilon^*(\mathscr{L}, \theta_{\mathscr{L}}) &= \limsup_{t\to\infty} \frac{t\theta_{\mathscr{L}}(t)}{\mathscr{L}(t)} \\ &= \limsup_{t\to\infty} \sum_{n=0}^{\infty} \left(\frac{t\Delta a_n}{a_n \Delta t_n + (t - t_n)\Delta a_n} \right) \mathbb{I}\{t \in [t_n, t_{n+1})\}\end{aligned}$$

and

$$\begin{aligned}\epsilon_*(\mathscr{L}, \theta_{\mathscr{L}}) &= \liminf_{t\to\infty} \frac{t\theta_{\mathscr{L}}(t)}{\mathscr{L}(t)} \\ &= \liminf_{t\to\infty} \sum_{n=0}^{\infty} \left(\frac{t\Delta a_n}{a_n \Delta t_n + (t - t_n)\Delta a_n} \right) \mathbb{I}\{t \in [t_n, t_{n+1})\}.\end{aligned}$$

This together with relation (4.80) implies that

$$\epsilon^*(\mathscr{L}, \theta_{\mathscr{L}}) = \limsup_{n\to\infty} \frac{t_n \Delta a_n}{a_n \Delta t_n} \tag{4.81}$$

and

$$\epsilon_*(\mathscr{L}, \theta_{\mathscr{L}}) = \liminf_{t\to\infty} \frac{t_n \Delta a_n}{a_n \Delta t_n}. \tag{4.82}$$

Now, Proposition 4.20 and relations (4.81) and (4.82) imply the following result.

Theorem 4.67 *Assume that condition (4.80) holds for the piecewise linear function $\mathscr{L}$. We have,*

1) if

$$\liminf_{n\to\infty} \frac{t_n \Delta a_n}{a_n \Delta t_n} > -\infty \quad \text{and} \quad \limsup_{n\to\infty} \frac{t_n \Delta a_n}{a_n \Delta t_n} < \infty,$$

then $\mathscr{L} \in \mathcal{PRV}$;

2) if the sequence $\{a_n\}$ is nondecreasing and

$$\limsup_{n\to\infty} \frac{t_n \Delta a_n}{a_n \Delta t_n} < \infty, \tag{4.83}$$

then $\mathscr{L} \in \mathcal{PRV}$;

3) if

$$\liminf_{n\to\infty} \frac{t_n \Delta a_n}{a_n \Delta t_n} > 0, \tag{4.84}$$

then $\mathscr{L} \in \mathcal{SQI}$;

4) if conditions (4.83) and (4.84) hold, then $\mathscr{L} \in \mathcal{POV}$;

5) if the limit

$$\lim_{n\to\infty} \frac{t_n \Delta a_n}{a_n \Delta t_n} = \rho \in (-\infty, \infty)$$

exists, then $\mathscr{L}$ is an RV-function of index ρ.

4.12 Comments

This chapter is based on the works of Buldygin et al. [63, 67, 70]. The results generalize the corresponding results for continuously differentiable functions obtained in Buldygin et al. [63, 64, 67].

The conditions obtained in this chapter allow us to prove corresponding regularity properties for various models, several of which are mentioned below.

Some questions for absolutely continuous functions are discussed by Bingham et al. [41]. For example, they prove the so-called *monotone density theorem* in Section 1.7.3 (see also Section 2.10.3 therein).

The theory of linear filters for stationary stochastic processes can be found, for example, in Lindgren [250]. Sufficient conditions for paths to be absolutely continuous are known for many important classes of stochastic processes. General conditions for a second-order stochastic process to possess absolutely continuous paths are obtained by Cambanis [77]. Janson et al. [200] give sufficient conditions on the filtration such that all totally inaccessible stopping times have hazard rates (being absolutely continuous, in other words).

Section 4.5 The whole Section 2.6 in Bingham et al. [41] (especially Theorem 2.6.1) is devoted to the asymptotic elasticity of functions, however the word "elasticity" is not used in [41]. Statement 4 of Proposition 4.17 is well-known for continuously differentiable functions f with $\theta = f'$ (see Seneta [324]).

For a function U being strictly increasing, strictly concave, continuously differentiable with $U'(0) = \lim_{x\to 0} U'(x) = \infty$, Kramkov and Schachermayer [241] consider asymptotic elasticity, elasticity and indices of elasticity for U with respect to its derivative U' (see, Definitions 4.16 and 4.19). Then, considering U as a utility function, these notions are used in a maximization problem for the mean utility of an agent in a financial market.

Section 4.7 The class of $\mathbb{ER}$-*functions* is implicitly treated in Matuszewska [268] (see also Bingham et al. [41]). The origin of the notion of almost increasing/decreasing functions is hard to trace back. It occurs, for example, in Bernstein [34]. The notion is independently used by Aljančić and Arandelović [8] in a discussion of the class $\mathcal{ORV}$ (see also Bingham et al. [41]). The Zygmund–Bari–Stechkin class, widely used in approximation theory, is a subset of the class of almost increasing functions (see Bari and Stechkin [26]; further details are given in Samko [313]).

Section 4.11 If a sequence $\{a_n\}_{n\geq 1}$ is such that $a_{n+1}/a_n \to 1$, $n \to \infty$, then the piecewise linear interpolation $\widehat{a}$, as defined in (4.53), is regularly varying if and only if $\{a_n\}_{n\geq 1}$ is regularly varying (see Galambos and Seneta [145] or Bojanic and Seneta [46]). In a similar manner, one can prove corresponding properties for sequences via the latter for their piecewise linear interpolations (each being an absolutely continuous function). Several "pathological" examples, where $a_{n+1}/a_n \not\to 1$, $n \to \infty$, are also mentioned in Galambos and Seneta [145].

Chapter 5
Nondegenerate Groups of Regular Points

5.1 Introduction

The defining property of an ORV-function f is that $f \in \mathbb{F}_+$ is measurable and the upper limit

$$f^*(\lambda) = \limsup_{t \to \infty} \frac{f(\lambda t)}{f(t)}$$

is positive and finite for all $\lambda > 0$ (see Definition 3.7).

The defining property of an RV-function f is that $f \in \mathbb{F}_+$ is measurable and the limit

$$\kappa_f(\lambda) = \lim_{t \to \infty} \frac{f(\lambda t)}{f(t)} \tag{5.1}$$

exists and is positive and finite for all $\lambda > 0$ (see Definition 3.2).

A natural extension of the class $\mathcal{RV}$ consists of those functions for which the limit in (5.1) exists for λ in a proper subset of the semiaxis $(0, \infty)$. Note that this limit exists for at least one point, namely for $\lambda = 1$. Moreover, it turns out that the set of those λ for which the limit in (5.1) exists is a multiplicative subgroup in $\mathbf{R}_+$.

The main aim of this chapter is to study a subclass of functions in $\mathcal{ORV}$ with "nondegenerate group of regular points", that is, those ORV-functions for which the limit $\kappa_f(\lambda)$ exists and is positive and finite for λ belonging to a certain multiplicative subgroup in $\mathbf{R}_+$. As a result, we discover different subclasses of ORV-functions which all contain RV-functions. Each of these classes possesses intrinsic characterization theorems, theorems on integral representations, etc. that differ from the corresponding counterparts in the theory of RV-functions. In particular, the limit functions in these cases can be represented as a product of an RV-function and

V. V. Buldygin et al., *Pseudo-Regularly Varying Functions and Generalized Renewal Processes*, Probability Theory and Stochastic Modelling 91, https://doi.org/10.1007/978-3-319-99537-3_5

a positive periodic function which extends the well-known representation of limit functions for the class $\mathcal{RV}$.

Chapter 5 is organized as follows. Section 5.2 contains some necessary definitions and auxiliary results concerning groups of regular points. Some typical examples of functions with nondegenerate groups of regular points are also given in Sect. 5.2.

A representative class of functions with nondegenerate group of regular points, the class of regular log-periodic functions, is studied in Sect. 5.3.

Section 5.4 treats some important properties of the mapping $f \mapsto f^*$.

The factorization representations of limit functions for several classes of functions with nondegenerate groups of regular points are obtained in Sects. 5.5 and 5.6.

These representations imply some new characterization theorems for functions of the class $\mathcal{RV}$ (see Sect. 5.7).

Section 5.8 contains the corresponding factorization representation for various classes with nondegenerate groups of regular points.

Finally, Sect. 5.9 is devoted to the proof of an appropriate uniform convergence theorem.

5.2 Groups of Regular Points

The main notions of this chapter are "*regular points*" and the "*group of regular points*".

5.2.1 Regular Points and Groups of Regular Points

Definition 5.1 Let $f \in \mathbb{F}_+$. A number $\lambda > 0$ is called a *regular point* of the function f if

$$f_*(\lambda) = f^*(\lambda) \in (0, \infty), \tag{5.2}$$

that is, if the limit

$$\kappa_f(\lambda) = \lim_{t\to\infty} \frac{f(\lambda t)}{f(t)}$$

exists and is positive and finite. The set of regular points of a function f is denoted by $\mathbb{G}_r(f)$.

The function $\kappa_f = (\kappa_f(\lambda), \lambda \in \mathbb{G}_r(f))$ is called the *index function* of f. The set of values of κ_f,

$$\mathbb{K}_r(f) = \kappa_f(\mathbb{G}_r(f)) = \{\kappa_f(\lambda), \lambda \in \mathbb{G}_r(f)\},$$

is called the *index map* of the set $\mathbb{G}_r(f)$.

The following two results present some important properties of the set of regular points and its index map.

Lemma 5.2 *The pair* $(\mathbb{G}_r(f), \times)$*, where* $\mathbb{G}_r(f)$ *is the set of regular points of a function* $f \in \mathbb{F}_+$ *and where the symbol "*$\times$*" denotes multiplication, is a multiplicative subgroup of the group* $(\mathbf{R}_+, \times)$.

Proof of Lemma 5.2 Definition 5.1 implies that $1 \in \mathbb{G}_r(f)$ and $\kappa_f(1) = 1$. Thus the set $\mathbb{G}_r(f)$ is nonempty.

If $\lambda_1, \lambda_2 \in \mathbb{G}_r(f)$, then

$$\lim_{t\to\infty} \frac{f(\lambda_1\lambda_2 t)}{f(t)} = \lim_{t\to\infty} \frac{f(\lambda_1\lambda_2 t)}{f(\lambda_2 t)} \lim_{t\to\infty} \frac{f(\lambda_2 t)}{f(t)} = \lim_{t\to\infty} \frac{f(\lambda_1 t)}{f(t)} \lim_{t\to\infty} \frac{f(\lambda_2 t)}{f(t)}.$$

This implies that $\kappa_f(\lambda_1\lambda_2)$ is well-defined and positive. Thus $\lambda_1\lambda_2 \in \mathbb{G}_r(f)$ and

$$\kappa_f(\lambda_1\lambda_2) = \kappa_f(\lambda_1)\kappa_f(\lambda_2). \tag{5.3}$$

If $\lambda \in \mathbb{G}_r(f)$, then

$$\lim_{t\to\infty} \frac{f(t/\lambda)}{f(t)} = \lim_{t\to\infty} \frac{f(t)}{f(\lambda t)} = 1 \Big/ \lim_{t\to\infty} \frac{f(\lambda t)}{f(t)}.$$

This implies that $\kappa_f(1/\lambda)$ is well-defined and positive as well. Hence $(1/\lambda) \in \mathbb{G}_r(f)$ and

$$\kappa_f(1/\lambda) = 1/\kappa_f(\lambda). \tag{5.4}$$

Therefore the set $\mathbb{G}_r(f)$ of regular points of a function f is a subgroup of the group $(\mathbf{R}_+, \times)$ with respect to multiplication. □

Relations (5.3) and (5.4) imply the following assertion.

Corollary 5.3 *The pair* $(\mathbb{K}_r(f), \times)$*, where* $\mathbb{K}_r(f)$ *is the index map of the set of regular points* $\mathbb{G}_r(f)$ *of a function* $f \in \mathbb{F}_+$*, is a multiplicative subgroup of the group* $(\mathbf{R}_+, \times)$.

It is worth mentioning that Theorems 5.34 and 5.40 below show that, under certain restrictions imposed on the function $f \in \mathbb{F}_+$, its index function κ_f is such that

$$\kappa_f(\lambda) = \lambda^\rho, \qquad \lambda \in \mathbb{G}_r(f),$$

where ρ is a real number depending on f, called the *index* of f, if f is an RV-function.

Along with the sets of positive regular points $\mathbb{G}_r(f)$, we also consider the sets

$$\mathbb{H}_r(f) = \ln(\mathbb{G}_r(f))$$

containing those real numbers u for which $\exp(u) \in \mathbb{G}_r(f)$. The sets $\mathbb{H}_r(f)$ are additive subgroups of the group $(\mathbf{R}, +)$ with respect to addition.

Moreover, we introduce an important property of the set of regular points.

Definition 5.4 If $\mathbb{G}_r(f) = \{1\}$, then the set $\mathbb{G}_r(f)$ as well as the corresponding group $(\mathbb{G}_r(f), \times)$ is called *degenerate*, otherwise $\mathbb{G}_r(f)$ and $(\mathbb{G}_r(f), \times)$ are called *nondegenerate*.

5.2.2 Full Sets of Regular Points for RV- (WRV)-Functions

In this subsection, we study various classes of functions with nondegenerate sets of regular points. The sets of regular points corresponding to these classes are different and, to some extent, determine the properties of the underlying functions. We start the survey of sets of regular points with "full" sets, where $\mathbb{G}_r(f) = \mathbf{R}_+$.

If

$$\mathbb{G}_r(f) = \mathbf{R}_+$$

for $f \in \mathbb{F}_+$, then f is a *weakly regularly varying function* (WRV) (see Definition 3.5). If, in addition, the function f is measurable, then f is *regularly varying* (RV) in Karamata's sense (see Definition 3.2).

Below we recall some characterization theorems for RV- and WRV-functions (see Karamata [206], Feller [129], de Haan [164], Seneta [324], Bingham et al. [41]). Another proof of these results will be given in Sect. 5.7 as a corollary of the theory of functions with nondegenerate sets of regular points. Recall that the notation "meas" stands for the Lebesgue measure.

We start with a result that follows from the following Steinhaus theorem

Proposition 5.5 *If $(T, \times)$ is a multiplicative subgroup of the group $(\mathbf{R}_+, \times)$ and T contains a set of positive Lebesgue measure, then $T = \mathbf{R}_+$.*

The proof of Proposition 5.5 can be found in Steinhaus [347] or Bingham et al. [41], Corollary 1.1.4.

5.2.3 Characterization Theorems

Proposition 5.5 means that, if $f \in \mathbb{F}_+$ and the set of its regular points $\mathbb{G}_r(f)$ contains a subset of positive Lebesgue measure, then $\mathbb{G}_r(f) = \mathbf{R}_+$, that is, $f \in \mathcal{WRV}$. The following two characterization theorems complement this result.

Theorem 5.6 (Characterization Theorem for RV-Functions) *Let f be a measurable function belonging to $\mathbb{F}_+$. If there is a measurable subset $\Lambda \subset \mathbb{G}_r(f)$ such*

that

$$\operatorname{meas}(\Lambda) > 0, \tag{5.5}$$

then

$$\mathbb{G}_r(f) = \mathbf{R}_+,$$

that is, f is an RV-function and there exists a real number $\rho = \rho_f$, called the index of f, such that

$$\kappa_f(\lambda) = \lambda^\rho, \qquad \lambda > 0. \tag{5.6}$$

Every RV-function f admits the following representation

$$f(t) = t^\rho \ell(t), \qquad t > 0, \tag{5.7}$$

where ℓ is a *slowly varying function* (SV), that is, ℓ is measurable and $\kappa_\ell(\lambda) \equiv 1$ (see Definition 3.3). Moreover, relations (5.6) and (5.7) are equivalent.

Theorem 5.7 (Characterization theorem for WRV-functions) *Let $f \in \mathbb{F}_+$ be such that both f and $1/f$ are bounded on each finite interval of $\mathbf{R}_+$ being sufficiently far away from the origin. If there exists a measurable set $\Lambda \subset \mathbb{G}_r(f)$ such that*

$$\operatorname{meas}(\Lambda) > 0,$$

then

$$\mathbb{G}_r(f) = \mathbf{R}_+,$$

that is, $f \in \mathcal{WRV}$ and there exists a real number $\rho = \rho_f$, called the index of f, such that (5.6) holds.

Note that every function f satisfying the assumptions of the characterization theorem for WRV-functions admits the following representation

$$f(t) = t^\rho w(t), \qquad t > 0, \tag{5.8}$$

where w is a *weakly slowly varying function* (WSV) (see Definition 3.5), that is, $\kappa_w(\lambda) \equiv 1$. Moreover, relations (5.8) and (5.6) are equivalent.

The above characterization theorems show that, if the set of regular points $\mathbb{G}_r(f)$ contains a subset of positive Lebesgue measure, then it coincides with $\mathbf{R}_+$ under some additional assumptions imposed on the function f. In this case, relations (5.6), (5.7) and (5.8) hold and f is a regularly varying function or it belongs to $\mathcal{WRV}$.

5.2.4 Null Sets of Regular Points

Next we consider several examples of functions f with nondegenerate but "thin" sets of regular points $\mathbb{G}_r(f)$, that is, with nondegenerate sets of regular points whose Lebesgue measure equals zero. Later we will see that the limit functions appearing in these examples are typical in a general case.

Example 5.8 Let

$$f(t) = r(t)\exp\{\sin(\ln t)\}, \qquad t > 0,$$

where $(r(t), t > 0)$ is an RV-function of index ρ. Then

$$f^*(\lambda) = \lambda^\rho \exp\{2|\sin(\ln\sqrt{\lambda})|\} \tag{5.9}$$

and

$$f_*(\lambda) = \lambda^\rho \exp\{-2|\sin(\ln\sqrt{\lambda})|\} \tag{5.10}$$

for all $\lambda > 0$. Relations (5.9) and (5.10) imply that

$$\mathbb{G}_r(f) = \{e^{2\pi n}; n \in \mathbf{Z}\}.$$

Thus the set $\mathbb{G}_r(f)$ is nondegenerate and

$$\kappa_f(\lambda) = \lambda^\rho, \qquad \lambda \in \mathbb{G}_r(f).$$

Note that

$$\exp\{2|\sin(u/2)|\}, \qquad u \in \mathbf{R},$$

is a positive periodic function whose set of periods is

$$\mathbb{H}_r(f) = \{2\pi n; n \in \mathbf{Z}\}.$$

Example 5.9 Let

$$f(t) = r(t)\exp\{\operatorname{sign}(\sin(\ln t))\}, \qquad t > 0,$$

where $(r(t), t > 0)$ is an RV-function of index ρ and $\operatorname{sign}(t) = 1$, for $t > 0$, $\operatorname{sign}(t) = -1$, for $t < 0$, and $\operatorname{sign}(0) = 0$. Then

$$f^*(\lambda) = \lambda^\rho \exp\{2\}, \qquad f_*(\lambda) = \lambda^\rho \exp\{-2\} \tag{5.11}$$

for every λ such that $\ln \lambda \neq 2\pi n$, $n \in \mathbf{Z}$, and

$$f^*(\lambda) = f_*(\lambda) = \lambda^\rho \tag{5.12}$$

for every λ such that $\ln \lambda = 2\pi n$, $n \in \mathbf{Z}$. Relations (5.11) and (5.12) imply that

$$\mathbb{G}_r(f) = \{e^{2\pi n}; n \in \mathbf{Z}\}.$$

Thus the set $\mathbb{G}_r(f)$ is nondegenerate and

$$\kappa_f(\lambda) = \lambda^\rho, \qquad \lambda \in \mathbb{G}_r(f).$$

The upper limit function is

$$f^*(\lambda) = \lambda^\rho \exp\{2(1 - I_{\mathbb{H}_r(f)}(\ln \lambda))\},$$

where $I_{\mathbb{H}_r(f)}$ denotes the indicator function of the set $\mathbb{H}_r(f)$. Note that

$$\exp\{2(1 - I_{\mathbb{H}_r(f)}(u)\}, \qquad u \in \mathbf{R},$$

is a positive periodic function whose set of periods is

$$\mathbb{H}_r(f) = \{2\pi n; n \in \mathbf{Z}\}.$$

Example 5.10 Let

$$f(t) = r(t) \exp\{d(t)\}, \qquad t > 0,$$

where $(r(t), t > 0)$ is an RV-function of index ρ and where $(d(t), t > 0)$ is the Dirichlet function, that is, $d(t) = 1$ for $t \in \mathbf{Q}$, and $d(t) = 0$ otherwise. Then

$$f^*(\lambda) = \lambda^\rho \exp\{1 - d(\lambda)\} \tag{5.13}$$

and

$$f_*(\lambda) = \lambda^\rho \exp\{d(\lambda) - 1\} \tag{5.14}$$

for all $\lambda > 0$. Relations (5.13) and (5.14) imply that

$$\mathbb{G}_r(f) = \mathbf{Q} \cap \mathbf{R}_+.$$

Thus the set $\mathbb{G}_r(f)$ is nondegenerate and

$$\kappa_f(\lambda) = \lambda^\rho, \qquad \lambda \in \mathbb{G}_r(f).$$

Note that the set $\mathbb{G}_r(f)$ is everywhere dense in $\mathbf{R}_+$ but $\operatorname{meas}(\mathbb{G}_r(f)) = 0$. The upper limit function f^* defined by (5.13) can be written as follows:

$$f^*(\lambda) = \lambda^\rho \exp\{1 - d(e^{\ln\lambda})\}.$$

Note that

$$\exp\{1 - d(e^u)\}, \qquad u \in \mathbf{R},$$

is a positive periodic function whose set of periods is

$$\mathbb{H}_r(f) = \{u \in \mathbf{R} : \exp(u) \in \mathbf{Q} \cap \mathbf{R}_+\}.$$

5.2.5 An Arbitrary Group of Regular Points

The following example shows that, given an arbitrary nondegenerate multiplicative subgroup $(\mathbb{G}, \times)$ of the group $(\mathbf{R}_+, \times)$, there exists a function $f \in \mathbb{F}_+$ such that $\mathbb{G}_r(f) = \mathbb{G}$.

Example 5.11 Let $\mathbb{G} \in \mathbf{R}_+$, $\mathbb{G}_r(f) \neq \{1\}$ and let $(\mathbb{G}, \times)$ be a multiplicative subgroup of $(\mathbf{R}_+, \times)$. Consider the function

$$f(t) = r(t) \exp\{I_{\mathbb{G}}(t)\}, \qquad t > 0,$$

where $(r(t), t > 0)$ is an RV-function of index ρ and where $I_{\mathbb{G}}$ denotes the indicator function of the set $\mathbb{G}$. Then

$$f^*(\lambda) = \lambda^\rho \exp\{1 - I_{\mathbb{G}}(\lambda)\}, \tag{5.15}$$

and

$$f_*(\lambda) = \lambda^\rho \exp\{I_{\mathbb{G}}(\lambda) - 1\} \tag{5.16}$$

for all $\lambda > 0$. Relations (5.15) and (5.16) imply that

$$\mathbb{G}_r(f) = \mathbb{G}.$$

Thus the set $\mathbb{G}_r(f)$ is nondegenerate and

$$\kappa_f(\lambda) = \lambda^\rho, \qquad \lambda \in \mathbb{G}_r(f).$$

The upper limit function f^* defined by (5.15) can be written as follows:

$$f^*(\lambda) = \lambda^\rho \exp\{1 - I_{\mathbb{G}}(e^{\ln\lambda})\}.$$

Note that

$$\exp\{1 - I_{\mathbb{G}}(\exp\{u\})\}, \qquad u \in \mathbf{R},$$

is a positive periodic function whose set of periods is

$$\mathbb{H}_r(f) = \{u \in \mathbf{R} : \exp(u) \in \mathbb{G}\}.$$

5.2.6 A Degenerate Group of Regular Points

Now we consider an example of a function with a degenerate set of regular points.

Example 5.12 Let

$$f(t) = r(t)(2 + \sin t), \qquad t > 0,$$

where $(r(t), t > 0)$ is an RV-function. Since

$$f_*(\lambda) < f^*(\lambda)$$

for all $\lambda \neq 1$, we conclude that $\mathbb{G}_r(f) = \{1\}$.

5.3 Regularly Periodic and Regularly Log-Periodic Functions

5.3.1 Definitions and Some Properties of Regularly Periodic and Regularly Log-Periodic Functions

Now we look at Examples 5.8 and 5.12 from a more general point of view. This will, in particular, allow us to increase the number of similar examples. For this, we introduce the classes of *regularly periodic* and *regularly log-periodic* functions.

Definition 5.13 A function $f \in \mathbb{F}_+$ is called *regularly periodic* if

$$f(t) = r(t)V(t), \qquad t > 0, \tag{5.17}$$

and *regularly log-periodic* if

$$f(t) = r(t)V(\ln t), \qquad t > 0, \tag{5.18}$$

where $(r(t) = t^\rho \ell(t),\ t > 0)$ is an RV-function of index $\rho \in \mathbf{R}$, $\ell = (\ell(t),\ t > 0)$ is an SV-function, and where $V = (V(t),\ t \in \mathbf{R})$ is a positive, continuous, periodic function. The family of regularly periodic functions is denoted by $\mathfrak{RP}$ and that of regularly log-periodic functions by $\mathfrak{RLP}$.

The aim of this subsection is to prove that, if a periodic function V is not constant (we say in this case that V is nondegenerate), the corresponding regularly periodic functions have degenerate sets of regular points, while the corresponding regularly log-periodic functions have nondegenerate sets of regular points. To achieve this goal, we determine the sets of regular points for regularly periodic and regularly log-periodic functions.

The function

$$r(t) = t^\rho \ell(t), \qquad t > 0,$$

is called the *regular component* (RV component) of the function f from (5.17) and (5.18), and its index ρ is called the *index* of f. The function ℓ is called the *slowly varying component* (SV component) of f. Note that the slowly varying component is not necessarily continuous.

The periodic function V is called the *periodic component* of the function $f \in \mathfrak{RP}$ and that of the function $f \in \mathfrak{RLP}$. The function $V \circ \ln = (V(\ln t),\ t > 0)$ is called the *log-periodic component* of $f \in \mathfrak{RLP}$.

Note that the function f introduced in Example 5.8 belongs to the class $\mathfrak{RLP}$, while the function f considered in Example 5.12 belongs to the class $\mathfrak{RP}$.

Definition 5.14 If $U = (U(t),\ t \in \mathbf{R})$ is a periodic function, then its set of periods is denoted by $S_{\mathrm{per}}(U)$; the set of its positive periods is denoted by $S^+_{\mathrm{per}}(U)$. The number

$$T(U) = \inf S^+_{\mathrm{per}}(U)$$

is called the *oscillation characteristic* of the function U.

If U is a periodic component of the function $f \in \mathfrak{RP}$ or that of the function $f \in \mathfrak{RLP}$, then its oscillation characteristic $T(U)$ is also called the *oscillation characteristic* of the function f.

Since

$$\mathbb{G}_r(f) = \mathbb{G}_r(V), \qquad f \in \mathfrak{RP}, \tag{5.19}$$

and

$$\mathbb{G}_r(f) = \mathbb{G}_r(V \circ \ln), \qquad f \in \mathfrak{RLP}, \tag{5.20}$$

determining the sets of regular points for functions of the classes $\mathfrak{RP}$ and $\mathfrak{RLP}$ can be reduced to determining the sets of regular points for their periodic or log-periodic components. To do so, we need some auxiliary results concerning periodic functions. In what follows the symbol $\{a\}$ stands for the *fractional part* of a number $a \in \mathbf{R}$.

Recall that $\{a\} = a - \lfloor a \rfloor$, where $\lfloor a \rfloor$ denotes the *integer part* of a, that is, $\lfloor a \rfloor$ is the largest integer not exceeding a.

Lemma 5.15 *Let $U = (U(t),\ t \in \mathbf{R})$ be a periodic function. If U possesses an arbitrarily small period and U is continuous at at least one point, then U is a constant (in particular, every real number is a period of U).*

Proof of Lemma 5.15 Let t_0 be a point of continuity of the function U. Then, for every $\varepsilon > 0$, there exists a number $\delta = \delta(\varepsilon, t_0) > 0$ such that $|U(t_0+\theta) - U(t_0)| \le \varepsilon$ if $|\theta| \le \delta$.

Fix an arbitrary $\varepsilon > 0$ and consider a positive period τ of the function U such that $\tau \le \delta$. Since

$$U(t) = U(t_0 + (t - t_0)) = U(t_0 + \theta + \lfloor (t - t_0)/\tau \rfloor \tau) = U(t_0 + \theta)$$

for all $t \in \mathbf{R}$ where $\theta = \{(t - t_0)/\tau\}\tau$, we obtain

$$|U(t) - U(t_0)| = |U(t_0 + \theta) - U(t_0)| \le \varepsilon,$$

whence Lemma 5.15 follows. □

Example 5.10 with the Dirichlet function shows that, in Lemma 5.15, one cannot drop the assumption on the continuity of U at at least one point.

Remark 5.16 If U is a continuous periodic function with oscillation characteristic $T(U)$, then Lemma 5.15 implies that either $T(U) = 0$ and the function U is constant or $T(U) > 0$ and $T(U)$ is the minimal positive period (also called the *principal period*) of the function U, that is,

$$S_{\text{per}}(U) = \{nT(U),\ n \in \mathbf{Z}\}\,.$$

The following result, known as *Kronecker's theorem*, will often be used throughout this chapter.

Proposition 5.17 *The following two statements are equivalent:*

1) *for all irrational numbers $\alpha > 0$, the set of fractional parts $\{n\alpha\}$, $n \in \mathbf{N}$, is everywhere dense in the interval $[0, 1]$;*
2) *if two real numbers x and y are not commensurable, that is their ratio x/y is an irrational number, then the set of numbers $nx + my$, where $n, m \in \mathbf{Z}$, is everywhere dense in $\mathbf{R}$.*

Kronecker's theorem implies that periodic functions coincide if they are asymptotically equivalent.

Lemma 5.18 *Let* $U_1 = (U_1(t), t \in \mathbf{R})$ *and* $U_2 = (U_2(t), t \in \mathbf{R})$ *be two positive periodic functions. Assume that one of them is continuous. If* $U_1 \sim U_2$, *then* $U_1 = U_2$.

Proof of Lemma 5.18 Let T_1 and T_2 be positive periods of the functions U_1 and U_2, respectively. Without loss of generality, assume that U_1 is continuous.

Assume that the numbers T_1 and T_2 are commensurable, that is, there exist natural numbers p and q such that

$$\frac{T_2}{T_1} = \frac{p}{q}.$$

Then

$$\frac{U_1(t + nq\,T_2)}{U_2(t + nq\,T_2)} = \frac{U_1(t + nq\,T_2)}{U_2(t)} = \frac{U_1(t + nq(T_2/T_1)\,T_1)}{U_2(t)}$$

$$= \frac{U_1(t + np\,T_1)}{U_2(t)} = \frac{U_1(t)}{U_2(t)}$$

for all $t \in \mathbf{R}$ and $n \in \mathbf{N}$, that is

$$\frac{U_1(t)}{U_2(t)} = \frac{U_1(t + nq\,T_2)}{U_2(t + nq\,T_2)}.$$

Thus

$$\frac{U_1(t)}{U_2(t)} = \lim_{n\to\infty} \frac{U_1(t + nq\,T_2)}{U_2(t + nq\,T_2)} = 1$$

for all $t \in \mathbf{R}$, since $U_1 \sim U_2$, whence $U_1 = U_2$.

Now assume that the numbers T_1 and T_2 are not commensurable, that is,

$$\frac{T_2}{T_1} = \alpha,$$

where α is a positive irrational number. Then

$$\frac{U_1(t + n\,T_2)}{U_2(t + n\,T_2)} = \frac{U_1(t + n\,T_2)}{U_2(t)} = \frac{U_1(t + n(T_2/T_1)\,T_1)}{U_2(t)} = \frac{U_1(t + \{n\alpha\}\,T_1)}{U_2(t)}$$

for all $t \in \mathbf{R}$ and $n \in \mathbf{N}$. Hence

$$\lim_{k\to\infty} U_1(t + \{n_k\alpha\}\,T_1) = U_2(t) \tag{5.21}$$

for all $t \in \mathbf{R}$ and any increasing to infinity sequence of natural numbers n_k, $k \geq 1$, since $U_1 \sim U_2$.

By Kronecker's theorem (see Proposition 5.17), there exists a sequence of natural numbers n_k, $k \geq 1$, increasing to infinity, such that

$$\lim_{k \to \infty} \{n_k \alpha\} T_1 = 0.$$

Considering (5.21) and taking into account the continuity of the function U_1, we conclude that

$$U_1(t) = \lim_{k \to \infty} U_1(t + \{n_k \alpha\} T_1) = U_2(t)$$

for all $t \in \mathbf{R}$. Therefore $U_1 = U_2$. □

Corollary 5.19 *Let $V = (V(t),\ t \in \mathbf{R})$ be a positive, continuous, periodic function with oscillation characteristic $T(V)$. Then the set of regular points $\mathbb{G}_r(V)$ is degenerate if and only if V is nonconstant, that is, if $T(V) > 0$ (see Remark 5.16).*

Proof of Corollary 5.19 Let the function V be nonconstant. Assume that the set $\mathbb{G}_r(V)$ is nondegenerate. Then there exists a constant $\lambda > 1$ such that

$$\lim_{t \to \infty} \frac{V(\lambda t)}{V(t)} = c \in (0, \infty). \tag{5.22}$$

Since the functions $V_\lambda = (V(\lambda t),\ t \in \mathbf{R})$ and $cV = (cV(t),\ t \in \mathbf{R})$ are positive, continuous, and periodic, Lemma 5.18 together with relation (5.22) implies that

$$V(\lambda t) = cV(t), \qquad t \in \mathbf{R}.$$

Choosing $t = 0$, we see that $c = 1$, whence

$$V_\lambda = V. \tag{5.23}$$

Since the function V is nonconstant, it possesses a principal period $T = T(V)$. At the same time, (5.23) implies that

$$V(t + (T/\lambda)) = V(\lambda(t + (T/\lambda))) = V(\lambda t + T) = V(\lambda t) = V(t)$$

for all $t \in \mathbf{R}$, that is, the number $T/\lambda < T$ is also a positive period of the function V. This contradiction proves that $\mathbb{G}_r(V) = \{1\}$.

To complete the proof of Corollary 5.19 assume that the set $\mathbb{G}_r(V)$ is degenerate, but the function V is nonconstant. Then, $\mathbb{G}_r(V) = \{1\}$ on one hand, but $\mathbb{G}_r(V) = \mathbf{R}$ on the other hand. This contradiction proves that V cannot be constant if the set $\mathbb{G}_r(V)$ is degenerate. □

The following result shows that if a periodic function belonging to $\mathfrak{RP}$ is nonconstant, then the set of regular points for this function is degenerate.

Corollary 5.20 *Let the function f belong to the class $\mathfrak{RP}$ and possess a periodic component V with oscillation characteristic $T(V)$. Then:*

1) *the set of regular points $\mathbb{G}_r(f)$ is degenerate if and only if the function V is nonconstant, that is, if $T(V) > 0$;*
2) *$\mathbb{G}_r(f) = \mathbf{R}_+$ if and only if the function V is constant, that is, if $T(V) = 0$.*

Corollary 5.20 follows from Corollary 5.19 together with relation (5.19). Note also that Corollary 5.20 generalizes Example 5.12.

5.3.2 Sets of Regular Points for Regularly Log-Periodic Functions

First we consider sets of regular points for regularly log-periodic functions $V \circ \ln = (V(\ln t),\ t > 0)$.

Lemma 5.21 *Let $V = (V(t), t \in \mathbf{R})$ be a positive, continuous, periodic function with oscillation characteristic $T(V)$. Then $T(V) > 0$, that is, the function V is nonconstant if and only if*

$$\mathbb{G}_r(V \circ \ln) = \left\{e^{nT(V)},\ n \in \mathbf{Z}\right\}. \tag{5.24}$$

Proof of Lemma 5.21 Let $T(V) > 0$. If $\lambda \in \left\{e^{nT(V)},\ n \in \mathbf{Z}\right\}$, then $\ln \lambda \in S_{\mathrm{per}}(V)$, and thus

$$\lim_{t\to\infty} \frac{V(\ln(\lambda t))}{V(\ln t)} = \lim_{t\to\infty} \frac{V(\ln \lambda + \ln t)}{V(\ln t)} = 1,$$

that is, $\lambda \in \mathbb{G}_r(V \circ \ln)$. Hence, if $T(V) > 0$, then

$$\mathbb{G}_r(V \circ \ln) \supset \left\{e^{nT(V)},\ n \in \mathbf{Z}\right\}. \tag{5.25}$$

Assume, as before, that $T(V) > 0$. Suppose also that $\lambda \in \mathbb{G}_r(V \circ \ln)$ and $\lambda \neq 1$. Then

$$\lim_{t\to\infty} \frac{V(\ln \lambda + \ln t)}{V(\ln t)} = \lim_{t\to\infty} \frac{V(\ln \lambda t)}{V(\ln t)} = c \in (0, \infty).$$

Moreover,

$$\lim_{t\to\infty} \frac{V(\ln \lambda + \ln t)}{V(\ln t)} = \lim_{t\to\infty} \frac{V(\ln \lambda + t)}{V(t)}.$$

Thus

$$\lim_{t\to\infty} \frac{V(\ln \lambda + t)}{cV(t)} = 1.$$

This together with Lemma 5.18 implies

$$V(\ln \lambda + t) = cV(t), \qquad t \in \mathbf{R}.$$

Repeating this procedure we obtain

$$V(n \ln \lambda) = c^n V(0), \qquad n \in \mathbf{N}.$$

If $c \neq 1$, then $\inf_{t\in\mathbf{R}} V(t) = 0$ or $\sup_{t\in\mathbf{R}} V(t) = \infty$. This contradicts the assumption that V is a positive, continuous, and periodic function. This proves $c = 1$. Therefore

$$V(\ln \lambda + t) = V(t), \qquad t \in \mathbf{R},$$

that is, $\ln \lambda \in S_{\text{per}}(U)$. Now, if $T(V) > 0$, then

$$\mathbb{G}_r(V \circ \ln) \subset \left\{ e^{nT(V)},\ n \in \mathbf{Z} \right\}. \tag{5.26}$$

Relations (5.25) and (5.26) imply that (5.24) holds provided $T(V) > 0$.

To complete the proof of Lemma 5.21 we assume that (5.24) is satisfied. If $T(V) = 0$, then V is a constant (see Remark 5.16) and thus $\mathbb{G}_r(V \circ \ln) = \mathbf{R}_+$. On the other hand, if $T(V) = 0$, the set $\left\{e^{nT(V)},\ n \in \mathbf{Z}\right\}$ contains a single element, which contradicts (5.24). This proves that $T(V) > 0$. □

For a constant function V,

$$\mathbb{G}_r(V \circ \ln) = \mathbb{G}_r(V) = \mathbf{R}_+$$

and Lemma 5.21 can be sharpened to some extent.

Lemma 5.22 *Let $V = (V(t), t \in \mathbf{R})$ be a positive, continuous, periodic function with oscillation characteristic $T(V)$.*

1) If $T(V) > 0$, then

$$\mathbb{G}_r(V \circ \ln) = \left\{ e^{nT(V)},\ n \in \mathbf{Z} \right\}.$$

2) If $T(V) = 0$, then

$$\mathbb{G}_r(V \circ \ln) = \mathbf{R}_+.$$

Lemma 5.22 and relation (5.20) provide a generalization of Example 5.8.

Corollary 5.23 *Let the function f belong to the class $\mathfrak{RLP}$ and possess the periodic component V with oscillation characteristic $T(V)$.*

1) If $T(V) > 0$, that is, if the function V is periodic with principal period $T(V)$, then

$$\mathbb{G}_r(f) = \left\{e^{nT(V)},\ n \in \mathbf{Z}\right\} \qquad \text{and} \qquad \mathbb{H}_r(f) = \{nT(V),\ n \in \mathbf{Z}\}.$$

2) If $T(V) = 0$, that is, if the function V is a constant, then

$$\mathbb{G}_r(f) = \mathbf{R}_+$$

and $f \in \mathcal{RV}$.

Remark 5.24 All assertions of this subsection concerning regularly periodic and regularly log-periodic functions remain true if ℓ in Definition 5.13 is a positive WSV-function (see Definition 3.5).

5.4 $*$-Invariant Limit Functions

For WORV-functions f, the transformations $f \mapsto f^*$ and $f \mapsto f_*$ are mappings from $\mathbb{F}_+$ to $\mathbb{F}_+$. In this subsection, we study *invariants* of these transformations and some of their properties. Recall that (see (3.7)), if $f \in \mathbb{F}_+$ is an WORV-function, then, for all $c > 0$,

$$0 < f_*(c) \le f^*(c) < \infty.$$

Definition 5.25 A function $f \in \mathbb{F}_+$ is called an *upper $*$-invariant function* if

$$f^*(\lambda) = f(\lambda) \qquad \text{for all} \qquad \lambda > 0,$$

and a *lower $*$-invariant function* if

$$f_*(\lambda) = f(\lambda) \qquad \text{for all} \qquad \lambda > 0.$$

Remark 5.26 Definition 5.25 implies, in particular, that any $*$-invariant WORV-function is positive.

As shown in the following result, a function which is a limit of a WORV-function is a $*$-invariant function.

Proposition 5.27 *Let f be a WORV-function with nondegenerate set of regular points $\mathbb{G}_r(f)$. Then its upper limit function f^* is an upper $*$-invariant function,* *that is, $f^{**} = (f^*)^* = f^*$, and its lower limit function f_* is a lower $*$-invariant function, that is, $f_{**} = (f_*)_* = f_*$. Moreover, $(f_*)^* = f^*$ and $(f^*)_* = f_*$.*

To prove Proposition 5.27 we need the following auxiliary result.

Lemma 5.28 *Let $f \in \mathcal{WORV}$. If $c \in \mathbb{G}_r(f)$, then*

$$f^*(c^n\lambda) = f^*(c^n)f^*(\lambda) = \kappa_f(c^n)f^*(\lambda) \tag{5.27}$$

and

$$f_*(c^n\lambda) = f_*(c^n)f_*(\lambda) = \kappa_f(c^n)f_*(\lambda) \tag{5.28}$$

for all $\lambda > 0$ and $n \in \mathbf{Z}$.

Proof of Lemma 5.28 Since $(\mathbb{G}_r(f), \times)$ is a multiplicative group, it suffices to prove relation (5.27) for $n = 1$.

Now, by Lemma 3.1, (5.2) implies that

$$f^*(c)f^*(\lambda) = f_*(c)f^*(\lambda) \le f^*(c\lambda) \le f^*(c)f^*(\lambda) = \kappa_f(c)f^*(\lambda)$$

for all $\lambda > 0$. Thus (5.27) is proved for $n = 1$ and all $\lambda > 0$.

Relation (5.28) follows from (5.27), since

$$\begin{aligned} f_*(c^n\lambda) &= \frac{1}{f^*(c^{-n}\lambda^{-1})} = \frac{1}{f^*(c^{-n})f^*(\lambda^{-1})} \\ &= f_*(c^n)f_*(\lambda) = \kappa_f(c^n)f_*(\lambda), \end{aligned}$$

so that the proof is complete. □

Proof of Proposition 5.27 Since $f \in \mathcal{WORV}$, we conclude that $f^* \in \mathbb{F}_+$ and, by Lemma 3.1,

$$f^{**}(\lambda) = \limsup_{x\to\infty} \frac{f^*(\lambda x)}{f^*(x)} \le f^*(\lambda) \tag{5.29}$$

for all $\lambda > 0$. Note that relation (5.29) holds for an arbitrary WORV-function.

Since the set $\mathbb{G}_r(f)$ is nondegenerate, we find a number

$$z \in \mathbb{G}_r(f) \setminus \{1\}.$$

Hence $c = \max\{z, 1/z\} \in \mathbb{G}_r(f)$, $c > 1$, and, by Lemma 5.28,

$$\frac{f^*(c^n\lambda)}{f^*(c^n)} = f^*(\lambda)$$

for all $\lambda > 0$ and $n \in \mathbf{N}$ (see (5.28)). This implies that

$$f^{**}(\lambda) = \limsup_{x\to\infty} \frac{f^*(\lambda x)}{f^*(x)} \ge \limsup_{n\to\infty} \frac{f^*(\lambda c^n)}{f^*(c^n)} = f^*(\lambda), \qquad \lambda > 0. \tag{5.30}$$

Relations (5.29) and (5.30) show that $f^{**} = f^*$ and thus f^* is an upper $*$-invariant function.

The equality $f_{**} = f_*$ is proved in the same way with (5.29) instead of (5.28). This proves that f_* is a lower $*$-invariant function.

It remains to note that, by Lemma 3.1,

$$(f_*)^*(\lambda) = \frac{1}{f_{**}(1/\lambda)} = \frac{1}{f_*(1/\lambda)} = f^*(\lambda),$$

that is, $(f_*)^* = f^*$. The proof of the equality $(f^*)_* = f_*$ is the same. □

Remark 5.29 Let f be a WORV-function with a nondegenerate set of regular points. If f is an upper $*$-invariant function, then it is positive according to Remark 5.26. Lemma 3.1 and Proposition 5.27 imply that

$$f_{**}(\lambda) = \frac{1}{(f_*)^*(1/\lambda)} = \frac{1}{f^*(1/\lambda)} = \frac{1}{f(1/\lambda)}$$

for all $\lambda > 0$. Therefore

$$f_{**}(\lambda) = \frac{1}{f(1/\lambda)}, \qquad \lambda > 0.$$

Similarly, if f is a lower $*$-invariant function, then

$$f^{**}(\lambda) = \frac{1}{f(1/\lambda)}, \qquad \lambda > 0.$$

Below are two examples of $*$-invariant functions.

Example 5.30 The function $f(t) = t^a$, $t > 0$, is both an upper $*$-invariant function and a lower $*$-invariant function for any $a \in \mathbf{R}$. Note that this property characterizes power functions in the class of WORV-functions.

Example 5.31 In view of Example 5.8 and Proposition 5.27, it follows that

$$f(t) = \exp\{2|\sin(\ln\sqrt{t})|\}, \qquad t > 0,$$

is an upper $*$-invariant function and

$$f(t) = \exp\{-2|\sin(\ln\sqrt{t})|\}, \qquad t > 0,$$

is a lower $*$-invariant function.

Corollary 5.32 *Let f be an WORV-function with nondegenerate set of regular points.*

1) If f is an upper $*$-invariant function, then $g = (1/f(1/t), t > 0)$ is a lower $*$-invariant function.
2) If f is a lower $*$-invariant function, then $g = (1/f(1/t), t > 0)$ is an upper $*$-invariant function.

Proof of Corollary 5.32 Let f be an upper $*$-invariant function. In view of Remark 5.26, it is positive and, moreover, Lemma 3.1 implies that

$$f_*(\lambda) = \frac{1}{f^*(1/\lambda)} = \frac{1}{f(1/\lambda)} = g(\lambda)$$

for all $\lambda > 0$. Thus $f_* = g$ and $f_{**} = g_*$. Further, taking Remark 5.29 into account, we see that

$$f_{**}(\lambda) = \frac{1}{f(1/\lambda)} = g(\lambda)$$

for all $\lambda > 0$. Thus $f_{**} = g$. So $g_* = g$, that is, the first statement of Corollary 5.32 is proved. The second statement can be proved similarly. □

Remark 5.33 If $(\psi(u), u \in \mathbf{R})$ is a positive and finite function, then $(\psi(\ln\lambda), \lambda > 0)$ is an upper $*$-invariant function if and only if

$$\limsup_{x\to\infty} \frac{\psi(u+x)}{\psi(x)} = \psi(u), \qquad u \in \mathbf{R}.$$

The latter condition is equivalent to

$$\liminf_{x\to\infty} \frac{\psi(u+x)}{\psi(x)} = \frac{1}{\psi(-u)}, \qquad u \in \mathbf{R}.$$

Moreover, $(\psi(\ln\lambda), \lambda > 0)$ is a lower $*$-invariant function if and only if

$$\liminf_{x\to\infty} \frac{\psi(u+x)}{\psi(x)} = \psi(u), \qquad u \in \mathbf{R}.$$

The latter condition is equivalent to

$$\limsup_{x\to\infty} \frac{\psi(u+x)}{\psi(x)} = \frac{1}{\psi(-u)}, \qquad u \in \mathbf{R}.$$

5.5 Weak Factorization Representations for Limit Functions

In this section, we study representations for upper and lower limit functions of WORV-functions f with nondegenerate sets of regular points $\mathbb{G}_r(f)$. Recall that (see Definition 5.14) we denote by $S_{\text{per}}(U)$ the set of periods of a periodic function $(U(t),\ t \in \mathbf{R})$.

Theorem 5.34 *Let f be a WORV-function with nondegenerate set of regular points $\mathbb{G}_r(f)$ and index function κ_f. Assume that $c \in \mathbb{G}_r(f) \setminus \{1\}$. Then the limit functions f^* and f_* can be represented as follows:*

$$f^*(\lambda) = \lambda^\alpha P(\ln \lambda), \qquad \lambda > 0, \tag{5.31}$$

and

$$f_*(\lambda) = \frac{\lambda^\alpha}{P(-\ln \lambda)}, \qquad \lambda > 0, \tag{5.32}$$

where

$$\alpha = \alpha(c) = \ln_c \kappa_f(c) \tag{5.33}$$

and where $(P(u), u \in \mathbf{R})$ is a positive periodic function with $P(0) = 1$ and

$$\{nu_0;\, n \in \mathbf{Z}\} \subset S_{\text{per}}(P) \subset \mathbb{H}_r(f), \tag{5.34}$$

where

$$u_0 = \ln c \neq 0.$$

Moreover, $(P(\ln \lambda), \lambda > 0)$ is upper $*$-invariant and $(1/P(-\ln \lambda), \lambda > 0)$ is lower $*$-invariant, that is,

$$\limsup_{x\to\infty} \frac{P(u+x)}{P(x)} = P(u) \qquad \text{and} \qquad \liminf_{x\to\infty} \frac{P(u+x)}{P(x)} = \frac{1}{P(-u)} \tag{5.35}$$

for all $u \in \mathbf{R}$.

Proof of Theorem 5.34 The function

$$h(u) = \ln(f^*(e^u)), \qquad u \in \mathbf{R},$$

is well-defined, since $f \in \mathcal{WORV}$ and thus the argument of the logarithm is positive. Now, Lemma 5.28 implies that

$$h(u + u_0) = h(u) + h(u_0) = h(u) + h_0, \qquad u \in \mathbf{R}, \tag{5.36}$$

where $h_0 = h(u_0) = \ln \kappa_f(c)$. Note that $h(0) = 0$, since $f^*(1) = 1$. Put

$$p(u) = h(u) - \frac{h_0}{u_0} u, \qquad u \in \mathbf{R}. \tag{5.37}$$

Relation (5.36) yields

$$p(u + u_0) = h(u + u_0) - \frac{h_0}{u_0}(u + u_0) = h(u) - \frac{h_0}{u_0} u = p(u)$$

for all $u \in \mathbf{R}$. This means that $(p(u), u \in \mathbf{R})$ is a periodic function and the set of its periods contains the set $\{nu_0; n \in \mathbf{Z}\}$. Moreover,

$$h(u) = \frac{h_0}{u_0} u + p(u), \qquad u \in \mathbf{R}.$$

Also,

$$f^*(\lambda) = \exp\{h(\ln \lambda)\}, \qquad \lambda > 0,$$

whence

$$f^*(\lambda) = \lambda^\alpha P(\ln \lambda), \qquad \lambda > 0,$$

where

$$\alpha = \frac{h_0}{u_0} = \frac{\ln \kappa_f(c)}{\ln c} = \ln_c \kappa_f(c),$$

and

$$P(u) = \exp\{p(u)\}, \qquad u \in \mathbf{R}.$$

This proves (5.31). Relation (5.32) follows from (5.31) and Lemma 3.1.

Since the function in (5.37) is periodic, $P = (P(u), u \in \mathbf{R})$ is also periodic with the same period. Thus $\{nu_0; n \in \mathbf{Z}\} \subset S_{\mathrm{per}}(P)$. Moreover, $P(0) = 1$, since $p(0) = 0$. Now (5.31) and (5.32) imply that $S_{\mathrm{per}}(P) \subset \mathbb{H}_r(f)$. Note also that the function P is positive.

Next we show that $(P(\ln \lambda), \lambda > 0)$ is upper $*$-invariant. Indeed, we conclude from (5.31) that

$$f^*(\lambda) = \lambda^\alpha P_l(\lambda), \qquad \lambda > 0,$$

where $P_l(\lambda) = P(\ln \lambda), \lambda > 0$. Then

$$f^{**}(\lambda) = \lambda^\alpha P_l^*(\lambda), \qquad \lambda > 0,$$

and we obtain from Proposition 5.27 that $f^{**} = f^*$. Thus

$$\lambda^\alpha P_l^*(\lambda) = \lambda^\alpha P_l(\lambda)$$

for all $\lambda > 0$.

Therefore, $P_l^* = P_l$, that is, $(P(\ln\lambda), \lambda > 0)$ is an upper $*$-invariant function. In turn, Corollary 5.32 shows that $(1/P(-\ln\lambda), \lambda > 0)$ is a lower $*$-invariant function. Relation (5.35) follows from Remark 5.33, which completes the proof. □

5.5.1 Some Corollaries and Remarks

Remark 5.35 Assume $f \in \mathbb{F}_+$ and let $(P(u), u \in \mathbf{R})$ be a positive, finite, and periodic function with $P(0) = 1$. If, for some $\alpha \in \mathbf{R}$, (5.31) holds, then f is a WORV-function with nondegenerate set of regular points $\mathbb{G}_r(f)$ and $S_{\text{per}}(P) \subset \mathbb{H}_r(f)$. Moreover,

$$\kappa_f(\lambda) = \lambda^\alpha, \qquad \lambda \in \exp(S_{\text{per}}(P)) = \{\lambda > 0 : \ln\lambda \in S_{\text{per}}(P)\}.$$

Remark 5.36 If f is a WORV-function with nondegenerate set of regular points, then Theorem 5.34 implies that

$$\{u : P(u)P(-u) = 1\} = \mathbb{H}_r(f).$$

Thus f is a WRV-function and (5.6) holds if and only if $P(u)P(-u) = 1$ for all $u \in \mathbf{R}$.

Corollary 5.37 *Let f be a WORV-function with nondegenerate set of regular points and let $(P(u), u \in \mathbf{R})$ be as in Theorem 5.34. Then:*

(a) $P(u)P(-u) \ge 1$ *for all* $u \in \mathbf{R}$*;*
(b) $\inf_{u\in\mathbf{R}} P(u) \sup_{u\in\mathbf{R}} P(u) \ge 1$, *where we set* $0 \cdot \infty = \infty$.

Proof of Corollary 5.37 Since $f^*(\lambda) \ge f_*(\lambda)$ holds for all $\lambda > 0$, relations (5.31) and (5.32) imply that

$$P(\ln\lambda) \ge \frac{1}{P(-\ln\lambda)}$$

for all $\lambda > 0$. Since P is positive, this implies that $P(u)P(-u) \ge 1$ for all $u \in \mathbf{R}$. Hence statement (a) is proved. Statement (b) follows immediately from (a). □

Remark 5.38 Since the representations (5.31) and (5.32) depend on c, they are not unique, which is related to the (possible) unboundedness of the function P. This means that the class $\mathcal{WORV}$ is too wide to provide unique representations. An appropriate restriction of $\mathcal{WORV}$ will be discussed in the next subsection.

Remark 5.38 explains why the representations (5.31) and (5.32) are not unique. As shown in the following result, the representation may be unique if the set of regular points has a rather simple structure.

Corollary 5.39 *Let f be a WORV-function with set of regular points $\mathbb{G}_r(f) = \{d^n, n \in \mathbf{Z}\}$, where $d > 1$ or, equivalently, $d \in (0, 1)$. Then there are a real number $\rho \in \mathbf{R}$ and a unique positive, periodic function $(P(u), u \in \mathbf{R})$ such that $\rho = \ln_c \kappa_f(c)$ for all $c \in \mathbb{G}_r(f) \setminus \{1\}$, $P(0) = 1$, and*

$$S_{\text{per}}(P) = \mathbb{H}_r(f) = \{n|\ln d|; n \in \mathbf{Z}\}. \tag{5.38}$$

Moreover, relations (5.31) and (5.32) hold in this case.

Proof of Corollary 5.39 Relation (5.3) implies that

$$\ln_c \kappa_f(c) = \ln_{d^n} \kappa_f(d^n) = \ln_{d^n} (\kappa_f(d))^n = \ln_d (\kappa_f(d))$$

for all $c = d^n$. Thus $\rho = \ln_d (\kappa_f(d))$, whence the uniqueness of the representations (5.31) and (5.32) follows. Relation (5.38) follows from (5.34) on observing that $u_0 = |\ln d|$. □

5.6 Strong Factorization Representations for Limit Functions

By Theorem 5.34, the limit functions f^* and f_* of a WORV-function f with nondegenerate set of regular points can be represented as a product of a power function and a positive periodic function with logarithmic argument. However, in general, these representations cannot be unique (see Remark 5.38). As shown in the following result, the corresponding representations are unique for WOURV-functions with nondegenerate set of regular points. Moreover, we obtain the form of the periodic component in the representations of the limit functions. Recall that (see Definition 3.12 and Remark 3.13), if $f \in \mathbb{F}_+$ belongs to $\mathcal{WOURV}$, then

$$0 < \inf_{c \in [a,b]} f_*(c) \le \sup_{c \in [a,b]} f^*(c) < \infty$$

for any interval $[a, b] \subset (0, \infty)$.

Theorem 5.40 *Let f be a WOURV-function with nondegenerate set of regular points $\mathbb{G}_r(f)$ and index function κ_f. Then,*

(i) there exists a unique real number $\rho \in \mathbf{R}$ such that

$$\rho = \ln_c \kappa_f(c), \qquad c \in \mathbb{G}_r(f) \setminus \{1\};$$

(ii) *if* $1 \in \{\kappa_f(c), c \in \mathbb{G}_r(f) \setminus \{1\}\}$, *then* $\rho = 0$;

(iii) *the index function* κ_f *is such that*

$$\kappa_f(\lambda) = \lambda^\rho, \qquad \lambda \in \mathbb{G}_r(f); \tag{5.39}$$

(iv) *the limit functions* f^* *and* f_* *are such that*

$$f^*(\lambda) = \lambda^\rho \mathscr{P}(\ln \lambda), \qquad \lambda > 0, \tag{5.40}$$

and

$$f_*(\lambda) = \frac{\lambda^\rho}{\mathscr{P}(-\ln \lambda)}, \qquad \lambda > 0, \tag{5.41}$$

where $(\mathscr{P}(u), u \in \mathbf{R})$ *is a positive periodic function with* $\mathscr{P}(0) = 1$,

$$0 < m = \inf_{u \in \mathbf{R}} \mathscr{P}(u) \leq \sup_{u \in \mathbf{R}} \mathscr{P}(u) = M < \infty, \tag{5.42}$$

and

$$S_{\text{per}}(\mathscr{P}) = \mathbb{H}_r(f),$$

where $S_{\text{per}}(\mathscr{P})$ *is the set of periods* $\mathscr{P}$;

(v) $(\mathscr{P}(\ln \lambda), \lambda > 0)$ *is an upper* $*$*-invariant function and* $(1/\mathscr{P}(-\ln \lambda), \lambda > 0)$ *is a lower* $*$*-invariant function, that is,*

$$\limsup_{x \to \infty} \frac{\mathscr{P}(u+x)}{\mathscr{P}(x)} = \mathscr{P}(u) \qquad \text{and} \qquad \liminf_{x \to \infty} \frac{\mathscr{P}(u+x)}{\mathscr{P}(x)} = \frac{1}{\mathscr{P}(-u)}$$

for all $u \in \mathbf{R}$;

(vi) $m = 1$ *in relation (5.42), that is,*

$$1 = \min_{u \in \mathbf{R}} \mathscr{P}(u) \leq \sup_{u \in \mathbf{R}} \mathscr{P}(u) < \infty;$$

(vii) $\mathfrak{p} = \ln \mathscr{P}$ *is a subadditive function, that is,*

$$\mathfrak{p}(u+x) \leq \mathfrak{p}(u) + \mathfrak{p}(x)$$

or, equivalently,

$$\mathscr{P}(u+x) \leq \mathscr{P}(u)\mathscr{P}(x)$$

for all $u, x \in \mathbf{R}$;

(viii) *given a function* f, *the representations (5.39), (5.40) and (5.41) are unique.*

Proof of Theorem 5.40 (i) First we consider the case where the group $(\mathbb{G}_r(f), \times)$ contains a unique generating number. By the assumption of the theorem, the group is nondegenerate and thus there exists a number $d > 1$ such that $\mathbb{G}_r(f) = \{d^n, n \in \mathbf{Z}\}$. Relation (5.3) implies that

$$\ln_c \kappa_f(c) = \ln_{d^n} \kappa_f(d^n) = \ln_{d^n} (\kappa_f(d))^n = \ln_d (\kappa_f(d))$$

for all $c = d^n$. Thus $\rho = \ln_d (\kappa_f(d))$ and statement (i) is proved in the case under consideration.

Now assume that the group $(\mathbb{G}_r(f), \times)$ has more than one generating number. Then, Theorem 5.34 implies that

$$f^*(\lambda) = \lambda^{\alpha_c} P_c(\ln \lambda), \qquad \lambda > 0,$$

for all $c \in \mathbb{G}_r(f) \setminus \{1\}$, where

$$\alpha_c = \ln_c \kappa_f(c)$$

and $(P_c(u), u \in \mathbf{R})$ is a positive periodic function with period $u_c = |\ln c|$ such that $P_c(0) = 1$. Thus

$$P_c(u) = f^*(e^u) \exp\{-\alpha_c u\}, \qquad u \in \mathbf{R}, \tag{5.43}$$

for all $c \in \mathbb{G}_r(f) \setminus \{1\}$. By the assumption of the theorem, f is a WOURV-function, and thus (3.8) holds. As P_c is periodic,

$$0 < m_c = \inf_{0 \le u \le u_c} P_c(u) = \inf_{-\infty < u < \infty} P_c(u) \tag{5.44}$$

$$\le \sup_{-\infty < u < \infty} P_c(u) = \sup_{0 \le u \le u_c} P_c(u) = M_c < \infty.$$

Since the group $(\mathbb{G}_r(f), \times)$ has two different generating numbers, there are numbers $c_1 \neq c_2 \in \mathbb{G}_r(f) \setminus \{1\}$ such that

$$\alpha_1 = \ln_{c_1} \kappa_f(c_1) \neq \alpha_2 = \ln_{c_2} \kappa_f(c_2).$$

Then (5.43) and (5.44) imply that

$$0 < \min\left\{\frac{m_{c_1}}{M_{c_2}}, \frac{m_{c_2}}{M_{c_1}}\right\} \le \liminf_{u\to\infty} \min\left\{\frac{P_{c_1}(u)}{P_{c_2}(u)}, \frac{P_{c_2}(u)}{P_{c_1}(u)}\right\}$$

$$= \liminf_{u\to\infty} \exp\{-|\alpha_1 - \alpha_2|u\} = 0.$$

This contradiction proves that all numbers α_c, $c \in \mathbb{G}_r(f) \setminus \{1\}$, are the same and thus $\rho = \ln_c \kappa_f(c)$ does not depend on $c \in \mathbb{G}_r(f) \setminus \{1\}$, which completes the proof of statement (i).

(ii) If $1 \in \{\kappa_f(c), c \in \mathbb{G}_r(f) \setminus \{1\}\}$, there exists a $c_1 \in \mathbb{G}_r(f) \setminus \{1\}$ such that $\kappa_f(c_1) = 1$. Taking statement (i) into account, we see that $\rho = 0$. This proves statement (ii).

(iii) Relation (iii) follows from (i).

(iv) Put

$$\mathscr{P}(u) = f^*(e^u) \exp\{-\rho u\}, \qquad u \in \mathbf{R}.$$

It is clear that

$$f^*(\lambda) = \lambda^\rho \mathscr{P}(\ln \lambda)$$

and

$$f_*(\lambda) = \frac{1}{f^*(1/\lambda)} = \frac{1}{(1/\lambda)^\rho \mathscr{P}(\ln (1/\lambda))} = \frac{\lambda^\rho}{\mathscr{P}(-\ln \lambda)}$$

for all $\lambda > 0$. Hence (5.40) and (5.41) are proved.

Statement (i) together with (5.43) implies that

$$\mathscr{P}(u) = P_c(u), \qquad u \in \mathbf{R}$$

for all $c \in \mathbb{G}_r(f) \setminus \{1\}$. Thus $(\mathscr{P}(u), u \in \mathbf{R})$ is a positive periodic function with $\mathscr{P}(0) = 1$, and, by Theorem 5.34,

$$\mathbb{H}_r(f) \subset \bigcup_{c \in \mathbb{G}_r(f) \setminus \{1\}} S_{\text{per}}(P_c) = S_{\text{per}}(\mathscr{P}).$$

Relations (5.40) and (5.41) imply that $S_{\text{per}}(\mathscr{P}) \subset \mathbb{H}_r(f)$. Therefore

$$\mathbb{H}_r(f) = S_{\text{per}}(\mathscr{P}).$$

Moreover, according to (5.44),

$$0 < \sup_{c \in \mathbb{G}_r(f) \setminus \{1\}} m_c \le \mathscr{P}(u) \le \inf_{c \in \mathbb{G}_r(f) \setminus \{1\}} M_c < \infty, \qquad u \in \mathbf{R}.$$

This proves (iv).

(v) Statement (v) follows from Theorem 5.34.

(vi) First we note that $0 < m \le \mathscr{P}(0) = 1$. Let $0 < m < 1$. This means that there exists a real number u' such that $\mathscr{P}(u') \in (0, 1)$. The number u' can be chosen arbitrarily, since the function $\mathscr{P}$ is positive. By statement (v), there exists a number $x' > 0$ such that

$$\frac{\mathscr{P}(x)}{\mathscr{P}(u'+x)} > \frac{1}{\gamma}$$

for all $x \geq x'$ and all numbers $\gamma \in (\mathscr{P}(u'), 1)$. This implies that

$$\mathscr{P}(x') > \frac{\mathscr{P}(u'+x')}{\gamma} > \frac{\mathscr{P}(2u'+x')}{\gamma^2} > \ldots > \frac{\mathscr{P}(nu'+x')}{\gamma^n} \geq \frac{m}{\gamma^n}$$

for all $n \in \mathbf{N}$. Using (5.42) we get a contradiction, i.e.

$$\infty = \sup_{n \in \mathbf{N}} \frac{m}{\gamma^n} \leq \mathscr{P}(x') < \infty,$$

whence we conclude that $m = 1$.

(vii) Assume that

$$\mathscr{P}(u'+x') = \gamma' \mathscr{P}(u')\mathscr{P}(x')$$

for some $u', x' \in \mathbf{R}$ and $\gamma' > 1$. The latter equality implies that

$$\frac{\mathscr{P}(u'+x'+nu_0)}{\mathscr{P}(x'+nu_0)} = \gamma' \mathscr{P}(u')$$

for all $n \in \mathbf{N}$, where $u_0 > 0$ is a fixed period of the function $\mathscr{P}$. According to (v) and (vi), we conclude that

$$1 = \limsup_{x \to \infty} \frac{\mathscr{P}(u'+x)}{\mathscr{P}(x)\mathscr{P}(u')} \geq \limsup_{n \to \infty} \frac{\mathscr{P}(u'+x'+nu_0)}{\mathscr{P}(x'+nu_0)\mathscr{P}(u')} = \gamma' > 1.$$

This contradiction proves (vii).

(viii) The uniqueness of the representations (5.39), (5.40) and (5.41) follows from statement (i).

So, the proof of Theorem 5.40 is complete. □

5.6.1 *Some Corollaries and Remarks*

Theorem 5.40 is one of the main results of Chap. 5. Below are some corollaries, which follow immediately from this result.

Definition 5.41 The index ρ in statement (iii) of Theorem 5.40 is called the *index of the limit functions* f^* and f_*, while the function $\mathscr{P}$ in statement (iv) is called the *periodic component of the limit functions.*

By Theorem 5.40, the limit functions of a WOURV-function with nondegenerate set of regular points are uniquely determined by its index and periodic component.

Remark 5.42 Let $f \in \mathbb{F}_+$, $\rho_1 \in \mathbf{R}$, and $(P_1(u), u \in \mathbf{R})$ be a positive periodic function such that $P_1(0) = 1$ and

$$0 < m \le P_1(u) \le M < \infty, \qquad u \in \mathbf{R}.$$

If, for all $\lambda > 0$,

$$f^*(\lambda) = \lambda^{\rho_1} P_1(\ln \lambda)$$

or, equivalently,

$$f_*(\lambda) = \frac{\lambda^{\rho_1}}{P_1(-\ln \lambda)},$$

then f is a WOURV-function with a nondegenerate set of regular points for which the index of the limit functions coincides with ρ_1 and the periodic component of these functions coincides with P_1.

By statement (vii) of Theorem 5.40, the function $\mathfrak{p} = \ln \mathscr{P}$ is subadditive, and thus

$$|\mathfrak{p}(u+x) - \mathfrak{p}(u)| \le \max\{\mathfrak{p}(x), \mathfrak{p}(-x)\}$$

for all $u, x \in \mathbf{R}$. Using this result we establish a relationship between the local and global continuity of the limit functions and their periodic component.

Corollary 5.43 *Let f be a WOURV-function with nondegenerate set of regular points and periodic component $\mathscr{P}$ of the limit functions. Then the following statements are equivalent:*

(a) f^* (f_*) *is continuous on* $\mathbf{R}_+$;
(b) $\mathscr{P}$ *is uniformly continuous on* $\mathbf{R}$;
(c) $\mathscr{P}$ *is continuous at the point* 0;
(d) f^* (f_*) *is continuous at the point* 1, *that is,* f *is a WPRV-function (see Definition 3.16 and Proposition 3.21).*

The following result contains conditions under which a WOURV-function with nondegenerate set of regular points is a WRV-function.

Corollary 5.44 *Let f be a WOURV-function with nondegenerate set of regular points $\mathbb{G}_r(f)$ and periodic component $\mathscr{P}$ of the limit functions. Then the following nine statements are equivalent:*

(a) *relation (5.6) holds, that is, $f \in \mathcal{WRV}$;*
(b) $\mathbb{G}_r(f) = \mathbf{R}_+$;
(c) $S_{\text{per}}(\mathscr{P}) = \mathbf{R}$;
(d) $\mathscr{P}(u) = 1$ *for all $u \in \mathbf{R}$;*
(e) $\mathscr{P}(u)\mathscr{P}(-u) = 1$ *for all $u \in \mathbf{R}$;*
(f) *the set $\mathbb{G}_r(f)$ is everywhere dense in $\mathbf{R}_+$ and the function f^* is continuous on $\mathbf{R}_+$;*
(g) *the set $\mathbb{G}_r(f)$ is everywhere dense in $\mathbf{R}_+$ and the function f^* is continuous at least at one point $\lambda \in \mathbf{R}_+$;*
(h) *the set of periods $S_{\text{per}}(\mathscr{P})$ is everywhere dense in $\mathbf{R}$ and the function $\mathscr{P}$ is continuous on $\mathbf{R}$;*
(i) *the set of periods $S_{\text{per}}(\mathscr{P})$ is everywhere dense in $\mathbf{R}$ and the function $\mathscr{P}$ is continuous at least at one point $u \in \mathbf{R}$.*

Corollary 5.44 follows from Theorem 5.40 and Lemma 5.15.

Remark 5.45 Let f be a WOURV-function with nondegenerate set of regular points $\mathbb{G}_r(f)$. If the set $\mathbb{G}_r(f)$ is everywhere dense in $\mathbf{R}_+$, then, by Corollary 5.44, either $f \in \mathcal{WRV}$ or the limit function f^* is discontinuous at every point of $\mathbf{R}$ (see Example 5.10).

In turn, by Kronecker's theorem (see Proposition 5.17), either the set $\mathbb{G}_r(f)$ is everywhere dense in $\mathbf{R}_+$ or $\mathbb{G}_r(f) = \{\lambda_0^n;\, n \in \mathbf{Z}\}$ for some $\lambda_0 > 1$. Therefore, if the function f^* (or $\mathscr{P}$) is continuous at least at one point, then either $f \in \mathcal{WRV}$ or

$$\mathbb{H}_r(f) = S_{\text{per}}(\mathscr{P}) = \{nu_0;\, n \in \mathbf{Z}\} \tag{5.45}$$

for some $u_0 > 0$ (see Examples 5.8 and 5.9 and Corollary 5.23).

Remark 5.46 Let f be a WORV-function with nondegenerate set of regular points. Assume that its limit function f^* is continuous at the point 1. This is equivalent to the condition that $f \in \mathcal{WPRV}$ possesses a nondegenerate set of regular points (see Definition 3.16 and Proposition 3.21). Using Example 3.14, Corollary 5.43, and Remark 5.45, we conclude that either $f \in \mathcal{WRV}$ or the limit function f^* is continuous on $\mathbf{R}_+$ and (5.45) holds for some $u_0 > 0$ (see Example 5.8). Moreover, Example 5.9 shows that the assumption of f^* to be continuous at the point 1 in statement (f) of Corollary 5.43 cannot be substituted by the condition that f^* is continuous at least at one point $\lambda \in \mathbf{R}_+$. Similarly, the condition that the function $\mathscr{P}$ is continuous at the point 0 in statement (e) of Corollary 5.43 cannot be substituted by the condition that $\mathscr{P}$ is continuous at least at one point $\lambda \in \mathbf{R}$.

Now we establish a method to evaluate the index and periodic components via the limit functions.

Corollary 5.47 *Let f be a WOURV-function with nondegenerate set of regular points $\mathbb{G}_r(f)$ and index ρ. Then*

$$\lim_{\lambda\to 0+} \frac{\ln f^*(\lambda)}{\ln \lambda} = \lim_{\lambda\to\infty} \frac{\ln f^*(\lambda)}{\ln \lambda} = \rho$$

and

$$\lim_{\lambda\to 0+} \frac{\ln f_*(\lambda)}{\ln \lambda} = \lim_{\lambda\to\infty} \frac{\ln f_*(\lambda)}{\ln \lambda} = \rho.$$

Corollary 5.47 follows from Theorem 5.40.

5.6.2 *Factorization Representations of Limit Functions for ORV-Functions with Nondegenerate Sets of Regular Points*

Since every ORV-function belongs to $\mathcal{WOURV}$ (see Remark 3.15), we obtain the following result for the class $\mathcal{ORV}$.

Theorem 5.48 *All statements of Theorem 5.40 as well as of its Corollaries 5.43–5.47 and Remarks 5.42–5.46 hold for any ORV-function f with nondegenerate set of regular points.*

5.7 Characterization Theorems

In this section, we show how Theorem 5.40 together with the Steinhaus theorem (see Proposition 5.5) can be used to derive well-known characterization theorems as well as some of their generalizations for RV- and WRV-functions.

If $f \in \mathcal{WOURV}$ is a function with nondegenerate set of regular points $\mathbb{G}_r(f)$, then, by Theorem 5.40, statement (iii), the index function κ_f is given by

$$\kappa_f(\lambda) = \lambda^\rho, \qquad \lambda \in \mathbb{G}_r(f), \tag{5.46}$$

where $\rho = \rho(f)$ is a real number called the index of the limit functions (see Definition 5.41). Relation (5.46) is a starting point to establish the results of this section.

We start with the following general assertion.

Proposition 5.49 *Let $f \in \mathbb{F}_+$. If the set of regular points $\mathbb{G}_r(f)$ contains a subset of positive Lebesgue measure, then $\mathbb{G}_r(f) = \mathbf{R}_+$, that is, $f \in \mathcal{WRV}$. If, in addition,*

$f \in \mathcal{WOURV}$, then relation (5.6) holds for some real number ρ. Moreover,

$$f(t) = t^{\rho} w(t), \qquad t > 0,$$

where $(w(t), t > 0)$ is a WSV-function, that is, $\kappa_w(\lambda) = 1$, $\lambda > 0$.

Proof of Proposition 5.49 Since the set $\mathbb{G}_r(f)$ contains a subset of positive Lebesgue measure, Proposition 5.5 guarantees that $\mathbb{G}_r(f) = \mathbf{R}_+$. Thus (5.46) yields $\kappa_f(\lambda) = \lambda^{\rho}$ for all $\lambda > 0$, that is, relation (5.6) holds.

Put

$$w(t) = f(t)t^{-\rho}, \qquad t > 0.$$

It is clear that

$$\begin{aligned} \kappa_w(\lambda) &= \lim_{t\to\infty} \frac{w(\lambda t)}{w(t)} = \lim_{t\to\infty} \frac{f(\lambda t)t^{\rho}}{(\lambda t)^{\rho} f(t)} \\ &= \lambda^{-\rho} \lim_{t\to\infty} \frac{f(\lambda t)}{f(t)} = \lambda^{-\rho}\kappa_f(\lambda) = \lambda^{-\rho} \cdot \lambda^{\rho} = 1 \end{aligned}$$

for all $\lambda > 0$, that is, $w \in \mathcal{WSV}$. Therefore Proposition 5.49 is proved. □

Now, we make use of Proposition 5.49 to prove the characterization theorems of Sect. 5.2.

Proof of the Characterization Theorem 5.6 for RV-functions. Let $f \in \mathbb{F}_+$ be measurable and let Λ be a measurable set such that $\Lambda \subset \mathbb{G}_r(f)$ and $\text{meas}(\Lambda) > 0$. Then Proposition 5.49 implies that $f \in \mathcal{WRV}$ and, moreover, $f \in \mathcal{RV}$, since f is measurable. Since $\mathcal{RV} \subset \mathcal{ORV}$, Remark 3.15 shows that $f \in \mathcal{OURV}$ and f is a function with nondegenerate set of regular points. This allows us to apply Proposition 5.49 once again. □

Proof of the Characterization Theorem 5.7 for WRV-functions. Let $f \in \mathbb{F}_+$ be such that both f and $1/f$ are bounded on any finite interval that is sufficiently far away from the origin. Also let Λ be a measurable set such that $\Lambda \subset \mathbb{G}_r(f)$ and $\text{meas}(\Lambda) > 0$. Then Proposition 5.49 implies that $f \in \mathcal{WRV}$ and thus $f \in \mathcal{WORV}$. Since both functions f and $1/f$ are bounded on finite intervals that are sufficiently far away from the origin, Lemma A.4 in Seneta [324] implies that there exists an interval $[a, b]$, $1 \le a < b < \infty$, such that $\sup_{c\in[a,b]} f^*(c) < \infty$. Recalling Remark 3.13, we conclude that $f \in \mathcal{WOURV}$ is a function with nondegenerate set of regular points and thus Proposition 5.49 can be applied once again. □

The following corollary coincides with another well-known result (see, e.g., Bingham et al. [41], Theorem 1.4.3).

Corollary 5.50 *Let $f \in \mathbb{F}_+$ and*

$$\limsup_{\lambda\downarrow 1} f^*(\lambda) \le 1 \qquad \text{or} \qquad \limsup_{\lambda\uparrow 1} f^*(\lambda) \le 1. \tag{5.47}$$

Then the following statements are equivalent:

(i) *relation (5.6) holds for some real number* ρ;
(ii) $f \in \mathcal{WRV}$;
(iii) $\mathbb{G}_r(f)$ *contains a subset of positive Lebesgue measure;*
(iv) $\mathbb{G}_r(f)$ *is everywhere dense in* $\mathbf{R}_+$;
(v) *there are two positive numbers* $\lambda_1, \lambda_2 \in \mathbb{G}_r(f) \setminus \{1\}$ *such that* $\ln\lambda_1/\ln\lambda_2$ *is an irrational number.*

Moreover, (i) implies (5.47).

The proof of Corollary 5.50 is based on the following auxiliary result.

Lemma 5.51 *Let* f *be a WOURV-function with nondegenerate set of regular points* $\mathbb{G}_r(f)$ *and with a periodic component* $(\mathscr{P}(u), u \in \mathbf{R})$ *of the limit functions. If there exist a number* $M \geq 1$, *an interval* $[a, b] \subset \mathbf{R}$, *and a number* $u_0 \in \mathbb{H}_r(f)$ *such that* $0 < u_0 < b - a$ *and* $\sup_{a \leq u \leq b} \mathscr{P}(u) \leq M$, *then*

$$1 = \min_{-\infty<u<\infty} \mathscr{P}(u) \leq \sup_{-\infty<u<\infty} \mathscr{P}(u) \leq M.$$

Proof of Lemma 5.51 Lemma 5.51 follows from statement (vi) of Theorem 5.40, since the function $\mathscr{P}$ is periodic with $S_{\text{per}}(\mathscr{P}) = \mathbb{H}_r(f)$. □

Proof of Corollary 5.50 The implications

$$\text{(i)} \Longrightarrow \text{(ii)} \Longrightarrow \text{(iii)} \Longrightarrow \text{(iv)} \Longrightarrow \text{(v)}$$

follow directly from Proposition 5.5.

If statement (v) holds, then Kronecker's theorem (see Proposition 5.17) guarantees that the set $\mathbb{G}_r(f)$ is everywhere dense in $\mathbf{R}_+$. Thus (v) $\Longrightarrow$ (iv), which proves that statements (v) and (iv) are equivalent.

Assume that (iv) is satisfied. Then the set $\mathbb{G}_r(f)$ is nondegenerate. Moreover, condition (5.47) and Remark 3.13 imply that $f \in \mathcal{WOURV}$. Now, Theorem 5.40 yields relations (5.40) and (5.41). Next we deduce from (5.47) that, given an arbitrary $\varepsilon > 0$, there exists a number $a_\varepsilon > 1$ such that

$$\sup_{\lambda\in[1,a_\varepsilon]} f^*(\lambda) \leq 1 + \varepsilon$$

or

$$\sup_{\lambda\in[1/a_\varepsilon,1]} f^*(\lambda) \leq 1 + \varepsilon.$$

Hence

$$\sup_{0\leq u\leq \ln b} \mathscr{P}(u) \leq b^{|\rho|}(1 + \varepsilon)$$

or

$$\sup_{-\ln b \le u \le 0} \mathscr{P}(u) \le b^{|\rho|}(1+\varepsilon)$$

for all $b \in (1, a_\varepsilon]$. Since the set $\mathbb{G}_r(f)$ is everywhere dense in $\mathbf{R}_+$, the set $\mathbb{H}_r(f)$ is everywhere dense in $\mathbf{R}$ and Lemma 5.51 implies that

$$1 = \min_{-\infty<u<\infty} \mathscr{P}(u) \le \sup_{-\infty<u<\infty} \mathscr{P}(u) \le b^{|\rho|}(1+\varepsilon)$$

for all $b \in (1, a_\varepsilon]$. Passing to the limit as $b \to 1$,

$$1 = \min_{-\infty<u<\infty} \mathscr{P}(u) \le \sup_{-\infty<u<\infty} \mathscr{P}(u) \le 1+\varepsilon$$

for all $\varepsilon > 0$. Thus $\mathscr{P}(u) \equiv 1$ and, by (5.40) and (5.41), we obtain

$$f^*(\lambda) = f_*(\lambda) = \lambda^\rho$$

for all $\lambda > 0$. This means that (i) holds. Therefore (iv) $\Rightarrow$ (i) and this means that statements (i), (ii), (iii), (iv) and (v) are equivalent. In order to complete the proof of Corollary 5.50, just observe that the implication (i) $\Rightarrow$ (5.47) is immediate from (5.6). □

Remark 5.52 Corollary 5.50 and Remark 5.45 show that either $\mathbb{G}_r(f) = \mathbf{R}_+$, or $\mathbb{G}_r(f) = \{\lambda_0^n;\, n \in \mathbf{Z}\}$ for some $\lambda_0 > 1$, provided condition (5.47).

Remark 5.53 Example 5.10 shows that condition (5.47) cannot be substituted, in general, by the following one:

$$\limsup_{\lambda \downarrow 1} f^*(\lambda) < \infty \qquad \text{or} \qquad \limsup_{\lambda \uparrow 1} f^*(\lambda) < \infty.$$

Moreover, Remark 5.45 and Example 5.8 show that assumption (iv), that is, the condition about the everywhere denseness of the set $\mathbb{G}_r(f)$ in $\mathbf{R}_+$ cannot, in general, be weakened for the implication (iv) $\Longrightarrow$ (i).

The proof of the following result is similar to that of Corollary 5.50.

Corollary 5.54 *Let $f \in \mathbb{F}_+$ and let there exist a $\lambda_0 >$ such that*

$$\lim_{\lambda \to \lambda_0} f^*(\lambda) = 1. \tag{5.48}$$

Then the following five statements are equivalent:

(i) there exists a real number ρ such that (5.6) holds;
(ii) $f \in \mathcal{WRV}$;
(iii) $\mathbb{G}_r(f)$ contains a subset of positive Lebesgue measure;

(iv) *the set* $\mathbb{G}_r(f)$ *is everywhere dense in* $\mathbf{R}_+$;
(v) *there are two positive numbers* $\lambda_1, \lambda_2 \in \mathbb{G}_r(f) \setminus \{1\}$ *such that* $\ln \lambda_1 / \ln \lambda_2$ *is an irrational number.*

Moreover, if (iv) and (5.48) hold with $\lambda_0 \neq 1$, *then* $\rho = 0$, *that is,* $f \in \mathcal{WSV}$.

Remark 5.55 Condition (5.48) is also a necessary condition for statement (i) in Corollary 5.54.

5.8 Factorization Representations of Functions

In Sects. 5.5 and 5.6, we considered the factorization representations for the limit functions in the subclasses $\mathcal{WORV}$ and $\mathcal{WOURV}$ of functions with nondegenerate sets of regular points. In this section, we study the factorization representations for the functions themselves.

The following proposition shows that an arbitrary WOURV-function with nondegenerate set of regular points can be represented as a product of a power function and a member of $\mathcal{WOSV}$. Recall that (see Definition 3.10) a function $f \in \mathbb{F}_+$ is called *O-weakly slowly varying* (WOSV) if

$$\sup_{c>0} f^*(c) < \infty.$$

A measurable WOSV-function $f \in \mathbb{F}_+$ is called *O-slowly varying* (OSV).

Proposition 5.56 *Let* f *be a WOURV-function with nondegenerate set of regular points and with an index* ρ *and periodic component* $\mathcal{P}$ *of its limit functions. Then*

$$f(t) = t^\rho s(t), \qquad t > 0, \tag{5.49}$$

where $(s(t), t > 0)$ *is a WOSV-function such that*

$$s^*(\lambda) = \mathcal{P}(\ln \lambda), \qquad \lambda > 0. \tag{5.50}$$

Proof of Proposition 5.56 Put

$$s(t) = \frac{f(t)}{t^\rho}, \qquad t > 0. \tag{5.51}$$

The assumptions of Proposition 5.56 and Theorem 5.40 imply relation (5.40) which, in turn, yields

$$s^*(\lambda) = \lambda^{-\rho} f^*(\lambda) = \mathcal{P}(\ln \lambda)$$

for all $\lambda > 0$. According to (5.42)

$$\sup_{\lambda>0} s^*(\lambda) = \sup_{u\in\mathbf{R}} \mathscr{P}(u) < \infty$$

and hence $(s(t), t > 0)$ is a WOSV-function. □

Corollary 5.57 *Let f be an ORV-function with nondegenerate set of regular points and with an index ρ and periodic component $\mathscr{P}$ of its limit functions. Then there exists an OSV-function $(s(t), t > 0)$ such that relations (5.49) and (5.50) hold.*

Proof of Corollary 5.57 Remark 3.15 shows that f is an OURV-function with nondegenerate set of regular points. According to Proposition 5.56, relations (5.49) and (5.50) hold with the WOSV-function $(s(t), t > 0)$ defined in (5.51). Since f is a measurable function, (5.51) implies that the function $(s(t), t > 0)$ is also measurable and hence $s \in \mathcal{OSV}$. □

Relation (5.49) justifies the following definition.

Definition 5.58 The index ρ in relation (5.49) is called the *index of the function f*.

Remark 5.59 Proposition 5.56 shows that the index of a WOURV-function with nondegenerate set of regular points coincides with the index of its limit functions (see Definition 5.41).

The possible form of an OSV-function is described in the following proposition.

Proposition 5.60 *A measurable function $(\psi(t), t > 0)$ belonging to $\mathbb{F}_+$ is an OSV-function if and only if*

$$\psi(t) = \ell(t)\theta(t), \qquad t > 0,$$

where $\ell \in \mathcal{SV}$ and θ is a positive measurable function such that both θ and $1/\theta$ are bounded on $(0, \infty)$.

The proof of Proposition 5.60 can be found in Drasin and Seneta [114].

Next we derive a factorization representation for ORV-functions with nondegenerate sets of regular points.

Theorem 5.61 *Let $f \in \mathbb{F}_+$ be a measurable function. Then, f belongs to $\mathcal{ORV}$ and possesses a nondegenerate set of regular points $\mathbb{G}_r(f)$ if and only if*

$$f(t) = r(t)\theta(t), \qquad t > 0, \tag{5.52}$$

where the functions r and θ are such that

(A1) *$(r(t), t > 0)$ is an RV-function;*
(A2) *$(\theta(t), t > 0)$ is a positive measurable function;*
(A3) *θ and $1/\theta$ are bounded on $(0, \infty)$;*

(A4) $\theta^*(\lambda) = \mathscr{P}(\ln \lambda)$, $\lambda > 0$, *where*
(A5) $(\mathscr{P}(u),\ u \in \mathbf{R})$ *is a positive periodic function with* $\mathscr{P}(0) = 1$ *and*

$$0 < \inf_{u \in \mathbf{R}} \mathscr{P}(u) \le \sup_{u \in \mathbf{R}} \mathscr{P}(u) < \infty.$$

Moreover, if relation (5.52) and conditions (A1)–(A5) are satisfied, then

(A6) *the index of the function r coincides with the index of the function f, and $\mathscr{P}$ coincides with the periodic component of the limit functions related to f;*
(A7) *the set of periods of the function $\mathscr{P}$ coincides with the set $\mathbb{H}_r(f)$;*
(A8) $(\mathscr{P}(\ln \lambda),\ \lambda > 0)$ *is an upper $*$-invariant function, that is, for all $u \in \mathbf{R}$,*

$$\limsup_{x \to \infty} \frac{\mathscr{P}(u+x)}{\mathscr{P}(x)} = \mathscr{P}(u) \qquad \textit{and} \qquad \liminf_{x \to \infty} \frac{\mathscr{P}(u+x)}{\mathscr{P}(x)} = \frac{1}{\mathscr{P}(-u)};$$

(A9) $\min_{u \in \mathbf{R}} \mathscr{P}(u) = 1$;
(A10) $\ln \mathscr{P}$ *is a subadditive function;*
(A11) $\mathbb{G}_r(f) = \mathbb{G}_r(\theta)$.

Proof of Theorem 5.61 Let f be an ORV-function with nondegenerate group of regular points. Then (5.52) and (A1)–(A11) follow from Corollary 5.57, Proposition 5.60, Theorem 5.40, and Theorem 5.48.

On the other hand, if (5.52) and (A1)–(A5) are satisfied, then $f \in \mathcal{ORV}$ and f possesses a nondegenerate set of regular points. The parts of Theorem 5.61 already proved imply relations (A6)–(A11). □

Remark 5.62 Relation (5.52) can be rewritten as follows:

$$f(t) = r(t)\psi(\ln t), \qquad t > 0,$$

where $\psi(x) = \theta(e^x)$, $x \in \mathbf{R}$. According to Theorem 5.61, the function ψ is positive, bounded, and measurable. The corresponding conditions on f are satisfied for the functions discussed in Examples 5.8–5.11.

It is known that (see Remark 3.4)

$$\lim_{t \to \infty} \frac{\ln r(t)}{\ln t} = \rho$$

for any RV-function $(r(t), t > 0)$ of index ρ. Below is a generalization of this result for ORV-function with nondegenerate sets of regular points.

Corollary 5.63 *Let f be an ORV-function with nondegenerate set of regular points and index ρ. Then*

$$\lim_{t \to \infty} \frac{\ln f(t)}{\ln t} = \rho.$$

Proof of Corollary 5.63 Using (5.52), (A1), (A3) and (A6), we get

$$\lim_{t\to\infty}\frac{\ln f(t)}{\ln t}=\lim_{t\to\infty}\frac{\ln r(t)}{\ln t}+\lim_{t\to\infty}\frac{\ln\theta(t)}{\ln t}$$
$$=\lim_{t\to\infty}\frac{\ln r(t)}{\ln t}=\rho,$$

which completes the proof. □

The representation (5.52) can be rewritten in an extended form.

Theorem 5.64 *Let $f\in\mathbb{F}_+$ be a measurable function. Then $f\in\mathcal{ORV}$ possesses a nondegenerate set of regular points $\mathbb{G}_r(f)$ if and only if*

$$f(t)=t^{\rho}\ell(t)\exp\{h(\ln t)\},\qquad t>0,\tag{5.53}$$

where

(B1) $\rho\in\mathbf{R}$;
(B2) $(\ell(t),t>0)$ *is an SV-function;*
(B3) $(h(u),u\in\mathbf{R})$ *is a measurable function such that* $\sup_{u\in\mathbf{R}}|h(u)|<\infty$;
(B4) *for all* $u\in\mathbf{R}$,

$$\limsup_{x\to\infty}[h(u+x)-h(x)]=\mathfrak{p}(u)\quad\text{and}\quad\liminf_{x\to\infty}[h(u+x)-h(x)]=-\mathfrak{p}(-u),$$

where $(\mathfrak{p}(u),u\in\mathbf{R})$ *is a periodic function such that*
(B5) $\mathfrak{p}(0)=0$ *and* $\sup_{u\in\mathbf{R}}|\mathfrak{p}(u)|<\infty$.

Moreover, if relation (5.53) as well as conditions (B1)–(B5) are satisfied, then

(B6) ρ *is the index of the function* f *and* $\mathfrak{p}=\ln\mathscr{P}(u)$, $u\in\mathbf{R}$, *where* $\mathscr{P}$ *is the periodic component of the limit functions corresponding to* f;
(B7) *the set of periods of the function* $\mathfrak{p}$ *coincides with* $\mathbb{H}_r(f)$;
(B8) *for all* $u\in\mathbf{R}$,

$$\limsup_{x\to\infty}[\mathfrak{p}(u+x)-\mathfrak{p}(x)]=\mathfrak{p}(u)\quad\text{and}\quad\liminf_{x\to\infty}[\mathfrak{p}(u+x)-\mathfrak{p}(x)]=-\mathfrak{p}(-u);$$

(B9) $\mathfrak{p}$ *is a nonnegative function and* $\min_{u\in\mathbf{R}}\mathfrak{p}(u)=0$;
(B10) $\mathfrak{p}$ *is a subadditive function, that is,* $\mathfrak{p}(u+x)\le\mathfrak{p}(u)+\mathfrak{p}(x)$ *for all* $u,x\in\mathbf{R}$.

Proof of Theorem 5.64 Theorem 5.64 is a corollary to Theorem 5.61 with

$$h(u)=\ln\theta(e^u),\qquad u\in\mathbf{R},$$

and

$$\mathfrak{p}(u) = \ln \mathscr{P}(u), \qquad u \in \mathbf{R}.$$

□

Corollary 5.65 *Let f be an ORV-function with nondegenerate set of regular points $\mathbb{G}_r(f)$, with index ρ and with periodic component $\mathscr{P}$ of its limit functions. Then $f \sim \breve{f}$, where*

$$\breve{f}(t) = ct^{\rho}\breve{\ell}(t)\exp\{h(\ln t)\}, \qquad t > 0,$$

$c > 0$, the function h is the same as in Theorem 5.64, $\mathfrak{p} = \ln \mathscr{P}$, and $\breve{\ell}$ is an infinitely often differentiable SV-function such that

$$\lim_{u\to\infty} \frac{d^n\breve{h}}{du^n}(u) = 0$$

for all $n \in \mathbf{N}$, where $\breve{h}(u) = \ln \breve{\ell}(e^u)$, $u \in \mathbf{R}$.

Corollary 5.65 follows from Theorem 5.64 and from a known result on the infinitely often differentiable versions of SV-functions (see, e.g., de Bruijn [54] and Bingham et al. [41], Theorem 1.3.3).

In order to look deeper into the result of Theorem 5.64, we introduce some new definitions.

Definition 5.66 We say that a function $(g(u), u \in \mathbf{R})$ is uniformly continuous at infinity if, given an arbitrary $\varepsilon > 0$, there are numbers $x' = x'(\varepsilon) > 0$ and $\delta = \delta(\varepsilon) > 0$ such that $|g(x_1) - g(x_2)| < \varepsilon$ if $x_1, x_2 \geq x'$ and $|x_1 - x_2| < \delta$.

It is clear that a function g is uniformly continuous at infinity if it is uniformly continuous on an interval $[A, \infty)$ for some $A \in \mathbf{R}$.

Definition 5.67 We say that a function $(g(u), u \in \mathbf{R})$ is almost periodic at infinity if, for arbitrary $\varepsilon > 0$ and $x_1, x_2 \in \mathbf{R}$, there exists a sequence of positive numbers $u_n = u_n(\varepsilon, x_1, x_2)$, $n \geq 1$, such that $u_n \to \infty$

$$\limsup_{n\to\infty} |g(x_i + u_n) - g(x_i)| < \varepsilon, \qquad i = 1, 2,$$

as $n \to \infty$.

Note that every almost periodic function in Bohr's sense (see Bohr [43]) is almost periodic at infinity.

The following proposition exhibits some new properties of the functions h, $\mathscr{P}$, and $\mathfrak{p}$ which hold under the assumption of uniform continuity at infinity or the assumption of almost periodicity at infinity of the function h (see Theorem 5.64).

Proposition 5.68 *Let f be an ORV-function with nondegenerate set of regular points, with an index ρ and with periodic component $\mathscr{P}$ of its limit functions and $\mathfrak{p} = \ln \mathscr{P}$. Then*

(i) if the function $\mathscr{P}$ is continuous at the point 0 or, equivalently, if the function f^* is continuous at the point 1, then the functions $\mathscr{P}$ and $\mathfrak{p}$ are uniformly continuous on $\mathbf{R}$ and f^* is continuous on $\mathbf{R}_+$;

(ii) if the function h is uniformly continuous at infinity, then $\mathscr{P}$ and $\mathfrak{p}$ are uniformly continuous on $\mathbf{R}$, where h is as defined in (5.53).

Moreover, if the function h is almost periodic at infinity, then

(iii) for all $u, x \in \mathbf{R}$,

$$h(u + x) \leq h(x) + \mathfrak{p}(u);$$

(iv) for all $u, x \in \mathbf{R}$,

$$|h(u + x) - h(x)| \leq \max\{|\mathfrak{p}(u)|, |\mathfrak{p}(-u)|\};$$

(v) if the function $\mathfrak{p}$ is continuous at the point 0 or, equivalently, if the function f^* is continuous at the point 1, then h is uniformly continuous on $\mathbf{R}$;

(vi) h is a periodic function and $S_{\text{per}}(h) = S_{\text{per}}(\mathfrak{p})$.

Proof of Proposition 5.68 Clearly, statement (i) follows from Corollary 5.43.

To prove (ii) note that the assumptions of Proposition 5.68 imply condition (B4) of Theorem 5.64. Thus

$$|\mathfrak{p}(u)| \leq \limsup_{x \to \infty} |h(u + x) - h(x)|$$

for all $u \in \mathbf{R}$. Since h is uniformly continuous at infinity, the function $\mathfrak{p}$ is continuous at the point 0. Now statement (i) implies that $\mathfrak{p}$ is uniformly continuous on $\mathbf{R}$. So, statement (ii) is proved.

Turning to the proof of statement (iii), assume that the function h is almost periodic at infinity, but there are real numbers $u', x' \in \mathbf{R}$, and $\delta' > 0$ such that

$$h(u' + x') - h(x') - \mathfrak{p}(u') = \delta'. \tag{5.54}$$

Since the function h is almost periodic at infinity, there exists a sequence $\{v_n\}_{n \geq 1}$ such that $\lim_{n \to \infty} v_n = \infty$,

$$\limsup_{n \to \infty} |h(u' + x' + v_n) - h(u' + x')| < \frac{\delta}{4}$$

and

$$\limsup_{n \to \infty} |h(x' + v_n) - h(x')| < \frac{\delta}{4}.$$

Hence (5.54) implies that

$$\limsup_{n\to\infty}[h(u'+x'+v_n)-h(x'+v_n)-\mathfrak{p}(u')]\geq\frac{\delta'}{2}.$$

Using condition (B4) of Theorem 5.64, we get a contradiction, i.e.

$$\begin{aligned}0&=\limsup_{x\to\infty}[h(u'+x)-h(x)-\mathfrak{p}(u')]\\&\geq\limsup_{n\to\infty}[h(u'+x'+v_n)-h(x'+v_n)-\mathfrak{p}(u')]\geq\frac{\delta'}{2}>0,\end{aligned}$$

whence statement (iii) follows.

Now (iii) implies that, for all $u, x \in \mathbf{R}$,

$$h(u+x)-h(x)\leq\mathfrak{p}(u)\leq\max\{|\mathfrak{p}(u)|,|\mathfrak{p}(-u)|\}.$$

Moreover,

$$h(u+x)-h(x)=-(h(u+x-u)-h(u+x))\geq -p(-u)\geq -\max\{|\mathfrak{p}(u)|,|\mathfrak{p}(-u)|\}$$

for all $u, x \in \mathbf{R}$. Thus statement (iv) holds and implies statement (v).

By Theorem 5.64, the function $\mathfrak{p}$ is periodic. Then statement (iv) implies that the function h is periodic and $S_{\mathrm{per}}(h) \supset S_{\mathrm{per}}(\mathfrak{p})$. In turn, condition (B4) of Theorem 5.64 implies that $S_{\mathrm{per}}(h) \subset S_{\mathrm{per}}(\mathfrak{p})$. Therefore statement (vi) is proved.

□

Remark 5.69 Let f be an ORV-function of the form (5.53), where h is a periodic function. Proposition 5.68 shows that h is continuous, that is, f is a regularly log-periodic function (see Definition 5.13) if and only if the function f^* is continuous at the point 1, that is, f is a PRV-function (see Definition 3.16 and Proposition 3.21).

Moreover, Example 5.9 shows that, in statement (v), the condition on the continuity of the function $\mathfrak{p}$ at the point 0 or, equivalently, the condition on the continuity of the function f^* at the point 1 cannot, in general, be exchanged with the assumption that the function $\mathfrak{p}$ is continuous at a fixed point $u \in \mathbf{R}$ or, equivalently, by the condition that the function f^* is continuous at a fixed point $\mathbf{R}_+$.

Remark 5.70 Statement (vi) implies that the class of functions h in the representation (5.53), which are almost periodic at infinity, belong to $\mathcal{ORV}$ and possess nondegenerate sets of regular points, contains only periodic functions.

Theorems 5.61, 5.64 combined with Proposition 5.68 and with the integral representations for RV-functions (see Seneta [324], Bingham et al. [41], and Sect. 3.5) and for PRV-functions (see Theorem 3.49 and 3.51), respectively, allow us to establish various integral representations for ORV-functions with nondegenerate sets of regular points.

5.9 Uniform Convergence Theorems

In this section, we consider uniform convergence theorems for WOURV- and ORV-functions with nondegenerate sets of regular points. These theorems complement the results of Chap. 3 (see Remark 3.48) and, together with the uniform convergence theorem for RV-functions, provide additional versions of uniform convergence theorems for general ORV-functions (see Aljančić and Arandelović [8], Theorem 1).

If f is a WOURV-function with nondegenerate set of regular points and if $\mathscr{P}$ is the periodic component of its limit functions, then, according to statement (v) of Theorem 5.40 (see also statements (A8) of Theorem 5.61 and (B8) of Theorem 5.64),

$$\limsup_{x\to\infty} \frac{\mathscr{P}(u+x)}{\mathscr{P}(x)} = \mathscr{P}(u) \qquad \text{and} \qquad \liminf_{x\to\infty} \frac{\mathscr{P}(u+x)}{\mathscr{P}(x)} = \frac{1}{\mathscr{P}(-u)}$$

for all $u \in \mathbf{R}$.

The following proposition improves the above relations.

Proposition 5.71 *Let f be a WOURV-function with a nondegenerate set of regular points. Then*

$$\limsup_{x\to\infty} \sup_{u\in\mathbf{R}} \left(\frac{\mathscr{P}(u+x)}{\mathscr{P}(x)} - \mathscr{P}(u) \right) = 0 \tag{5.55}$$

and

$$\liminf_{x\to\infty} \inf_{u\in\mathbf{R}} \left(\frac{\mathscr{P}(u+x)}{\mathscr{P}(x)} - \frac{1}{\mathscr{P}(-u)} \right) = 0, \tag{5.56}$$

where $\mathscr{P}$ is the periodic component of the limit functions corresponding to f. Moreover, (5.55) and (5.56) are equivalent.

Proof of Proposition 5.71 Applying statement (A5) of Theorem 5.61, we see that equality (5.55) is equivalent to

$$\limsup_{x\to\infty} \sup_{u\in\mathbf{R}} [\mathfrak{p}(u+x) - \mathfrak{p}(x) - \mathfrak{p}(u)] = 0, \tag{5.57}$$

while (5.56) is equivalent to

$$\liminf_{x\to\infty} \inf_{u\in\mathbf{R}} [\mathfrak{p}(u+x) - \mathfrak{p}(x) + \mathfrak{p}(-u)] = 0, \tag{5.58}$$

where $\mathfrak{p} = \ln \mathscr{P}$. Since

$$\begin{aligned}\liminf_{x\to\infty} \inf_{u\in\mathbf{R}}[\mathfrak{p}(u+x) - \mathfrak{p}(x) + \mathfrak{p}(-u)] \\ &= \liminf_{x\to\infty} \inf_{u\in\mathbf{R}}[\mathfrak{p}(x-u) - \mathfrak{p}(x) + \mathfrak{p}(u)] \\ &= -\limsup_{x\to\infty} \sup_{u\in\mathbf{R}}[\mathfrak{p}(u+(x-u)) - \mathfrak{p}(x-u) - \mathfrak{p}(u)]\end{aligned}$$

and

$$\begin{aligned}\limsup_{x\to\infty} \sup_{u\in\mathbf{R}}[\mathfrak{p}(u+x) - \mathfrak{p}(x) - \mathfrak{p}(u)] \\ &= \limsup_{x\to\infty} \sup_{u\in\mathbf{R}}[\mathfrak{p}(x-u) - \mathfrak{p}(x) - \mathfrak{p}(-u)] \\ &= -\liminf_{x\to\infty} \inf_{u\in\mathbf{R}}[\mathfrak{p}(u+(x-u)) - \mathfrak{p}(x-u) + \mathfrak{p}(-u)],\end{aligned}$$

relations (5.57) and (5.58), as well as relations (5.55) and (5.56), are equivalent.

So, it suffices to prove (5.57). To do so, put

$$\delta = \limsup_{x\to\infty} \sup_{u\in\mathbf{R}}[\mathfrak{p}(u+x) - \mathfrak{p}(x) - \mathfrak{p}(u)].$$

Note that $\delta \le 0$ by statement (B10) of Theorem 5.64. Assume that $\delta < 0$. Then relation (B8) of Theorem 5.64 implies that

$$0 = \limsup_{x\to\infty}[\mathfrak{p}(u'+x) - \mathfrak{p}(x) - \mathfrak{p}(u')] \le \limsup_{x\to\infty} \sup_{u\in\mathbf{R}}[\mathfrak{p}(u+x) - \mathfrak{p}(x) - \mathfrak{p}(u)] < 0$$

for all $u' \in \mathbf{R}$. This contradiction proves (5.57). □

The following is the main result of this section for functions which are *uniformly continuous at infinity* (see Definition 5.66).

Theorem 5.72 *Let f be an ORV-function with nondegenerate set of regular points, with an index ρ, and with periodic component $\mathscr{P}$ of its limit functions. If the function θ (see (5.52)) or, equivalently, if the function h (see (5.53)) is uniformly continuous at infinity, then, for an arbitrary interval $[a, b] \subset \mathbf{R}_+$,*

$$\limsup_{t\to\infty} \sup_{\lambda\in[a,b]} \left(\frac{f(\lambda t)}{f(t)} - \lambda^\rho \mathscr{P}(\ln \lambda)\right) = 0 \tag{5.59}$$

and

$$\liminf_{t\to\infty} \inf_{\lambda\in[a,b]} \left(\frac{f(\lambda t)}{f(t)} - \frac{\lambda^\rho}{\mathscr{P}(-\ln \lambda)}\right) = 0. \tag{5.60}$$

Moreover, (5.59) holds for an arbitrary interval $[a, b] \subset \mathbf{R}_+$ *if and only if (5.60) holds for all* $[a, b] \subset \mathbf{R}_+$.

Remark 5.73 Taking into account the representations for the limit functions f^* and f_* derived in statement (iv) of Theorem 5.40 (see (5.40) and (5.41)), we see that relation (5.59) can be rewritten as

$$\limsup_{t\to\infty} \sup_{\lambda\in[a,b]} \left(\frac{f(\lambda t)}{f(t)} - f^*(\lambda) \right) = 0,$$

while relation (5.60) takes the form

$$\liminf_{t\to\infty} \inf_{\lambda\in[a,b]} \left(\frac{f(\lambda t)}{f(t)} - f_*(\lambda) \right) = 0.$$

Proof of Theorem 5.72 Under the assumptions of Theorem 5.72, the assumptions of Theorem 5.61 are also satisfied and thus relation (5.52) holds. Assume that the function θ in this relation is uniformly continuous at infinity. This is equivalent to the assumption that the function h in (5.53) is uniformly continuous at infinity. Under this assumption, we prove relations (5.59) and (5.60).

Relation (5.52) implies that

$$\begin{aligned} \frac{f(\lambda t)}{f(t)} - \lambda^\rho \mathscr{P}(\ln \lambda) &= \frac{r(\lambda t)\theta(\lambda t)}{r(t)\theta(t)} - \lambda^\rho \mathscr{P}(\ln \lambda) \\ &= \frac{\theta(\lambda t)}{\theta(t)} \left(\frac{r(\lambda t)}{r(t)} - \lambda^\rho \right) + \lambda^\rho \left(\frac{\theta(\lambda t)}{\theta(t)} - \mathscr{P}(\ln \lambda) \right). \end{aligned}$$

Since conditions (A2) and (A3) of Theorem 5.61 are satisfied for the function θ, and relation (3.25) holds for the RV-function r, it is clear that relation (5.59) holds for an arbitrary interval $[a, b] \subset \mathbf{R}_+$ if and only if

$$\limsup_{t\to\infty} \sup_{\lambda\in[a,b]} \left(\frac{\theta(\lambda t)}{\theta(t)} - \mathscr{P}(\ln \lambda) \right) = 0$$

for an arbitrary interval $[a, b] \subset \mathbf{R}_+$. Taking condition (A5) of Theorem 5.61 into account, we see that the latter relation is equivalent to

$$\limsup_{t\to\infty} \sup_{\lambda\in[a,b]} \left(\frac{\theta(\lambda t)}{\theta(t)\mathscr{P}(\ln \lambda)} - 1 \right) = 0$$

for an arbitrary interval $[a, b] \subset \mathbf{R}_+$. This, in turn, is equivalent to

$$\limsup_{x\to\infty} \sup_{u\in[c,d]} [h(u+x) - h(x) - \mathfrak{p}(u)] = 0 \tag{5.61}$$

for an arbitrary interval $[c, d] \subset \mathbf{R}$, where $h(u) = \ln \theta(e^u)$, $u \in \mathbf{R}$ (see (5.53)) and $\mathfrak{p} = \ln \mathscr{P}$.

Therefore, in order to prove that (5.59) holds for an arbitrary interval $[a, b] \subset \mathbf{R}_+$, it suffices to prove (5.61) for an arbitrary interval $[c, d] \subset \mathbf{R}$.

For a fixed interval $[c, d] \subset \mathbf{R}$, put

$$\delta = \limsup_{x\to\infty} \sup_{u\in[c,d]} [h(u + x) - h(x) - \mathfrak{p}(u)].$$

Under the assumptions of Theorem 5.72, the assumptions of Theorem 5.64 are also satisfied and thus statement (B4) of Theorem 5.64 holds. Hence

$$\delta = \limsup_{x\to\infty} \sup_{u\in[c,d]} [h(u+x) - h(x) - \mathfrak{p}(u)] \ge \limsup_{x\to\infty} [h(c+x) - h(x) - \mathfrak{p}(c)] = 0.$$

Assume that $\delta > 0$. Then there are sequences $\{u_n\}_{n\ge1}$ and $\{x_n\}_{n\ge1}$ such that $\{u_n\}_{n\ge1} \subset [a, b]$ and $x_n \to \infty$ as $n \to \infty$ and

$$h(u_n + x_n) - h(x_n) - \mathfrak{p}(u_n) > \delta/2$$

for all $n \ge 1$. Thus there is a sequence $\{n_k\}_{k\ge1} \subset \mathbf{N}$ such that $u_{n_k} \to u' \in [a, b]$ as $k \to \infty$. Now we conclude that

$$\lim_{k\to\infty} |h(x_{n_k}) - h((u_{n_k} - u') + x_{n_k})| = 0,$$

since the function θ, as well as the function h, is uniformly continuous at infinity. Moreover,

$$\mathfrak{p}(u_{n_k}) \to \mathfrak{p}(u'), \qquad k \to \infty,$$

since the function $\mathfrak{p}$ is continuous in view of Proposition 5.68. Therefore

$$\limsup_{k\to\infty} [h(u_{n_k} + x_{n_k}) - h(x_{n_k})] - \mathfrak{p}(u') \ge \delta/2 > 0.$$

Applying statement (B4) of Theorem 5.64, we see that

$$\begin{aligned}
0 &= \mathfrak{p}(u') - \mathfrak{p}(u') = \limsup_{y\to\infty} [h(u' + y) - h(y)] - \mathfrak{p}(u') \\
&\ge \limsup_{k\to\infty} [h(u' + (u_{n_k} - u') + x_{n_k}) - h((u_{n_k} - u') + x_{n_k})] - \mathfrak{p}(u') \\
&= \limsup_{k\to\infty} [h(u' + (u_{n_k} - u') + x_{n_k}) - h(x_{n_k}) + h(x_{n_k}) - h((u_{n_k} - u') + x_{n_k})] - \mathfrak{p}(u') \\
&= \limsup_{k\to\infty} [h(u_{n_k} + x_{n_k}) - h(x_{n_k})] - \mathfrak{p}(u') \ge \delta/2 > 0.
\end{aligned}$$

This contradiction proves that $\delta = 0$.

Summarizing, relation (5.61) is proved for an arbitrary $[c, d] \subset \mathbf{R}$ and relation (5.59) is proved for an arbitrary interval $[a, b] \subset \mathbf{R}_+$.

To complete the proof of Theorem 5.72, we show that

$$\liminf_{x\to\infty} \inf_{u\in[c,d]} [h(u + x) - h(x) + \mathfrak{p}(-u)] = 0 \tag{5.62}$$

holds. In the same way as above, we prove that relation (5.60) holds for an arbitrary interval $[a, b] \subset \mathbf{R}_+$ if and only if relation (5.62) holds for all $[c, d] \subset \mathbf{R}$. Since

$$\begin{aligned}
&\liminf_{x\to\infty} \inf_{u\in[c,d]} [h(x + u) - h(x) - \mathfrak{p}(-u)] \\
&\quad = -\limsup_{x\to\infty} \sup_{u\in[c,d]} [h((x + u) - u) - h(x + u) - \mathfrak{p}(-u)] \\
&\qquad - \limsup_{x\to\infty} \sup_{u\in[c,d]} [h((x + u) - u) - h(x + u) - \mathfrak{p}(-u)] \\
&\quad = -\limsup_{x\to\infty} \sup_{u\in[-d,-c]} [h(x + u) - h(x) - \mathfrak{p}(u)],
\end{aligned}$$

it is clear that relation (5.62) holds for an arbitrary interval $[c, d] \subset \mathbf{R}$ if and only if relation (5.61) holds for an arbitrary interval $[c, d] \subset \mathbf{R}$.

This implies that relation (5.60) holds for an arbitrary interval $[a, b] \subset \mathbf{R}_+$ if and only if (5.59) holds for an arbitrary interval $[a, b] \subset \mathbf{R}_+$. Therefore Theorem 5.72 is proved. □

It is reasonable to separate relations (5.61) and (5.62) appearing in the proof of Theorem 5.72.

Corollary 5.74 *Let f be an ORV-function with nondegenerate set of regular points and with periodic component $\mathscr{P}$ of its limit functions. If the function h defined in (5.53) is uniformly continuous at infinity, then, for all intervals $[c, d] \subset \mathbf{R}$,*

$$\limsup_{x\to\infty} \sup_{u\in[c,d]} [h(u + x) - h(x) - \mathfrak{p}(u)] = 0$$

and

$$\liminf_{x\to\infty} \inf_{u\in[c,d]} [h(u + x) - h(x) + \mathfrak{p}(-u)] = 0,$$

where $\mathfrak{p} = \ln \mathscr{P}$. Moreover, the first of these relations holds for all $[c, d] \subset \mathbf{R}$ if and only if the second relation holds for all $[c, d] \subset \mathbf{R}$.

The following result follows from Corollary 5.74 and Proposition 5.68.

Corollary 5.75 *Let f be an ORV-function with nondegenerate set of regular points and with periodic component $\mathscr{P}$ of its limit functions. If the function h defined* *in (5.53) is almost periodic at infinity (see Definition 5.67) and if the limit function f^* is continuous at the point* 1 *or, equivalently, if the function* $\mathfrak{p} = \ln \mathscr{P}$ *is continuous at the point* 0*, then the functions* $\mathfrak{p}$ *and* h *are uniformly continuous on* $\mathbf{R}$*, the function* h *is periodic,*

$$\limsup_{x\to\infty} \sup_{u\in\mathbf{R}} [h(u+x) - h(x) - \mathfrak{p}(u)] = 0,$$

and

$$\liminf_{x\to\infty} \inf_{u\in\mathbf{R}} [h(u+x) - h(x) + \mathfrak{p}(-u)] = 0.$$

5.10 Comments

This chapter is based on Buldygin et al. [58, 60].

From a general point of view, some sections of this chapter study the solutions of the functional equation

$$\phi(cx) = \phi(c)\phi(x), \qquad x > 0, \tag{5.63}$$

where $\phi(1) = 1$ and the variable c belongs to a nondegenerate multiplicative subgroup $\mathbb{G}$ of the group $\mathbf{R}_+$. If $\mathbb{G} = \mathbf{R}_+$, then this is known as the *Hamel functional equation* (see Hamel [171]). Clearly, (5.63) is closely related to the *Cauchy functional equation*

$$\phi(x+y) = \phi(x) + \phi(y), \qquad x, y \in \mathbf{R}. \tag{5.64}$$

As is well known, any measurable solution of (5.64) is a linear function.

Section 5.2 In 1821, Cauchy [79] was able to prove that any continuous solution of (5.64) is a linear function. Many extensions of this result have been obtained since then. For example, Darboux [98] proved the same result by assuming continuity at a single point. Theorem 5.7 provides the well-known result that any RV function which is bounded on a finite interval is linear.

Cauchy treated not only the functional equation (5.64) but also (5.63) for $x, y \in \mathbf{R}$ as well as $\phi(xy) = \phi(x) + \phi(y)$ and $\phi(x+y) = \phi(x)\phi(y)$ on pages 98–113 and 220–229 of [79]. His arguments are not convincing, since he used a statement like "*an infinite sum of continuous functions is continuous*". Abel [2] gave a counterexample to the above "statement" and clarified some details. At the International Congress of Mathematicians in Paris in 1900, Hilbert [183] asked whether a weaker assumption on ϕ (compared to its continuity) in equation (5.64) forces ϕ to be linear (see Aczél [5] for a detailed discussion). Banach [23] and Sierpiński [330] in 1920 proved that measurability suffices. It is also worth mentioning that Hamel in 1905 showed that there could be other solutions of (5.64) different from linear ones.

A nice introduction to the modern treatment of the Cauchy function equation is given by Kuczma [247] (classical texts on this topic are Aczél [4] and Kuczma [246]; also see Sahoo and Kanappan [312]). A treatment of the topic for semigroups of linear operators is given by Engel and Nagel [117]. Drasin and Shea [115] consider functions with exceptional sets (a prototype of functions with sets of regular points).

Functions with nondegenerate groups of regular points (without mentioning the name) are used by Grinevich and Khokhlov [156] in integral representations for semi-stable laws.

Section 5.3 The proof of Proposition 5.17 can be found in Hardy and Wright [174, Theorems 438 and 439] or Yadrenko [368, Theorem 11].

Considering Lemma 5.15 we conclude that the limit functions f^* and f_* of a WORV-function f are solutions of equation (5.63), where $\mathbb{G}$ is the group of regular points of the function f.

Log-periodic functions are popular in several applied models of economics (see, for instance, Rodríguez-Caballero and Knapik [304]), statistical physics (cf. Aguiar and Guedes [7]), chaos (see Barra et al. [27]), or the theory of art (e.g. Brissaud [53]) etc. The so-called *log-periodic power law* is a useful tool in forecasting financial crashes (see Brée and Joseph [51]). Some other authors doubt that it really helps in forecasting the crashes (cf. Wosnitza and Leker [367]).

Section 5.5 Given a fixed group $\mathbb{G}$, the set $S_+(\mathbb{G})$ of all positive solutions ϕ of equation (5.55) is also a multiplicative group. Moreover, for every solution $\phi \in S_+(\mathbb{G})$, there exists a dual solution $\phi'(x) = 1/\phi(1/x)$, $x > 0$. Now, Theorem 5.34 shows that every solution $\phi \in S_+(\mathbb{G})$ can be represented as a product of a power function and a positive periodic function having a logarithmic argument.

Section 5.6 In turn, Theorem 5.40 shows that this representation is unique for functions $\phi \in S_+(\mathbb{G})$ such that ϕ and $1/\phi$ are locally bounded. Moreover, Theorem 5.40 describes the properties of the periodic components. Note that similar results are valid for measurable functions $\phi \in S_+(\mathbb{G})$.

Section 5.7 The proof of the characterization theorem for the class $\mathcal{RV}$ can be found in Seneta [324, Theorem 1.3, p. 9] or Bingham et al. [41, Theorem 1.4.1, p. 17].

Section 5.8 O-weakly slowly varying functions used in Proposition 5.56 are introduced by Drasin and Seneta [114]. (These functions are named "S–O functions" therein.)

Section 5.9 The proof of the uniform convergence theorem for the class $\mathcal{RV}$ can be found in Seneta [324, Theorem 1.2, p. 2] or Bingham et al. [41, Theorem 1.2.1, p. 6].

Chapter 6
Karamata's Theorem for Integrals

6.1 Introduction

In this chapter, we continue to study some functions with nondegenerate groups of regular points, namely the regularly log-periodic functions considered in Chap. 5. The main aim of this chapter is to extend the well-known Karamata theorem on the asymptotic behavior of integrals of RV-functions to the case of log-periodic functions. For this class of functions, we are able to obtain the precise asymptotic behavior of integrals similarly to the class $\mathcal{RV}$ and determine the limit values. This becomes possible in view of the well-defined notion of an index for a log-periodic function. The essence of the Karamata theorem for regularly varying functions is that

$$\text{if} \quad f \in \mathcal{RV}_\rho, \quad \text{then} \quad \int_A^{(\cdot)} f(u)\,du \in \mathcal{RV}_{\rho+1}.$$

Another (more precise) way to state this is that

$$\int_A^x f(u)\,du \sim \frac{xf(x)}{\rho+1}, \qquad x \to \infty,$$

provided $\rho \neq -1$. The asymptotic behavior of integrals of functions considered in this chapter involves an additional log-periodic factor.

Chapter 6 is organized as follows. The direct and converse parts of the Karamata theorem for RV-functions are stated in Sect. 6.2.

Some necessary facts about regularly log-periodic functions are given in Sect. 6.3.

V. V. Buldygin et al., *Pseudo-Regularly Varying Functions and Generalized Renewal Processes*, Probability Theory and Stochastic Modelling 91, https://doi.org/10.1007/978-3-319-99537-3_6

Section 6.4 contains a generalization of the Karamata theorem to the class of regularly log-periodic functions.

The proof of this result is given in Sects. 6.5–6.7.

6.2 Karamata's Theorem for Integrals of RV-Functions

Given $A > 0$, we consider a family $\mathbb{FM}_+(A)$ of positive and (Lebesgue) measurable functions $f = (f(t), t \geq A)$. Note that

$$\mathbb{FM}_+(A) \subset \mathbb{F}_+(A) \subset \mathbb{F}_+.$$

Moreover, we assume that all RV-functions considered in this section belong to the class $\mathbb{FM}_+(A)$.

Recall that a measurable, real-valued function $(\varphi(t), t \geq A)$ is called *locally integrable* if it is (Lebesgue) integrable on any interval $[a, b] \subset [A, \infty)$.

Karamata's theorem on the asymptotic behavior of integrals of RV-functions consists of two parts, a direct and a converse one.

6.2.1 *Karamata's Theorem: Direct Part*

Theorem 6.1 (Karamata's Theorem, Direct Part) *Let f be a locally integrable RV-function of index $\rho \in \mathbf{R}$.*

1) If $\rho > -1$, then

$$\frac{xf(x)}{\int_A^x f(t)dt} \to \rho + 1, \qquad x \to \infty. \tag{6.1}$$

2) If $\rho < -1$, then

$$\frac{xf(x)}{\int_x^\infty f(t)\,dt} \to |\rho + 1|, \qquad x \to \infty. \tag{6.2}$$

3) If $\rho = -1$ and $I_f(\infty) = \infty$, then

$$\frac{xf(x)}{\int_A^x f(t)\,dt} \to 0, \qquad x \to \infty, \tag{6.3}$$

where

$$I_f(\infty) = \int_A^\infty f(t)\,dt.$$

4) If $\rho = -1$ *and* $I_f(\infty) < \infty$*, then*

$$\frac{xf(x)}{\int_x^\infty f(t)dt} \to 0, \qquad x \to \infty. \tag{6.4}$$

6.2.2 Karamata's Theorem: Converse Part

Theorem 6.2 (Karamata's Theorem, Converse Part) *Let* f *be a locally integrable function in the class* $\mathbb{FM}_+(A)$*.*

5) If there exists a number $\gamma \in (0, \infty)$ *such that*

$$\frac{xf(x)}{\int_A^x f(t)dt} \to \gamma, \qquad x \to \infty, \tag{6.5}$$

then f *is an RV-function of index* $\rho = \gamma - 1$*.*

6) If $I_f(\infty) < \infty$ *and there exists a number* $\gamma \in (0, \infty)$ *such that*

$$\frac{xf(x)}{\int_x^\infty f(t)dt} \to \gamma, \qquad x \to \infty, \tag{6.6}$$

then f *is an RV-function of index* $\rho = -\gamma - 1$*.*

6.3 Regularly Log-Periodic and Regularly Log-Bounded Functions

Let $A > 0$. Recall that (see (5.18)) a function f in the class $\mathbb{FM}_+(A)$ is called *regularly log-periodic* if

$$f(t) = t^\rho \ell(t) V(\ln t), \qquad t \geq A, \tag{6.7}$$

where $\rho \in \mathbf{R}$, $\ell = (\ell(t),\ t \geq A)$ is an SV-function and $V = (V(t),\ t \in \mathbf{R})$ is a positive continuous periodic function. The class of functions represented in the form (6.7) is denoted by $\mathfrak{RLP}(A)$.

The function

$$r(t) = t^\rho \ell(t), \qquad t \geq A,$$

related to (6.7) is called the *regular component* or *RV-component* of the function f. The index ρ of r is called the *index* of the function f, while ℓ is called the *slowly varying component* or *SV-component* of the function f. The slowly-varying component may be a discontinuous function.

Moreover, the function V in (6.7) is called the *periodic component* of the function f. Finally,

$$V \circ \ln = (V(\ln t),\ t > 0)$$

is called the *log-periodic component* of the function f.

Recall that (see Definition 5.13) $S_{\text{per}}(U)$ denotes the set of periods of a periodic function U and $S^+_{\text{per}}(U)$ denotes the set of its positive periods. The number

$$T(U) = \inf S^+_{\text{per}}(U)$$

is called the *oscillation characteristic* of the function U. If U is the periodic component of a function $f \in \mathfrak{RLP}(A)$, then its oscillation characteristic $T(U)$ is called the *oscillation characteristic of the function* f and is sometimes denoted by T_f.

The function V in (6.7) is positive, continuous and periodic. This function is constant, that is, there exists a $c > 0$ such that $V(t) \equiv c$ if and only if f is an RV-function (see Corollary 5.23). In this case, the function $V \circ \ln$ is also constant and it can be merged with ℓ. Otherwise, the function $V \circ \ln$ is not an SV-function and it cannot be merged with ℓ.

In what follows we assume, unless otherwise stated, that

$$V(0) = 1. \tag{6.8}$$

Condition (6.8) does not restrict the class of regularly log-periodic functions, since every member f in this class can be represented in the form of

$$f(t) = t^\rho \ell_0(t) V_0(\ln t), \qquad t \geq A,$$

where $V_0(t) = V(t)/V(0)$ and $\ell_0(t) = V(0)\ell(t)$. It is clear that $V_0(0) = 1$ and that ℓ_0 is an SV-function.

The representation (6.7) is called *canonical* if its periodic component is *canonical*, that is, if condition (6.8) holds.

The class of regularly log-periodic functions possessing a positive oscillation characteristic T_f is denoted by $\mathfrak{RLP}_+(A)$. Note that the functions f in this class possess nondegenerate sets of regular points

$$\mathbb{G}_r(f) = \left\{ e^{nT_f},\ n \in \mathbf{Z} \right\} \tag{6.9}$$

(see Corollary 5.23). The class of functions

$$\mathfrak{RLP}_0(A) = \mathfrak{RLP}(A) \setminus \mathfrak{RLP}_+(A) \tag{6.10}$$

coincides with the class of regularly log-periodic functions satisfying $T_f = 0$ and whose periodic components V are constant. Thus condition (6.8) implies that

$V(x) \equiv 1$. Therefore, the class $\mathfrak{RLP}_0(A)$ coincides with the class of RV-functions defined on $[A, \infty)$.

The properties of the functions in $\mathfrak{RLP}_+(A)$ justify the notion "periodic" used for this class.

If f is a locally integrable, regularly log-periodic function of index $\rho > -1$, then

$$\int_A^\infty f(t)dt = \infty. \tag{6.11}$$

Correspondingly, if f is a locally integrable regularly log-periodic function of index $\rho < -1$, then

$$\int_A^\infty f(t)dt < \infty. \tag{6.12}$$

Indeed, for an arbitrary locally integrable RV-function $(r(t), t \geq A)$ with index $\rho > -1$,

$$\int_A^\infty r(t)dt = \infty$$

(see, for example, Bingham et al. [41], de Haan [164], Seneta [324]). Thus (6.11) follows from the relation

$$\int_A^\infty f(t)dt = \int_A^\infty t^\rho \ell(t) V(\ln t)dt \geq \kappa \int_A^\infty t^\rho \ell(t)\, dt = \infty,$$

where $\kappa = \inf_{x \in \mathbf{R}} V(t) > 0$. Assertion (6.12) can be proved in the same way.

6.3.1 *Regularly Log-Bounded Functions*

Along with the class of regularly log-periodic functions we consider the wider class of *regularly log-bounded* functions.

Definition 6.3 A function $f \in \mathbb{FM}_+(A)$ is called *regularly log-bounded* if

$$f(t) = t^\rho \ell(t) V(\ln t), \qquad t \geq A, \tag{6.13}$$

where $\rho \in \mathbf{R}$, $(\ell(t),\ t \geq A)$ is an SV-function and $V = (V(t),\ t \in \mathbf{R})$ is a positive, measurable function such that

$$\kappa = \inf_{t \in \mathbf{R}} V(t) > 0 \qquad \text{and} \qquad K = \sup_{t \in \mathbf{R}} V(t) < \infty. \tag{6.14}$$

The class of regularly log-bounded functions is denoted by $\mathfrak{RLB}(A)$.

It is clear that

$$\mathfrak{RLP}(A) \subset \mathfrak{RLB}(A) \qquad \text{and} \qquad \mathfrak{RLB}(A) \setminus \mathfrak{RLP}(A) \neq \varnothing.$$

If f is a regularly log-bounded function, then similarly to the case of log-periodic functions, $r(t) = t^\rho \ell(t)$, $t \geq A$, is called the *regularly varying component* of f, its index ρ is called the *index* of f, ℓ is called the *slowly varying component* of f, and V is called the *bounded component* of f.

Remark 6.4 Similarly to the case of regularly log-periodic functions, one can prove that relations (6.11) and (6.12) are satisfied for a locally integrable and regularly log-bounded function f with index $\rho \neq 1$ if $\rho > -1$ or $\rho < -1$, respectively.

Remark 6.5 Comparing the definitions of regularly log-bounded functions and OSV-functions (see Definition 6.3 and 3.10) and recalling Proposition 5.60, we see that the class of regularly log-bounded functions with index $\rho = 0$ coincides with the class of OSV-functions. Thus, a regularly log-bounded function f can be represented as

$$f(t) = t^\rho s(t), \qquad t > 0,$$

where $(s(t), t > 0)$ is an OSV-function.

6.4 Generalizations of Karamata's Theorem

Like the original Karamata theorem on the asymptotic behavior of integrals of RV-functions (see Sect. 6.2), its generalizations for the class of regularly log-periodic functions $\mathfrak{RLP}(A)$ consist of a direct and a converse part.

6.4.1 Direct Part: The Case $\rho \neq -1$

We start with the case when $\rho \neq -1$ (see (6.1) and (6.2)).

Theorem 6.6 *Let f be a locally integrable function in the class $\mathfrak{RLP}(A)$ with index $\rho \neq -1$, periodic component V, and oscillation characteristic $T(V) = T$. Then there exists a positive, continuous, periodic function $D = (D(t), t \in \mathbf{R})$, depending on ρ and V, such that*

$$\int_A^x f(t)\, dtt \sim x D(\ln x) f(x), \qquad x \to \infty, \tag{6.15}$$

if $\rho > -1$, *and*

$$\int_x^\infty f(t)\,dt \sim xD(\ln x)f(x), \qquad x \to \infty, \tag{6.16}$$

if $\rho < -1$. *Moreover,*

1) *if* $f \in \mathcal{RV}$ *(that is, if* $T = 0$, $V(t) \equiv 1$*), then*

$$D(t) = D_{V,\rho}(t) \equiv \frac{1}{|\rho + 1|}; \tag{6.17}$$

2) *if* $f \in \mathfrak{RLP}_+(A)$ *(that is, if* $T > 0$*) and* $\rho > -1$*, then*

$$D(t) = D_{V,\rho}(t) = \frac{\int_0^{T\{t/T\}} e^{(\rho+1)y} V(y)dy + C}{V(t)\exp\{T(\rho+1)\{t/T\}\}}, \qquad t \geq 0, \tag{6.18}$$

where $\{t/T\}$ *is the fractional part of* t/T *and*

$$C = C_{V,\rho} = \frac{\int_0^T e^{(\rho+1)y} V(y)\,dy}{e^{T(\rho+1)} - 1};$$

3) *if* $f \in \mathfrak{RLP}_+(A)$ *(that is, if* $T > 0$*) and* $\rho < -1$*, then*

$$D(t) = D_{V,\rho}(t) = \frac{\int_{T\{t/T\}}^{T} e^{(\rho+1)y} V(y)dy + C}{V(t)\exp\{T(\rho+1)\{t/T\}\}}, \qquad t \geq 0, \tag{6.19}$$

where

$$C = C_{V,\rho} = \frac{\int_0^T e^{(\rho+1)y} V(y)dy}{e^{T|\rho+1|} - 1};$$

4) $T(DV) = T(D) = T(V)$*, where* $T(DV)$ *is the oscillation characteristic of the positive, continuous, periodic function* $DV = (D(t)V(t),\ t \in \mathbf{R})$.

6.4.2 Direct Part: The Case $\rho = -1$

The generalization of Karamata's theorem (see (6.3) and (6.4)) for regularly log-periodic functions of index $\rho = -1$ follows from the next result, where a wider class is treated compared to the class of regularly log-bounded functions $\mathfrak{RLB}(A)$ (see Definition 6.3).

Proposition 6.7 *Let f be a locally integrable function in the class $\mathfrak{RLB}(A)$ with index $\rho = -1$.*

1) If $\int_A^\infty f(t)\,dt = \infty$, then

$$\lim_{x\to\infty} \frac{xf(x)}{\int_A^x f(t)\,dt} = 0.$$

2) If $\int_A^\infty f(t)dt < \infty$, then

$$\lim_{x\to\infty} \frac{xf(x)}{\int_x^\infty f(t)dt} = 0.$$

Proof of Proposition 6.7 First we rewrite the corresponding part of Karamata's Theorem 6.1 in a form being convenient for our purposes. Recall that any RV-function of index $\rho = -1$ is given by $(\ell(t)/t, t \geq A)$, where ℓ is an SV-function. Thus, if ℓ is a locally integrable SV function, then Theorem 6.1 implies that (see (6.3) and (6.4))

$$\lim_{x\to\infty} \frac{\ell(x)}{\int_A^x (\ell(t)/t)dt} = 0, \tag{6.20}$$

if $\int_A^\infty (\ell(t)/t)\,dt = \infty$, and

$$\lim_{x\to\infty} \frac{\ell(x)}{\int_x^\infty (\ell(t)/t)\,dt} = 0, \tag{6.21}$$

if $\int_A^\infty (\ell(t)/t)dt < \infty$.

By assumption, f belongs to the class $\mathfrak{RLB}(A)$ and has index $\rho = -1$. Hence (see (6.13)) $f(t) = \ell(t)V(\ln t)/t, t \geq A$, where V is a positive, measurable function for which condition (6.14) is satisfied. Thus

$$\kappa \int_A^\infty (\ell(t)/t)dt \leq \int_A^\infty f(t\,dt \leq K \int_A^\infty (\ell(t)/t)\,dt$$

and, for arbitrary $x > A$,

$$0 \leq \frac{xf(x)}{\int_A^x f(t)dt} \leq \frac{K\ell(x)}{\kappa \int_A^x (\ell(t)/t)\,dt}$$

and

$$\frac{xf(x)}{\int_x^\infty f(t)dt} \leq \frac{K\ell(x)}{\kappa \int_x^\infty (\ell(t)/t)\,dt}.$$

Hence statements 1) and 2) follow from relations (6.20) and (6.21). □

6.4.3 Converse Part

The generalization of the converse part of Karamata's Theorem 6.2 (see (6.5) and (6.6)) to the class of regularly log-periodic functions is considered only for the case when $\rho \neq -1$, since Proposition 6.7 shows that statements 1) and 2) for $\rho = -1$ hold for the class of functions $\mathfrak{RLB}(A)$, which is wider than the class $\mathfrak{RLP}(A)$.

Theorem 6.8 *Let $f \in \mathbb{FM}_+(A)$ be a locally integrable function such that either*

$$\int_A^x f(t)dt \sim xB(\ln x)f(x), \qquad x \to \infty, \tag{6.22}$$

or

$$\int_A^\infty f(t)\,dt < \infty \qquad \text{and} \qquad \int_x^\infty f(t)\,dt \sim xB(\ln x)f(x), \qquad x \to \infty, \tag{6.23}$$

where $(B(t),\ t \in \mathbf{R})$ is a positive, continuous, periodic function. Then f is a regularly log-periodic function ($f \in \mathfrak{RLP}(A)$) with periodic component V and index $\rho > -1$, if (6.22) holds, and $\rho < -1$. if (6.23) holds. Moreover,

1) if $T(B) = 0$, that is, if $B(t) \equiv \beta > 0$, then

$$|\rho + 1| = \frac{1}{\beta} \tag{6.24}$$

and $V(t) \equiv 1$, that is, f is an RV-function of index ρ;

2) if $T(B) > 0$, then f belongs to the class $\mathfrak{RLP}_+(A)$ and possesses the index

$$|\rho + 1| = \left(\frac{1}{B}\right)_{av} \tag{6.25}$$

and periodic component

$$V(x) = \frac{B(0)}{B(x)} \exp\left\{\pm \int_0^x \left(\frac{1}{B(t)} - \left(\frac{1}{B}\right)_{av}\right) dt\right\}, \qquad x \geq 0, \tag{6.26}$$

where

$$\left(\frac{1}{B}\right)_{av} = \frac{1}{T(B)} \int_0^{T(B)} \frac{dt}{B(t)} \tag{6.27}$$

and the "+"-sign applies in relation (6.26), if $\rho + 1 > 0$, whereas "−" applies, if $\rho + 1 < 0$;

3) *$T(BV) = T(V) = T(B)$, where $T(BV)$ is the oscillation characteristic of a positive, continuous, periodic function $BV = (B(t)V(t),\ t \in \mathbf{R})$.*

Remark 6.9 Relation (6.22) implies that $\liminf_{x\to\infty} xf(x) > 0$, whence $\int_A^\infty f(t)\,dt = \infty$. Thus the latter condition is omitted for the statement of (6.22).

Remark 6.10 Relations (6.15), (6.16) and (6.22), (6.23) uniquely determine the functions D and B, respectively. Namely, if (6.15) holds for two positive continuous periodic functions D_1 and D_2, then $D_1 \sim D_2$, whence $D_1 = D_2$ by Lemma 5.18. Thus $D = B$ in Theorems 6.6 and 6.8, hence we conclude that, if the function B possesses a property, then the function D possesses the same property, and *vice versa*.

Remark 6.11 Passing to the limits in (6.18) and (6.19) as $T \to 0$ and if the function V tends pointwise to 1, then the Lebesgue dominated convergence theorem yields

$$D_{V,\rho}(t) \to \frac{1}{\rho+1},$$

that is, (6.17) is "compatible" with (6.18) and (6.19). Similarly, (6.24) and (6.25) are also "compatible".

Remark 6.12 If f is a locally integrable, regularly log-periodic function ($f(t) = t^\rho \ell(t)V(\ln t)$) with $\rho > -1$, then, by Theorem 6.6,

$$\int_A^x f(t)\,dt = xD(\ln x)f(x)\ell_1(x)$$
$$= x^{\rho+1}\ell(x)\ell_1(x)D(\ln x)V(\ln x) = x^{\rho+1}\ell_2(x)DV(\ln x),$$

where ℓ_1 and $\ell_2 = \ell\ell_1$ are SV-functions, DV is a positive, continuous, periodic function, and $T(DV) = T(V)$. It is clear that the function $(\int_A^x f(t)\,dt, x \ge A)$ is continuous. Therefore the mapping $f(\cdot) \mapsto \int_A^{(\cdot)} f(t)\,dt$ transforms a locally integrable, regularly log-periodic function of index $\rho > -1$ into a continuous regularly log-periodic function. The oscillation characteristic does not change under this mapping. In view of Theorem 6.6, the same properties are satisfied for regularly log-periodic functions of index $\rho < -1$ in the case of the mapping $f(\cdot) \mapsto \int_{(\cdot)}^\infty f(t)dt$.

6.5 Asymptotic Properties of Integrals

We start our study of asymptotic properties of integrals with varying limits with a result (see Lemma 6.14) showing that one can omit (in the sense of the asymptotic behavior) the slowly varying component in the integral of a regularly log-bounded function of index $\rho \ne -1$ (see Definition 6.3).

The method of the proof of Lemma 6.14 is similar to that used in the proof of the direct part of Karamata' theorem for RV-functions (see Bingham et al. [41]). An important role in the proof is played by Potter's theorem for SV-functions (see Potter [293], Bingham et al. [41]).

Theorem 6.13 (Potter's Theorem) *If $\ell = (\ell(t), t \geq A)$ is an SV-function, then, given an arbitrary $c > 1$ and $\delta > 0$, there exists an $X = X(c, \delta) \geq A$ such that*

$$\frac{\ell(z)}{\ell(u)} \leq c \max\left\{\left(\frac{z}{u}\right)^{\delta}, \left(\frac{u}{z}\right)^{\delta}\right\}, \qquad z \geq X, \quad u \geq X. \tag{6.28}$$

Now we are ready to present Lemma 6.14, where we put $V_l(t) = V(\ln t)$, whence $f(t) = t^{\rho}\ell(t)V_l(t)$.

Lemma 6.14 *Let f be a locally integrable function in the class $\mathfrak{RLB}(A)$ with index $\rho \neq -1$, SV-component ℓ, and bounded component V. If $\rho > -1$, then*

$$\int_A^x f(t)\,dt \sim \ell(x) \int_A^x t^{\rho} V_l(t)\,dt, \qquad x \to \infty. \tag{6.29}$$

If $\rho < -1$, then

$$\int_x^\infty f(t)\,dt \sim \ell(x) \int_x^\infty t^{\rho} V_l(t)\,dt, \qquad x \to \infty. \tag{6.30}$$

Proof of Lemma 6.14 The cases $\rho > -1$ and $\rho < -1$ are considered separately. First we treat the case of $\rho > -1$ and show that relation (6.29) holds in this case.

Put

$$J(x) = \int_{A/x}^1 y^{\rho}\left(\frac{\ell(xy)}{\ell(x)} - 1\right) V_l(xy)dy, \qquad x \geq A,$$

and

$$J_\varepsilon(x) = \int_\varepsilon^1 y^{\rho}\left(\frac{\ell(xy)}{\ell(x)} - 1\right) V_l(xy)dy, \qquad x \geq (A/\varepsilon)$$

for $\varepsilon \in (0, 1)$.

Since $f \in \mathfrak{RLB}$ and $\rho > -1$, we conclude from (6.14) that

$$d = \frac{\sup_{x \geq A} V_l(x)}{\rho + 1} < \infty.$$

Applying the uniform convergence for SV-functions (see Remark 3.48) we obtain

$$\lim_{x\to\infty} J_\varepsilon(x) = 0 \tag{6.31}$$

for all $\varepsilon \in (0, 1)$. Simple algebra shows that, for all $\varepsilon \in (0, 1)$ and $x \geq (A/\varepsilon)$,

$$\begin{aligned}|J(x) - J_\varepsilon(x)| &\leq \int_{A/x}^{\varepsilon} y^\rho \left|\frac{\ell(xy)}{\ell(x)} - 1\right| V_l(xy)dy \\ &\leq d\left(\varepsilon^{\rho+1} + (\rho+1)\int_{A/x}^{\varepsilon} y^\rho \frac{\ell(xy)}{\ell(x)} dy\right).\end{aligned}$$

Fix $\delta \in (0, \rho+1)$. According to Potter's Theorem 6.13 (see (6.28)), there exists an $x(\delta) > 0$ such that

$$\int_{A/x}^{\varepsilon} y^\rho \frac{\ell(xy)}{\ell(x)} dy \leq 2\int_{A/x}^{\varepsilon} y^{\rho-\delta} dy \leq \frac{2\varepsilon^{\rho+1-\delta}}{\rho+1-\delta}$$

for $x \geq \max\{x(\delta), (A/\varepsilon)\}$. Hence

$$\limsup_{x\to\infty} |J(x) - J_\varepsilon(x)| \leq d\left(\varepsilon^{\rho+1} + \frac{2(\rho+1)\varepsilon^{\rho+1-\delta}}{\rho+1-\delta}\right)$$

for all $\varepsilon \in (0, 1)$. Since $\rho + 1 - \delta > 0$, we get

$$\limsup_{\varepsilon\to 0} \limsup_{x\to\infty} |J(x) - J_\varepsilon(x)| = 0. \tag{6.32}$$

Relations (6.31) and (6.32) imply that

$$\lim_{x\to\infty} J(x) = 0.$$

Moreover, it follows from (6.14) that

$$\liminf_{x\to\infty} \int_{A/x}^{1} y^\rho V_l(xy)dy > 0.$$

Thus

$$\int_{A/x}^{1} y^\rho \ell(xy) V_l(xy)\, dy \sim \ell(x) \int_{A/x}^{1} y^\rho V_l(xy)\, dy, \qquad x \to \infty,$$

whence (6.29) follows, since

$$\int_A^x f(t)\, dt = \int_A^x t^\rho \ell(t) V_l(t) dt = x^{\rho+1} \int_{A/x}^{1} y^\rho \ell(xy) V_l(xy)\, dy$$

and

$$x^{\rho+1}\ell(x)\int_{A/x}^{1} y^{\rho}V_l(xy)\,dy = \ell(x)\int_{A}^{x} t^{\rho}V_l(t)\,dt,$$

$x \geq A$.

Now we consider the case when $\rho < -1$ and prove that (6.30) holds. Remark 6.4 shows that the corresponding integrals in (6.30) converge.

For $a \in [1, \infty)$, put

$$J_{a,\infty}(x) = \int_a^{\infty} y^{\rho}\left(\frac{\ell(xy)}{\ell(x)} - 1\right) V_l(xy)dy, \qquad x \geq A,$$

and

$$J_{1,a}(x) = \int_1^{a} y^{\rho}\left(\frac{\ell(xy)}{\ell(x)} - 1\right) V_l(xy)dy, \qquad x \geq A.$$

The uniform convergence theorem for SV-functions (see Remark 3.48) implies that

$$\lim_{x\to\infty} J_{1,a}(x) = 0 \tag{6.33}$$

for all $a \in [1, \infty)$. Moreover, since $\rho < -1$,

$$\left|J_{a,\infty}(x)\right| \leq d_1\left(\int_a^{\infty} y^{\rho}\frac{\ell(xy)}{\ell(x)}dy + \frac{a^{1+\rho}}{|1+\rho|}\right)$$

for all $a \in [1, \infty)$ and $x \geq A$, where $d_1 = \sup_{x\geq A} V_l(x)$. By condition (6.14), $d_1 < \infty$. Fix $\delta \in (0, |1+\rho|)$ and use Potter's Theorem 6.13. Then there exists an $x(\delta) > 0$ such that

$$\int_a^{\infty} y^{\rho}\frac{\ell(xy)}{\ell(x)}\,dy \leq 2\int_a^{\infty} y^{\rho+\delta}\,dy \leq \frac{2a^{1+\rho+\delta}}{|1+\rho+\delta|}$$

for $x \geq \max\{x(\delta), A\}$ (see (6.28)). Thus

$$\limsup_{x\to\infty}\left|J_{a,\infty}(x)\right| \leq d_1\Big(\frac{2a^{1+\rho+\delta}}{|1+\rho+\delta|} + \frac{a^{1+\rho}}{|1+\rho|}\Big)$$

for all $a \in [1, \infty)$. Since $1+\rho+\delta < 0$, we obtain

$$\limsup_{a\to\infty}\limsup_{x\to\infty}\left|J_{a,\infty}(x)\right| = 0. \tag{6.34}$$

Relations (6.33) and (6.34) imply that

$$\lim_{x\to\infty} J_{1,\infty}(x) = 0.$$

Moreover, we conclude from (6.14) that

$$\liminf_{x\to\infty} \int_1^\infty y^\rho V_l(xy)\,dy > 0.$$

Thus

$$\int_1^\infty y^\rho \ell(xy) V_l(xy)\,dy \sim \ell(x) \int_1^\infty y^\rho V_l(xy)\,dy, \qquad x \to \infty,$$

whence (6.30) follows, since

$$\int_x^\infty f(t)dt = \int_x^\infty t^\rho \ell(t) V_l(t)\,dt = x^{\rho+1} \int_1^\infty y^\rho \ell(xy) V_l(xy)\,dy$$

and

$$x^{\rho+1}\ell(x) \int_1^\infty y^\rho V_l(xy)\,dy = \ell(x) \int_x^\infty t^\rho V_l(t)\,dt$$

for $x \ge A$.

So, the proof of Lemma 6.14 is complete. □

Remark 6.15 For $V(x) \equiv \text{const} > 0$, Lemma 6.14 coincides with the direct part of Karamata's theorem for RV-functions of index $\rho \ne -1$ (see relations (6.1) and (6.2) in Theorem 6.1).

Remark 6.16 Relation (6.29) holds if $\rho > -1$, in which case $\int_A^\infty f(t)\,dt = \infty$ (see Remark 6.4). If $\rho > -1$, then

$$\int_{x_1}^x f(t)\,dt \sim \ell(x) \int_{x_2}^x t^\rho V(\ln t)\,dt, \qquad x \to \infty,$$

for all $x_1 \ge A$ and $x_2 > 0$.

Remark 6.17 Relations (6.29) and (6.30) can be rewritten as follows:

$$\int_A^x t^\rho \ell(t) V(\ln t)\,dt \sim \ell(x) \int_A^x t^\rho V(\ln t)\,dt, \qquad x \to \infty,$$

and

$$\int_x^\infty t^\rho \ell(t) V(\ln t)\,dt \sim \ell(x) \int_x^\infty t^\rho V(\ln t)\,dt, \qquad x \to \infty.$$

Remark 6.18 Lemma 6.14 and Remarks 6.15–6.17 hold for regularly log-periodic functions f, since this class is included in the class of regularly log-bounded functions.

6.6 An Auxiliary Result

Relation (6.10) shows that, in the proof of Theorem 6.6 for the class $\mathfrak{RLP}(A)$ of regularly log-periodic functions, one can confine oneself to the class $\mathfrak{RLP}_+(A)$ of regularly log-periodic functions with positive oscillation characteristic.

The following result is the first step toward proving Theorem 6.6.

Lemma 6.19 *Let $f \in \mathfrak{RLP}_+(A)$ be a locally integrable function with index $\rho \neq -1$, periodic component V, and oscillation characteristic $T(V) = T$. Then there exists a positive, continuous, periodic function $D = (D(t),\ t \in \mathbf{R})$, depending on ρ and V, such that*

1) if $\rho > -1$, then

$$\int_A^x f(t)\,dt \sim xD(\ln x)f(x), \qquad x \to \infty, \tag{6.35}$$

where $D(t) = D_{V,\rho}(t)$ is given by

$$D(t) = \frac{1}{V(t)\exp\{T(\rho+1)\{t/T\}\}}\left(\int_0^{T\{t/T\}} e^{(\rho+1)y}V(y)\,dy + C\right) \qquad t \geq 0,$$

$$\text{with} \qquad C = C_{V,\rho} = \frac{1}{e^{T(\rho+1)}-1}\int_0^T e^{(\rho+1)y}V(y)\,dy$$

and $\{t/T\}$ denoting the fractional part of t/T;

2) if $\rho < -1$, then

$$\int_x^\infty f(t)\,dt \sim xD(\ln x)f(x), \qquad x \to \infty, \tag{6.36}$$

where $D(t) = D_{V,\rho}(t)$ is given by

$$D(t) = \frac{1}{V(t)\exp\{T(\rho+1)\{t/T\}\}}\left(\int_{T\{t/T\}}^T e^{(\rho+1)y}V(y)\,dy + C\right), \qquad t \geq 0,$$

$$\text{with} \qquad C = C_{V,\rho} = \frac{1}{e^{T|\rho+1|}-1}\int_0^T e^{(\rho+1)y}V(y)\,dy;$$

3)

$$T(D) \le T(V), \tag{6.37}$$

where $T(D)$ is the oscillation characteristic of the function D.

Proof of Lemma 6.19 The cases $\rho > -1$ and $\rho < -1$ are considered separately. First we treat the case of $\rho > -1$ and show that (6.35) holds.

Relations (6.29) and (6.11) together with Remark 6.16 imply that

$$\int_A^x f(t)\,dt \sim \ell(x)Y_1(x), \qquad x \to \infty, \tag{6.38}$$

where ℓ is the SV-component of the function f (see (6.7)) and

$$Y_1(x) = \int_1^x t^\rho V(\ln t)\,dt, \qquad x \ge 1.$$

It is clear that

$$Y_1(x) = \int_0^{\ln x} e^{(\rho+1)t} V(t)\,dt = Z^\bullet(\ln x),$$

where

$$Z^\bullet(x) = \int_0^x e^{(\rho+1)t} V(t)\,dt = \sum_{k=0}^{\lfloor x/T\rfloor -1} \int_{kT}^{(k+1)T} e^{(\rho+1)t} V(t)\,dt + R(x),$$

$$\text{with} \qquad R(x) = \int_{T\lfloor x/T\rfloor}^x e^{(\rho+1)t} V(t)\,dt,$$

and where $\lfloor x/T\rfloor$ is the integer part of x/T.

Since V is a periodic function with period T,

$$\int_{kT}^{(k+1)T} e^{(\rho+1)t} V(t)\,dt = \int_0^T e^{(\rho+1)(y+kT)} V(y)\,dy$$
$$= e^{kT(\rho+1)} \int_0^T e^{(\rho+1)y} V(y)\,dy$$

for all $k \in \mathbf{N}$. Thus

$$Z^\bullet(x) = C\left(e^{T(\rho+1)\lfloor x/T\rfloor} - 1\right) + R(x),$$

where

$$C = \frac{1}{e^{T(\rho+1)} - 1} \int_0^T e^{(\rho+1)y} V(y)\,dy > 0.$$

Moreover,

$$R(x) = \int_{T\lfloor x/T \rfloor}^{x} e^{(\rho+1)u} V(u)\,du = e^{T(\rho+1)\lfloor x/T \rfloor} \int_0^{T\{x/T\}} e^{(\rho+1)u} V(u)\,du.$$

After some simple algebra we get

$$Y_1(x) = x^{\rho+1} \Theta^{\bullet}(\ln x) - C, \qquad x \ge 1, \tag{6.39}$$

where

$$\Theta^{\bullet}(t) = \frac{\int_0^{T\{t/T\}} e^{(\rho+1)y} V(y)dy + C}{\exp\{T(\rho+1)\{t/T\}\}}, \qquad t \ge 0.$$

For $t \ge 0$, the function $\Theta^{\bullet}$ is positive and periodic with period T. It is straightforward to check that $\Theta^{\bullet}$ is continuous. Denote by $\tilde{\Theta}^{\bullet} = (\tilde{\Theta}^{\bullet}(t),\ t \in \mathbf{R})$ the continuous, periodic continuation of the function $\Theta^{\bullet}$ to $\mathbf{R}$. Since $\rho > -1$, relation (6.39) implies that $\lim_{x\to\infty} Y_1(x) = \infty$ and hence

$$Y_1(x) \sim x^{\rho+1} \tilde{\Theta}^{\bullet}(\ln x), \qquad x \to \infty.$$

Putting

$$D(t) = \frac{\tilde{\Theta}^{\bullet}(t)}{V(t)},$$

we see that (6.35) holds in view of (6.38). It remains to note that $(D(t),\ t \in \mathbf{R})$ is a positive continuous periodic function with period T, whence relation (6.37) follows. Thus, the first statement and a part of the third statement are proved.

Now we treat the case of $\rho < -1$ and show that (6.36) holds. Note that (see (6.12)) the integral on the left-hand side of (6.36) converges in this case.

Relation (6.30) implies that

$$\int_x^{\infty} f(t)\,dt \sim \ell(x) Y(x), \qquad x \to \infty, \tag{6.40}$$

where ℓ is the SV-component of the function f (see (6.7)) and

$$Y(x) = \int_x^{\infty} t^{\rho} V(\ln t)\,dt, \qquad x \ge A.$$

It is clear that $Y(x) = Z_\bullet(\ln x)$, where

$$Z_\bullet(x) = \int_x^\infty e^{(\rho+1)t} V(t)\,dt = \sum_{k=1+\lfloor x/T\rfloor}^{\infty} \int_{kT}^{(k+1)T} e^{(\rho+1)t} V(t)\,dt + R(x)$$

with

$$R(x) = \int_x^{T(1+\lfloor x/T\rfloor)} e^{(\rho+1)t} V(t)\,dt.$$

Since the function V is periodic with period T,

$$\int_{kT}^{(k+1)T} e^{(\rho+1)t} V(t\,dt = \int_0^T e^{(\rho+1)(y+kT)} V(y)\,dy = e^{kT(\rho+1)} \int_0^T e^{(\rho+1)y} V(y)\,dy$$

for all $k \in \mathbf{N}$. Thus

$$Z_\bullet(x) = Ce^{T(\rho+1)\lfloor x/T\rfloor} + R(x),$$

where

$$C = \frac{e^{T(\rho+1)} \int_0^T e^{(\rho+1)y} V(y)\,dy}{1 - e^{T(\rho+1)}} = \frac{\int_0^T e^{(\rho+1)y} V(y)\,dy}{e^{T|\rho+1|} - 1} > 0.$$

Moreover,

$$R(x) = \int_x^{T(1+\lfloor x/T\rfloor)} e^{(\rho+1)u} V(u)\,du = e^{T(\rho+1)\lfloor x/T\rfloor} \int_{T\{x/T\}}^{T} e^{(\rho+1)u} V(u)\,du.$$

After some simple algebra we get

$$Y(x) = x^{\rho+1} \Theta_\bullet(\ln x), \qquad x \ge 1, \tag{6.41}$$

where

$$\Theta_\bullet(t) = \frac{\int_{T\{t/T\}}^{T} e^{(\rho+1)y} V(y)dy + C}{\exp\{T(\rho+1)\{t/T\}\}}, \qquad t \ge 0.$$

The function $\Theta_\bullet$ is positive and periodic with period T for $t \ge 0$. It is straightforward to check that $\Theta_\bullet$ is continuous. Denote by $\tilde{\Theta}_\bullet = (\tilde{\Theta}_\bullet(t),\ t \in \mathbf{R})$ the continuous periodic continuation of the function $\Theta_\bullet$ to $\mathbf{R}$. Putting

$$D(t) = \frac{\tilde{\Theta}_\bullet(t)}{V(t)},$$

we see that (6.36) holds in view of (6.40) and (6.41). It remains to note that $(D(t),\ t \in \mathbf{R})$ is a positive, continuous, periodic function with period T, whence relation (6.37) follows. Therefore the second and third statements are also proved. □

6.7 Proof of the Generalized Karamata Theorem

Some parts of the proof of Theorem 6.6 are already given in the previous sections. This section contains the final step of the proof of Theorem 6.6 and the proof of its converse part, i.e. Theorem 6.8. We make use of Lemma 6.19 as an intermediate result for Theorem 6.6. The complete proof Theorem 6.6 can be given after we have proved Theorem 6.8.

6.7.1 *Proof of the Converse Part of the Generalized Karamata Theorem*

The following result not only allows us to prove Theorem 6.8 but also contains some useful complements.

Theorem 6.20 *Let $f \in \mathbb{FM}_+(A)$ be a locally integrable function such that either*

$$\int_A^x f(t)dt \sim \frac{x\Gamma(\ln x)f(x)}{\gamma}, \qquad x \to \infty, \tag{6.42}$$

or

$$\int_A^\infty f(t)dt < \infty \qquad \text{and} \qquad \int_x^\infty f(t)dt \sim \frac{x\Gamma(\ln x)f(x)}{\gamma}, \qquad x \to \infty, \tag{6.43}$$

where $\gamma \in (0, \infty)$ and $(\Gamma(t),\ t \in \mathbf{R})$, $\Gamma(0) = 1$, is a positive, continuous, periodic function with oscillation characteristic $T(\Gamma) = T$. Then f is a regularly log-periodic function, i.e. $f \in \mathfrak{RLP}(A)$, whose periodic component and index are denoted by V and ρ, respectively. If (6.42) holds, then $\rho > -1$; otherwise, if (6.43) holds, then $\rho < -1$.

Moreover,

1) $T = 0$ if and only if $V(t) \equiv 1$ and

$$|\rho + 1| = \gamma,$$

that is, f is an RV-function with index ρ;

2) $T > 0$ if and only if f belongs to the class $\mathfrak{RLP}_+(A)$ with

$$|\rho + 1| = \gamma \left(\frac{1}{\Gamma}\right)_{av}$$

and

$$V(t) = \frac{1}{\Gamma(t)} \exp\left\{\pm\gamma \int_0^t \left(\frac{1}{\Gamma(u)} - \left(\frac{1}{\Gamma}\right)_{av}\right) du\right\}, \qquad t \geq 0, \tag{6.44}$$

where

$$\left(\frac{1}{\Gamma}\right)_{av} = \frac{1}{T}\int_0^T \frac{du}{\Gamma(u)}.$$

The "+"-sign in (6.44) is used if $\rho + 1 > 0$, otherwise, if $\rho + 1 < 0$, then "−" applies in (6.44);

3)

$$T(\Gamma V) = T(V) = T(\Gamma) = T, \tag{6.45}$$

where $T(V)$ is the oscillation characteristic of the function V and $T(\Gamma V)$ is the oscillation characteristic of the positive, continuous, periodic function $\Gamma V = (\Gamma(t)V(t),\ t \in \mathbf{R})$.

Proof of Theorem 6.20 First let condition (6.42) hold. Put

$$b(x) = \frac{xf(x)\Gamma(\ln x)}{\int_A^x f(t)dt}, \qquad x > A, \tag{6.46}$$

and note that

$$\lim_{x\to\infty} b(x) = \gamma \tag{6.47}$$

according to condition (6.42). Now (6.46) implies that

$$\frac{b(x)}{x\Gamma(\ln x)} = \frac{f(x)}{\int_A^x f(t)dt}.$$

Integrating both sides we prove that

$$\int_{A+1}^x \frac{b(t)dt}{t\Gamma(\ln t)} = \ln\left(\int_A^x f(t)dt\right) - \ln\left(\int_A^{A+1} f(t)dt\right)$$

for all $x \ge A + 1$. Hence

$$\int_A^x f(t)dt = a \exp\left\{\int_{A+1}^x \frac{b(t)dt}{t\Gamma(\ln t)}\right\}, \qquad x \ge A+1,$$

where $a = \int_A^{A+1} f(t)dt$. Then (6.46) implies

$$f(x) = \frac{ab(x)}{x\Gamma(\ln x)} \exp\left\{\int_{A+1}^x \frac{b(t)dt}{t\Gamma(\ln t)}\right\}.$$

It is convenient to rewrite f as

$$f(x) = \frac{ab(x)c(x)}{x\Gamma(\ln x)} \exp\left\{\gamma \int_{A+1}^x \frac{dt}{t\Gamma(\ln t)}\right\},$$

where

$$c(x) = \exp\left\{\int_{A+1}^x \frac{b(t) - \gamma}{t\Gamma(\ln t)} dt\right\}.$$

By (6.47),

$$\lim_{t\to\infty} \frac{b(t) - \gamma}{\Gamma(\ln t)} = 0,$$

whence we conclude that $(c(x),\ x \ge A+1)$ is an SV-function in view of the integral representation of SV-functions (see (3.26)). The function $ab(x)c(x)$ is also an SV-function as a product of two SV-functions. Moreover, for $x \ge A + 1$,

$$\int_{A+1}^x \frac{dt}{t\Gamma(\ln t)} = \int_{x_0}^{\ln x} \frac{dt}{\Gamma(t)} = \int_0^{\ln x} \frac{dt}{\Gamma(t)} - \int_0^{x_0} \frac{dt}{\Gamma(t)},$$

where $x_0 = \ln(A + 1)$. Thus

$$f(x) = \frac{\ell(x)}{x\Gamma(\ln x)} M(\ln x), \qquad x \ge A+1, \tag{6.48}$$

where

$$M(x) = \exp\left\{\gamma \int_0^x \frac{dt}{\Gamma(t)}\right\}, \qquad x \ge 0,$$

and

$$\ell(x) = ab(x)c(x) \exp\left\{-\gamma \int_0^{x_0} \frac{dt}{\Gamma(t)}\right\}, \qquad x \ge A+1,$$

is an SV-function.

Introduce the number $\left(\frac{1}{\Gamma}\right)_{av}$ and the positive, continuous, periodic function $\tilde{h} = (\tilde{h}(x),\ x \in \mathbf{R})$. Let

$$\left(\frac{1}{\Gamma}\right)_{av} = \frac{1}{\Gamma(0)} = 1 \tag{6.49}$$

if $T = 0$, that is, if the function Γ is constant, and

$$\left(\frac{1}{\Gamma}\right)_{av} = \frac{1}{T}\int_0^T \frac{dt}{\Gamma(t)} \tag{6.50}$$

if $T > 0$. Moreover, let $\tilde{h}(x) \equiv 0$ if $T = 0$, and

$$\tilde{h}(x) = \int_0^x \left(\frac{1}{\Gamma(t)} - \left(\frac{1}{\Gamma}\right)_{av}\right) dt, \qquad x \geq 0,$$

if $T > 0$. Note that the number T is a period of the function $\tilde{h}$.

Since, for $x \geq 0$,

$$\int_0^x \frac{dt}{\Gamma(t)} = x\left(\frac{1}{\Gamma}\right)_{av} + \int_0^x \left(\frac{1}{\Gamma(t)} - \left(\frac{1}{\Gamma}\right)_{av}\right) dt = x\left(\frac{1}{\Gamma}\right)_{av} + \tilde{h}(x)$$

and

$$M(x) = \exp\left\{x\gamma\left(\frac{1}{\Gamma}\right)_{av} + \gamma\tilde{h}(x)\right\},$$

we conclude from (6.48) the following representation of the function f:

$$f(x) = x^{\rho}\ell(x)V(\ln x), \qquad x \geq A+1, \tag{6.51}$$

where

$$\rho = \gamma\left(\frac{1}{\Gamma}\right)_{av} - 1, \qquad V(x) = \exp\{h(x)\}, \qquad x \in \mathbf{R},$$

and

$$h(x) = \gamma\tilde{h}(x) - \ln\Gamma(x), \qquad x \in \mathbf{R}.$$

Since, for $x \geq 0$,

$$V(x) = \frac{1}{\Gamma(x)}\exp\left\{\gamma\int_0^x \left(\frac{1}{\Gamma(t)} - \left(\frac{1}{\Gamma}\right)_{av}\right) dt\right\}, \tag{6.52}$$

we see that V is a positive, continuous, periodic function with period T and $V(0) = 1$. Relation (6.51) shows that f is a regularly log-periodic function of index

$$\rho = \gamma \left(\frac{1}{\Gamma}\right)_{av} - 1 > -1. \tag{6.53}$$

Now we prove the first statement of Theorem 6.20. If $T = 0$, then $\Gamma(x) \equiv 1$. Thus $V(x) \equiv 1$, that is, f is an RV-function of index $\rho = \gamma - 1$. On the other hand, if $V(x) \equiv 1$, then (6.52) implies that the function Γ is a continuously differentiable solution of the following linear differential equation of first order with constant coefficients, i.e.

$$\frac{d\Gamma(x)}{dx} = -\gamma \left(\frac{1}{\Gamma}\right)_{av} \Gamma(x) + \gamma, \qquad x \in \mathbf{R},$$

with initial condition $\Gamma(0) = 1$. Since Γ is a continuous, positive, periodic function, this is satisfied if and only if $\Gamma(x) \equiv 1$, that is, if $T = 0$. Thus $T = 0$ occurs if and only if f is an RV-function of index $\rho = \gamma - 1$. This proves the first statement of Theorem 6.20.

Correspondingly, the inequality $T > 0$ is valid only if V is a positive, continuous, periodic function with period T. Thus

$$0 < T(V) \le T. \tag{6.54}$$

This together with relations (6.51)–(6.53) implies the second statement of Theorem 6.20.

To complete the proof in the case of condition (6.42), one needs to prove relation (6.45).

Let $f \in \mathcal{RV}$. Then $V(x) \equiv \Gamma(x) \equiv 1$, whence $T(\Gamma V) = T(V) = T(\Gamma) = 0$. Now let $f \in \mathfrak{RLP}_+$. Since V is the periodic component of the function f, relation (6.53) and Lemma 6.19 imply that there is a positive, continuous, periodic function D, for which relation (6.35) and inequality

$$T(D) \le T(V) \tag{6.55}$$

hold. Condition (6.42) for the function $D_1 = \gamma^{-1}\Gamma$ implies (6.35), whence we conclude that the positive, continuous, periodic functions D and D_1 are asymptotically equivalent. By Lemma 5.18, these functions coincide and

$$T(D) = T(D_1) = T.$$

Taking relations (6.54) and (6.55) into account, we obtain

$$T(V) = T. \tag{6.56}$$

Consider the function $\Gamma V = (\Gamma(x)V(x),\ x \in \mathbf{R})$, for which

$$\Gamma(x)V(x) = \exp\left\{\gamma \int_0^x \left(\frac{1}{\Gamma(t)} - \left(\frac{1}{\Gamma}\right)_{av}\right) dt\right\}, \qquad x \ge 0,$$

by (6.52). Since the right-hand side of this relation represents a periodic function with period $T > 0$, the function ΓV has period T. Now we show that T is the principal period of the function ΓV. It is sufficient to show that there is no smaller period for the function

$$\Psi(x) = \frac{1}{\gamma} \ln(\Gamma(x)V(x)) = \int_0^x \left(\frac{1}{\Gamma(t)} - \left(\frac{1}{\Gamma}\right)_{av}\right) dt, \qquad x \ge 0.$$

Assume that there exists a number $\tau \in (0, T)$ such that

$$\Psi(x+\tau) = \Psi(x)$$

for all $x \ge 0$. The function Γ is positive and continuous and the latter equality yields

$$\Gamma(x+\tau) = \left(\frac{d\Psi(x+\tau)}{dx} + \left(\frac{1}{\Gamma}\right)_{av}\right)^{-1} = \left(\frac{d\Psi(x)}{dx} + \left(\frac{1}{\Gamma}\right)_{av}\right)^{-1} = \Gamma(x)$$

for all $x \ge 0$. This contradicts the fact that T is the minimal positive period of the function Γ. Thus $T(\Gamma V) = T$, which together with (6.56) proves (6.45). Hence, Theorem 6.20 is proved under condition (6.42).

Now we assume that condition (6.43) holds. Put

$$b(x) = \frac{x f(x) \Gamma(\ln x)}{\int_x^\infty f(t)dt}, \qquad x \ge A, \tag{6.57}$$

and note that, by (6.43),

$$\lim_{x\to\infty} b(x) = \gamma. \tag{6.58}$$

Now (6.57) implies that

$$\frac{b(x)}{x\Gamma(\ln x)} = \frac{f(x)}{\int_x^\infty f(t)dt}.$$

Integrating both sides we see that

$$\int_A^x \frac{b(t)dt}{t\Gamma(\ln t)} = -\ln\left(\int_x^\infty f(t)dt\right) + \ln\left(I_f(\infty)\right), \qquad x \ge A,$$

for all $x \ge A$, where $I_f(\infty) = \int_A^\infty f(t)\,dt$. Therefore

$$\int_x^\infty f(t)dt = I_f(\infty)\exp\left\{-\int_A^x \frac{b(t)\,dt}{t\Gamma(\ln t)}\right\}, \qquad x \ge A.$$

From (6.57) we obtain

$$f(x) = \frac{I_f(\infty)b(x)}{x\Gamma(\ln x)}\exp\left\{\int_A^x \frac{\gamma - b(t)}{t\Gamma(\ln t)}dt\right\}\exp\left\{-\gamma\int_A^x \frac{dt}{t\Gamma(\ln t)}\right\}.$$

Put $\ell_1(x) = I_f(\infty)b(x)c(x)$, $x \ge A$, where

$$c(x) = \exp\left\{\int_A^x \frac{\gamma - b(t)}{t\Gamma(\ln t)}\,dt\right\}.$$

Note that

$$\lim_{t\to\infty} \frac{b(t) - \gamma}{\Gamma(\ln t)} = 0$$

by relation (6.58), whence we conclude that $(c(x),\ x \ge A)$ is an SV-function in view of the integral representation theorem for SV-functions (see (3.26)). The function ℓ_1 is also an SV-function as a product of SV-functions. Thus

$$f(x) = \frac{\ell(x)}{x\Gamma(\ln x)}M(\ln x), \qquad x \ge A, \tag{6.59}$$

where

$$M(x) = \exp\left\{-\gamma\int_0^x \frac{dt}{\Gamma(t)}\right\}, \qquad x \ge 0,$$

and

$$\ell(x) = \ell_1(x)\exp\left\{\gamma\int_0^{\ln A} \frac{dt}{\Gamma(t)}\right\}, \qquad x \ge A.$$

It is clear that $\ell \in \mathcal{SV}$.

By the same reasoning as above, we obtain from (6.59) a representation of the function f similar to (6.51), i.e.

$$f(x) = x^\rho \ell(x) V(\ln x), \qquad x \ge A, \tag{6.60}$$

where

$$\rho = -\gamma \left(\frac{1}{\Gamma}\right)_{av} - 1,$$

$$V(x) = \exp\{h(x)\}, \qquad x \in \mathbf{R},$$

$$h(x) = -\gamma \tilde{h}(x) - \ln \Gamma(x), \qquad x \in \mathbf{R},$$

and the number $\left(\frac{1}{\Gamma}\right)_{av}$ and function $\tilde{h}$ are the same as above.

Since

$$V(x) = \frac{1}{\Gamma(x)} \exp\left\{-\gamma \int_0^x \left(\frac{1}{\Gamma(t)} - \left(\frac{1}{\Gamma}\right)_{av}\right) dt\right\}$$

for $x \ge 0$, the function V is positive, continuous, and periodic with period T and $V(0) = 1$. Hence representation (6.60) means that f is a regularly log-periodic function of index

$$\rho = -\gamma \left(\frac{1}{\Gamma}\right)_{av} - 1 < -1.$$

The rest of the proof coincides with that of Theorem 6.20 under condition (6.42). Therefore the proof of Theorem 6.20 is complete. □

Proof of Theorem 6.8 Theorem 6.8 follows from Theorem 6.20 by putting

$$\Gamma(x) = \frac{B(x)}{B(0)} \qquad \text{and} \qquad \gamma = \frac{1}{B(0)}.$$

□

6.7.2 The Proof of the Generalized Karamata Theorem, Direct Part

Now we are in position to complete the proof of Theorem 6.6.

Proof of Theorem 6.6 The first statement of Theorem 6.6 follows from the direct part of Karamata's theorem for RV-functions and also from Lemma 6.14. The second statement of Theorem 6.6 is a consequence of Lemma 6.19, while the third statement results from statement 3) of Theorem 6.8. □

6.8 Comments

This chapter is based on Buldygin and Pavlenkov [72]. The results presented here complement those in Chaps. 4 and 5.

Section 6.2 Theorems 6.1 and 6.2 are called *Karamata's theorem (direct part)* and *Karamata's theorem (converse part)*, respectively, after Bingham et al. [41, Theorems 1.5.8 and 1.6.1]. Combined together they constitute the original *Karamata integral theorem* (see Karamata [205, Theorem IV] for continuous functions and de Haan [164] for measurable functions). Another proof of Karamata's integral theorem is presented in Seneta [324, Theorem 2.1] stated in the form that only functions of the class $\mathcal{RV}$ possess both properties described in the direct and converse parts. Thus, Theorems 6.1 and 6.2 provide another characterization of the class of regularly varying functions. Yet another characterization of the class $\mathcal{RV}$, also involving integrals of regularly varying functions, is due to Aljančić and Karamata [10] (see also Bingham et al. [41, Theorem 1.6.2] or Seneta [324, Theorem 2.2]).

The case $\rho = -1$ in the Karamata theorem is special, since the limits in (6.3) and (6.4) are zero and the integrals are slowly varying (see Parameswaran [290]) rather than regularly varying as in the other cases. Moreover, the integrals are members of the *de Haan class* of functions (see de Haan [164]). A simple, yet general proof of both parts of Karamata's theorem is presented in Geluk and de Haan [148] (see also de Haan and Ferreira [167]).

Related to Karamata's theorem are the notions of *quasi-monotonicity* and *near-monotonicity* (see Bojanić and Karamata [44]) and gauge functions (see Bingham et al. [41, Section 2.8]). For example, the left gauge function for $\ell \in \mathcal{SV}$ is defined as

$$\kappa(\eta) = \limsup_{x\to\infty} \frac{1}{x^\eta \ell(\eta)} \int_0^\eta t^\eta \, |d\ell(t)|.$$

Section 6.4 Generalizations of the Karamata theorem (direct part) for the classes of functions $\mathcal{ER}$ and $\mathcal{PRV}$ are obtained in Cline [90, Theorem 4.3]. The statements in Cline [90, Theorem 4.3] are of the following nature: "... *if a function f belongs to the class $\mathcal{PRV}$, then the integral of f also belongs to $\mathcal{PRV}$* ...". The most general presentation of results of such kind can be found in Bingham et al. [41, Section 2.6]. A similar statement for rapidly varying functions is proved by Elez and Djurčić [116]. Much earlier, de Haan [164] extended Karamata's theorem to monotone rapidly varying functions multiplied by powers. A further extension is due to Bingham et al. [41, Section 2.6.3].

The case of functions of the class $\mathcal{ORV}$ is considered by Seneta [324, Theorem A.3] and Maller [259] (see also Bingham and Goldie [40] for more results concerning the class $\mathcal{ORV}$). Karamata's theorem for regularly log-periodic functions including the precise form of the limit value is proved by Buldygin and Pavlenkov [73]. An analog of Karamata's theorem for functions of dominated variation is given in Seneta [324, Theorem A.5]. Moreover, a characterization of functions of dominated variation can be provided in terms of Karamata's theorem (see Seneta [324, Theorems A.5 and A.6]. Some related results are considered by Bingham and Ostaszewski [42] for Beurling regularly varying functions.

A related result to Theorem 6.1 is obtained by Aljančić and Karamata [10]. Bingham and Goldie [40] further discuss an application of this result to Frullani integrals (see also Bingham et al. [41, Section 1.6.4]).

Soni and Soni [341] study the asymptotic behavior of the integral transform $\int_0^\infty k(xt)f(t)\,dt$ as $x \downarrow 0$ for $f \in \mathcal{RV}$ and for some kernels k. Their results correspond to the Karamata theorem (direct part) if $k(s) = \mathbb{I}_{[0,1]}(s)$, $s \geq 0$. The asymptotic behavior of the integral $\int_0^\infty \ell(xt)f(t)\,dt$ for $\ell \in \mathcal{SV}$ and general f is considered by Aljančić et al. [9] (see Vuilleumier [363] for a generalization).

We would also like to mention a classical problem for trigonometric series related to the topic of this chapter. Hardy [172, 173] was the first to find the asymptotic behavior of the series $\sum_{n=1}^\infty w(n)\cos(nx)$ as $x \to \infty$ for a regularly varying function w. This result has been generalised by many authors since then and indeed has become classical. Chen and Chen [82] were able to cover the case of $w \in \mathcal{ORV}$.

Section 6.5 In Embrechts and Goldie [118], the asymptotic behavior of the tail of an infinitely divisible distribution is compared with integrals of its Lévy measure. Some closely related results under multivariate regular variation are obtained by Hult and Lindskog [186].

Chapter 7
Asymptotically Quasi-inverse Functions

7.1 Introduction

Let f be a real function f defined and locally bounded on $[t_0, \infty)$ for some $t_0 \ge 0$. Assume that $f(t) \to \infty$ as $t \to \infty$. Then its *generalized inverse function*

$$f^{\leftarrow}(s) = \inf\{t \in [t_0, \infty) : f(t) > s\}$$

is defined on the set $[\inf_{t \ge t_0} f(t_0), \infty)$. The function $f^{\leftarrow}$ is nondecreasing and tends to ∞ as $s \to \infty$. Generalized inverses, known as *quantile functions*, play an important role in probability theory and statistics. The *value-at-Risk* or *return period*, being generalized inverses of corresponding functions, are useful tools in financial and insurance mathematics.

If f is continuous and (strictly) increasing, then its inverse function f^{-1} exists and coincides with $f^{\leftarrow}$. The inverse function is characterized by the following two properties:

(i) $f(f^{-1}(t)) = t$, $t \in [f(t_0), \infty)$,
(ii) $f^{-1}(f(t)) = t$, $t \in [t_0, \infty)$.

The inverse function f^{-1} may not exist if f is discontinuous or f is not increasing. In many of these cases the generalized inverse function is a natural substitution for the inverse function but $f^{\leftarrow}$ cannot satisfy both conditions (i) and (ii). For example, if the function f is continuous and $f(t) \to \infty$ as $t \to \infty$, then

$$f(f^{\leftarrow}(t)) = t, \qquad t \ge f(t_0), \tag{7.1}$$

that is, condition (i) is satisfied for $f^{\leftarrow}$, but (ii) may fail. Along with generalized inverse functions, one may consider other "inverse" functions possessing

V. V. Buldygin et al., *Pseudo-Regularly Varying Functions and Generalized Renewal Processes*, Probability Theory and Stochastic Modelling 91, https://doi.org/10.1007/978-3-319-99537-3_7

property (7.1). If f is not monotone, then the first and last exit times from an interval $(0, x)$ are different and similarly the generalized inverse function has many other analogs defined by considering subsequent exit times.

RV-functions f with positive indices provide another appropriate example. In this case, the generalized inverse functions satisfy conditions (i) and (ii) in an "asymptotic sense", namely

$$f(f^{\leftarrow}(t)) \sim f^{\leftarrow}(f(t)) \sim t, \qquad t \to \infty. \tag{7.2}$$

There are various known definitions of "*quasi-inverse*" functions. Each of them weakens conditions (i) and (ii) in one way or another. We shall consider *asymptotically quasi-inverse functions* $f^{\sim}$ defined by the following asymptotic analogue of (i):

$$f(f^{\sim}(t)) \sim t, \qquad t \to \infty. \tag{7.3}$$

We also consider *asymptotically inverse functions* for which, in addition to the latter condition, an asymptotic analogue of (ii) is satisfied, i.e.

$$f^{\sim}(f(t)) \sim t, \qquad t \to \infty. \tag{7.4}$$

Therefore conditions (i) and (ii) are satisfied for asymptotically inverse functions, but only in an asymptotic sense. In view of (7.2), every generalized inverse function is an asymptotically inverse function as well in the class $\mathcal{RV}$. This statement fails in the general case even if f is increasing and left continuous (see Example 7.7 below).

It is sometimes convenient to say that a function $f^{\sim}$ is *a right asymptotically inverse function* to f if (7.3) holds or *a left asymptotically inverse function* to f if (7.4) holds.

Note that an asymptotically quasi-inverse function $f^{\sim}$ may not be unique, even if f possesses an inverse function. On the other hand, not all functions f possess an asymptotically quasi-inverse function, which raises the question about the class of functions f for which asymptotically quasi-inverse functions exist.

In this chapter, we consider four problems concerning asymptotically quasi-inverse functions:

(A) For which functions are their generalized inverse functions also asymptotically quasi-inverse?
(B) For which functions are their asymptotically quasi-inverse functions also asymptotically inverse?
(C) For which functions are their asymptotically quasi-inverse functions asymptotically equivalent?
(D) For which functions is the asymptotic behavior of their asymptotically quasi-inverse functions determined by the asymptotic behavior of the original functions (and *vice versa*)?

Problems (A)–(D) are closely related to each other and will be considered separately for the classes $\mathcal{PRV}$, $\mathcal{SQI}$, and $\mathcal{POV}$ (recall Sect. 3.3).

Note that problem (D) is related to the following question: which conditions yield the implications

$$\lim_{t\to\infty} \frac{g(t)}{f(t)} = a \quad \Longrightarrow \quad \lim_{t\to\infty} \frac{g_{\text{inv}}(t)}{f_{\text{inv}}(t)} = b, \tag{7.5}$$

$$\lim_{t\to\infty} \frac{g(t)}{f(t)} = a \quad \Longleftarrow \quad \lim_{t\to\infty} \frac{g_{\text{inv}}(t)}{f_{\text{inv}}(t)} = b, \tag{7.6}$$

$$\lim_{t\to\infty} \frac{g(t)}{f(t)} = a \quad \Longleftrightarrow \quad \lim_{t\to\infty} \frac{g_{\text{inv}}(t)}{f_{\text{inv}}(t)} = b, \tag{7.7}$$

where $f(t) \to \infty$ and $g(t) \to \infty$ as $t \to \infty$, and where f_{inv} and g_{inv} are "inverse functions" in some sense for f and g, respectively, and $a, b \in [0, \infty]$?

Such implications are well known in probability theory. They appear, for example, when studying the relationship between strong laws of large numbers for sums of independent, identically distributed random variables and their renewal process counterparts (see Chaps. 1 and 2). In Chaps. 8 and 9, some applications of the general theory developed in this chapter will be given to study the asymptotic behavior of renewal processes or that of the solutions of stochastic differential equations.

A general result proved in this chapter (see Sect. 7.5) shows that, under certain assumptions,

$$\lim_{t\to\infty} \frac{g(t)}{f(t)} = a \in (0, \infty) \quad \Longrightarrow \quad \lim_{t\to\infty} \frac{g^{\sim}(t)}{f^{\sim}(t/a)} = 1, \tag{7.8}$$

where $f^{\sim}$ and $g^{\sim}$ are *asymptotically quasi-inverse functions* for f and g, respectively. Note that the right-hand side in (7.8) differs from the right-hand side of (7.5). If f is a regularly varying function with index $\rho > 0$ (see (3.5)), then the right-hand side in (7.8) transforms to

$$\lim_{t\to\infty} \frac{g^{\sim}(t)}{f^{\sim}(t)} = \left(\frac{1}{a}\right)^{\frac{1}{\rho}} \tag{7.9}$$

which is similar to the right-hand side of (7.5).

In Sect. 7.6, we prove that relation (7.5) holds under certain assumptions for $a \neq 1$, $b \neq 1$, f_{inv} and g_{inv} being asymptotically quasi-inverse functions only if f belongs to the class of functions with nondegenerate set of regular points (recall Sect. 5.2).

The main aim of this chapter is to study properties of asymptotically quasi-inverse functions and to solve problems (A)–(D). To do so, we need a more careful consideration of the classes $\mathcal{PRV}$, $\mathcal{SQI}$, and $\mathcal{POV}$.

Another general problem is discussed in this chapter, namely, let f be a nondecreasing function and let $f^{\sim}$ be its asymptotically quasi-inverse function, then the question is about conditions to be imposed on the function f (or on $f^{\sim}$) to imply that $f^{\sim}$ (or f) belongs to a certain class of functions with some generalized regular variation? An answer to this question allows us to determine the classes of functions containing both f ($f^{\sim}$) and $f^{\sim}$ (f). In other words, we determine the classes of functions which are invariant under the transformations $f \mapsto f^{\sim}$ and $f^{\sim} \mapsto f$. A simple example of such a class is given by the class of continuous, monotone, regularly varying functions with positive indices.

Chapter 7 is organized as follows. First, asymptotically quasi-inverse and asymptotically inverse functions are considered in Sect. 7.2.

In Sect. 7.3, conditions are given under which quasi-inverse functions preserve the asymptotic equivalence of functions.

Properties and characterizations of POV-functions and their asymptotically quasi-inverse functions are studied in Sect. 7.4.

The limit behavior of the ratio of asymptotically quasi-inverse functions is discussed in Sect. 7.5.

A relationship between limit points of the ratio of two functions and those of the corresponding quasi-inverse functions are studied in Sect. 7.6.

In Sect. 7.7, we deal with a relationship between the asymptotic properties of f and $f^{\sim}$.

Using the results obtained in the preceding sections, we study in Sect. 7.8 properties of piecewise linear interpolations and the ratio of functions being asymptotically quasi-inverse to the underlying sequences.

Finally, in Sect. 7.9, we apply the general results to investigate the asymptotic stability of solutions of the Cauchy problem for ordinary differential equations.

7.2 Asymptotically Quasi-inverse and Asymptotically Inverse Functions

In this section, we study asymptotically quasi-inverse and asymptotically inverse functions and find conditions for their existence. Theorem 7.16 shows that every PRV-function $f \in \mathbb{F}^{\infty}$ (see Definition 3.16) possesses an asymptotically quasi-inverse function. It follows from Theorem 7.18 that every POV-function (see Definition 3.32) possesses an asymptotically inverse function.

7.2.1 Generalized Inverse Functions

For $f \in \mathbb{F}^{(\infty)}$ and $s \geq 0$, we consider the following four *generalized inverse functions*:

(GIF1) $f^{\leftarrowtail}(s) = \inf\{t \geq 0 : f(t) \geq s\}$,
(GIF2) $f^{\leftarrow}(s) = \inf\{t \geq 0 : f(t) > s\}$,
(GIF3) $f^{\rightarrow}(s) = \sup\{t \geq 0 : f(t) < s\}$,
(GIF4) $f^{\rightarrowtail}(s) = \sup\{t \geq 0 : f(t) \leq s\}$,

where $\sup \varnothing = 0$. Note that $f^{\leftarrow}(\cdot)$ is the generalized inverse function for f. Obviously, all four functions $f^{\leftarrowtail}(t)$, $f^{\leftarrow}(t)$, $f^{\rightarrow}(t)$, and $f^{\rightarrowtail}(t)$ are analogues of the functionals $L(t)$, $M(t)$, and $T(t)$ studied in Chaps. 1 and 2.

All these functions are nondecreasing and tend to ∞ as $s \to \infty$. It is possible that $f^{\rightarrow}(s) = \infty$ or $f^{\rightarrowtail}(s) = \infty$ for all s. It is also clear that, for all $s > 0$ and $t > 0$,

$$f^{\leftarrowtail}(s) \leq f^{\leftarrow}(s) \leq f^{\rightarrowtail}(s), \qquad f^{\leftarrowtail}(s) \leq f^{\rightarrow}(s) \leq f^{\rightarrowtail}(s), \tag{7.10}$$

$$f^{\leftarrowtail}(f(t)) \leq t \leq f^{\rightarrowtail}(f(t)).$$

If $f \in \mathbb{F}^{\infty}_{\mathrm{inc}}$, then for all sufficiently large s,

$$f^{\leftarrowtail}(s) = f^{\leftarrow}(s) = f^{\rightarrow}(s) = f^{\rightarrowtail}(s) \tag{7.11}$$

and these functions belong to the class $\mathbb{C}^{\infty}_{\mathrm{ndec}}$. In addition, if $f \in \mathbb{C}^{\infty}_{\mathrm{inc}}$, then there exists the *inverse function* $f^{-1} \in \mathbb{C}^{\infty}_{\mathrm{inc}}$ such that, for all sufficiently large s and t,

$$f(f^{-1}(s)) = s \qquad \text{and} \qquad f^{-1}(f(t)) = t,$$

and, for all sufficiently large s,

$$f^{-1}(s) = f^{\leftarrowtail}(s) = f^{\leftarrow}(s) = f^{\rightarrow}(s) = f^{\rightarrowtail}(s).$$

7.2.2 Quasi-inverse Functions

Following Buldygin et al. [59], we recall the definition of a *quasi-inverse function*, which will be useful in the sequel.

Definition 7.1 Let $f \in \mathbb{F}^{(\infty)}$. A function $f^{(-1)} \in \mathbb{F}^{\infty}$ is called a *quasi-inverse function* for f if

$$f(f^{(-1)}(s)) = s$$

for all sufficiently large $s > 0$.

For any $f \in \mathbb{C}^{(\infty)}$, a quasi-inverse function exists, but may not be unique. Below are some appropriate examples.

Example 7.2 Let $x \in \mathbb{C}^{(\infty)}$ and $s_0 = x(0) > 0$. Consider the "first exit time" from the set $(-\infty, s)$, i.e.

$$x_F^{(-1)}(s) = \begin{cases} 0, & 0 \le s < s_0, \\ \inf\{t \ge 0 : x(t) = s\}, & s \ge s_0. \end{cases}$$

It is clear that $x_F^{(-1)} = x^{\leftarrowtail} = x^{\leftarrow}$ and $x(x_F^{(-1)}(s)) = s$ for all $s > s_0$. Thus, the functions $x_F^{(-1)}$, $x^{\leftarrowtail}$, and $x^{\leftarrow}$ are quasi-inverses for x.

Example 7.3 Let $x \in \mathbb{C}^{\infty}$ and $s_0 = x(0) > 0$. Consider the "last exit time" from $(-\infty, s)$, i.e.

$$x_L^{(-1)}(s) = \begin{cases} 0, & 0 \le s < s_0, \\ \sup\{t \ge 0 : x(t) = s\}, & s \ge s_0. \end{cases}$$

It is clear that $x_L^{(-1)} = x^{\rightarrow} = x^{\rightarrowtail}$ and $x(x_L^{(-1)}(s)) = s$ for all $s > s_0$. Thus, the functions $x_L^{(-1)}$, $x^{\rightarrow}$, and $x^{\rightarrowtail}$ are quasi-inverses for x. Note that $x_F^{(-1)}(s) \le x_L^{(-1)}(s)$, $s > 0$, and, in general, $x_F^{(-1)} \ne x_L^{(-1)}$.

Example 7.4 Let $x \in \mathbb{C}^{(\infty)}$ and $s_0 = x(0) > 0$. Assume that

$$\operatorname{card}\{t \ge 0 : x(t) = s\} \in (1, \infty)$$

if $s > s_0 = \max\{x(0), 0\}$. Consider the "second exit time", i.e.

$$x_2^{(-1)}(s) = \begin{cases} 0, & 0 \le s < s_0, \\ \inf\{t > x_F^{(-1)}(s) : x(t) = s\}, & s \ge s_0. \end{cases}$$

It is clear that $x(x_2^{(-1)}(s)) = s$ for all $s > s_0$. Thus, $x_2^{(-1)}$ is a quasi-inverse function for x. Note that

$$x_F^{(-1)}(s) \le x_2^{(-1)}(s) \le x_L^{(-1)}(s)$$

for sufficiently large s. Similarly to the case of $n = 2$, one can define the "n^{th} exit time" for $n > 2$.

Example 7.5 If $f \in POV$, then Theorem 3.79 implies that there exists an asymptotically equivalent function $g \in \mathbb{C}_{\text{inc}}^{\infty}$, which obviously possesses an inverse function g^{-1}. Since $f \sim g$ and $g^{-1} \in \mathbb{C}_{\text{inc}}^{\infty}$,

$$f(g^{-1}(s)) \sim g(g^{-1}(s)) \sim s \quad \text{as} \quad s \to \infty.$$

Thus, g^{-1} is a quasi-inverse function for f.

7.2.3 *Asymptotically Quasi-inverse and Asymptotically Inverse Functions*

Definition 7.6 Let $f \in \mathbb{F}^{(\infty)}$. We say that $f^{\sim}$ is an *asymptotically quasi-inverse function* for f if

(i) $f^{\sim} \in \mathbb{F}^{\infty}$;
(ii) $f(f^{\sim}(s)) \sim s \quad \text{as} \quad s \to \infty$.

We denote by $\mathbb{AQI}(f)$ the class of asymptotically quasi-inverse functions for f.

It is sometimes convenient to say that $f^{\sim}$ is a right asymptotically quasi-inverse function if it possesses properties (i) and (ii). One can also speak of a left asymptotically quasi-inverse function for which (ii) is substituted with

(ii′) $f^{\sim}(f(s)) \sim s \quad \text{as} \quad s \to \infty$.

First we show by an example that the generalized inverse function $f^{\leftarrow}(\cdot)$, even if f is increasing and left-continuous, does not necessarily belong to $\mathbb{AQI}(f)$.

Example 7.7 Let $\{x_n\}_{n\ge1}$, $\{v_n\}_{n\ge1}$, and $\{g_n\}_{n\ge1}$ be three increasing sequences such that

$$x_n \to \infty, \qquad g_n \to \infty, \qquad g_n < v_n < g_{n+1}, \qquad \frac{g_n}{v_n} \not\to 1, \qquad \frac{g_n}{g_{n+1}} \not\to 1.$$

Consider the function

$$g(x) = \sum_{n=1}^{\infty} g_n \mathbb{I}_{(x_n, x_{n+1}]}(x), \qquad x \ge x_1.$$

Then $g^{\leftarrow}(v_n) = x_{n+1}$ and thus $g(g^{\leftarrow}(v_n)) = g_n$. Therefore $g^{\leftarrow}(\cdot)$ is not a right asymptotically quasi-inverse function.

Further, $g^{\leftarrow}(g(x_{n+1})) = g^{\leftarrow}(g_n) = g_{n+1}$ and $g^{\leftarrow}(\cdot)$ is not a left asymptotically quasi-inverse function.

It is clear that every quasi-inverse function is an asymptotically quasi-inverse function.

Example 7.8 If $x \in \mathbb{F}^\infty$ and

$$x^{\leftarrow}(s) \sim x^{\rightarrowtail}(s) \quad \text{as} \quad s \to \infty,$$

then (7.10) implies that

$$x^{\leftarrow}(x(s)) \sim x^{\rightarrowtail}(x(s)) \sim s \quad \text{as} \quad s \to \infty,$$

that is, x is an asymptotically quasi-inverse function for $x^{\leftarrow}$ and $x^{\rightarrowtail}$.

Example 7.9 If $x \in \mathbb{F}^\infty_{\text{inc}}$, then Example 7.8 and relation (7.11) show that x is an asymptotically quasi-inverse function for $x^{\leftarrow}$ and $x^{\rightarrowtail}$.

Along with asymptotically quasi-inverse functions we consider asymptotically inverse functions (see, for example, Bingham et al. [41]).

Definition 7.10 Let $f \in \mathbb{F}^{(\infty)}$. A function $\tilde{f}^{-1}$ is called *asymptotically inverse* for f if $\tilde{f}^{-1}$ is asymptotically quasi-inverse for f, that is,

(i) $\tilde{f}^{-1} \in \mathbb{F}^\infty$;
(ii) $f(\tilde{f}^{-1}(s)) \sim s \quad \text{as} \quad s \to \infty$,

and, in addition,

(iii) $\tilde{f}^{-1}(f(t)) \sim t \quad \text{as} \quad t \to \infty$.

We denote by $\mathbb{AI}(f)$ the class of asymptotically inverse functions for f.

Every inverse function is asymptotically inverse. Definition 7.10 also implies that $\mathbb{AI}(f) \subset \mathbb{AQI}(f)$. If $f \in \mathbb{F}^\infty$ and an asymptotically inverse function $\tilde{f}^{-1}$ exists, then the function f is also asymptotically inverse for $\tilde{f}^{-1}$.

Example 7.11 If $x \in \mathbb{C}^\infty$ and

$$x^{\leftarrow}(s) \sim x^{\rightarrowtail}(s) \quad \text{as} \quad s \to \infty,$$

then Examples 7.2, 7.3, and 7.8 show that x is an asymptotically inverse function for both $x^{\leftarrow}$ and $x^{\rightarrowtail}$. In turn, the functions $x^{\leftarrow}$ and $x^{\rightarrowtail}$ are both asymptotically inverse for x.

In the following proposition, we list some simple, but useful properties of asymptotically quasi-inverse and asymptotically inverse functions.

Proposition 7.12 *Let* $f \in \mathbb{F}^{(\infty)}$.

(A) If $g \sim f$, *then* $\mathbb{AQI}(g) = \mathbb{AQI}(f)$.
(B) If $f \in \mathcal{WPRV}$, $\varphi \in \mathbb{AQI}(f)$, *and* $q \sim \varphi$, *then* $q \in \mathbb{AQI}(f)$.
(C) If $f \in \mathbb{F}^\infty$, $g \sim f$, *and* $\psi \in \mathbb{AI}(f) \cap \mathcal{WPRV}$, *then* $\psi \in \mathbb{AI}(g)$ *and* $g \in \mathbb{AI}(\psi)$.
(D) If $f \in \mathbb{F}^\infty$ *and* $g \sim f$, *then* $\mathbb{AI}(g) \cap \mathcal{WPRV} = \mathbb{AI}(f) \cap \mathcal{WPRV}$.

(E) *If* $f \in \mathbb{F}^\infty \cap \mathcal{WPRV}$, $\psi \in \mathbb{AI}(f)$ *and* $q \sim \psi$, *then* $q \in \mathbb{AI}(f)$ *(and* $f \in \mathbb{AI}(q)$ *as well).*

(F) *If* $f \in \mathbb{F}^\infty \cap \mathcal{WPRV}$, $\psi \in \mathbb{AI}(f) \cap \mathcal{WPRV}$, $g \sim f$, *and* $q \sim \psi$, *then* $g \in \mathbb{AI}(q)$ *and* $q \in \mathbb{AI}(g)$.

Proof of Proposition 7.12 If $g(t) \sim f(t)$, as $t \to \infty$, then $g(f^{\sim}(s)) \sim s$, as $s \to \infty$, since $f^{\sim}(s) \to \infty$ and $f(f^{\sim}(s)) \sim s$, as $s \to \infty$, by Definition 7.6. Thus, $f^{\sim} \in \mathbb{AQI}(g)$, that is, $\mathbb{AQI}(f) \subset \mathbb{AQI}(g)$. Since the relation of asymptotic equivalence $g \sim f$ is symmetric with respect to f and g, the same method proves $\mathbb{AQI}(g) \subset \mathbb{AQI}(f)$. Therefore, $\mathbb{AQI}(g) = \mathbb{AQI}(f)$ and (A) is proved.

If $f \in \mathcal{WPRV}$, $\varphi \in \mathbb{AQI}(f)$ and $q \sim \varphi$, then $f(q(s)) \sim f(\varphi(s)) \sim s$, as $s \to \infty$, since the function f preserves the asymptotic equivalence of functions in view of Theorem 3.42. This proves (B).

If $g \sim f$ and $\psi \in \mathbb{AI}(f)$, then (A) implies that $\psi \in \mathbb{AQI}(g)$. If, additionally, $f \in \mathbb{F}^\infty$ and $\psi \in \mathcal{WPRV}$, then (B) implies that $g \in \mathbb{AQI}(\psi)$, since $f \in \mathbb{AQI}(\psi)$ by Definition 7.10. Therefore, $\psi \in \mathbb{AI}(g)$ and $g \in \mathbb{AI}(\psi)$, since $g \in \mathbb{F}^\infty$. Thus (C) is proved.

Now (C) implies that

$$\mathbb{AI}(f) \cap \mathcal{WPRV} \subset \mathbb{AI}(g) \cap \mathcal{WPRV}.$$

Since the asymptotic equivalence $g \sim f$ is symmetric with respect to f and g, similarly

$$\mathbb{AI}(g) \cap \mathcal{WPRV} \subset \mathbb{AI}(f) \cap \mathcal{WPRV}.$$

Therefore, $\mathbb{AI}(g) \cap \mathcal{WPRV} = \mathbb{AI}(f) \cap \mathcal{WPRV}$, that is, (D) is proved.

Statement (E) follows from (C), since the function f is asymptotically inverse for $\psi \in \mathbb{AI}(f)$, if $f \in \mathbb{F}^\infty$ (see Definition 7.10). In turn, (F) follows from (E) and (C). □

Example 7.13 Let f be an RV-function with positive index ρ. Then (see Bingham et al. [41], p. 28) there exists an asymptotically inverse function $\tilde{f}^{-1}$, which is unique up to asymptotic equivalence. This asymptotically inverse function belongs to $\mathcal{RV}_{1/\rho}$. Moreover, one of the asymptotically equivalent versions of $\tilde{f}^{-1}$ is the generalized inverse

$$f^{\leftarrow}(s) = \inf\{t \ge 0 : f(t) > s\}.$$

7.2.4 Asymptotically Quasi-inverse Functions for PRV-Functions

An important property of PRV-functions is that their generalized inverses are, under certain natural assumptions, also asymptotically quasi-inverse.

Lemma 7.14 *Let $f \in \mathcal{PRV}$.*

1) If $f \in \mathbb{F}^{(\infty)}$, then the two generalized inverses $f^{\leftarrow}$ and $f^{\leftarrowtail}$ are asymptotically quasi-inverse for f.
2) If $f \in \mathbb{F}^{\infty}$, then the four generalized inverses $f^{\leftarrow}$, $f^{\leftarrowtail}$, $f^{\rightarrow}$, and $f^{\rightarrowtail}$ are asymptotically quasi-inverse for f.

Proof of Lemma 7.14 The proof of this result uses some ideas of the proof of Theorem 1.5.12 in Bingham et al. [41] together with Theorem 3.59.

Consider the function $f^{\leftarrow}$. It is clear that condition (i) in Definition 7.6 holds for $f^{\sim} = f^{\leftarrow}$. It remains to check condition (ii).

Choose $A > 1$. Then, in view of Theorem 3.59, there are $b > 0$, $\lambda_A > 1$, and $u_A > 0$ such that, for all $\lambda \in (1, \lambda_A)$, the inequalities

$$A^{-1}\lambda^{-b} f(v) \le f(u) \le A\lambda^{b} f(v)$$

hold for all $u \ge u_A$ and $v \in [\lambda^{-1}u, \lambda u]$. Fix $\lambda \in (1, \lambda_A)$ and choose s such that $f^{\leftarrow}(s) > u_A$. According to the definition of $f^{\leftarrow}$, there exists a $v' \in [f^{\leftarrow}(s), \lambda f^{\leftarrow}(s)]$ such that $f(v') > s$. Moreover, $f(v'') \le s$ for all $v'' \in [\lambda^{-1} f^{\leftarrow}(s), f^{\leftarrow}(s)]$. Put $u = f^{\leftarrow}(s)$. First we choose $v = v'$ and then $v = v''$. As a result,

$$A^{-1}\lambda^{-b} s \le A^{-1}\lambda^{-b} f(v') \le f(f^{\leftarrow}(s)) \le A\lambda^{b} f(v'') \le A\lambda^{b} s.$$

Thus,

$$A^{-1}\lambda^{-b} \le \liminf_{s\to\infty} \frac{f(f^{\leftarrow}(s))}{s} \le \limsup_{s\to\infty} \frac{f(f^{\leftarrow}(s))}{s} \le A\lambda^{b}.$$

Passing to the limit as $\lambda \downarrow 1$ and then as $A \downarrow 1$, we see that $f(f^{\leftarrow}(s)) \sim s$ as $s \to \infty$. Therefore, $f^{\leftarrow}$ is an asymptotically quasi-inverse function for f.

The proof for the functions $f^{\leftarrowtail}$, $f^{\rightarrow}$, and $f^{\rightarrowtail}$ is the same. □

The following result complements Examples 7.2, 7.3, and 7.8–7.11.

Lemma 7.15 *Let $f \in \mathbb{F}^{\infty}$.*

1) If $f^{\leftarrowtail} \sim f^{\rightarrowtail}$, then f is an asymptotically quasi-inverse function for each of the four generalized inverses $f^{\leftarrowtail}$, $f^{\leftarrow}$, $f^{\rightarrow}$, and $f^{\rightarrowtail}$.
2) If $f \in \mathbb{F}^{\infty}_{\mathrm{inc}}$, then f is an asymptotically quasi-inverse function for each of the four generalized inverses $f^{\leftarrowtail}$, $f^{\leftarrow}$, $f^{\rightarrow}$, and $f^{\rightarrowtail}$.
3) If $f \in \mathbb{C}^{\infty}$ and $f^{\leftarrowtail} \sim f^{\rightarrowtail}$, then f is an asymptotically inverse function for each of the four generalized inverses $f^{\leftarrowtail}$, $f^{\leftarrow}$, $f^{\rightarrow}$, and $f^{\rightarrowtail}$; in turn, each of these four functions is asymptotically inverse for f.

Proof of Lemma 7.15 Indeed, the definition of the functions $f^{\leftarrowtail}$ and $f^{\rightarrowtail}$ implies that

$$f^{\leftarrowtail}(f(t)) \le t \le f^{\rightarrowtail}(f(t))$$

for all sufficiently large t (see (7.10)). Thus,

$$\frac{f^{\leftarrowtail}(f(t))}{f^{\rightarrowtail}(f(t))} \le \frac{f^{\leftarrowtail}(f(t))}{t} \le \frac{f^{\rightarrowtail}(f(t))}{t} \le \frac{f^{\rightarrowtail}(f(t))}{f^{\leftarrowtail}(f(t))}$$

and

$$\lim_{t\to\infty} \frac{f^{\leftarrowtail}(f(t))}{t} = \lim_{t\to\infty} \frac{f^{\rightarrowtail}(f(t))}{t} = 1.$$

This together with (7.10) implies that

$$\lim_{t\to\infty} \frac{f^{\leftarrow}(f(t))}{t} = \lim_{t\to\infty} \frac{f^{\rightarrow}(f(t))}{t} = 1,$$

since $f^{\leftarrowtail}(s) \le f^{\leftarrow}(s) \le f^{\rightarrowtail}(s)$ and $f^{\leftarrowtail}(s) \le f^{\rightarrow}(s) \le f^{\rightarrowtail}(s)$ for all $s > 0$. Therefore, statement 1) is proved.

Statement 2) follows from 1) together with (7.11), and statement 3) follows from 1) combined with Examples 7.2 and 7.3. □

The following result contains conditions for the existence of asymptotically inverse functions.

Theorem 7.16 *Let $f \in \mathbb{F}^{(\infty)}$.*

1) *If $f \in \mathcal{PRV}$, then the generalized inverse functions $f^{\leftarrowtail}$ and $f^{\leftarrow}$ are both asymptotically quasi-inverse for f.*
2) *If $f \in \mathcal{PRV}$ and $f \in \mathbb{F}^{\infty}$, then the generalized inverse functions $f^{\leftarrowtail}$, $f^{\leftarrow}$, $f^{\rightarrow}$, and $f^{\rightarrowtail}$ are asymptotically quasi-inverse for f.*
3) *If $f \in \mathbb{C}^{(\infty)}$, then the generalized inverse functions $f^{\leftarrow}$ and $f^{\leftarrowtail}$ are quasi-inverse for f.*
4) *If $f \in \mathbb{C}^{\infty}$, then the generalized inverse functions $f^{\leftarrowtail}$, $f^{\leftarrow}$, $f^{\rightarrow}$, and $f^{\rightarrowtail}$ are quasi-inverses for f.*
5) *If $f \in \mathbb{F}^{\infty}$ and $f^{\leftarrowtail} \sim f^{\rightarrowtail}$, then f is an asymptotically quasi-inverse function for each of the generalized inverses $f^{\leftarrowtail}$, $f^{\leftarrow}$, $f^{\rightarrow}$, and $f^{\rightarrowtail}$.*
6) *If $f \in \mathbb{F}^{\infty}_{\mathrm{inc}}$, then f is an asymptotically quasi-inverse function for each of the generalized inverses $f^{\leftarrowtail}$, $f^{\leftarrow}$, $f^{\rightarrow}$, $f^{\rightarrowtail}$.*
7) *If $f \in \mathcal{PRV}$, $f \in \mathbb{F}^{\infty}$ and $f^{\leftarrowtail} \sim f^{\rightarrowtail}$, then the generalized inverse functions $f^{\leftarrowtail}$, $f^{\leftarrow}$, $f^{\rightarrow}$, and $f^{\rightarrowtail}$ are asymptotically inverse for f; on the other hand, f is asymptotically inverse for each of the four generalized inverses.*

8) *If* $f \in \mathcal{PRV}$ *and* $f \in \mathbb{F}^{\infty}_{\text{inc}}$, *then the generalized inverse functions* $f^{\leftarrowtail}$, $f^{\leftarrow}$, $f^{\rightarrow}$, *and* $f^{\rightarrowtail}$ *are asymptotically inverse for* f; *on the other hand,* f *is an asymptotically inverse function for each of the four generalized inverses.*
9) *If* $f \in \mathbb{C}^{\infty}$ *and* $f^{\leftarrowtail} \sim f^{\rightarrowtail}$, *then the generalized inverse functions* $f^{\leftarrowtail}$, $f^{\leftarrow}$, $f^{\rightarrow}$ *and,* $f^{\rightarrowtail}$ *are asymptotically inverse for* f; *on the other hand,* f *is an asymptotically inverse function for each of the four generalized inverses.*

Theorem 7.16 follows from Lemmas 7.14 and 7.15 combined with Examples 7.2 and 7.3

The following result shows how one can construct asymptotically quasi-inverse (inverse) functions with the help of generalized inverse functions $f^{\leftarrowtail}$ and $f^{\rightarrowtail}$.

Corollary 7.17 *Let* $f \in \mathbb{F}^{(\infty)} \cap \mathcal{PRV}$ *and assume there exists a nondecreasing function* h *such that* $f \sim h$. *Moreover, let* $q \in \mathbb{F}_{+}$ *and assume there exists an* $s_0 > 0$ *such that*

$$f^{\leftarrowtail}(s) \le q(s) \le f^{\rightarrowtail}(s) \quad \textit{as} \quad s \ge s_0.$$

Then:

1) *the function* q *is asymptotically quasi-inverse for* f;
2) *if* $f^{\leftarrowtail} \sim f^{\rightarrowtail}$, *then the function* q *is asymptotically inverse for* f *and*

$$q \sim f^{\leftarrowtail} \sim f^{\leftarrow} \sim f^{\rightarrow} \sim f^{\rightarrowtail}.$$

Proof of Corollary 7.17 Since $f \in \mathbb{F}^{(\infty)} \cap \mathcal{PRV}$, $f \sim h$, and h is a nondecreasing function, then $f \in \mathbb{F}^{\infty} \cap \mathcal{PRV}$ and $h \in \mathbb{F}^{\infty} \cap \mathcal{PRV}$.

Theorem 7.16 and Proposition 7.12 imply that $f^{\leftarrowtail}$ and $f^{\rightarrowtail}$ are asymptotically quasi-inverse functions for h, that is,

$$h(f^{\leftarrowtail}(s)) \sim s \sim h(f^{\rightarrowtail}(s)) \quad \text{as} \quad s \to \infty.$$

Thus $h(q(s)) \sim s$ as $s \to \infty$, since

$$h(f^{\leftarrowtail}(s)) \le h(q(s)) \le h(f^{\rightarrowtail}(s))$$

for all sufficiently large s. Hence, q is an asymptotically quasi-inverse function for h and, by Proposition 7.12, for f as well. Therefore, statement 1) is proved.

Further, let $f^{\leftarrowtail} \sim f^{\rightarrowtail}$. Then

$$f^{\leftarrowtail} \sim q \sim f^{\rightarrowtail},$$

since $f^{\leftarrowtail}(s) \le q(s) \le f^{\rightarrowtail}(s)$ for all sufficiently large s. Now Theorem 7.16 implies that

$$q(f(s)) \sim s \quad \text{as} \quad s \to \infty.$$

So, it is clear that statement 2) follows from statement 1). □

7.2.5 Asymptotically Inverse Functions for POV-Functions

We show in Theorem 7.18 that every POV-function possesses an asymptotically inverse function.

Theorem 7.18 *Let $f \in \mathcal{POV}$ and $q \in \mathbb{F}_+$. Then:*

1) *The generalized inverse functions $f^{\leftarrowtail}$, $f^{\leftarrow}$, $f^{\rightarrow}$, and $f^{\rightarrowtail}$ are asymptotically equivalent and are asymptotically inverse for f.*
2) *If there exists an $s_0 > 0$ such that*

$$f^{\leftarrowtail}(s) \le q(s) \le f^{\rightarrowtail}(s) \quad as \quad s \ge s_0,$$

then q is an asymptotically inverse function for f and

$$q \sim f^{\leftarrowtail} \sim f^{\leftarrow} \sim f^{\rightarrow} \sim f^{\rightarrowtail}.$$

Proof of Theorem 7.18 Since $f \in \mathcal{POV}$, we get $f \in \mathbb{F}^\infty$. In addition, there exists an increasing continuous POV-function h such that $f \sim h$ (see Theorem 3.79).

Recalling Theorem 7.16 and Corollary 7.17 we see that the asymptotic equivalence of functions $f^{\leftarrowtail}$ and $f^{\rightarrowtail}$ suffices for the proof of Theorem 7.18. First, note that Corollary 3.78 yields that condition (3.45) holds for all sequences of positive numbers $\{c_n\}_{n\ge1}$ and $\{t_n\}_{n\ge1}$ such that $\limsup_{n\to\infty} c_n > 1$ and $\lim_{n\to\infty} t_n = \infty$.

Assume that $f^{\leftarrowtail}$ and $f^{\rightarrowtail}$ are not asymptotically equivalent functions. Then there exists a sequence $\{s_n\}_{n\ge1}$ such that $\lim_{n\to\infty} s_n = \infty$ and

$$\limsup_{n\to\infty} \frac{f^{\rightarrowtail}(s_n)}{f^{\leftarrowtail}(s_n)} > 1. \tag{7.12}$$

The definition of the functions $f^{\leftarrowtail}$ and $f^{\rightarrowtail}$ implies that there exist two sequences $\{a_n\}_{n\ge1}$ and $\{b_n\}_{n\ge1}$ such that $a_n \uparrow 1$, $b_n \downarrow 1$, and $f(a_n f^{\rightarrowtail}(s_n)) \le s_n$ and $f(b_n f^{\leftarrowtail}(s_n)) \ge s_n$. Thus,

$$\limsup_{n\to\infty} \frac{f(a_n f^{\rightarrowtail}(s_n))}{f(b_n f^{\leftarrowtail}(s_n))} \le 1.$$

On the other hand, inequalities (7.12) and condition (3.45) result in

$$\limsup_{n\to\infty} \frac{f(a_n f^{\rightarrowtail}(s_n))}{f(b_n f^{\leftarrowtail}(s_n))} > 1.$$

This contradiction proves Theorem 7.18. □

To complete this section, we present some examples showing that, in general, the generalized inverse functions $f^{\leftarrowtail}$ and $f^{\rightarrowtail}$ are not asymptotically equivalent, but are quasi-inverse for the function f. Recall that $\lfloor x \rfloor$ denotes the integer part of a real number x.

Example 7.19 Let $f(t) = \lfloor \ln t \rfloor, t \geq 1$. The function f belongs to $\mathcal{SV}$ and thus to $\mathcal{PRV}$, but not to $\mathcal{POV}$ (see Remark 3.33). By Lemma 7.14, the functions $f^{\leftarrow}$ and $f^{\rightarrow}$ are asymptotically quasi-inverse for f. On the other hand, these functions are not asymptotically equivalent. Moreover, the function $(e^t,\ t \geq 0)$ is asymptotically quasi-inverse for f. Note that f does not have any asymptotically inverse functions. Indeed, if such a function g existed, then we would have $g(f(t)) \sim t$ as $t \to \infty$. This means that, for $\theta \in (0, 1)$ and $t_n = \exp\{n + \theta\}$,

$$g(f(t_n)) = g(n) \sim \exp\{n + \theta\} \quad \text{as} \quad n \to \infty,$$

which is not the case, since θ is arbitrary. Note that $\lfloor \ln t \rfloor \sim \ln t$ as $t \to \infty$ and the function $(\ln t, t \geq 1)$ does not belong to $\mathcal{POV}$. Nevertheless it possesses the inverse function $(e^t,\ t \geq 0)$, and $f^{\leftarrow}$ and $f^{\rightarrow}$ coincide with $(e^t,\ t \geq 0)$.

Example 7.20 Let $f(t) = e^{\lfloor t \rfloor},\ t \geq 0$. The generalized inverse functions $f^{\leftarrow}$ and $f^{\rightarrow}$ are not asymptotically quasi-inverse for f. On the other hand, the functions $f^{\leftarrow}$ and $f^{\rightarrow}$ are asymptotically equivalent. Note that f does not belong to $\mathcal{PRV}$.

The following examples demonstrate that asymptotically quasi-inverse functions may not exist, if the original functions do not belong to $\mathcal{PRV}$.

Example 7.21 The function $(e^{\lfloor t \rfloor},\ t \geq 0)$ does not belong to $\mathcal{PRV}$. We show that this function does not possess any asymptotically quasi-inverse functions. Indeed, given an arbitrary function $\varphi \in \mathbb{F}_+$ and a sequence $t_n = \exp\{n + \frac{1}{2}\}$, $n \geq 1$, we have either $\lfloor \varphi(t_n) \rfloor \geq n + 1$ or $\lfloor \varphi(t_n) \rfloor \leq n$. This implies that either

$$\begin{aligned}\limsup_{t\to\infty} \frac{\exp\{\lfloor \varphi(t) \rfloor\}}{t} &\geq \limsup_{n\to\infty} \frac{\exp\{\lfloor \varphi(t_n) \rfloor\}}{t_n} \\ &= \exp\left\{\limsup_{n\to\infty} \left(\lfloor \varphi(t_n) \rfloor - n - \frac{1}{2}\right)\right\} \geq \exp\left\{\frac{1}{2}\right\} > 1,\end{aligned}$$

or

$$\begin{aligned}\liminf_{t\to\infty} \frac{\exp\{\lfloor \varphi(t) \rfloor\}}{t} &\leq \liminf_{n\to\infty} \frac{\exp\{\lfloor \varphi(t_n) \rfloor\}}{t_n} \\ &= \exp\left\{\liminf_{n\to\infty} \left(\lfloor \varphi(t_n) \rfloor - n - \frac{1}{2}\right)\right\} \leq \exp\left\{-\frac{1}{2}\right\} < 1.\end{aligned}$$

Example 7.22 Example 7.21 and Proposition 7.12 imply that $f(t) = \frac{t}{\lfloor t \rfloor} \cdot e^{\lfloor t \rfloor}$, $t \geq 1$, does not have any asymptotically quasi-inverse functions, since $f(t) \sim e^{\lfloor t \rfloor}$ as $t \to \infty$. Note that f is an increasing function.

Example 7.23 Let $a \in \mathbb{F}_+$. Similar to Example 7.21, it can be shown that the function $(e^{\lfloor a(t) \rfloor}, t > 0)$ does not have any asymptotically quasi-inverse functions. Lemma 7.14 implies that $(e^{\lfloor a(t) \rfloor}, t > 0)$ does not belong to $\mathcal{PRV}$ for an arbitrary function $a \in \mathbb{F}^{(\infty)}$.

7.2.6 Asymptotically Right and Asymptotically Left Quasi-inverse Functions

Let $f \in \mathbb{F}^{(\infty)}$. An asymptotically quasi-inverse function $f^{\sim}$ is characterized by the property

$$(f \circ f^{\sim})(t) \sim t, \qquad t \to \infty,$$

(see Definition 7.6), where the symbol $\circ$ stands for the composition of functions.

Definition 7.24 Any function $f_r^{\sim}$ such that $f_r^{\sim} \in \mathbb{F}^{\infty}$ and

$$(f \circ f_r^{\sim})(t) \sim t, \qquad t \to \infty, \tag{7.13}$$

is called an *asymptotically right quasi-inverse* function for f.

It is clear that an "asymptotically right quasi-inverse function" is another name for an "asymptotically quasi-inverse function" (see Definition 7.10). To emphasize a certain "symmetry" property, we introduce one more notion.

Definition 7.25 Any function $f_\ell^{\sim}$ such that $f_\ell^{\sim} \in \mathbb{F}^{(\infty)}$ and

$$(f_\ell^{\sim} \circ f)(t) \sim t, \qquad t \to \infty. \tag{7.14}$$

is called an *asymptotically left quasi-inverse* function.

The generalized inverse function $f^{\leftarrow}$ is asymptotically right quasi-inverse for f, if f is continuous. Similarly, $f^{\leftarrow}$ is asymptotically left quasi-inverse for f, if f is increasing. In turn, an asymptotically inverse function is both asymptotically right quasi-inverse and asymptotically left quasi-inverse.

If g is an asymptotically right quasi-inverse function for $f \in \mathbb{F}^{\infty}$, then f is an asymptotically left quasi-inverse function for g, and *vice versa*. According to this symmetry, one can transform statements regarding right quasi-inverse functions to statements regarding left quasi-inverse functions, and *vice versa* (see Theorems 7.80–7.83).

7.3 Properties of Asymptotically Quasi-inverse Functions

In this section, we find conditions for a function f, under which their asymptotically quasi-inverse function $f^{\sim}$ and asymptotically inverse function $\tilde{f}^{-1}$ belong to $\mathcal{WPRV}$ or to $\mathcal{PRV}$, respectively (see Definition 3.16). In other words, we find conditions under which asymptotically quasi-inverse functions $f^{\sim}$ and asymptotically inverse functions $\tilde{f}^{-1}$ preserve the asymptotic equivalence of functions and sequences (see Definitions 3.39 and 3.41).

Among the conditions to be imposed on the function f, under which $f^{\sim} \in \mathcal{WPRV}$, the crucial one is that f should be a WSQI-function (see Definition 3.28), that is, condition (3.14) should be satisfied for f.

Considering the correspondence between f and $f^{\sim}$ (or $\tilde{f}^{-1}$), the WPRV- and WSQI-properties are "dual" in some sense. The corresponding statements will be proved in this section.

Recall that (see Definition 7.6 and 7.10) $\mathbb{AQI}(f)$ denotes the class of asymptotically quasi-inverse functions for f and $\mathbb{AI}(f)$ denotes the class of asymptotically inverse functions for f.

Lemma 7.26 *Let $f \in \mathbb{F}^{(\infty)}$ and $f^{\sim} \in \mathbb{AQI}(f)$. A function $f^{\sim}$ belongs to $\mathcal{WPRV}$ (thus, preserves the asymptotic equivalence of functions and sequences) if and only if*

$$\limsup_{c\to 1}\limsup_{t\to\infty}\frac{f^{\sim}(ct)}{f^{\sim}(t)} = 1. \tag{7.15}$$

Lemma 7.26 follows immediately from Definition 3.16 and Theorem 3.42.

The following result provides conditions on the function f, under which $f^{\sim} \in \mathcal{WPRV}$.

Proposition 7.27 *Let $f \in \mathcal{WSQI}$, that is, let condition (3.14) hold and $f^{\sim} \in \mathbb{AQI}(f)$. If there exists a nondecreasing function h such that $f \sim h$, then $f^{\sim} \in \mathcal{WPRV}$ (thus, $f^{\sim}$ preserves the asymptotic equivalence of functions and sequences).*

Proof of Proposition 7.27 Since $\mathcal{WSQI} \subset \mathcal{WPI}$ (see Remarks 3.22 and 3.33), Lemma 3.63 implies that $f \in \mathbb{F}^{(\infty)}$. Moreover, $f \in \mathbb{F}^{\infty}$ and $h \in \mathbb{F}^{\infty}_{\text{ndec}}$, since $f \sim h$, where h is a nondecreasing function (see Corollary 3.64).

Assume that condition (3.14) holds, but condition (7.15) does not. Then there exist a number $\delta > 0$ and two sequences $\{c_n\}_{n\geq 1}$ and $\{s_n\}_{n\geq 1}$ such that $c_n \to 1$ and $s_n \to \infty$ as $n \to \infty$ and

$$f^{\sim}(c_n s_n) \geq (1+\delta) f^{\sim}(s_n) \quad \text{for all} \quad n \geq 1.$$

According to (3.14),

$$\begin{aligned}
1 &= \lim_{n\to\infty} \frac{f(f^{\sim}(c_n s_n))}{c_n s_n} = \lim_{n\to\infty} \frac{f(f^{\sim}(c_n s_n))}{s_n} = \lim_{n\to\infty} \frac{f(f^{\sim}(c_n s_n))}{f(f^{\sim}(s_n))} \\
&= \lim_{n\to\infty} \frac{h(f^{\sim}(c_n s_n))}{h(f^{\sim}(s_n))} \geq \liminf_{n\to\infty} \frac{h((1+\delta) f^{\sim}(s_n))}{h(f^{\sim}(s_n))} \\
&\geq \liminf_{t\to\infty} \frac{h((1+\delta)t)}{h(t)} = \liminf_{t\to\infty} \frac{f((1+\delta)t)}{f(t)} > 1.
\end{aligned}$$

This contradiction proves the implication (3.14) $\Longrightarrow$ (7.15). An application of Lemma 7.26 completes the proof of Proposition 7.27. □

Proposition 7.27 shows that

$$f \in \mathcal{WSQI} \quad \Longrightarrow \quad f^{\sim} \in \mathcal{WPRV}$$

under certain assumptions. The following result complements Proposition 7.27 and states that the same assumptions yield

$$f \in \mathcal{WPRV} \quad \Longrightarrow \quad f^{\sim} \in \mathcal{WSQI}.$$

Proposition 7.28 *Let $f \in \mathbb{F}^{(\infty)} \cap \mathcal{WPRV}$ and $f^{\sim} \in \mathbb{AQI}(f)$.*

1) *If there exists a nondecreasing function h such that $f \sim h$, then $f^{\sim} \in \mathcal{WSQI}$, that is,*

$$(f^{\sim})_*(c) = \liminf_{t\to\infty} \frac{f^{\sim}(ct)}{f^{\sim}(t)} > 1 \quad \textit{for all} \quad c > 1 \tag{7.16}$$

(see condition (3.14)).

2) *If there exists a nondecreasing function h such that $f^{\sim} \sim h$, then $f^{\sim} \in \mathcal{WSQI}$, that is, condition (7.16) holds.*

Proof of Proposition 7.28 We first prove statement 1). If condition (7.16) does not hold, then there exist a number $c_0 > 1$ and a sequence of real numbers $\{t_n\}_{n\geq 1}$ such that $t_n \to \infty$ as $n \to \infty$ and

$$\lim_{n\to\infty} \frac{f^{\sim}(t_n)}{f^{\sim}(c_0 t_n)} = \beta \in [1, \infty]. \tag{7.17}$$

If $\beta = 1$, then (7.17) implies that

$$1 > \frac{1}{c_0} = \lim_{n\to\infty} \frac{f(f^{\sim}(t_n))}{f(f^{\sim}(c_0 t_n))} = 1, \tag{7.18}$$

since, by Theorem 3.42, f preserves the asymptotic equivalence of sequences. If $\beta \in (1, \infty]$, then

$$1 > \frac{1}{c_0} = \lim_{n\to\infty} \frac{f(f^{\sim}(t_n))}{f(f^{\sim}(c_0 t_n))} = \lim_{n\to\infty} \frac{h(f^{\sim}(t_n))}{h(f^{\sim}(c_0 t_n))} \geq 1,$$

since $f \sim h$ and h is nondecreasing. This contradiction proves statement 1).

We now prove statement 2). If condition (7.16) does not hold, then relation (7.17) is satisfied, whence

$$1 \geq \lim_{n\to\infty} \frac{h(t_n)}{h(c_0 t_n)} = \lim_{n\to\infty} \frac{f^{\sim}(t_n)}{f^{\sim}(c_0 t_n)} = \beta \geq 1,$$

since $f^{\sim} \sim h$, where the function h is nondecreasing. Thus, $\beta = 1$ and (7.18) holds. This contradiction proves statement 2). □

Example 7.29 The generalized inverse functions $f^{\leftarrow}$, $f^{\leftarrowtail}$, $f^{\rightarrow}$, and $f^{\rightarrowtail}$ are nondecreasing, thus these functions are measurable. If $f \in \mathbb{F}^{(\infty)} \cap \mathcal{PRV}$, then Proposition 7.28 and Lemma 7.14 imply that $f^{\leftarrow}$ and $f^{\leftarrowtail}$ are SQI-functions. A similar reasoning proves that $f^{\leftarrow}$, $f^{\leftarrowtail}$, $f^{\rightarrow}$, and $f^{\rightarrowtail}$ are SQI-functions if $f \in \mathbb{F}^{\infty} \cap \mathcal{PRV}$.

Since every asymptotically inverse function $\tilde{f}^{-1}$ is asymptotically quasi-inverse for f and, if $f \in \mathbb{F}^{\infty}$, the function f is asymptotically quasi-inverse for $\tilde{f}^{-1}$ (see Definitions 7.6 and 7.10). This together with Proposition 7.28 yields the following result.

Corollary 7.30 *Let $f \in \mathbb{F}^{\infty}$ and $\psi \in \mathbb{AI}(f)$. If $\psi \in \mathcal{WPRV}$ and if there exists a nondecreasing function h such that either $f \sim h$ or $\psi \sim h$, then $f \in \mathcal{WSQI}$, that is, condition (3.14) holds.*

Corollary 7.30 and Propositions 7.27 and 7.28 imply the following two theorems.

Theorem 7.31 *Let $f \in \mathbb{F}^{(\infty)}$ and $\psi \in \mathbb{AI}(f)$. If there exists a nondecreasing function h such that $f \sim h$, then $\psi \in \mathcal{WPRV}$ (and therefore ψ preserves the asymptotic equivalence of functions and sequences) if and only if $f \in \mathcal{WSQI}$, that is, if condition (3.14) holds.*

Theorem 7.32 *Let $f \in \mathbb{F}^{\infty}$ and $\psi \in \mathbb{AI}(f)$. If there exists a nondecreasing function h such that $\psi \sim h$, then $f \in \mathcal{WPRV}$ (and therefore f preserves the asymptotic equivalence of functions and sequences) if and only if $\psi \in \mathcal{WSQI}$, that is, if*

$$(\tilde{f}^{-1})_*(c) = \liminf_{t\to\infty} \frac{\tilde{f}^{-1}(ct)}{\tilde{f}^{-1}(t)} > 1 \quad \textit{for all} \quad c > 1. \tag{7.19}$$

If $f \in \mathbb{C}^{\infty}_{\text{inc}}$, then the inverse function exists for f. Then, $f^{-1} \in \mathbb{C}^{\infty}_{\text{inc}}$ and Theorems 7.31 and 7.32 imply the following result.

Corollary 7.33 *Let $f \in \mathbb{C}^{\infty}_{\text{inc}}$ and f^{-1} be the inverse function for f. The function f^{-1} belongs to $\mathcal{PRV}$ (and therefore preserves the asymptotic equivalence of functions and sequences) if and only if $f \in \mathcal{SQI}$, that is, if condition (3.14) holds. Moreover, $f \in \mathcal{PRV}$ (and therefore f preserves the asymptotic equivalence of functions and sequences) if and only if $f^{-1} \in \mathcal{SQI}$, that is, if*

$$(f^{-1})_*(c) = \liminf_{t\to\infty} \frac{f^{-1}(ct)}{f^{-1}(t)} > 1 \quad \textit{for all} \quad c > 1$$

(see condition (3.14)).

Since $f \in \mathcal{POV}$ is equivalent to $f \in \mathcal{PRV} \cap \mathcal{SQI}$ (see Definition 3.32), Corollary 7.33 implies the following important characterization, which can even be generalized (see Corollary 7.42).

Corollary 7.34 *Let $f \in \mathbb{C}^{\infty}_{\text{inc}}$ and f^{-1} be the inverse function for f. The following statements are equivalent:*

1) $f \in \mathcal{POV}$;
2) $f^{-1} \in \mathcal{POV}$;
3) $f \in \mathcal{SQI}$ and $f^{-1} \in \mathcal{SQI}$;
4) $f \in \mathcal{PRV}$ and $f^{-1} \in \mathcal{PRV}$;
5) f and f^{-1} preserve the asymptotic equivalence of functions and sequences.

Some additional properties of WSQI-functions are listed in Theorem 7.35. This result complements Proposition 7.27 and, in particular, shows that the classes of asymptotically quasi-inverse functions and of asymptotically inverse functions coincide with the WSQI-functions under some additional assumptions. Similar results for POV-functions are considered in the next section; some applications for obtaining relationships between the asymptotic behavior of a ratio of WSQI-functions and that of their asymptotically quasi-inverse functions are contained in Sect. 7.5 (see Theorem 7.50).

Theorem 7.35 *Let $f \in \mathcal{WSQI}$ and*

$$f \sim f_0 \in \mathbb{F}_{\text{ndec}}. \tag{7.20}$$

Moreover, let

$$\mathbb{AI}(f_0) \neq \varnothing \tag{7.21}$$

and $\psi_0 \in \mathbb{AI}(f_0)$. Then:

1) $\psi_0 \in \mathbb{AI}(f)$ and $\psi_0 \in \mathcal{WPRV}$;
2) if $g \sim f$, then

$$\mathbb{AQI}(g) = \mathbb{AI}(g) = \mathbb{AI}(f) = \mathbb{AQI}(f) \subset \mathcal{WPRV}; \tag{7.22}$$

3) if $g \sim f$, $q \in \mathbb{AQI}(g)$ and $\varphi \in \mathbb{AQI}(f)$, then $q \sim \varphi \sim \psi_0$.

Proof of Theorem 7.35 Since $f \sim f_0 \in \mathbb{F}_{\mathrm{ndec}}$ and $f \in \mathcal{WSQI}$, we get $f_0 \in \mathcal{WSQI}$, $f_0 \in \mathbb{F}^{\infty}_{\mathrm{ndec}}$, and $f \in \mathbb{F}^{\infty}$ (see Lemma 3.63 and the beginning of the proof of Proposition 7.27).

Proposition 7.27 implies that $\psi_0 \in \mathcal{WPRV}$ and ψ_0 preserves the equivalence of functions by Theorem 3.42. Thus

$$\psi_0(f(s)) \sim \psi_0(f_0(s)) \sim s \quad \text{as} \quad s \to \infty.$$

This yields

$$f \in \mathbb{AQI}(\psi_0). \tag{7.23}$$

Moreover, since $f \sim f_0$,

$$f(\psi_0(s)) \sim f_0(\psi_0(s)) \sim s \quad \text{as} \quad s \to \infty.$$

Thus, $\psi_0 \in \mathbb{AQI}(f)$ and from (7.23) we see that $\psi_0 \in \mathbb{AI}(f)$. So, the first statement is proved.

To prove the second statement assume that $g \sim f$ and $q \in \mathbb{AQI}(g)$. Then

$$f(q(s)) \sim g(q(s)) \sim s \quad \text{as} \quad s \to \infty.$$

Hence

$$q \in \mathbb{AQI}(f). \tag{7.24}$$

According to statement 1), ψ_0 preserves the asymptotic equivalence of functions, whence

$$q(s) \sim \psi_0(f_0((q(s))) \sim \psi_0(f(q(s))) \sim \psi_0(s) \quad \text{as} \quad s \to \infty.$$

Thus,

$$q \sim \psi_0 \tag{7.25}$$

and we conclude that

$$q(f(s)) \sim \psi_0(f(s)) \sim s \quad \text{as} \quad s \to \infty.$$

This proves that $f \in \mathbb{AQI}(q)$. This together with (7.24) implies that $q \in \mathbb{AI}(f)$. Therefore, the inclusion

$$\mathbb{AQI}(g) \subset \mathbb{AI}(f)$$

is proved from which we immediately obtain that

$$\mathbb{AI}(g) \subset \mathbb{AQI}(g) \subset \mathbb{AI}(f). \tag{7.26}$$

The assumptions of statement 2) in Theorem 7.35 are symmetric with respect to the functions f and g and we conclude that, along with (7.26), the further inclusion

$$\mathbb{AI}(f) \subset \mathbb{AQI}(f) \subset \mathbb{AI}(g)$$

also holds. From the latter together with (7.26) we obtain

$$\mathbb{AQI}(g) = \mathbb{AI}(g) = \mathbb{AI}(f) = \mathbb{AQI}(f).$$

Relation (7.25) and statement 1) imply that $q \in \mathcal{WPRV}$, whence

$$\mathbb{AQI}(g) \subset \mathcal{WPRV}$$

and statement 2) is also proved.

To prove statement 3) assume that $g \sim f$, $q \in \mathbb{AQI}(g)$, and $\varphi \in \mathbb{AQI}(f)$. Recall that $q \sim \psi_0$ (see (7.25)). Since the assumptions of statement 3) in Theorem 7.35 are symmetric with respect to the functions f and g, we conclude that, along with (7.26), the relation $\varphi \sim \psi_0$ also holds. This implies that $q \sim \varphi$ and statement 3) is proved. □

7.4 Asymptotically Quasi-inverse Functions and the POV-Property

The results obtained in the previous two sections together with Theorem 3.79 on equivalent increasing and continuous versions for POV-functions allow us to study the properties of asymptotically quasi-inverse and asymptotically inverse functions for the class $\mathcal{POV}$ rather completely (see Definitions 3.32, 7.6 and 7.10).

The existence of asymptotically quasi-inverse functions and asymptotically inverse functions for the class $\mathcal{POV}$ has been proved in Theorem 7.18. This result also shows that the generalized inverse functions $f^{\hookleftarrow}$, $f^{\leftarrow}$, $f^{\rightarrow}$, and $f^{\rightarrowtail}$ are asymptotically equivalent and asymptotically inverse functions for f.

The following result describes the class of functions that are asymptotically quasi-inverse for WPOV- and POV-functions. In particular, we prove that only WPOV-functions can be asymptotically quasi-inverse for members of the class $\mathcal{POV}$. Therefore, only POV-functions can be measurable, asymptotically quasi-inverses functions for members of the class $\mathcal{POV}$.

Proposition 7.36 *Let $f \in \mathcal{WPOV}$ and $f^{\sim} \in \mathbb{AQI}(f)$.*

1) *If there exists a nondecreasing function h such that $f \sim h$, then $f^{\sim} \in \mathcal{WPOV}$.*
2) *If $f \in \mathcal{POV}$, then $f^{\sim} \in \mathcal{WPOV}$.*
3) *If $f \in \mathcal{POV}$ and $f^{\sim}$ is measurable, then $f^{\sim} \in \mathcal{POV}$.*

Proof of Proposition 7.36 Let $f \in \mathcal{WPOV}$ and let h be a nondecreasing function such that $f \sim h$. Since

$$f \in \mathcal{WPOV} \Longleftrightarrow f \in \mathcal{WPRV} \cap \mathcal{WSQI},$$

we have $f \in \mathcal{WSQI}$. This implies that $f \in \mathbb{F}^{\infty}$ (see the proof of Proposition 7.27). Thus statement 1) follows from Propositions 7.27 and 7.28.

Statement 2) follows from statement 1) in view of Theorem 3.79 on the equivalent increasing version of a POV-function.

In turn, statement 3) follows from statement 2) by Definition 3.32.

An asymptotically inverse function $\tilde{f}^{-1}$ is asymptotically quasi-inverse for f and f is asymptotically quasi-inverse for $\tilde{f}^{-1}$, if $f \in \mathbb{F}^{\infty}$ (see Definitions 7.6 and 7.10). Thus Proposition 7.36 follows from the following result. □

Corollary 7.37 *Let $f \in \mathbb{F}^{\infty}$ and $\psi \in \mathbb{AI}(f)$.*

1) *If $f \in \mathcal{POV}$ and ψ is measurable, then $\psi \in \mathcal{POV}$.*
2) *If $\psi \in \mathcal{POV}$ and f is measurable, then $f \in \mathcal{POV}$.*
3) *If f and ψ are measurable, then $f \in \mathcal{POV}$ if and only if $\psi \in \mathcal{POV}$.*

If $f \in \mathcal{POV}$, then f is a PRV-function. Moreover, according to Proposition 7.36, $\tilde{f}^{-1}$ is a WPOV-function and thus a WPRV-function. This is an important property, since, by Theorem 3.42, the functions f and $\tilde{f}^{-1}$ preserve the asymptotic equivalence of functions and sequences. It turns out that this property characterizes POV-functions under certain extra conditions.

Proposition 7.38 *Let $f \in \mathbb{F}^{\infty}$ and $\psi \in \mathbb{AI}(f)$.*

1) *If $f \in \mathcal{POV}$, then $\psi \in \mathcal{WPRV}$ and both functions f and ψ preserve the asymptotic equivalence of functions and sequences.*
2) *If $f \in \mathcal{POV}$ and ψ is measurable, then $\psi \in \mathcal{PRV}$.*
3) *If $f \in \mathcal{PRV}$, $\psi \in \mathcal{WPRV}$ and there exists a nondecreasing function h such that $f \sim h$ or $\psi \sim h$, then $f \in \mathcal{POV}$.*

Proof of Proposition 7.38 As mentioned above, statement 1) follows from Proposition 7.36 and Theorem 3.42. Statement 2) is immediate from statement 1). Finally, statement 3) follows from Corollary 7.27, since $\mathcal{POV} = \mathcal{PRV} \cap \mathcal{WSQI}$. □

A result similar to Proposition 7.38 is valid if one exchanges the PRV-condition for the SQI-condition.

Proposition 7.39 *Let $f \in \mathbb{F}^\infty$ and $\psi \in \mathbb{AI}(f)$.*

1) *If $f \in \mathcal{POV}$, then $f \in \mathcal{SQI}$ and $\psi \in \mathcal{WSQI}$.*
2) *If $f \in \mathcal{POV}$ and ψ is measurable, then $f \in \mathcal{SQI}$ and $\psi \in \mathcal{SQI}$.*
3) *If $f \in \mathcal{SQI}$, $\psi \in \mathcal{WSQI}$, and there exists a nondecreasing function h such that $\psi \sim h$, then $f \in \mathcal{POV}$.*

Proof of Proposition 7.39 If $f \in \mathcal{POV}$, then $f \in \mathcal{SQI}$. Moreover, according to Proposition 7.36, ψ is a WPOV-function and thus a WSQI-function. This means that statement 1) holds, whence statement 2) follows immediately. Finally, if the assumptions of statement 3) are satisfied, then $f \in \mathcal{WPRV}$ by applying Proposition 7.27 to the function ψ. To complete the proof, just note that $\mathcal{POV} = \mathcal{SQI} \cap \mathcal{WPRV}$. □

Theorem 7.40 *Let $f \in \mathbb{F}^\infty$ and $\psi \in \mathbb{AI}(f)$. Also let the functions f and ψ be measurable. Then the following two statements are equivalent:*

(a) $f \in \mathcal{POV}$;
(b) $\psi \in \mathcal{POV}$.

If there exists a nondecreasing function h such that $f \sim h$ or $\psi \sim h$, then each of the statements (a) and (b) is equivalent to each of the following two statements:

(c) $f \in \mathcal{PRV}$ and $\psi \in \mathcal{PRV}$;
(d) f and ψ preserve the asymptotic equivalence of functions and sequences.

If there exists a nondecreasing function h such that $\psi \sim h$, then each of the statements (a), (b), (c), and (d) is equivalent to

(e) $f \in \mathcal{SQI}$ and $\psi \in \mathcal{SQI}$.

Theorem 7.40 follows from Propositions 7.36–7.39, Corollary 7.37, and Theorem 3.42.

The following result is a direct corollary of Theorem 7.40.

Corollary 7.41 *Let $f \in \mathbb{F}^\infty$ and $\psi \in \mathbb{AI}(f)$. Also let f and ψ be measurable. If there exists a nondecreasing function h such that $\psi \sim h$, then the following conditions are equivalent:*

1) *$f \in \mathcal{POV}$;*
2) *$\psi \in \mathcal{POV}$;*
3) *$f \in \mathcal{SQI}$ and $\psi \in \mathcal{SQI}$;*
4) *$f \in \mathcal{PRV}$ and $\psi \in \mathcal{PRV}$;*
5) *f and ψ preserve the asymptotic equivalence of functions and sequences.*

Since any nondecreasing function is measurable, Corollary 7.41 for nondecreasing asymptotically inverse functions can be stated in the following way.

Corollary 7.42 *Let $f \in \mathbb{F}^{\infty}$ and $\psi \in \mathbb{AI}(f)$. If f is measurable and ψ is nondecreasing, then the following conditions are equivalent:*

1) $f \in \mathcal{POV}$;
2) $\psi \in \mathcal{POV}$;
3) $f \in \mathcal{SQI}$ and $\psi \in \mathcal{SQI}$;
4) $f \in \mathcal{PRV}$ and $\psi \in \mathcal{PRV}$;
5) f and ψ preserve the asymptotic equivalence of functions and sequences.

Proposition 7.36 enables us to extend Theorem 7.18 as follows.

Theorem 7.43 *Let $f \in \mathcal{POV}$ and $f^{\leftarrowtail}$, $f^{\leftarrow}$, $f^{\rightarrow}$, and $f^{\rightarrowtail}$ be its generalized inverse functions (see Sect. 7.2). Then:*

1) $f^{\leftarrowtail}$, $f^{\leftarrow}$, $f^{\rightarrow}$, and $f^{\rightarrowtail}$ are nondecreasing POV-functions, asymptotically inverse for f, and are asymptotically equivalent to each other;
2) if a function $q \in \mathbb{F}_{+}$ is such that $f^{\leftarrowtail}(s) \leq q(s) \leq f^{\rightarrowtail}(s)$ for sufficiently large s, then q is a WPOV-function, asymptotically inverse for f, and asymptotically equivalent to each of the four generalized inverses $f^{\leftarrowtail}$, $f^{\leftarrow}$, $f^{\rightarrow}$, and $f^{\rightarrowtail}$;
3) if q is a nondecreasing function such that $f^{\leftarrowtail}(s) \leq q(s) \leq f^{\rightarrowtail}(s)$ for sufficiently large s, then q is a POV-function, nondecreasing and asymptotically inverse for f, and it is asymptotically equivalent to each of the four generalized inverses $f^{\leftarrowtail}$, $f^{\leftarrow}$, $f^{\rightarrow}$, and $f^{\rightarrowtail}$.

The following assertion complements Proposition 7.12 and shows that all asymptotically inverse functions for a POV-function f are the same for all asymptotically equivalent versions of f. Moreover, the class of asymptotically inverse functions for an arbitrary function being asymptotically equivalent to a POV-function f coincides with the class of asymptotically inverse functions for the function f.

Proposition 7.44 *Let $f \in \mathbb{F}^{(\infty)}$, $\psi \in \mathbb{AI}(f)$, $g \sim f$, and $q \sim \psi$.*

1) If $f \in \mathcal{WSQI}$ and there exists a nondecreasing function h such that $f \sim h$, then $\mathbb{AI}(g) = \mathbb{AI}(g) \cap \mathcal{WPRV} = \mathbb{AI}(f) \cap \mathcal{WPRV} = \mathbb{AI}(f)$.
2) If $f \in \mathcal{WPOV}$ and there exists a nondecreasing function h such that $f \sim h$, then $\mathbb{AI}(g) = \mathbb{AI}(g) \cap \mathcal{WPOV} = \mathbb{AI}(f) \cap \mathcal{WPOV} = \mathbb{AI}(f)$.
3) If $f \in \mathcal{WPOV}$ and there exists a nondecreasing function h such that $f \sim h$, then $q \in \mathbb{AI}(g) \cap \mathcal{WPOV} = \mathbb{AI}(f) \cap \mathcal{WPOV}$ and $g \in \mathbb{AI}(q) = \mathbb{AI}(\psi)$.
4) If $f \in \mathcal{POV}$, then $q \in \mathbb{AI}(g) = \mathbb{AI}(f)$ and $g \in \mathbb{AI}(q) = \mathbb{AI}(\psi)$.

Proof of Proposition 7.44 Statement 1) follows from statement 2) of Theorem 7.35. In turn, statement 2) follows from statement 1) and Proposition 7.36. Statement 3) follows from statement 2) and Propositions 7.12 (E) and 7.36. Statement 4) follows from statement 3) and Theorem 3.79 on the equivalent increasing versions for POV-functions. □

The following result shows that a function q being asymptotically quasi-inverse for a POV-function f, or for an asymptotically equivalent version of f, is asymptotically inverse for f and belongs to $\mathcal{WPOV}$. Moreover, such a function is unique up to asymptotic equivalence. Finally, if q measurable, then $q \in \mathcal{POV}$.

Theorem 7.45 *Let $f \in \mathcal{POV}$, $\psi \in \mathbb{AI}(f)$, and $\varphi \in \mathbb{AQI}(f)$. Then:*

1) $\varphi \sim \psi$, $\varphi \in \mathcal{WPOV}$, and $\varphi \in \mathcal{POV}$, provided φ is measurable;
2) $\varphi \in \mathbb{AI}(f)$.

Proof of Theorem 7.45 Recall that $\psi \in \mathcal{WPOV}$ and thus, by Proposition 7.36, $\psi \in \mathcal{WPRV}$. Therefore the function ψ preserves the equivalence of functions and sequences (see Theorem 3.42).

Since the function φ is asymptotically quasi-inverse for f (see Definition 7.6), we get $\varphi \in \mathbb{F}^\infty$ and $f(\varphi(s)) \sim s$ as $s \to \infty$. The function ψ is asymptotically inverse for f and preserves the asymptotic equivalence of functions and sequences, hence

$$\varphi(s) \sim \psi(f(\varphi(s)) \sim \psi(s) \quad \text{as} \quad s \to \infty.$$

Thus, $\varphi \sim \psi$, whence (or, alternatively, from Proposition 7.36) we conclude that

$$\varphi \in \mathcal{WPOV}.$$

Therefore, statement 1) is proved.

By the assumptions of the theorem, $f \in \mathcal{POV}$ and thus $f \in \mathbb{F}^\infty$ (see Corollary 3.75). Statement 1) implies that

$$\varphi(f(s)) \sim \psi(f(s)) \sim s \quad \text{as} \quad s \to \infty.$$

This completes the proof of statement 2). □

For RV-functions, Theorem 7.45 takes the following form.

Corollary 7.46 *Let f be an RV-function with index $\rho > 0$. Then*

$$\mathbb{AQI}(f) = \mathbb{AI}(f)$$

and there exists an SV-function L such that

$$f^{\sim}(s) \sim s^{1/\rho} L(s) \quad \text{as} \quad s \to \infty.$$

Remark 7.47 Corollary 7.46 shows that

$$f \in \mathcal{RV}_\rho \Longleftrightarrow f^{\sim} \in \mathcal{RV}_{1/\rho},$$

for $\rho > 0$, that is, $f \in \mathcal{RV}_\rho$ if and only if $f^{\sim} \in \mathcal{RV}_{1/\rho}$.

The following theorem combines the above results regarding asymptotically inverse and asymptotically quasi-inverse functions for POV-functions.

Theorem 7.48 *Let $f \in \mathcal{POV}$. Then:*

1) there exists a continuous POV-function f_1, increasing to infinity, such that $f_1 \sim f$ and the inverse f_1^{-1} is a continuous POV-function, increasing to infinity;
2) $f_1^{-1} \in \mathbb{AI}(f)$;
3) if $\varphi \in \mathbb{AQI}(f)$ and $q \sim \varphi$, then $q \in \mathbb{AI}(f)$, $q \in \mathcal{WPOV}$, and $q \sim f_1^{-1}$;
4) if $g \sim f$ and $q \in \mathbb{AQI}(g)$, then $q \in \mathbb{AI}(g) = \mathbb{AI}(f)$, $q \in \mathcal{WPOV}$, and the function q is asymptotically equivalent to any asymptotically quasi-inverse function for f.

Proof of Theorem 7.48 Statement 1) of Theorem 7.48 follows from Theorems 3.79 and 7.40, while statement 2) follows from Proposition 7.44.

Further, if $\varphi \in \mathbb{AQI}(f)$, then we conclude from statement 2) and Theorem 7.45 that $\varphi \sim f_1^{-1}$, $\varphi \in \mathcal{WPOV}$, and $\varphi \in \mathbb{AI}(f)$. Thus, $q \sim \varphi$ implies that $q \sim f_1^{-1}$ and $q \in \mathcal{WPOV}$. Moreover, Proposition 7.44 yields $q \sim f_1^{-1}$. Therefore, statement 3) is proved.

Let $g \sim f$ and $q \in \mathbb{AQI}(g)$. Then

$$f(q(s)) \sim g(q(s)) \sim s \quad \text{as} \quad s \to \infty.$$

Thus, $q \in \mathbb{AQI}(f)$ and statement 3) gives $q \in \mathbb{AI}(f)$, $q \in \mathcal{WPOV}$, and $q \sim f_1^{-1}$. Note that the equality $\mathbb{AI}(g) = \mathbb{AI}(f)$ has been proved in Proposition 7.44. Note also that the relation $q \sim f_1^{-1}$ implies the asymptotic equivalence of q to an arbitrary asymptotically quasi-inverse function for f. This completes the proof of statement 4). □

7.5 Properties of a Ratio of Asymptotically Quasi-inverse Functions

Below (see Theorem 7.50) we show that, under certain assumptions, the asymptotic behavior of a ratio of two functions, being asymptotically quasi-inverse for WSQI-functions, is the same as the asymptotic behavior of the ratio of the original functions. This result together with its corollaries describe conditions for (7.5)–(7.7) and play an important role in applications of asymptotically quasi-inverse functions.

First we provide an example showing that, if two functions are asymptotically equivalent, then this does not mean that all pairs of their asymptotically quasi-inverse functions are also equivalent. In particular, it may well happen that the generalized inverse functions are not asymptotically equivalent, while some other AQIFs are.

Example 7.49 Following Example 7.19 let

$$f(t) = \log(t), \qquad t \geq 1,$$
$$g(t) = \lfloor \log(t) \rfloor, \qquad t \geq 1.$$

Then $f \sim g$ and $h(t) = e^t$, $t \geq 1$, is an asymptotically quasi-inverse function for both f and g. Thus there exists a pair $f^{\sim}$ and $g^{\sim}$ of asymptotically equivalent functions. On the other hand, $f^{\leftarrow}(t) = e^t$ but

$$g^{\leftarrow}(n) = e^{n+1}, \qquad n \geq 1.$$

Therefore $f^{\leftarrow}(\cdot)$ and $g^{\leftarrow}(\cdot)$ are not asymptotically equivalent.

7.5.1 A General Approach to the Proof

Consider the following general approach to investigating the relationship between the asymptotic behavior of the ratio of two functions and that of their asymptotically quasi-inverse functions.

Let $f \in \mathcal{WSQI}$, $f \in \mathbb{C}^{\infty}_{\text{inc}}$, $x \in \mathbb{F}_+$, and $x^{\sim} \in \mathbb{AQI}(x)$. This yields the implication

$$\text{(A)} \qquad \lim_{t\to\infty} \frac{x(t)}{f(t)} = 1 \qquad \Longrightarrow \qquad \lim_{s\to\infty} \frac{x^{\sim}(s)}{f^{-1}(s)} = 1.$$

The proof of (A) consists of two steps:

$$\lim_{t\to\infty} \frac{x(t)}{f(t)} = 1 \qquad \Longrightarrow \qquad \lim_{s\to\infty} \frac{f(x^{\sim}(s))}{s} = 1 \qquad \Longrightarrow \qquad \lim_{s\to\infty} \frac{x^{\sim}(s)}{f^{-1}(s)} = 1.$$

The first of these implications holds, since $x^{\sim}$ is an asymptotically quasi-inverse function for x, that is, $x(x^{\sim}(s)) \sim s$ as $s \to \infty$. The second implication holds, since f is a WSQI-function and, by Corollary 7.33, its inverse f^{-1} is a PRV-function and thus preserves the asymptotic equivalence of functions.

Note that the conditions $f \in \mathcal{WSQI}$ and $f^{-1} \in \mathcal{PRV}$ are equivalent, so that it suffices to check one of them in any particular problem.

If $x^{\sim}$ is asymptotically inverse for x and $f^{-1} \in \mathcal{WSQI}$ or, equivalently, if $f \in \mathcal{PRV}$, then the "inverse" implication holds as follows:

$$\text{(B)} \qquad \lim_{t\to\infty} \frac{x(t)}{f(t)} = 1 \qquad \Longleftarrow \qquad \lim_{s\to\infty} \frac{x^{\sim}(s)}{f^{-1}(s)} = 1.$$

The proof of implication (B) also consists of two steps:

$$\lim_{t\to\infty} \frac{x(t)}{f(t)} = 1 \qquad \Longleftarrow \qquad \lim_{t\to\infty} \frac{f^{-1}(x(t))}{t} = 1 \qquad \Longleftarrow \qquad \lim_{s\to\infty} \frac{x^{\sim}(s)}{f^{-1}(s)} = 1.$$

If f is a POV-function, then both implications (A) and (B) hold.

This general approach with necessary refinements will be applied throughout this section.

7.5.2 The Limit Behavior of the Ratio of Functions Which Are Asymptotically Quasi-inverses for SQI-Functions

Theorem 7.50 *Let $f \in \mathcal{WSQI}$, $f \sim f_0 \in \mathbb{F}_{\text{ndec}}$, $\mathbb{AI}(f_0) \neq \varnothing$, $\psi_0 \in \mathbb{AI}(f_0)$, and $x \in \mathbb{F}_+$. If*

$$\lim_{t\to\infty} \frac{x(t)}{f(t)} = a, \tag{7.27}$$

with $a \in (0, \infty)$, then, for all $x^{\sim} \in \mathbb{AQI}(x)$ and $f^{\sim} \in \mathbb{AQI}(f)$,

$$\lim_{s\to\infty} \frac{x^{\sim}(s)}{f^{\sim}(s/a)} = \lim_{s\to\infty} \frac{x^{\sim}(s)}{\psi_0(s/a)} = 1. \tag{7.28}$$

In particular, if

$$\lim_{t\to\infty} \frac{x(t)}{f(t)} = 1, \tag{7.29}$$

then, for all $x^{\sim} \in \mathbb{AQI}(x)$ and $f^{\sim} \in \mathbb{AQI}(f)$,

$$\lim_{s\to\infty} \frac{x^{\sim}(s)}{f^{\sim}(s)} = \lim_{s\to\infty} \frac{x^{\sim}(s)}{\psi_0(s)} = 1. \tag{7.30}$$

Theorem 7.50 follows immediately from Theorem 7.35 by using the general approach described in Sect. 7.5.1.

Corollary 7.51 *Let $f \in \mathcal{WSQI}$, $f \sim f_0 \in \mathbb{C}_{\text{inc}}$, and $x \in \mathbb{F}_+$. If relation (7.27) holds, with $a \in (0, \infty)$, then*

$$\lim_{s\to\infty} \frac{x^{\sim}(s)}{f^{\sim}(s/a)} = \lim_{s\to\infty} \frac{x^{\sim}(s)}{f_0^{-1}(s/a)} = 1$$

for all $x^{\sim} \in \mathbb{AQI}(x)$ and $f^{\sim} \in \mathbb{AQI}(f)$, where f_0^{-1} is the inverse function of f_0.

In particular, if relation (7.29) holds, then

$$\lim_{s\to\infty} \frac{x^{\sim}(s)}{f^{\sim}(s)} = \lim_{s\to\infty} \frac{x^{\sim}(s)}{f_0^{-1}(s)} = 1$$

for all $x^{\sim} \in \mathbb{AQI}(x)$ *and* $f^{\sim} \in \mathbb{AQI}(f)$.

Under some extra assumptions imposed on the function f, Corollary 7.51 takes the following form.

Corollary 7.52 *Let* $f \in \mathbb{C}_{\text{inc}} \cap \mathcal{WSQI}$ *and* $x \in \mathbb{F}_{+}$. *If relation (7.27) holds, with* $a \in (0, \infty)$, *then*

$$\lim_{s\to\infty} \frac{x^{\sim}(s)}{f^{\sim}(s/a)} = \lim_{s\to\infty} \frac{x^{\sim}(s)}{f^{-1}(s/a)} = 1$$

for all $x^{\sim} \in \mathbb{AQI}(x)$ *and* $f^{\sim} \in \mathbb{AQI}(f)$, *where* f^{-1} *is the inverse function of* f.

In particular, if relation (7.29) holds, then

$$\lim_{s\to\infty} \frac{x^{\sim}(s)}{f^{\sim}(s)} = \lim_{s\to\infty} \frac{x^{\sim}(s)}{f^{-1}(s)} = 1$$

for all $x^{\sim} \in \mathbb{AQI}(x)$ *and* $f^{\sim} \in \mathbb{AQI}(f)$.

Theorem 7.50 can be rewritten for the generalized inverses of the function x.

Theorem 7.53 *Let* $f \in \mathbb{F}^{\infty}$, $f \sim f_0 \in \mathbb{F}_{\text{ndec}}$, $\mathbb{AI}(f_0) \neq \varnothing$, $\psi_0 \in \mathbb{AI}(f_0)$, *and* $a \in (0, \infty)$.

1) If $f \in \mathcal{WSQI}$ *and* $x \in \mathbb{C}^{\infty}$, *then*

$$\lim_{t\to\infty} \frac{x(t)}{f(t)} = a \quad \Longrightarrow \quad \lim_{s\to\infty} \frac{x^{\leftarrow}(s)}{\psi_0(s/a)} = \lim_{s\to\infty} \frac{x^{\rightarrowtail}(s)}{\psi_0(s/a)} = 1$$

and, for an arbitrary function $f^{\sim} \in \mathbb{AQI}(f)$,

$$\lim_{t\to\infty} \frac{x(t)}{f(t)} = a \quad \Longrightarrow \quad \lim_{s\to\infty} \frac{x^{\leftarrow}(s)}{f^{\sim}(s/a)} = \lim_{s\to\infty} \frac{x^{\rightarrowtail}(s)}{f^{\sim}(s/a)} = 1.$$

2) If $f \in \mathcal{WSQI}$, $f \in \mathbb{C}^{\infty}$, *and* $x \in \mathbb{C}^{\infty}$, *then*

$$\lim_{t\to\infty} \frac{x(t)}{f(t)} = a \quad \Longrightarrow \quad \lim_{s\to\infty} \frac{x^{\leftarrow}(s)}{f^{\sim}(s/a)} = \lim_{s\to\infty} \frac{x^{\rightarrowtail}(s)}{f^{\sim}(s/a)} = 1,$$

where $f^{\sim} \in \{f^{\leftarrow}, f^{\rightarrowtail}\}$.

3) If $f \in \mathcal{WPRV}$, *then*

$$\lim_{t\to\infty} \frac{x(t)}{f(t)} = a \quad \Longleftarrow \quad \lim_{s\to\infty} \frac{x^{\leftarrow}(s)}{\psi_0(s/a)} = \lim_{s\to\infty} \frac{x^{\rightarrowtail}(s)}{\psi_0(s/a)} = 1.$$

4) If $f \sim f_0 \in \mathbb{C}^{\infty}_{\text{inc}}$, then statements 1) and 3) hold if

$$\psi_0 = f_0^{-1},$$

where f_0^{-1} is the inverse function of f_0.

5) If $f \in \mathcal{PRV}$ and $f^{\leftarrow} \sim f^{\rightarrowtail}$, then

$$\lim_{t\to\infty}\frac{x(t)}{f(t)} = a \qquad \Longleftarrow \qquad \lim_{s\to\infty}\frac{x^{\leftarrow}(s)}{f^{\sim}(s/a)} = \lim_{s\to\infty}\frac{x^{\rightarrowtail}(s)}{f^{\sim}(s/a)} = 1,$$

where $f^{\sim} \in \{f^{\leftarrow}, f^{\rightarrowtail}\}$.

6) If $f \in \mathcal{WPRV}$ and $x \in \mathbb{F}^{\infty}_{\text{inc}}$, then

$$\lim_{t\to\infty}\frac{x(t)}{f(t)} = a \qquad \Longleftarrow \qquad \lim_{s\to\infty}\frac{x^{\leftarrow}(s)}{\psi_0(s/a)} = 1.$$

7) If $f \in \mathcal{WPRV}$, $f \in \mathbb{F}^{\infty}_{\text{inc}}$ and $x \in \mathbb{F}^{\infty}_{\text{inc}}$, then

$$\lim_{t\to\infty}\frac{x(t)}{f(t)} = a \qquad \Longleftarrow \qquad \lim_{s\to\infty}\frac{x^{\leftarrow}(s)}{f^{\leftarrow}(s)} = 1.$$

Proof of Theorem 7.53 Statement 1) follows from Theorem 7.50 and statement 2) follows from statement 1) combined with statement 4) of Theorem 7.16.

Let $f \in \mathcal{WPRV}$ and

$$\lim_{s\to\infty}\frac{x^{\leftarrow}(s)}{\psi_0(s/a)} = \frac{x^{\rightarrowtail}(s)}{\psi_0(s/a)} = 1.$$

Then

$$\lim_{s\to\infty}\frac{x^{\leftarrow}(s)}{x^{\rightarrowtail}(s)} = 1,$$

whence $x \in \mathbb{AQI}(x^{\leftarrow})$ in view of statement 1) of Lemma 7.15. Since $x^{\leftarrow} \in \mathbb{F}_{\text{ndec}}$, the function ψ_0 is asymptotically equivalent to a function from the class $\mathbb{F}_{\text{ndec}}$. Moreover, $f_0 \in \mathcal{WPRV}$, since $f \in \mathcal{WPRV}$ and $f \sim f_0$. This together with the second part of Proposition 7.28 implies that $\psi_0 \in \mathcal{WSQI}$. Note that $f_0 \in \mathbb{AI}(\psi_0)$. Thus, all assumptions of Theorem 7.50 are satisfied for the functions ψ_0, f_0, $x^{\leftarrow}$, and x. Therefore

$$\lim_{t\to\infty}\frac{x(t)}{af_0(t)} = 1.$$

Since $f \sim f_0$,

$$\lim_{t\to\infty} \frac{x(t)}{f(t)} = a.$$

So, statement 3) is proved.

Statement 4) follows from the first three statements, while statement 5) results from statement 3), since $f^{\leftarrow} \in \mathbb{AI}(f)$ and $f^{\rightarrowtail} \in \mathbb{AI}(f)$ in view of $f \in \mathcal{PRV}$, $f \in \mathbb{F}^{\infty}$, and $f^{\leftarrow} \sim f^{\rightarrowtail}$ according to statement 7) of Theorem 7.16. In turn, statement 6) follows from statement 3) and relation (7.11), while statement 7) is a corollary of statement 5) and relation (7.11). □

Remark 7.54 The condition $x \in \mathbb{C}^{\infty}$ in statements 1) and 2) of Theorem 7.53 is needed for the generalized inverse functions $x^{\leftarrow}$ and $x^{\rightarrowtail}$ to be asymptotically quasi-inverse functions for x. If x is not a continuous function, then one has to use another condition, for example $x \in \mathcal{PRV}$, under which $x^{\leftarrow}$ and $x^{\rightarrowtail}$ are still asymptotically quasi-inverse for x (see Theorem 7.16). In contrast, if $x \in \mathcal{PRV}$ and $\lim_{t\to\infty}(x(t)/f(t)) = a \in (0,\infty)$, then $f \in \mathcal{WPRV}$ which together with $f \in \mathcal{WSQI}$ gives $f \in \mathcal{WPOV}$. This means that, for the case of general discontinuous functions, statements 1) and 2) can be proved for $f \in \mathcal{WPOV}$ (see Theorem 7.57).

7.5.3 A Characterization of SQI- and PRV-Functions

If $g \in \mathbb{C}^{\infty}_{\mathrm{inc}}$ and $g \in \mathcal{SQI}$, then

$$x \sim g \quad \Longrightarrow \quad x^{\sim} \sim g^{-1} \tag{7.31}$$

holds for all $x \in \mathbb{F}^{(\infty)}$, which possess asymptotically quasi-inverse functions $x^{\sim}$ (recall Corollary 7.52). It turns out that the converse is also true for $g \in \mathbb{C}^{\infty}_{\mathrm{inc}}$.

Theorem 7.55 *Let $g \in \mathbb{C}^{\infty}_{\mathrm{inc}}$. Then implication (7.31) holds for all $x \in \mathbb{F}^{(\infty)}$ such that $\mathbb{AQI}(x) \neq \varnothing$ if and only if $g \in \mathcal{SQI}$.*

Proof of Theorem 7.55 Let relation (7.31) hold for all $x \in \mathbb{F}^{(\infty)}$ such that $\mathbb{AQI}(x) \neq \varnothing$.

We show that $g^{-1} \in \mathcal{PRV}$. Consider the function $I(t) = t$, $t \ge 0$. Let $u, v \in \mathbb{C}^{\infty}_{\mathrm{inc}}$ and $u \sim v$. Then

$$u \sim v \quad \Longleftrightarrow \quad u \circ v^{-1} \sim I \quad \Longleftrightarrow \quad u \circ v^{-1} \circ g \sim g.$$

Taking (7.31) into account, we obtain the equivalences

$$(u \circ v^{-1} \circ g)^{-1} \sim g^{-1} \quad \Longleftrightarrow \quad g^{-1} \circ v \circ u^{-1} \sim g^{-1} \quad \Longleftrightarrow \quad g^{-1} \circ v \sim g^{-1} \circ u.$$

The last of these relations shows that g^{-1} preserves the asymptotic equivalence of functions in the class $\mathbb{C}_{\text{inc}}^{\infty}$. Now, Theorem 3.42 implies that $g^{-1} \in \mathcal{PRV}$. Then $g \in \mathcal{SQI}$ by Corollary 7.33. It remains to recall that Corollary 7.52 also contains the converse statement. □

Along with relation (7.31) for functions $g \in \mathbb{C}_{\text{inc}}^{\infty}$ we consider the following one:

$$x \sim g^{-1} \quad \Longrightarrow \quad x^{\sim} \sim g. \tag{7.32}$$

Theorem 7.55 and Corollary 7.33 provide a characterization of PRV-functions.

Theorem 7.56 *Let $g \in \mathbb{C}_{\text{inc}}^{\infty}$. Then implication (7.32) holds for all $x \in \mathbb{F}^{(\infty)}$ such that $\mathbb{AQI}(x) \neq \varnothing$ if and only if $g \in \mathcal{PRV}$.*

7.5.4 The Limit Behavior of the Ratio of Functions Which Are Asymptotically Quasi-inverse for POV-Functions

For POV-functions f, Theorem 7.50 and its corollaries become more complete.

Theorem 7.57 *Let $f \in \mathcal{POV}$, $f^{\sim} \in \mathbb{AQI}(f)$, $x \in \mathbb{F}^{(\infty)}$, $x^{\sim} \in \mathbb{AQI}(x)$, and $a \in (0, \infty)$.*

1) If relation (7.27) holds, that is, if

$$\lim_{t\to\infty} \frac{x(t)}{f(t)} = a,$$

then

$$\lim_{s\to\infty} \frac{x^{\sim}(s)}{f^{\sim}(s/a)} = 1;$$

in particular, if relation (7.29) holds, that is, if

$$\lim_{t\to\infty} \frac{x(t)}{f(t)} = 1,$$

then

$$\lim_{s\to\infty} \frac{x^{\sim}(s)}{f^{\sim}(s)} = 1.$$

2) *If* $\tilde{x}^{-1} \in \mathbb{AI}(x)$*, then*

$$\lim_{t\to\infty} \frac{x(t)}{f(t)} = a \qquad \Longleftrightarrow \qquad \lim_{s\to\infty} \frac{\tilde{x}^{-1}(s)}{f^{\sim}(s/a)} = 1;$$

in particular,

$$\lim_{t\to\infty} \frac{x(t)}{f(t)} = 1 \qquad \Longleftrightarrow \qquad \lim_{s\to\infty} \frac{\tilde{x}^{-1}(s)}{f^{\sim}(s)} = 1.$$

3) *If* $x \in \mathcal{WPRV}$*, then*

$$\lim_{t\to\infty} \frac{x(t)}{f(t)} = a \qquad \Longleftrightarrow \qquad \lim_{s\to\infty} \frac{x^{\sim}(s)}{f^{\sim}(s/a)} = 1;$$

in particular,

$$\lim_{t\to\infty} \frac{x(t)}{f(t)} = 1 \qquad \Longleftrightarrow \qquad \lim_{s\to\infty} \frac{x^{\sim}(s)}{f^{\sim}(s)} = 1.$$

4) *If the function* x *is measurable, then statement 1) holds with*

$$x^{\sim} \in \{x^{\leftarrowtail}, x^{\leftarrow}, x^{\rightarrow}, x^{\rightarrowtail}\},$$

where $x^{\leftarrowtail}$, $x^{\leftarrow}$, $x^{\rightarrow}$, *and* $x^{\rightarrowtail}$ *are the generalized inverses for* x.

5) *If the function* x *is measurable and relation (7.27) holds, then*

$$\lim_{s\to\infty} \frac{z(s)}{f^{\sim}(s/a)} = 1,$$

for an arbitrary function z *such that* $x^{\leftarrowtail}(s) \le z(s) \le x^{\rightarrowtail}(s)$ *for all sufficiently large* s.

6) *If* $x \in \mathcal{POV}$*, then statement 2) holds for*

$$\tilde{x}^{-1} \in \{x^{\leftarrowtail}, x^{\leftarrow}, x^{\rightarrow}, x^{\rightarrowtail}\}.$$

7) *If the function* x *is measurable, then*

$$\lim_{t\to\infty} \frac{x(t)}{f(t)} = a \qquad \Longleftrightarrow \qquad \lim_{s\to\infty} \frac{x^{\leftarrowtail}(s)}{f^{\sim}(s/a)} = \frac{x^{\rightarrowtail}(s)}{f^{\sim}(s/a)} = 1;$$

in particular,

$$\lim_{t\to\infty} \frac{x(t)}{f(t)} = 1 \qquad \Longleftrightarrow \qquad \lim_{s\to\infty} \frac{x^{\leftarrowtail}(s)}{f^{\sim}(s)} = \frac{x^{\rightarrowtail}(s)}{f^{\sim}(s)} = 1.$$

8) *If* $x \in \mathbb{F}^{\infty}_{\text{inc}}$, *then*

$$\lim_{t\to\infty} \frac{x(t)}{f(t)} = a \qquad \Longleftrightarrow \qquad \lim_{s\to\infty} \frac{x^{\leftarrowtail}(s)}{f^{\sim}(s/a)} = 1;$$

in particular,

$$\lim_{t\to\infty} \frac{x(t)}{f(t)} = 1 \qquad \Longleftrightarrow \qquad \lim_{s\to\infty} \frac{x^{\leftarrowtail}(s)}{f^{\sim}(s)} = 1.$$

9) *If* $x \in \mathbb{C}^{\infty}$, *then*

$$\lim_{t\to\infty} \frac{x(t)}{f(t)} = a \qquad \Longleftrightarrow \qquad \lim_{s\to\infty} \frac{x^{\leftarrowtail}(s)}{f^{\sim}(s/a)} = 1;$$

in particular,

$$\lim_{t\to\infty} \frac{x(t)}{f(t)} = 1 \qquad \Longleftrightarrow \qquad \lim_{s\to\infty} \frac{x^{\leftarrowtail}(s)}{f^{\sim}(s)} = 1.$$

10) *Statements 1)–9) hold for*

$$f^{\sim} \in \{f^{\leftarrowtail}, f^{\leftarrow}, f^{\rightarrow}, f^{\rightarrowtail}\},$$

where $f^{\leftarrowtail}$, $f^{\leftarrow}$, $f^{\rightarrow}$, *and* $f^{\rightarrowtail}$ *are the generalized inverses for* f.

11) *If* $f \sim f_0 \in \mathbb{C}^{\infty}_{\text{inc}}$, *then statements 1)–9) hold for*

$$f^{\sim} = f_0^{-1},$$

where f_0^{-1} *is the inverse function of* f_0.

Proof of Theorem 7.57 Statement 1) follows from Theorems 7.50 and 3.79 (see also Theorem 7.48).

To prove statement 2) one only needs to show the implication

$$\lim_{s\to\infty} \frac{\tilde{x}^{-1}(s)}{f^{\sim}(s/a)} = 1 \qquad \Longrightarrow \qquad \lim_{t\to\infty} \frac{x(t)}{f(t)} = a.$$

Let

$$\lim_{s\to\infty} \frac{\tilde{x}^{-1}(s)}{f^{\sim}(s/a)} = 1,$$

with $a \in (0, \infty)$. Since $f \in \mathcal{POV}$, Theorem 7.48 implies that

$$f^{\sim} \sim f_1^{-1},$$

where f_1^{-1} is the continuous POV-function, increasing to infinity, as given there. Thus

$$\lim_{s\to\infty} \frac{\tilde{x}^{-1}(s)}{f_1^{-1}(s/a)} = 1.$$

Applying statement 1), we get

$$\lim_{t\to\infty} \frac{x(t)}{f_1(t)} = a,$$

whence

$$\lim_{t\to\infty} \frac{x(t)}{f(t)} = a,$$

since $f \sim f_1$. So, statement 2) is proved.

In turn, one needs to prove the implication

$$\lim_{s\to\infty} \frac{x^{\sim}(s)}{f^{\sim}(s/a)} = 1 \qquad \Longrightarrow \qquad \lim_{t\to\infty} \frac{x(t)}{f(t)} = a$$

in order to get statement 3). For this statement the assumption $x \in \mathcal{WPRV}$ is needed.

Assume that $x \in \mathcal{WPRV}$ and

$$\lim_{s\to\infty} \frac{x^{\sim}(s)}{f^{\sim}(s/a)} = 1.$$

As in the proof of statement 2), we obtain

$$\lim_{s\to\infty} \frac{x^{\sim}(s)}{f_1^{-1}(s/a)} = 1,$$

where f_1^{-1} is the inverse of the continuous function f_1, increasing to infinity, defined in Theorem 7.48.

Since $x \in \mathcal{WPRV}$, Theorem 3.42 implies that x preserves the asymptotic equivalence of functions. Hence

$$\begin{aligned} 1 = \lim_{s\to\infty} \frac{x(x^{\sim}(s))}{x(f_1^{-1}(s/a))} &= \lim_{s\to\infty} \frac{s}{x(f_1^{-1}(s/a))} \\ = \lim_{s\to\infty} \frac{af(s)}{x(f_1^{-1}(af(s)/a)} &= \lim_{s\to\infty} \frac{af(s)}{x(s)} \end{aligned}$$

and

$$\lim_{t\to\infty} \frac{x(t)}{f(t)} = a.$$

So, statement 3) is proved.

If the function x is measurable and relation (7.27) holds, then $x \in \mathcal{POV}$, since $f \in \mathcal{POV}$. Now, statement 1) of Theorem 7.43 implies statement 4), whence statement 5) follows.

In turn, statement 6) follows from statements 1) and 2) of Theorem 7.43 and statement 7) from statements 3) and 4) of Theorem 7.53. Next, statement 8) follows from statement 7) and relation (7.11) and statement 9) is a corollary of statement 2) of Theorem 7.48 combined with statement 4) of Theorem 7.16.

Finally, statement 10) follows from statements 1)–9) together with Theorem 7.43, while statement 11) is a consequence of statements 1)–9) and Proposition 7.44. □

7.5.5 *Zero and Infinite Limit Points of the Ratio of Asymptotically Quasi-inverse Functions*

Consider the relationship between the limit behavior of the ratio of asymptotically quasi-inverse functions for the case when the ratio of the underlying functions tends to 0 or to ∞. To extend Theorem 7.50 to these cases, one needs to introduce some extra conditions. One of these conditions is that the function ψ_0 is a WPI-function, that is,

$$\liminf_{s\to\infty} \frac{\psi_0(c_0 s)}{\psi_0(s)} > 1, \qquad \text{for some } c_0 > 1 \tag{7.33}$$

(recall (3.13)).

Theorem 7.58 *Let $f \in \mathcal{WSQI}$, $f \sim f_0 \in \mathbb{F}_{\text{ndec}}$, $\mathbb{AI}(f_0) \neq \varnothing$, $\psi_0 \in \mathbb{AI}(f_0) \cap \mathcal{WPI}$, $\psi_0 \sim q \in \mathbb{F}^{\infty}_{\text{ndec}}$, and $x \in \mathbb{F}^{(\infty)}$. Then, for all $x^{\sim} \in \mathbb{AQI}(x)$ and $f^{\sim} \in \mathbb{AQI}(f)$,*

$$\lim_{t\to\infty} \frac{x(t)}{f(t)} = \infty \qquad \Longrightarrow \qquad \lim_{s\to\infty} \frac{x^{\sim}(s)}{f^{\sim}(s)} = \lim_{s\to\infty} \frac{x^{\sim}(s)}{\psi_0(s)} = 0, \tag{7.34}$$

$$\lim_{t\to\infty} \frac{x(t)}{f(t)} = 0 \qquad \Longrightarrow \qquad \lim_{s\to\infty} \frac{x^{\sim}(s)}{f^{\sim}(s)} = \lim_{s\to\infty} \frac{x^{\sim}(s)}{\psi_0(s)} = \infty. \tag{7.35}$$

Proof of Theorem 7.58 First we show that

$$\lim_{c\to\infty} l(c) = \infty \qquad \text{and} \qquad \lim_{c\to 0} r(c) = 0, \tag{7.36}$$

where

$$l(c) = \liminf_{s\to\infty} \frac{\psi_0(cs)}{\psi_0(s)}, \qquad r(c) = \limsup_{s\to\infty} \frac{\psi_0(cs)}{\psi_0(s)}.$$

Since

$$l(c) = \liminf_{s\to\infty} \frac{\psi_0(cs)}{\psi_0(s)} = \liminf_{s\to\infty} \frac{q(cs)}{q(s)}$$

and $q \in \mathbb{F}^{\infty}_{\mathrm{ndec}}$, we conclude that $l \in \mathbb{F}_{\mathrm{ndec}}$. Taking (7.33) into account, we see that

$$l_\infty = \liminf_{c\to\infty} l(c) \ge 1.$$

Moreover, Lemma 3.1 implies that

$$l_\infty = \liminf_{c\to\infty} l(c) = \liminf_{c\to\infty} l\left(c^2\right) \ge \left(\liminf_{c\to\infty} l(c)\right)^2 = l_\infty^2.$$

Hence $l_\infty = \infty$ and the first relation in (7.36) is proved.

Since $r(c) = 1/l(1/c)$, the second relation in (7.36) follows from the first one.

Before we begin with the proof of relation (7.34), note that Theorem 7.35 yields $\psi_0 \in \mathcal{WPRV}$. Hence ψ_0 preserves the asymptotic equivalence of functions (recall Theorem 3.42). Thus,

$$1 = \lim_{s\to\infty} \frac{\psi_0(x(x^{\sim}(s)))}{\psi_0(s)} = \lim_{s\to\infty} \frac{\psi_0(x(x^{\sim}(s)))}{x^{\sim}(s)} \cdot \frac{x^{\sim}(s)}{\psi_0(s)}, \tag{7.37}$$

since $x(x^{\sim}(s)) \sim s$ as $s \to \infty$.

Assume that the left-hand side of relation (7.34) holds. Put, for $t > 0$,

$$a(t) = x(t)/f(t).$$

The left-hand side of relation (7.34) implies that $a(t) \to \infty$ as $t \to \infty$, whence

$$\begin{aligned}
\liminf_{s\to\infty} \frac{\psi_0(x(x^{\sim}(s)))}{x^{\sim}(s)} &\ge \liminf_{t\to\infty} \frac{\psi_0(x(t))}{t} = \liminf_{t\to\infty} \frac{\psi_0(a(t)f(t))}{t} \\
&= \liminf_{t\to\infty} \frac{q(a(t)f(t))}{t} \ge \liminf_{c\to\infty} \liminf_{t\to\infty} \frac{q(cf(t))}{t} \\
&= \liminf_{c\to\infty} \liminf_{t\to\infty} \frac{\psi_0(cf(t))}{t} = \liminf_{c\to\infty} \liminf_{t\to\infty} \frac{\psi_0(cf(t))}{\psi_0(f(t))} \\
&\ge \liminf_{c\to\infty} \liminf_{t\to\infty} \frac{\psi_0(ct)}{\psi_0(t)} = \lim_{c\to\infty} l(c),
\end{aligned}$$

since $\psi_0(f(t)) \sim t$ as $t \to \infty$. We conclude from this together with (7.36) that

$$\liminf_{s\to\infty} \frac{\psi_0(x(x^{\sim}(s)))}{x^{\sim}(s)} = \infty,$$

whence

$$\limsup_{s\to\infty} \frac{x^{\sim}(s)}{\psi_0(s)} = 0$$

in view of (7.37). It remains to recall that, by Theorem 7.35, $f^{\sim} \sim \psi_0$. Thus,

$$\limsup_{s\to\infty} \frac{x^{\sim}(s)}{f^{\sim}(s)} = \limsup_{s\to\infty} \frac{x^{\sim}(s)}{\psi_0(s)} = 0,$$

which proves relation (7.34).

Relation (7.35) follows in the same way. □

Below are two immediate corollaries of Theorem 7.58.

Corollary 7.59 *Let $f \in \mathcal{WSQI}$, $f \sim f_0 \in \mathbb{C}_{\mathrm{inc}}$, and $f_0^{-1} \in \mathcal{WPI}$, where f_0^{-1} is the inverse function of f_0, and $x \in \mathbb{F}^{(\infty)}$. Then relations (7.34) and (7.35) hold if $\psi_0 = f_0^{-1}$ for all $x^{\sim} \in \mathbb{AQI}(x)$ and $f^{\sim} \in \mathbb{AQI}(f)$.*

Corollary 7.60 *Let $f \in \mathcal{WSQI} \cap \mathbb{C}_{\mathrm{inc}}$ and $f^{-1} \in \mathcal{WPI}$, where f^{-1} is the inverse function for f, and $x \in \mathbb{F}^{(\infty)}$. Then relations (7.34) and (7.35) hold for all $x^{\sim} \in \mathbb{AQI}(x)$ and $f^{\sim} \in \mathbb{AQI}(f)$ if $\psi_0 = f^{-1}$.*

Theorems 7.58 and 7.16 imply the following two statements on the asymptotic behavior of generalized inverse functions $x^{\leftarrowtail}$, $x^{\leftarrow}$, $x^{\rightarrow}$, and $x^{\rightarrowtail}$.

Corollary 7.61 *Let $f \in \mathcal{WSQI}$, $f \sim f_0 \in \mathbb{F}_{\mathrm{ndec}}$, $\mathbb{AI}(f_0) \neq \varnothing$, $\psi_0 \in \mathbb{AI}(f_0) \cap \mathcal{WPI}$, $\psi_0 \sim q \in \mathbb{F}^{\infty}_{\mathrm{ndec}}$, and $x \in \mathbb{C}^{\infty}$. Then relations (7.34) and (7.35) hold for all $f^{\sim} \in \mathbb{AQI}(f)$ and $x^{\sim} \in \{x^{\leftarrowtail}, x^{\leftarrow}, x^{\rightarrow}, x^{\rightarrowtail}\}$.*

Corollary 7.62 *Let $f \in \mathcal{WSQI} \cap \mathbb{C}_{\mathrm{inc}}$ and $f^{-1} \in \mathcal{WPI}$, where f^{-1} is the inverse function of f, and $x \in \mathbb{C}^{\infty}$. Then relations (7.34) and (7.35) hold for all $f^{\sim} \in \mathbb{AQI}(f)$ and $x^{\sim} \in \{x^{\leftarrowtail}, x^{\leftarrow}, x^{\rightarrow}, x^{\rightarrowtail}\}$ if $\psi_0 = f^{-1}$.*

7.5.6 *Zero and Infinite Limit Points of the Ratio of Functions Which Are Asymptotically Quasi-inverse for POV-Functions*

Applying Theorems 7.48 and 7.16 we are now in a position to state Theorem 7.58 for POV-functions in a complete form.

Theorem 7.63 *Let $f \in \mathcal{POV}$, $x \in \mathbb{F}^{(\infty)}$, and $f^{\sim} \in \mathbb{AQI}(f)$.*

1) If $x^{\sim} \in \mathbb{AQI}(x)$, then

$$\lim_{t\to\infty} \frac{x(t)}{f(t)} = \infty \qquad \Longrightarrow \qquad \lim_{s\to\infty} \frac{x^{\sim}(s)}{f^{\sim}(s)} = 0,$$

$$\lim_{t\to\infty} \frac{x(t)}{f(t)} = 0 \qquad \Longrightarrow \qquad \lim_{s\to\infty} \frac{x^{\sim}(s)}{f^{\sim}(s)} = \infty.$$

2) If $\tilde{x}^{-1} \in \mathbb{AI}(x)$, then

$$\lim_{t\to\infty} \frac{x(t)}{f(t)} = \infty \qquad \Longleftrightarrow \qquad \lim_{s\to\infty} \frac{\tilde{x}^{-1}(s)}{f^{\sim}(s)} = 0,$$

$$\lim_{t\to\infty} \frac{x(t)}{f(t)} = 0 \qquad \Longleftrightarrow \qquad \lim_{s\to\infty} \frac{\tilde{x}^{-1}(s)}{f^{\sim}(s)} = \infty.$$

3) If $x \in \mathbb{C}^{\infty}$, then statement 1) holds for

$$x^{\sim} \in \{x^{\leftarrowtail}, x^{\leftarrow}, x^{\rightarrow}, x^{\rightarrowtail}\},$$

where $x^{\leftarrowtail}$, $x^{\leftarrow}$, $x^{\rightarrow}$, $x^{\rightarrowtail}$ are the generalized inverses for x.

4) If $x \in \mathcal{PRV}$, then statement 1) holds for

$$x^{\sim} \in \{x^{\leftarrowtail}, x^{\leftarrow}, x^{\rightarrow}, x^{\rightarrowtail}\}.$$

5) If $x \in \mathbb{C}^{\infty}$ and $x^{\leftarrowtail} \sim x^{\rightarrowtail}$, then statement 2) holds for

$$\tilde{x}^{-1} \in \{x^{\leftarrowtail}, x^{\leftarrow}, x^{\rightarrow}, x^{\rightarrowtail}\}.$$

6) If $x \in \mathcal{PRV}$ and $x^{\leftarrowtail} \sim x^{\rightarrowtail}$, then statement 2) holds for

$$\tilde{x}^{-1} \in \{x^{\leftarrowtail}, x^{\leftarrow}, x^{\rightarrow}, x^{\rightarrowtail}\}.$$

7) If $x \in \mathcal{POV}$, then statement 2) holds for

$$\tilde{x}^{-1} \in \{x^{\leftarrowtail}, x^{\leftarrow}, x^{\rightarrow}, x^{\rightarrowtail}\}.$$

8) Statements 1)–7) hold for

$$f^{\sim} \in \{f^{\leftarrowtail}, f^{\leftarrow}, f^{\rightarrow}, f^{\rightarrowtail}\},$$

where $f^{\leftarrowtail}$, $f^{\leftarrow}$, $f^{\rightarrow}$, $f^{\rightarrowtail}$ are the generalized inverses for f.

Proof of Theorem 7.63 Statement 1) follows from Theorems 7.58 and 7.48.

To prove statement 2), one needs to show that

$$\lim_{s\to\infty} \frac{\tilde{x}^{-1}(s)}{f^{\sim}(s)} = 0 \qquad \Longrightarrow \qquad \lim_{t\to\infty} \frac{x(t)}{f(t)} = \infty, \tag{7.38}$$

$$\lim_{s\to\infty} \frac{\tilde{x}^{-1}(s)}{f^{\sim}(s)} = \infty \qquad \Longrightarrow \qquad \lim_{t\to\infty} \frac{x(t)}{f(t)} = 0. \tag{7.39}$$

For the proof of relation (7.38) we assume that

$$\lim_{s\to\infty} \frac{\tilde{x}^{-1}(s)}{f^{\sim}(s)} = 0.$$

Since $f \in \mathcal{POV}$, Theorem 7.48 implies that

$$f^{\sim} \sim f_1^{-1},$$

where f_1^{-1} is the continuous POV-function, increasing to infinity, as defined there. Thus,

$$\lim_{s\to\infty} \frac{\tilde{x}^{-1}(s)}{f_1^{-1}(s/a)} = 0.$$

Applying statement 1), we obtain

$$\lim_{t\to\infty} \frac{x(t)}{f_1(t)} = \infty,$$

whence

$$\lim_{t\to\infty} \frac{x(t)}{f(t)} = \infty,$$

since $f \sim f_1$. Therefore, relation (7.38) is proved. The proof of (7.39) is the same, so that the proof of statement 2) is complete.

Statements 3) and 4) follow from 1) and 4) of Theorem 7.16. Statements 5) and 6) follow from 2) together with statements 7) and 9) of Theorem 7.16. Statement 7) follows from 2) and Examples 7.2 and 7.4. Finally, statement 8) follows immediately from Theorem 7.43. $\square$

7.5.7 The Limit Behavior of the Ratio of Functions Which Are Asymptotically Quasi-inverse for RV-Functions

Some statements of Theorems 7.57 and 7.63 can be made more precise and simple. In what follows we set $1/\infty := 0$ and $1/0 := \infty$.

Theorem 7.64 *Let f be an RV-function with index $\rho > 0$, $f^{\sim} \in \mathbb{AQI}(f)$, $x \in \mathbb{F}^{(\infty)}$, and $a \in [0, \infty]$.*

1) If $x^{\sim} \in \mathbb{AQI}(x)$, then

$$\lim_{t\to\infty} \frac{x(t)}{f(t)} = a \qquad \Longrightarrow \qquad \lim_{s\to\infty} \frac{x^{\sim}(s)}{f^{\sim}(s)} = \left(\frac{1}{a}\right)^{\frac{1}{\rho}}. \tag{7.40}$$

2) If $\tilde{x}^{-1} \in \mathbb{AI}(x)$, then

$$\lim_{t\to\infty} \frac{x(t)}{f(t)} = a \qquad \Longleftrightarrow \qquad \lim_{s\to\infty} \frac{\tilde{x}^{-1}(s)}{f^{\sim}(s)} = \left(\frac{1}{a}\right)^{\frac{1}{\rho}}. \tag{7.41}$$

3) If $x \in \mathbb{C}^{\infty}$, then statement 1) holds for

$$x^{\sim} \in \{x^{\leftarrowtail}, x^{\leftarrow}, x^{\rightarrow}, x^{\rightarrowtail}\},$$

where $x^{\leftarrowtail}$, $x^{\leftarrow}$, $x^{\rightarrow}$, and $x^{\rightarrowtail}$ are the generalized inverses for x.

4) If $x \in \mathbb{C}^{\infty}$, then

$$\lim_{t\to\infty} \frac{x(t)}{f(t)} = a \qquad \Longrightarrow \qquad \lim_{s\to\infty} \frac{z^{\sim}(s)}{f^{\sim}(s)} = \left(\frac{1}{a}\right)^{\frac{1}{\rho}} \tag{7.42}$$

for any function z such that $x^{\leftarrowtail}(s) \le z(s) \le x^{\rightarrowtail}(s)$ for all sufficiently large s.

5) If $x \in \mathcal{POV}$, then statement 2) holds for

$$\tilde{x}^{-1} \in \{x^{\leftarrowtail}, x^{\leftarrow}, x^{\rightarrow}, x^{\rightarrowtail}\};$$

in particular, relation (7.42) holds.

6) If $x \in \mathcal{WPRV}$, then statement 2) holds for $a \in (0, \infty)$.

7) If x is measurable, then statement 1) holds for $a \in (0, \infty)$ and

$$\tilde{x}^{-1} \in \{x^{\leftarrowtail}, x^{\leftarrow}, x^{\rightarrow}, x^{\rightarrowtail}\};$$

in particular, relation (7.42) holds for $a \in (0, \infty)$.

8) If $x \in \mathbb{C}^{\infty}$ and $a \in (0, \infty)$, then

$$\lim_{t\to\infty} \frac{x(t)}{f(t)} = a \qquad \Longleftrightarrow \qquad \lim_{s\to\infty} \frac{x^{\leftarrowtail}(s)}{f^{\sim}(s)} = \frac{x^{\rightarrowtail}(s)}{f^{\sim}(s)} = \left(\frac{1}{a}\right)^{\frac{1}{\rho}}.$$

9) If $x \in \mathbb{C}^\infty$ and $x^{\leftarrowtail} \sim x^{\rightarrowtail}$, then statement 2) holds for

$$\tilde{x}^{-1} \in \{x^{\leftarrowtail}, x^{\leftarrow}, x^{\rightarrow}, x^{\rightarrowtail}\}.$$

10) Statements 1)–9) hold for

$$f^{\sim} \in \{f^{\leftarrowtail}, f^{\leftarrow}, f^{\rightarrow}, f^{\rightarrowtail}\},$$

where $f^{\leftarrowtail}$, $f^{\leftarrow}$, $f^{\rightarrow}$, and $f^{\rightarrowtail}$ are the generalized inverses for f.

Proof of Theorem 7.64 Theorem 7.64 follows from Theorems 7.57 and 7.63 together with Corollary 7.46. Since $f \in \mathcal{RV}_\rho$ and $\rho > 0$, we apply Corollary 7.46 for $c > 0$ and pass to the limit as $s \to \infty$, which gives

$$f^{\sim}(cs) \sim (cs)^{1/\rho} L^{\sim}(cs) \sim (cs)^{1/\rho} L^{\sim}(s) \sim c^{1/\rho} f^{\sim}(s)$$

for an arbitrary asymptotically quasi-inverse function $f^{\sim}$. Thus

$$\lim_{s\to\infty} \frac{x^{\sim}(s)}{f^{\sim}(s/a)} = 1$$

for $a \in (0, \infty)$ is equivalent to

$$\lim_{s\to\infty} \frac{x^{\sim}(s)}{f^{\sim}(s)} = \left(\frac{1}{a}\right)^{\frac{1}{\rho}},$$

which completes the proof. □

7.6 Limit Points of the Ratio of Asymptotically Quasi-inverse Functions

We have shown above that the implication

$$\lim_{t\to\infty} \frac{g(t)}{f(t)} = a \in (0, \infty) \quad \Longrightarrow \quad \lim_{t\to\infty} \frac{g^{\sim}(t)}{f^{\sim}(t/a)} = 1 \tag{7.43}$$

holds under certain assumptions, where $f^{\sim}$ and $g^{\sim}$ are asymptotically quasi-inverse functions for f and g, respectively. If $a \neq 1$, then the right-hand side has a complicated form. On the other hand, Theorem 7.64 shows that, for RV-functions with index $\rho > 0$, relation (7.43) turns into the "classical" form

$$\lim_{t\to\infty} \frac{g(t)}{f(t)} = a \in (0, \infty) \quad \Longrightarrow \quad \lim_{t\to\infty} \frac{g^{\sim}(t)}{f^{\sim}(t)} = b \in (0, \infty), \tag{7.44}$$

where

$$b = \left(\frac{1}{a}\right)^{\frac{1}{\rho}}. \tag{7.45}$$

So, the natural question arises about conditions on the function f under which relation (7.44) holds with $a \neq 1$. To answer this and similar questions, we consider some relations between limit points of the ratio of functions f and g belonging to the class $\mathbb{F}^{(\infty)}$ and limit points of the ratio of their asymptotically quasi-inverse functions $f^{\sim}$ and $g^{\sim}$.

Let $\mathbb{LP}(\varphi)$ be the set of limit points, as $t \to \infty$, of a real function $\varphi = (\varphi(t), t > 0)$ and let $\mathbb{LP}(\{z_n\})$ be the set of limit points of a sequence $\{z_n\}$. For $f \in \mathbb{F}^{(\infty)}$, put

$$l_*(c; f, \{z_n\}) = \liminf_{n\to\infty} \frac{z_n}{f(cf^{\sim}(z_n))}, \qquad c > 0,$$

and

$$l^*(c; f, \{z_n\}) = \limsup_{n\to\infty} \frac{z_n}{f(cf^{\sim}(z_n))}, \qquad c > 0,$$

where $f^{\sim} \in \mathbb{AQI}(f)$, that is, $f^{\sim}$ is an asymptotically quasi-inverse function for f.

Lemma 7.65 *Let $f, g \in \mathbb{F}^{(\infty)}$, $f \in \mathcal{WPRV}$, $f^{\sim} \in \mathbb{AQI}(f)$, and $g^{\sim} \in \mathbb{AQI}(g)$. If*

$$b \in \mathbb{LP}(g^{\sim}/f^{\sim}) \cap \mathbf{R}_+,$$

then

$$l_*(b; f, \{s_n\}) \in \mathbb{LP}(g/f)$$

and

$$l^*(b; f, \{s_n\}) \in \mathbb{LP}(g/f)$$

for an arbitrary sequence of positive numbers $\{s_n\}$ such that $s_n \to \infty$ and

$$\lim_{n\to\infty} \frac{g^{\sim}(s_n)}{f^{\sim}(s_n)} = b. \tag{7.46}$$

Proof of Lemma 7.65 According to Theorem 3.42, the WPRV-function f preserves the asymptotic equivalence of sequences. Thus condition (7.46) implies that

$$\lim_{n\to\infty} \frac{g^{\sim}(s_n)}{bf^{\sim}(s_n)} = 1 \Longrightarrow \lim_{n\to\infty} \frac{f(g^{\sim}(s_n))}{f(bf^{\sim}(s_n))} = 1$$
$$\Longrightarrow \lim_{n\to\infty} \frac{f(bf^{\sim}(s_n))}{f(g^{\sim}(s_n))} = 1$$

$$\Longrightarrow \lim_{n\to\infty} \left(\frac{s_n}{f(g^{\sim}(s_n))}\right)\left(\frac{f(bf^{\sim}(s_n))}{s_n}\right) = 1$$

$$\Longrightarrow \lim_{n\to\infty} \left(\frac{g(t_n)}{f(t_n)}\right)\left(\frac{f(bf^{\sim}(s_n))}{s_n}\right) = 1,$$

where $t_n = g^{\sim}(s_n) \to \infty$. The latter equality is only possible if

$$\liminf_{n\to\infty} \frac{g(t_n)}{f(t_n)} = 1 \Big/ \limsup_{n\to\infty} \frac{f(bf^{\sim}(s_n))}{s_n} = l_*(b;\, f, \{s_n\})$$

and

$$\limsup_{n\to\infty} \frac{g(t_n)}{f(t_n)} = 1 \Big/ \liminf_{n\to\infty} \frac{f(bf^{\sim}(s_n))}{s_n} = l^*(b;\, f, \{s_n\}).$$

This proves Lemma 7.65, since

$$\liminf_{n\to\infty} \frac{g(t_n)}{f(t_n)} \in \mathbb{LP}(g/f)$$

and

$$\limsup_{n\to\infty} \frac{g(t_n)}{f(t_n)} \in \mathbb{LP}(g/f).$$

□

Corollary 7.66 *Let $f, g \in \mathbb{F}^{(\infty)}$, $f \in \mathcal{WPRV}$, $f^{\sim} \in \mathbb{AQI}(f)$, and $g^{\sim} \in \mathbb{AQI}(g)$. Then*

$$1 \in \mathbb{LP}(g^{\sim}/f^{\sim}) \quad \Longrightarrow \quad 1 \in \mathbb{LP}(g/f).$$

Proof of Corollary 7.66 Corollary 7.66 follows from Lemma 7.65, since

$$l_*(1;\, f, \{s_n\}) \in \mathbb{LP}(g/f) = 1$$

and

$$l^*(b;\, f, \{s_n\}) \in \mathbb{LP}(g/f) = 1$$

for an arbitrary sequence of positive numbers $\{s_n\}$ such that $s_n \to \infty$. □

Corollary 7.67 *Assume that $f, g \in \mathbb{F}^{(\infty)}$, $f \in \mathcal{WPRV}$, $f^{\sim} \in \mathbb{AQI}(f)$, $g^{\sim} \in \mathbb{AQI}(g)$, and $f^{\sim} \in \mathbb{C}^{\infty}$. If the limits*

$$\lim_{t\to\infty} \frac{g(t)}{f(t)} = a \in (0, \infty) \qquad \textit{and} \qquad \lim_{s\to\infty} \frac{g^{\sim}(s)}{f^{\sim}(s)} = b \in (0, \infty) \tag{7.47}$$

exist, then the limit

$$\lim_{t\to\infty} \frac{f(t/b)}{f(t)} \in (0,\infty) \tag{7.48}$$

also exists and

$$\lim_{t\to\infty} \frac{f(t/b)}{f(t)} = a. \tag{7.49}$$

Proof of Corollary 7.67 Let condition (7.47) hold. According to Lemma 7.65,

$$l_*(b;\ f, \{s_n\}) = l^*(b;\ f, \{s_n\}) = \lim_{t\to\infty} \frac{g(t)}{f(t)}$$

for an arbitrary sequence of positive numbers $\{s_n\}$ such that $s_n \to \infty$. This means that the limit

$$\lim_{t\to\infty} \frac{t}{f(bf^{\sim}(t))}$$

exists and

$$\lim_{t\to\infty} \frac{g(t)}{f(t)} = \lim_{t\to\infty} \frac{t}{f(bf^{\sim}(t))} = \lim_{t\to\infty} \frac{f(f^{\sim}(t))}{f(bf^{\sim}(t))}.$$

On the other hand, $f^{\sim} \in \mathbb{C}^{\infty}$ and thus

$$\lim_{t\to\infty} \frac{f(f^{\sim}(t))}{f(bf^{\sim}(t))} = \lim_{t\to\infty} \frac{f(t)}{f(bt)} = \lim_{t\to\infty} \frac{f(t/b)}{f(t)}. \tag{7.50}$$

Therefore, relation (7.48) and equality (7.49) are proved. □

The following result is similar to Corollary 7.67, but with other conditions on the functions f and $f^{\sim}$.

Theorem 7.68 *Let $f \in \mathbb{F}^{\infty}_{\text{ndec}} \cap \mathcal{WPRV}$, $f^{\sim} \in \mathbb{AQI}(f) \cap \mathbb{F}_{\text{ndec}}$, $g \in \mathbb{F}^{(\infty)}$, and $g^{\sim} \in \mathbb{AQI}(g)$. Then condition (7.47) implies relation (7.48) and equality (7.49).*

We need the following auxiliary result to prove Theorem 7.68.

Lemma 7.69 *Let $f \in \mathbb{F}^{\infty}_{\text{ndec}}$, $f^{\sim} \in \mathbb{AQI}(f) \cap \mathbb{F}_{\text{ndec}}$, and $c > 0$. The limit*

$$l_1(c) = \lim_{t\to\infty} \frac{f(cf^{\sim}(t))}{t} \in [0,\infty]$$

exists if and only if this holds for the limit

$$l_2(c) = \lim_{t\to\infty} \frac{f(ct)}{f(t)} \in [0, \infty].$$

Moreover, $l_1(c) = l_2(c)$.

Proof of Lemma 7.69 If $f^{\sim} \in \mathbb{AQI}(f)$, then $f^{\sim} \in \mathbb{F}^{\infty}$ (see Definition 7.1). Thus the condition that $f^{\sim} \in \mathbb{AQI}(f) \cap \mathbb{F}_{\text{ndec}}$ is equivalent to

$$f^{\sim} \in \mathbb{AQI}(f) \cap \mathbb{F}^{\infty}_{\text{ndec}}.$$

Assume that the limit $l_2(c)$ exists. Then

$$\lim_{t\to\infty} \frac{f(ct)}{f(t)} = \lim_{t\to\infty} \frac{f(cf^{\sim}(t))}{f(f^{\sim}(t))} = \lim_{t\to\infty} \frac{f(cf^{\sim}(t))}{t},$$

that is, the limit $l_1(c)$ exists and $l_1(c) = l_2(c)$.

Now we assume that the limit $l_1(c)$ exists. Consider a sequence of positive numbers $\{s_n\}$ tending to ∞. Since $f^{\sim}$ is nondecreasing, there exist, for any $n \ge 1$, positive numbers t_n and τ_n such that $t_n \to \infty$, $t_n/\tau_n \to 1$, and $f^{\sim}(t_n) \le s_n \le f^{\sim}(\tau_n)$. Then

$$\begin{aligned} l_1(c) &= \liminf_{n\to\infty} \frac{f(cf^{\sim}(t_n))}{t_n} = \liminf_{n\to\infty} \frac{f(cf^{\sim}(t_n))}{\tau_n} = \liminf_{n\to\infty} \frac{f(cf^{\sim}(t_n))}{f(f^{\sim}(\tau_n))} \\ &\le \liminf_{n\to\infty} \frac{f(cs_n)}{f(s_n)} \le \limsup_{n\to\infty} \frac{f(cs_n)}{f(s_n)} \\ &\le \limsup_{n\to\infty} \frac{f(cf^{\sim}(\tau_n))}{f(f^{\sim}(t_n))} = \limsup_{n\to\infty} \frac{f(cf^{\sim}(\tau_n))}{t_n} = \limsup_{n\to\infty} \frac{f(cf^{\sim}(\tau_n))}{\tau_n} = l_1(c), \end{aligned}$$

since f is a nondecreasing function. This implies that, for any sequence of positive numbers $\{s_n\}$ tending to ∞,

$$\lim_{n\to\infty} \frac{f(cs_n)}{f(s_n)} = l_1(c).$$

Thus the limit $l_2(c)$ exists and $l_1(c) = l_2(c)$, which proves Lemma 7.69. □

Proof of Theorem 7.68 The proof is similar to that of Corollary 7.67 and is based on Lemma 7.69. □

7.6.1 Necessary Conditions for Relation (7.44)

Corollary 7.67 or Theorem 7.68 allow us to derive relation (7.48) and equality (7.49) from (7.47).

If condition (7.47) holds with $b = 1$, then $f(t/b) = f(t)$ and relation (7.48) is trivial and equality (7.49) implies that $a = 1$. Therefore, $b \neq 1$ if $a \neq 1$.

On the other hand, if $b > 0$ and $b \neq 1$, then relation (7.48) is not trivial and means that the limit

$$\lim_{t\to\infty} \frac{f(t/b)}{f(t)} = a \in (0, \infty) \tag{7.51}$$

exists.

According to Definitions 5.1 and 5.4, condition (7.51) means that f is a function with a nondegenerate set of regular points $\mathbb{G}_r(f)$ and that $1/b \in \mathbb{G}_r(f)$. Moreover,

$$\kappa_f(1/b) = a,$$

where $(\kappa_f(\lambda), \lambda \in \mathbb{G}_r(f))$ is the index function for f.

Let f be a WOURV-function (see Definition 3.12 and Remark 3.13). As proved above, f is a WOURV-function with a nondegenerate set of regular points and the assumptions of Theorem 5.40 are satisfied for f. Relation (5.39) shows that

$$\kappa_f(\lambda) = \lambda^\rho, \qquad \lambda \in \mathbb{G}_r(f),$$

where ρ is the index of the limit functions for f and ρ coincides with the index of f, which can be evaluated as

$$\rho = \lim_{t\to\infty} \frac{\ln f(t)}{\ln t}$$

(see Definition 5.41 and 5.58, Remark 5.59, and Corollary 5.63).

Therefore

$$a = \kappa_f(1/b) = \left(\frac{1}{b}\right)^\rho, \tag{7.52}$$

whence, by passing to the limit and since $\rho \neq 0$,

$$b = \left(\frac{1}{a}\right)^{\frac{1}{\rho}}.$$

The latter equality coincides with (7.45). Also note that $\rho > 0$ only if $f \in \mathbb{F}^\infty$.

On summarizing the above, we obtain the following proposition.

Proposition 7.70 *Let $f, g \in \mathbb{F}^{(\infty)}$, $f \in \mathcal{WOURV} \cap \mathcal{WPRV}$, $f^{\sim} \in \mathbb{AQI}(f)$, and $g^{\sim} \in \mathbb{AQI}(g)$. Further, let $f^{\sim} \in \mathbb{C}^{\infty}$ or $f \in \mathbb{F}_{\text{ndec}}$ and $f^{\sim} \in \mathbb{F}_{\text{ndec}}$. If the limits*

$$\lim_{t\to\infty} \frac{g(t)}{f(t)} = a \in (0, \infty) \qquad \text{and} \qquad \lim_{s\to\infty} \frac{g^{\sim}(s)}{f^{\sim}(s)} = b \in (0, \infty)$$

exist, where $b \neq 1$, then f is a WOURV-function with a nondegenerate set of regular points $\mathbb{G}_r(f)$ and with index ρ. Moreover, $1/b \in \mathbb{G}_r(f)$ and equality (7.52) holds.

Proposition 7.70 provides important necessary conditions for relation (7.44) with $a \neq 1$, since then also $b \neq 1$.

Corollary 7.71 *Let $f, g \in \mathbb{F}^{(\infty)}$, $f \in \mathcal{WOURV} \cap \mathcal{WPRV}$, $f^{\sim} \in \mathbb{AQI}(f)$, and $g^{\sim} \in \mathbb{AQI}(g)$. Further, let $f^{\sim} \in \mathbb{C}^{\infty}$ or $f \in \mathbb{F}_{\text{ndec}}$ and $f^{\sim} \in \mathbb{F}_{\text{ndec}}$. If relation (7.44) holds with $a \neq 1$, then f is a WOURV-function with a nondegenerate set of regular points $\mathbb{G}_r(f)$ and with index ρ. Moreover, $1/b \in \mathbb{G}_r(f)$ and equality (7.52) holds.*

Remark 7.72 Remarks 3.15 and 3.17 imply that

$$\mathcal{PRV} \subset \mathcal{ORV} = \mathcal{OURV}.$$

Thus the condition $f \in \mathcal{WOURV} \cap \mathcal{WPRV}$ can be exchanged with the condition $f \in \mathcal{PRV}$ in Corollary 7.71 under measurability of the function f in Proposition 7.70.

7.7 Duality Between Certain Classes of Functions

Let $f \in \mathbb{F}^{(\infty)}$ and let $f^{\sim} \in \mathbb{AQI}(f)$, that is, $f^{\sim}$ is an asymptotically quasi-inverse function for f. An important question is about conditions under which the functions f and $f^{\sim}$ belong to certain classes of functions studied in the theory of regular variation and its generalizations.

An answer to this question allows one to determine the classes of functions for which both f ($f^{\sim}$) and $f^{\sim}$ (f) belong to the same class. In other words, we are interested in describing the classes which are invariant under the transformations $f \mapsto f^{\sim}$ and $f^{\sim} \mapsto f$.

Throughout this section, we assume that f and $f^{\sim}$ are measurable functions.

7.7.1 Some Auxiliary Results

It is known (see Remark 7.47 and Bingham et al. [41, Theorem 1.5.12]), that, if $\rho > 0$, then every asymptotically quasi-inverse function $f^{\sim}$ for $f \in \mathcal{RV}_{\rho}$ is an asymptotically inverse function for f, and that

$$f \in \mathcal{RV}_\rho \qquad \Longleftrightarrow \qquad f^{\sim} \in \mathcal{RV}_{1/\rho}.$$

Rewriting this relation as

$$f \in \mathcal{RV}_+ \qquad \Longleftrightarrow \qquad f^{\sim} \in \mathcal{RV}_+, \tag{7.53}$$

where $\mathcal{RV}_+$ is the class of RV-functions with positive indices, we see that $\mathcal{RV}_+$ is an invariant class under the transformations $f \mapsto f^{\sim}$ and $f^{\sim} \mapsto f$.

Moreover, if $f \in \mathbb{F}_{\text{ndec}}$, that is, if the function f is nondecreasing, then

$$f \in \mathcal{SQI} \qquad \Longrightarrow \qquad f^{\sim} \in \mathcal{PRV}$$

and if, additionally, $f^{\sim} \in \mathbb{AI}(f)$, that is, if $f^{\sim}$ is an asymptotically inverse function for f, then

$$f \in \mathcal{SQI} \qquad \Longleftrightarrow \qquad f^{\sim} \in \mathcal{PRV}$$

(see Proposition 7.27 and Theorem 7.31).

On the other hand, if $f \in \mathbb{F}_{\text{ndec}}$ or $f^{\sim} \in \mathbb{F}_{\text{ndec}}$, then

$$f \in \mathcal{PRV} \qquad \Longrightarrow \qquad f^{\sim} \in \mathcal{SQI},$$

and, if $f^{\sim} \in \mathbb{AI}(f)$, then

$$f \in \mathcal{PRV} \qquad \Longleftrightarrow \qquad f^{\sim} \in \mathcal{SQI} \tag{7.54}$$

(see Proposition 7.36 and Corollary 7.37).

The following relation is similar to (7.53):

$$f \in \mathcal{POV} \qquad \Longleftrightarrow \qquad f^{\sim} \in \mathcal{POV}. \tag{7.55}$$

If $f \in \mathcal{POV}$, then $f^{\sim}$ is an asymptotically inverse function for f (see Theorems 7.40 and 7.45).

Our aim is to find similar characterizations for the classes $\mathcal{ORV}$, $\mathcal{SV}$, $\mathcal{RV}_\infty$, and some other classes of functions (see Sect. 3.2).

7.7.2 The Images of Some Classes of Functions Under the Transformations $f \mapsto f^{\sim}$ and $f^{\sim} \mapsto f$

First we state the main results of this section and highlight some common features. The proofs are given at the end of the section.

Theorem 7.73 *Let $f \in \mathbb{F}^{\infty}_{\text{ndec}}$. Then,*

$$f \in \mathcal{PI} \quad \Longrightarrow \quad f^{\sim} \in \mathcal{ORV}.$$

Theorem 7.74 *Let $f \in \mathbb{F}^{\infty}_{\text{ndec}}$ and $f^{\sim} \in \mathbb{F}_{\text{ndec}}$. Then,*

(i) $f \in \mathcal{PI} \quad \Longleftrightarrow \quad f^{\sim} \in \mathcal{ORV}$;
(ii) $f \in \mathcal{ORV} \quad \Longleftrightarrow \quad f^{\sim} \in \mathcal{PI}$.

Theorem 7.75 *Let $f \in \mathbb{F}^{\infty}_{\text{ndec}}$ and $f^{\sim} \in \mathbb{F}_{\text{ndec}}$. Then,*

(iii) $f \in \mathcal{RV}_{\infty} \quad \Longleftrightarrow \quad f^{\sim} \in \mathcal{SV}$;
(iv) $f \in \mathcal{SV} \quad \Longleftrightarrow \quad f^{\sim} \in \mathcal{RV}_{\infty}$.

Theorem 7.74 implies the following result.

Theorem 7.76 *Let $f \in \mathbb{F}^{\infty}_{\text{ndec}}$ and $f^{\sim} \in \mathbb{F}_{\text{ndec}}$. Then,*

$$f \in \mathcal{ORV} \cap \mathcal{PI} \quad \Longleftrightarrow \quad f^{\sim} \in \mathcal{ORV} \cap \mathcal{PI}.$$

From the definition of the classes $\mathcal{RV}$, $\mathcal{PRV}$, and $\mathcal{ORV}$ (see Sects. 3.2 and 3.3) we obtain the inclusions

$$\mathcal{RV} \subset \mathcal{PRV} \subset \mathcal{ORV}$$

(see Remark 3.17). Moreover, the definition of $\mathcal{RV}_{+}$, $\mathcal{RV}_{\infty}$, $\mathcal{SQI}$, and $\mathcal{PI}$ (recall Sects. 3.2 and 3.3) further implies that

$$\mathcal{RV}_{+} \cup \mathcal{RV}_{\infty} \subset \mathcal{SQI} \subset \mathcal{PI}.$$

An intersection of the corresponding classes in these inclusions leads to

$$\mathcal{RV}_{+} \subset \mathcal{PRV} \cap \mathcal{SQI} \subset \mathcal{ORV} \cap \mathcal{PI}.$$

It is worth mentioning that all three classes of functions $\mathcal{RV}_{+}$, $\mathcal{PRV} \cap \mathcal{SQI} = \mathcal{POV}$, and $\mathcal{ORV} \cap \mathcal{PI}$ are invariant under the transformations $f \mapsto f^{\sim}$ and $f^{\sim} \mapsto f$ in view of relations (7.53), (7.55) and Theorem 7.76.

Consider the conditions

$$\lim_{t\to\infty} \frac{f(c_0 f^{\sim}(t))}{t} = \infty, \quad \text{for some} \quad c_0 > 1 \tag{7.56}$$

and

$$\lim_{t\to\infty} \frac{f(c_0 t)}{f(t)} = \infty, \qquad \text{for some} \quad c_0 > 1. \tag{7.57}$$

It is clear that conditions (7.56) and (7.57) are equivalent if the function $f^{\sim}$ is continuous. Moreover, Lemma 7.69 shows that these conditions are equivalent if both functions f and $f^{\sim}$ are nondecreasing.

Let $\mathcal{QV}$ and $\mathcal{QV}_1$ be the two subsets of measurable functions in the class $\mathbb{F}^{(\infty)}$ satisfying conditions (7.56) and (7.57), respectively.

Theorem 7.77 *Let* $f \in \mathbb{F}^{\infty}_{\text{ndec}}$. *Then,*

$$f \in \mathcal{QV} \qquad \Longleftrightarrow \qquad f^{\sim} \in \mathcal{OSV}.$$

Theorem 7.78 *Let* $f \in \mathbb{F}^{\infty}_{\text{ndec}}$ *and* $f^{\sim} \in \mathbb{F}_{\text{ndec}}$. *Then,*

(v) $f \in \mathcal{QV}_1 \qquad \Longleftrightarrow \qquad f^{\sim} \in \mathcal{OSV};$
(vi) $f \in \mathcal{OSV} \qquad \Longleftrightarrow \qquad f^{\sim} \in \mathcal{QV}_1.$

Note that Theorems 7.75 and 7.78, in contrast to Theorem 7.74, do not lead to new classes of functions, which are invariant under the transformations $f \mapsto f^{\sim}$ and $f^{\sim} \mapsto f$, since

$$\mathcal{RV}_\infty \cap \mathcal{SV} = \varnothing$$

and

$$\mathcal{OSV} \cap \mathcal{QV}_1 = \varnothing.$$

7.7.3 *Proof of Theorems 7.73–7.78*

Proof of Theorem 7.73 Let $f \in \mathcal{PI}$, but $f^{\sim} \notin \mathcal{ORV}$.

Since $f^{\sim}$ is not an ORV-function, there exists a number $a > 0$ such that

$$(f^{\sim})^*(a) = \limsup_{t\to\infty} \frac{f^{\sim}(at)}{f^{\sim}(t)} = \infty. \tag{7.58}$$

Thus, given an arbitrary $\varepsilon > 0$, there exists a number $a = a(\varepsilon) \in (0, 1+\varepsilon)$ such that (7.58) holds, since $(f^{\sim})^*(\sqrt{u}) \geq \sqrt{(f^{\sim})^*(u)}$ for all $u > 1$. This implies that, for any $\varepsilon > 0$, there exist an $a = a(\varepsilon) \in (0, 1+\varepsilon)$ and a sequence of positive numbers $\{s_n\}$, increasing to ∞, such that

$$f^{\sim}(a(\varepsilon)s_n) \geq A f^{\sim}(s_n) \tag{7.59}$$

for all $A > 0$ and all sufficiently large n.

Since $f \in \mathcal{PI}$, there exists a $\delta > 0$ such that

$$f(c_0 t_n) > (1+\delta) f(t_n) \tag{7.60}$$

for all sufficiently large n and an arbitrary sequence $\{t_n\}$ increasing to ∞, where c_0 is the constant from condition (3.13).

Since $f \in \mathbb{F}^{\infty}_{\text{ndec}}$, the inequalities (7.59) and (7.60), with $\varepsilon = \delta$ and $A = c_0$, imply

$$\begin{aligned} 1+\delta > a(\delta) &= \lim_{n\to\infty} \frac{f(f^{\sim}(a(\delta)s_n))}{s_n} \ge \limsup_{n\to\infty} \frac{f(c_0 f^{\sim}(s_n))}{s_n} \\ &\ge (1+\delta) \lim_{n\to\infty} \frac{f(f^{\sim}(s_n))}{s_n} = 1+\delta. \end{aligned}$$

This contradiction proves Theorem 7.73. □

To prove Theorem 7.74 we need an auxiliary result.

Lemma 7.79 *Let* $f \in \mathbb{F}^{\infty}_{\text{ndec}}$ *and* $f^{\sim} \in \mathbb{F}_{\text{ndec}}$. *Then,*

$$f^{\sim} \in \mathcal{ORV} \qquad \Longrightarrow \qquad f \in \mathcal{PI}.$$

Proof of Lemma 7.79 Let $f^{\sim} \in \mathcal{ORV}$, but $f \notin \mathcal{PI}$.

Since $f^{\sim} \in \mathcal{ORV}$, for an arbitrary $a > 1$, there exists a number $A > 1$ such that

$$f^{\sim}(au) \le A f^{\sim}(u) \tag{7.61}$$

for all sufficiently large s.

Since $f \notin \mathcal{PI}$ and $f \in \mathbb{F}^{\infty}_{\text{ndec}}$, there exists an increasing sequence of positive numbers $\{s_n\}$ such that $s_n \to \infty$ and

$$\lim_{n\to\infty} \frac{f(As_n)}{f(s_n)} = 1.$$

Put $t_n = \sup\{t \ge 0 : f^{\sim}(t) \le s_n\}$ and $\tau_n = \inf\{t \ge 0 : f^{\sim}(t) > s_n\}$, $n \ge 1$, where $\sup \varnothing := 0$. Then $t_n \to \infty$ and $t_n = \tau_n$, $n \ge 1$, since the function $f^{\sim}$ is nondecreasing and unbounded.

Putting $u_n = t_n - 1$ and $v_n = t_n + 1$, $n \ge 1$, we get

$$f^{\sim}(u_n) \le s_n \le f^{\sim}(v_n), \qquad n \ge 1, \tag{7.62}$$

and

$$u_n \sim f(s_n) \sim v_n, \qquad n \to \infty, \tag{7.63}$$

since the functions $f^{\sim}$ and f are nondecreasing.

Therefore, relations (7.61)–(7.63) yield

$$1 < a = \lim_{n\to\infty} \frac{f(f^{\sim}(au_n))}{u_n} \le \limsup_{n\to\infty} \frac{f(Af^{\sim}(u_n))}{u_n} \le \limsup_{n\to\infty} \frac{f(As_n)}{u_n}$$
$$= \limsup_{n\to\infty} \frac{f(As_n)}{f(s_n)} \frac{f(s_n)}{u_n} = \lim_{n\to\infty} \frac{f(As_n)}{f(s_n)} = 1.$$

This contradiction proves Lemma 7.79. □

Proof of Theorem 7.74 Statement (i) follows from Theorem 7.73 and Lemma 7.79.

To prove statement (ii) assume that $f \notin \mathcal{ORV}$, but $f^{\sim} \in \mathcal{PI}$. Since $f^{\sim} \in \mathcal{PI}$, there exist numbers $c_0 > 1$ and $b > 1$ such that

$$f^{\sim}(c_0 u) \ge b f^{\sim}(u)$$

for all sufficiently large u.

Since the function f is nondecreasing and $f \notin \mathcal{ORV}$, $f^*(a) = \infty$ for all $a > 1$ (see the proof of Theorem 7.73). Thus, for all numbers $a > 1$ and $A > 1$, there exists an increasing sequence of positive numbers $\{s_n\}$ such that $s_n \to \infty$ and

$$f(as_n) \ge Af(s_n), \qquad n \ge 1.$$

As in the proof of Lemma 7.79, choose sequences $\{u_n\}$ and $\{v_n\}$ and put $a = b$ and $A = 2c_0$. Since the function f is nondecreasing,

$$c_0 = \lim_{n\to\infty} \frac{f(f^{\sim}(c_0 v_n))}{v_n} \ge \limsup_{n\to\infty} \frac{f(bf^{\sim}(v_n))}{v_n} \ge \limsup_{n\to\infty} \frac{f(bs_n)}{v_n}$$
$$\ge 2c_0 \limsup_{n\to\infty} \frac{f(s_n)}{v_n} = 2c_0.$$

This contradiction proves that $f \in \mathcal{ORV}$. Thus

$$f \in \mathcal{ORV} \quad \Longleftarrow \quad f^{\sim} \in \mathcal{PI}.$$

Now, assume that $f \in \mathcal{ORV}$, but $f^{\sim} \notin \mathcal{PI}$. Since $f \in \mathcal{ORV}$, for any $a > 1$, there exists an $A > 1$ such that

$$f(at) \le Af(t)$$

for all sufficiently large t.

Further, since $f^{\sim} \notin \mathcal{PI}$, for all $c > 1$ and $b > 1$, there exists an increasing sequence of positive numbers $\{s_n\}$ such that $s_n \to \infty$ and

$$f^{\sim}(cs_n) \le bf^{\sim}(s_n), \qquad n \ge 1.$$

Fix $a > 1$ and put $b = a$ and $c = 2A$. Since the function f is nondecreasing,

$$2A = \lim_{n\to\infty} \frac{f(f^{\sim}(cs_n))}{s_n} \le \limsup_{n\to\infty} \frac{f(af^{\sim}(s_n))}{s_n}$$
$$\le \limsup_{n\to\infty} \frac{Af(f^{\sim}(s_n))}{s_n} = A \limsup_{n\to\infty} \frac{f(f^{\sim}(s_n))}{s_n} = A.$$

This contradiction proves that

$$f^{\sim} \in \mathcal{ORV} \quad \Longrightarrow \quad f^{\sim} \in \mathcal{PI},$$

which completes the proof of statement (ii). So, Theorem 7.74 is proved. □

Proof of Theorem 7.75 To prove statement (iii) assume that $f \in \mathcal{RV}_\infty$, but $f^{\sim} \notin \mathcal{SV}$.

Since $f^{\sim} \notin \mathcal{SV}$, there exist numbers $c_0 > 1$ and $a > 1$ and an increasing sequence of positive numbers $\{t_n\}$ such that $t_n \to \infty$ and

$$af^{\sim}(t_n) \le f^{\sim}(c_0 t_n), \qquad n \ge 1.$$

Since $f \in \mathcal{RV}_\infty$, this implies that

$$\infty = \lim_{n\to\infty} \frac{f\,(af^{\sim}(t_n))}{f\,(f^{\sim}(t_n))} \le \lim_{n\to\infty} \frac{f\,(f^{\sim}(c_0 t_n))}{f\,(f^{\sim}(t_n))} = c_0.$$

This contradiction proves that

$$f \in \mathcal{RV}_\infty \quad \Longrightarrow \quad f^{\sim} \in \mathcal{SV}.$$

Now, assume that $f^{\sim} \in \mathcal{SV}$, but $f \notin \mathcal{RV}_\infty$.

Since $f^{\sim} \in \mathcal{SV}$, for all $c > 1$ and $a > 1$,

$$f^{\sim}(ct) \le af^{\sim}(t)$$

for all sufficiently large t.

Further, since $f \notin \mathcal{RV}_\infty$, there exist numbers $\lambda > 1$ and $A > 0$ and an increasing sequence of positive numbers $\{s_n\}$ such that $s_n \to \infty$ and

$$f(\lambda s_n) \le Af(s_n), \qquad n \ge 1.$$

Put $a = \lambda$, $c > A$ and choose sequences $\{u_n\}$ and $\{v_n\}$ as in the proof of Lemma 7.79. Since the function f is nondecreasing,

$$c = \lim_{n\to\infty} \frac{f(f^{\sim}(cu_n))}{u_n} \le \limsup_{n\to\infty} \frac{f(af^{\sim}(u_n))}{u_n} \le \limsup_{n\to\infty} \frac{f(as_n)}{u_n}$$
$$\le A \limsup_{n\to\infty} \frac{f(s_n)}{u_n} = A.$$

This contradiction proves statement (iii).

To prove statement (iv) assume that $f \in \mathcal{SV}$, but $f^{\sim} \notin \mathcal{RV}_\infty$.

Since $f^{\sim} \notin \mathcal{RV}_\infty$, there are numbers $c > 1$ and $a > 0$ and an increasing sequence of positive numbers $\{t_n\}$ such that $t_n \to \infty$ and

$$f^{\sim}(ct_n) \le af^{\sim}(t_n), \qquad n \ge 1.$$

This implies that $f(f^{\sim}(ct_n)) \le f(af^{\sim}(t_n))$ for all $n \ge 1$.

Moreover, since $f \in \mathcal{SV}$,

$$1 < c = \lim_{n\to\infty} \frac{f(f^{\sim}(ct_n))}{t_n} \le \lim_{n\to\infty} \frac{f(af^{\sim}(t_n))}{t_n} = \lim_{n\to\infty} \frac{f(f^{\sim}(t_n))}{t_n} = 1.$$

This contradiction proves the implication

$$f \in \mathcal{SV} \quad \Longrightarrow \quad f^{\sim} \in \mathcal{RV}_\infty.$$

Now, we assume that $f^{\sim} \in \mathcal{RV}_\infty$, but $f \notin \mathcal{SV}$.

Since $f \notin \mathcal{SV}$, there are numbers $c_0 > 1$ and $a > 1$ and an increasing sequence of positive numbers $\{s_n\}$ such that $s_n \to \infty$ and

$$f(c_0 s_n) \ge af(s_n), \qquad n \ge 1. \tag{7.64}$$

Let $c > c_0$. Since $f^{\sim} \in \mathcal{RV}_\infty$, for any $\lambda > 1$ and all sufficiently large t, we have

$$f^{\sim}(\lambda t) \ge cf^{\sim}(t). \tag{7.65}$$

Choose increasing sequences of positive numbers $\{u_n\}$ and $\{v_n\}$ such that $u_n \to \infty$ and $v_n \to \infty$, for which relations (7.62) and (7.63) hold (see the proof of Lemma 7.79).

Since the function f is nondecreasing, relations (7.62), (7.64), and (7.65) with $t = v_n$ imply that

$$f(f^{\sim}(\lambda v_n)) \ge f(cf^{\sim}(v_n)) \ge f(c_0 s_n) \ge af(s_n)$$

for all sufficiently large n. From this and (7.63) we obtain that

$$1 = \limsup_{n\to\infty} \frac{f(s_n)}{v_n} = \limsup_{n\to\infty} \frac{\lambda f(s_n)}{f(f^{\sim}(\lambda v_n))} \le \frac{\lambda}{a}$$

for all $\lambda > 1$. Passing to the limit as $\lambda \downarrow 1$, we get a contradiction which proves

$$f^{\sim} \in \mathcal{RV}_{\infty} \quad \Longrightarrow \quad f \in \mathcal{SV}$$

and completes the proof of statement (iv). So, the proof of Theorem 7.75 is now complete. □

Proof of Theorem 7.77 Let $f^{\sim} \in \mathcal{OSV}$. Then there exists a number $A > 0$ such that $f^{\sim}(ct) \le Af^{\sim}(t)$ for all $c > 0$ and all sufficiently large t. Since the function f is nondecreasing,

$$c = \lim_{t\to\infty} \frac{f(f^{\sim}(ct))}{t} \le \liminf_{t\to\infty} \frac{f(Af^{\sim}(t))}{t}$$

for all $c > 0$. This means that condition (7.56) with $c_0 = A$ is satisfied. Thus $f \in \mathcal{QV}$, whence

$$f^{\sim} \in \mathcal{OSV} \quad \Longrightarrow \quad f \in \mathcal{QV}.$$

Now, assume that $f \in \mathcal{QV}$, but $f^{\sim} \notin \mathcal{OSV}$.

Since $f^{\sim} \notin \mathcal{OSV}$, for any $A > 0$ there exist a number $c > 0$ and an increasing sequence of positive numbers $\{t_n\}$ such that $t_n \to \infty$ and

$$f^{\sim}(ct_n) \ge Af^{\sim}(t_n), \qquad n \ge 1.$$

In particular, for $A = c_0$, there exist a $c > 0$ and $\{t_n\}$ such that $t_n \to \infty$ and

$$f^{\sim}(ct_n) \ge c_0 f^{\sim}(t_n), \qquad n \ge 1.$$

Since the function f is nondecreasing,

$$c = \lim_{n\to\infty} \frac{f(f^{\sim}(ct_n))}{t_n} \ge \liminf_{n\to\infty} \frac{f(c_0 f^{\sim}(t_n))}{t_n} \ge \liminf_{t\to\infty} \frac{f(c_0 f^{\sim}(t))}{t},$$

which contradicts $f \in \mathcal{QV}$. Therefore

$$f \in \mathcal{QV} \quad \Longrightarrow \quad f^{\sim} \in \mathcal{OSV},$$

and the proof of Theorem 7.77 is complete. □

Proof of Theorem 7.78 Statement (v) follows from Theorem 7.77 and Lemma 7.69.

To prove statement (vi) assume that $f^{\sim} \in \mathcal{QV}_1$. Then there exists a number $c_0 > 1$ such that

$$\lim_{t\to\infty} \frac{f^{\sim}(c_0 t)}{f^{\sim}(t)} = \infty.$$

Thus, for all $A > 0$ and for all sufficiently large t,

$$f^{\sim}(c_0 t) \geq A f^{\sim}(t).$$

Since f is a nondecreasing function,

$$c_0 = \lim_{t\to\infty} \frac{f(f^{\sim}(c_0 t))}{t} \geq \limsup_{t\to\infty} \frac{f(Af^{\sim}(t))}{t} \tag{7.66}$$

for all $A > 0$. Next we show that

$$\sup_{A>0} f^*(A) \leq c_0. \tag{7.67}$$

Indeed, otherwise there exist numbers $A_0 > 0$ and $B > 0$ such that $f^*(A_0) > c_0$ and $c_0 < B < f^*(A_0)$. Now, choose an increasing sequence of positive numbers $\{s_n\}$ such that $s_n \to \infty$ and

$$f(A_0 s_n) \geq B f(s_n), \qquad n \geq 1.$$

Then we introduce sequences $\{u_n\}$ and $\{v_n\}$ in the same way as in the proof of Lemma 7.79. As a result,

$$f(A_0 f^{\sim}(v_n)) \geq f(A_0 s_n) \geq B f(s_n) \geq B f(f^{\sim}(u_n)).$$

Taking (7.63) and (7.66) into account, we arrive at the contradiction

$$c_0 \geq \limsup_{n\to\infty} \frac{f(A_0 f^{\sim}(v_n))}{v_n} \geq B \limsup_{n\to\infty} \frac{f(f^{\sim}(v_n))}{v_n} = B,$$

which proves (7.67). This inequality means that $f \in \mathcal{OSV}$, which yields

$$f^{\sim} \in \mathcal{QV}_1 \quad \Longrightarrow \quad f \in \mathcal{OSV}.$$

Now, assume that $f \in \mathcal{OSV}$, but $f^{\sim} \notin \mathcal{QV}_1$.

Since $f \in \mathcal{OSV}$, there exists a number $c_0 \geq 1$ such that $f^*(A) \leq c_0$ for all $A > 0$. Moreover, $f^{\sim} \notin \mathcal{QV}_1$ and thus, for all $c > 1$,

$$\liminf_{t\to\infty} \frac{f^{\sim}(ct)}{f^{\sim}(t)} < \infty.$$

This implies that, for all $c > 1$, there exists a number $B > 0$ such that $f^{\sim}(ct) \le Bf^{\sim}(t)$ for infinitely many t. Thus, with $c > c_0$ we obtain the contradiction

$$c = \lim_{t\to\infty} \frac{f(f^{\sim}(ct))}{t} \le \liminf_{t\to\infty} \frac{f(Bf^{\sim}(t))}{f(f^{\sim}(t))} \le \limsup_{s\to\infty} \frac{f(Bs)}{f(s)} \le c_0,$$

which yields

$$f \in \mathcal{OSV} \quad \Longrightarrow \quad f^{\sim} \in \mathcal{QV}_1.$$

So, statement (vi) is proved and the proof of Theorem 7.78 is complete. □

7.7.4 Equivalence Theorems for Asymptotically Left Quasi-inverse Functions

Asymptotically right quasi-inverse and asymptotically left quasi-inverse functions have been defined in Sect. 7.2 (see Definition 7.24 and 7.25).

Recall that an asymptotically quasi-inverse function is an asymptotically right quasi-inverse function. Moreover, for $f \in \mathbb{F}^{\infty}$, the function f is an asymptotically left quasi-inverse function for g, if g is an asymptotically right quasi-inverse function for f, and *vice versa*. Theorems 7.74–7.78 imply the following results for asymptotically left quasi-inverse functions.

Theorem 7.80 *Let $f \in \mathbb{F}^{\infty}_{\text{ndec}}$ and $f_{\ell}^{\sim} \in \mathbb{F}_{\text{ndec}}$. Then,*

(a) $f \in \mathcal{PI} \iff f_{\ell}^{\sim} \in \mathcal{ORV}$;
(b) $f \in \mathcal{ORV} \iff f_{\ell}^{\sim} \in \mathcal{PI}$;
(c) $f \in \mathcal{ORV} \cap \mathcal{PI} \iff f_{\ell}^{\sim} \in \mathcal{ORV} \cap \mathcal{PI}$.

Theorem 7.81 *Let $f \in \mathbb{F}^{\infty}_{\text{ndec}}$ and $f_{\ell}^{\sim} \in \mathbb{F}_{\text{ndec}}$. Then,*

(a) $f \in \mathcal{RV}_{\infty} \iff f_{\ell}^{\sim} \in \mathcal{SV}$;
(b) $f \in \mathcal{SV} \iff f_{\ell}^{\sim} \in \mathcal{RV}_{\infty}$.

Theorem 7.82 *Let $f \in \mathbb{F}^{\infty}$ and $f_{\ell}^{\sim} \in \mathbb{F}_{\text{ndec}}$. Then,*

$$f \in \mathcal{OSV} \iff f_{\ell}^{\sim} \in \mathcal{QV}.$$

Theorem 7.83 *Let $f \in \mathbb{F}^{\infty}_{\text{ndec}}$ and $f_{\ell}^{\sim} \in \mathbb{F}_{\text{ndec}}$. Then,*

(a) $f \in \mathcal{OSV} \iff f_{\ell}^{\sim} \in \mathcal{QV}_1$;
(b) $f \in \mathcal{QV}_1 \iff f_{\ell}^{\sim} \in \mathcal{OSV}$.

7.8 Generalized Inverse Functions Corresponding to Sequences

For a sequence of real numbers $\{x\} = \{x_n\}_{n\ge 0}$ such that $x_0 = 0$ and

$$\{x_n\}_{n\ge 0} \in \mathbb{S}^{\infty}, \qquad \text{that is,} \qquad \lim_{n\to\infty} x_n = \infty,$$

define the four *generalized inverse functions*, pointwise for all $s \ge 0$, as follows:

$$\begin{aligned}
\{x\}^{\leftarrowtail}(s) &= \min\{n \ge 0 : x(n) \ge s\},\\
\{x\}^{\leftarrow}(s) &= \min\{n \ge 0 : x(n) > s\},\\
\{x\}^{\rightarrow}(s) &= \max\{n \ge 0 : x(n) < s\},\\
\{x\}^{\rightarrowtail}(s) &= \max\{n \ge 0 : x(n) \le s\},
\end{aligned} \tag{7.68}$$

which all take nonnegative integer values. Since $x_0 = 0$,

$$\{x\}^{\leftarrowtail}(0) = 0, \qquad \{x\}^{\leftarrow}(0) \ge 0, \qquad \{x\}^{\rightarrow}(0) \ge 0, \qquad \{x\}^{\rightarrowtail}(0) \ge 0.$$

If $x_n \ge 0$, $n \ge 1$, we set $\{x\}^{\rightarrow}(0) = 0$.

The generalized inverse functions $\{x\}^{\leftarrowtail}$, $\{x\}^{\leftarrow}$, $\{x\}^{\rightarrow}$, and $\{x\}^{\rightarrowtail}$ are similar to the generalized inverse functions $f^{\leftarrowtail}$, $f^{\leftarrow}$, $f^{\rightarrow}$, and $f^{\rightarrowtail}$ defined in Sect. 7.2 for functions f with continuous argument. Some properties are also similar.

The functions $\{x\}^{\leftarrowtail}$, $\{x\}^{\leftarrow}$, $\{x\}^{\rightarrow}$, and $\{x\}^{\rightarrowtail}$ are nondecreasing and, provided $\{x_n\}_{n\ge 0} \in \mathbb{S}^{\infty}$, tend to ∞ as $s \to \infty$. For all $s \ge 0$,

$$\{x\}^{\leftarrowtail}(s) \le \{x\}^{\leftarrow}(s) \le \{x\}^{\rightarrowtail}(s) + 1, \qquad \{x\}^{\leftarrowtail}(s) \le \{x\}^{\rightarrow}(s) + 1, \tag{7.69}$$

$$\{x\}^{\rightarrow}(s) \le \{x\}^{\rightarrowtail}(s).$$

Moreover,

$$\{x\}^{\leftarrowtail}(x_n) \le n \le \{x\}^{\rightarrowtail}(x_n) \tag{7.70}$$

for all $n \ge 0$ such that $x_n \ge 0$.

If the sequence $\{x_n\}_{n\ge 0}$ is increasing, that is, if $x_{n+1} > x_n$, $n \ge 0$, then

$$\begin{aligned}
\{x\}^{\rightarrow}(s) &\le \{x\}^{\rightarrowtail}(s) \le \{x\}^{\leftarrowtail}(s) \le \{x\}^{\leftarrow}(s)\\
&\le \{x\}^{\rightarrowtail}(s) + 1 \le \{x\}^{\rightarrow}(s) + 2,
\end{aligned} \tag{7.71}$$

for all $s \ge 0$. The latter inequalities complement (7.69) and show that

$$\{x\}^{\leftarrowtail} \sim \{x\}^{\leftarrow} \sim \{x\}^{\rightarrow} \sim \{x\}^{\rightarrowtail} \tag{7.72}$$

if $\{x_n\}_{n\ge 0} \in \mathbb{S}^\infty$.

The generalized inverse functions for sequences have a clear meaning. If $\{x_n\}_{n\ge 0}$ is considered as values of a certain variable that changes at discrete times (say), then $\{x\}^{\leftarrowtail}(s)$ is the first exit time of this variable from the set $(-\infty, s)$ or the first hitting time of the set $[s, \infty)$. Correspondingly, $\{x\}^{\leftarrow}(s)$ is the first exit time from the set $(-\infty, s]$ or the first hitting time of the set (s, ∞). In turn, $\{x\}^{\rightarrow}(s)$ is the last sojourn time in the set $(-\infty, s)$, while $\{x\}^{\rightarrowtail}(s)$ is the last sojourn time in the set $(-\infty, s]$. Note that the function $\{x\}^{\rightarrowtail}$ constructed from a sequence of sums of random variables $\{S_n\}$ has been studied already in Chaps. 1 and 2 (see, e.g., the definition in (1.1)).

For the generalized inverse functions $\{x\}^{\leftarrowtail}$, $\{x\}^{\leftarrow}$, $\{x\}^{\rightarrow}$, and $\{x\}^{\rightarrowtail}$, the following characterizing duality relations are satisfied:

$$\begin{aligned}
\{x\}^{\leftarrowtail}(s) = n \ge 1 &\iff x_n \ge s \text{ and } x_k < s \text{ for } 0 \le k < n, \\
\{x\}^{\leftarrow}(s) = n \ge 1 &\iff x_n > s \text{ and } x_k \le s \text{ for } 0 \le k < n, \\
\{x\}^{\rightarrow}(s) = n \ge 0 &\iff x_n < s \text{ and } x_k \ge s \text{ for } k > n, \\
\{x\}^{\rightarrowtail}(s) = n \ge 0 &\iff x_n \le s \text{ and } x_k > s \text{ for } k > n.
\end{aligned} \tag{7.73}$$

These duality relations imply that, under certain additional assumptions, a generalized inverse function for a sequence $\{x_n\}_{n\ge 0}$ is asymptotically quasi-inverse for this sequence (see also Lemma 2.2).

Lemma 7.84 *Let* $\{x_n\}_{n\ge 0} \in \mathbb{S}^\infty$ *and*

$$\lim_{n\to\infty} \frac{x_{n+1}}{x_n} = 1. \tag{7.74}$$

Then

$$\lim_{s\to\infty} \frac{x_{\{x\}^{\leftarrowtail}(s)}}{s} = \lim_{s\to\infty} \frac{x_{\{x\}^{\leftarrow}(s)}}{s} = \lim_{s\to\infty} \frac{x_{\{x\}^{\rightarrow}(s)}}{s} \tag{7.75}$$

$$= \lim_{s\to\infty} \frac{x_{\{x\}^{\rightarrowtail}(s)}}{s} = 1. \tag{7.76}$$

Proof of Lemma 7.84 According to (7.73),

$$x_{(\{x\}^{\leftarrowtail}(s)-1)} < s \le x_{\{x\}^{\leftarrowtail}(s)}$$

and

$$\frac{x_{(\{x\}^{\leftarrowtail}(s)-1)}}{x_{\{x\}^{\leftarrowtail}(s)}} < \frac{s}{x_{\{x\}^{\leftarrowtail}(s)}} \le 1$$

if $\{x\}^{\leftarrowtail}(s) \ge 1$. Taking (7.74) into account and $\{x\}^{\leftarrowtail}(s) \to \infty$, as $s \to \infty$, we get

$$\lim_{s\to\infty} \frac{x_{\{x\}^{\leftarrowtail}(s)}}{s} = 1.$$

The other equalities in (7.75) can be proved analogously. □

The following result complements Lemma 7.84 and shows that the sequence $\{x_n\}_{n\ge0}$ is an "asymptotically quasi-inverse" for each of its generalized inverse functions if the functions $\{x\}^{\leftarrowtail}$ and $\{x\}^{\rightarrowtail}$ are asymptotically equivalent.

Lemma 7.85 *Let* $\{x_n\}_{n\ge0} \in \mathbb{S}^\infty$ *and* $\{x\}^{\leftarrowtail} \sim \{x\}^{\rightarrowtail}$. *Then*

$$\lim_{n\to\infty} \frac{\{x\}^{\leftarrowtail}(x_n)}{n} = \lim_{n\to\infty} \frac{\{x\}^{\leftarrow}(x_n)}{n} = \lim_{n\to\infty} \frac{\{x\}^{\rightarrow}(x_n)}{n} \tag{7.77}$$
$$= \lim_{n\to\infty} \frac{\{x\}^{\rightarrowtail}(x_n)}{n} = 1.$$

Proof of Lemma 7.85 According to (7.70),

$$\frac{\{x\}^{\leftarrowtail}(x_n)}{\{x\}^{\rightarrowtail}(x_n)} \le \frac{n}{\{x\}^{\rightarrowtail}(x_n)} \le 1 \le \frac{n}{\{x\}^{\leftarrowtail}(x_n)} \le \frac{\{x\}^{\rightarrowtail}(x_n)}{\{x\}^{\leftarrowtail}(x_n)}$$

if $x_n \ge 0$. The assumptions of the lemma imply that

$$\lim_{n\to\infty} \frac{\{x\}^{\leftarrowtail}(x_n)}{n} = \lim_{n\to\infty} \frac{\{x\}^{\rightarrowtail}(x_n)}{n} = 1.$$

Now (7.77) follows from relation (7.69). □

Remark 7.86 If there exist a number $a \in (0, \infty)$ and a function $f \in \mathbb{F}_+$ such that

$$\lim_{s\to\infty} \frac{\{x\}^{\leftarrowtail}(s)}{f(s)} = \lim_{s\to\infty} \frac{\{x\}^{\rightarrowtail}(s)}{f(s)} = a, \tag{7.78}$$

then $\{x\}^{\leftarrowtail} \sim \{x\}^{\rightarrowtail}$. Therefore, if $\{x_n\}_{n\ge0} \in \mathbb{S}^\infty$ and (7.78) holds, then so does (7.77).

7.8.1 The Limit Behavior of the Ratio of Asymptotically Quasi-inverse Functions Constructed from Sequences

Let $\{x\} = \{x_n\}_{n \geq 0}$ be a sequence of real numbers such that $x_0 = 0$ and $\{x_n\}_{n\geq 0} \in \mathbb{S}^\infty$ and let $\{p\} = \{p_n\}_{n\geq 0}$ be a sequence of real numbers such that $p_0 = 0$, $\{p_n\}_{n\geq 0} \in \mathbb{S}^\infty$, and $p_n > 0$, $n \geq 1$.

Assume that

$$\lim_{n\to\infty} \frac{x_n}{p_n} = a \in (0, \infty).$$

We consider the problem of finding conditions for

$$\lim_{s\to\infty} \frac{\{x\}^\sim(s)}{\{p\}^\sim(s/a)} = 1,$$

where $\{x\}^\sim$ and $\{p\}^\sim$ are "asymptotically quasi-inverse" functions for the sequences $\{x_n\}_{n\geq 0}$ and $\{p_n\}_{n\geq 0}$, respectively. In particular, this problem is considered for the functions

$$\{x\}^\sim \in \left\{\{x\}^{\leftarrowtail}, \{x\}^{\leftarrow}, \{x\}^{\rightarrow}, \{x\}^{\rightarrowtail}\right\}$$

and

$$\{p\}^\sim \in \left\{\{p\}^{\leftarrowtail}, \{p\}^{\leftarrow}, \{p\}^{\rightarrow}, \{p\}^{\rightarrowtail}\right\}.$$

More general problems are to find conditions under which, for a given positive number a,

$$\lim_{n\to\infty} \frac{x_n}{p_n} = a \quad \Longrightarrow \quad \lim_{s\to\infty} \frac{\{x\}^\sim(s)}{\{p\}^\sim(s/a)} = 1 \tag{7.79}$$

or the converse implication

$$\lim_{n\to\infty} \frac{x_n}{p_n} = a \quad \Longleftarrow \quad \lim_{s\to\infty} \frac{\{x\}^\sim(s)}{\{p\}^\sim(s/a)} = 1 \tag{7.80}$$

or the equivalence

$$\lim_{n\to\infty} \frac{x_n}{p_n} = a \quad \Longleftrightarrow \quad \lim_{s\to\infty} \frac{\{x\}^\sim(s)}{\{p\}^\sim(s/a)} = 1. \tag{7.81}$$

The cases $a = 0$ and $a = \infty$ are also of great interest.

For functions with continuous argument, the same problems have already been considered in Sect. 7.5. With the help of piecewise linear interpolations we reduce the problems for sequences to the corresponding problems for functions with continuous argument.

7.8.2 Piecewise Linear Interpolations of Sequences and Functions

Recall that the continuous function

$$\widehat{x}(t) = x_{\lfloor t \rfloor} + (x_{\lfloor t \rfloor+1} - x_{\lfloor t \rfloor})(t - \lfloor t \rfloor), \qquad t \ge 0, \tag{7.82}$$

where $\lfloor t \rfloor$ denotes the integer part of a number t, is called the *piecewise linear interpolation of a sequence* $\{x_n\} = \{x_n\}_{n\ge 0}$. Such functions and their properties have been considered in Sects. 3.4 (see Lemma 3.43) and 4.11.

Definition 7.87 We say that $\{x_n\}$ is a PRV-*sequence*, SQI-*sequence*, POV-*sequence*, ORV-*sequence*, RV-*sequence* etc., if the piecewise linear interpolation of this sequence is a PRV-*function*, SQI-*function*, POV-*function*, ORV-*function*, RV-*function* etc.

Theorems 4.57–4.63 provide some conditions under which piecewise linear interpolations of sequences belong to a certain class of functions and thus provide conditions under which the sequences themselves belong to the corresponding class.

The continuous function

$$\widehat{f}(t) = f(\lfloor t \rfloor) + (f(\lfloor t \rfloor + 1) - f(\lfloor t \rfloor))(t - \lfloor t \rfloor), \qquad t \ge 0, \tag{7.83}$$

is called the *piecewise linear interpolation of the function* $f = (f(t), t > 0)$.

Relation (7.83) implies that $\widehat{f}$ coincides with the piecewise linear interpolation of the sequence $\{f(n)\}_{n\ge 0}$. Throughout this section we assume that $f(0) = 0$.

The following result shows that, under some conditions, the piecewise linear interpolations preserve some important properties of the original functions.

Lemma 7.88 *The following statements hold true:*

(a) if $f \in \mathcal{WPRV}$, *then* $\widehat{f} \sim f$ *and* $\widehat{f} \in \mathcal{PRV}$;
(b) if $f \in \mathbb{C}^{\infty}_{\text{inc}}$, *then* $\hat{f}^{-1} \sim f^{-1}$, *where* $\hat{f}^{-1}$ *and* f^{-1} *are the inverse functions of* $\widehat{f}$ *and* f, *respectively;*
(c) if $f \in \mathbb{C}_{\text{inc}}$ *and* $f \in \mathcal{WSQI}$, *then* $\hat{f}^{-1} \sim f^{-1}$ *and* $\widehat{f} \in \mathcal{SQI}$;
(d) if $f \in \mathcal{WPOV}$, *then* $\widehat{f} \sim f$, $\hat{f}^{-1} \sim f^{\sim}$, *and* $\widehat{f} \in \mathcal{POV}$, *where* $f^{\sim}$ *is an arbitrary asymptotically quasi-inverse function for* f;
(e) if f *is an RV-function with positive index* ρ, *then* $\widehat{f} \sim f$, $\hat{f}^{-1} \sim f^{\sim}$, *and* $\widehat{f} \in \mathcal{RV}_\rho$, *where* $f^{\sim}$ *is an arbitrary asymptotically quasi-inverse function for* f.

Proof of Lemma 7.88 Let $f \in \mathcal{WPRV}$. Then

$$\lim_{t\to\infty} \frac{f(\lfloor t \rfloor)}{f(t)} = \lim_{t\to\infty} \frac{f(\lfloor t \rfloor + 1)}{f(t)} = 1,$$

since f preserves the equivalence of functions by Theorem 3.42. Moreover,

$$\left| \frac{\widehat{f}(t)}{f(t)} - 1 \right| \le \left| \frac{f(\lfloor t \rfloor)}{f(t)} - 1 \right| + \left| \frac{f(\lfloor t \rfloor + 1)}{f(t)} - 1 \right|,$$

whence

$$\lim_{t\to\infty} \left| \frac{\widehat{f}(t)}{f(t)} - 1 \right| = 0,$$

that is, $\widehat{f} \sim f$. Therefore, for all $c > 0$,

$$(\widehat{f})^*(c) = \limsup_{t\to\infty} \frac{\widehat{f}(ct)}{\widehat{f}(t)} = \limsup_{t\to\infty} \frac{f(ct)}{f(t)} = f^*(c). \tag{7.84}$$

Since $f \in \mathcal{WPRV}$, relation (7.84) implies that $\widehat{f} \in \mathcal{WPRV}$. The function $\widehat{f}$ is continuous and thus measurable. So, $\widehat{f} \in \mathcal{PRV}$, which proves statement (a).

Let $f \in \mathbb{C}_{\text{inc}}^{\infty}$. Then $\widehat{f} \in \mathbb{C}_{\text{inc}}^{\infty}$. The function f possesses the inverse function f^{-1} and thus $\widehat{f}$ also possesses the inverse function

$$\widehat{f}^{-1}(s) = n + \frac{s - f(n)}{f(n+1) - f(n)}, \qquad s \in [f(n), f(n+1)),\ n \ge 0.$$

Comparing the functions f^{-1} and $\widehat{f}^{-1}$, we see that

$$\frac{n}{n+1} \le \frac{\widehat{f}^{-1}(s)}{f^{-1}(s)} \le \frac{n+1}{n}, \qquad s \in [f(n), f(n+1)),\ n \ge 0.$$

Hence

$$\lim_{s\to\infty} \frac{\widehat{f}^{-1}(s)}{f^{-1}(s)} = 1,$$

and this proves statement (b).

Let $f \in \mathbb{C}_{\text{inc}} \cap \mathcal{WSQI}$. In view of Corollaries 3.64 and 3.70, $f \in \mathbb{C}_{\text{inc}}^{\infty}$ and $\widehat{f} \in \mathbb{C}_{\text{inc}}^{\infty}$. The function f possesses the inverse function f^{-1}, which, by Corollary 7.33, belongs to $\mathcal{PRV}$. Correspondingly, the function $\widehat{f}$ also possesses the inverse function $\widehat{f}^{-1}$ and $f^{-1} \sim \widehat{f}^{-1}$ by statement (b). This means that $\widehat{f}^{-1} \in \mathcal{PRV}$, since $f^{-1} \in \mathcal{PRV}$. Applying Corollary 7.33 again we get $\widehat{f} \in \mathcal{SQI}$. So, statement (c) is proved.

Let $f \in \mathcal{WPOV}$. Then $f \in \mathcal{WPRV}$ and $\widehat{f} \sim f$ according to statement (a). This implies that $\widehat{f} \in \mathcal{WPOV}$, since $f \in \mathcal{WPOV}$. The function $\widehat{f}$ is continuous and thus measurable, whence $\widehat{f} \in \mathcal{POV}$. Finally, statement 4) of Theorem 7.48 implies that $\widehat{f}^{-1} \sim f^{\sim}$ and statement (d) is proved. Statement (e), in turn, follows from statement (d). □

Remark 7.89 For quickly growing functions f, the relation $f \sim \widehat{f}$ does not hold. Indeed, consider, e.g., $f(t) = a^t$, $t \geq 0$, where $a > 1$. Then, for any natural number n,

$$\frac{\widehat{f}(n+\frac{1}{2})}{f(n+\frac{1}{2})} = \frac{a^{n+1}+a^n}{2a^{n+\frac{1}{2}}} = \frac{a+1}{2\sqrt{a}} > 1,$$

whence

$$\limsup_{t\to\infty} \frac{\widehat{f}(t)}{f(t)} > 1.$$

Moreover, if $f(t) = t^t$, $t > 0$, then

$$\limsup_{t\to\infty} \frac{\widehat{f}(t)}{f(t)} = \infty.$$

This example explains why the relation $f \sim \widehat{f}$ is omitted in statement (c) of Lemma 7.88.

The following result improves statement (v) of Lemma 3.43.

Lemma 7.90 *Let $\{x_n\}$ and $\{c_n\}$ be two sequences of real numbers, with $c_n > 0$ for all sufficiently large n. Then, for all $a \in [0, \infty]$,*

$$\lim_{n\to\infty} \frac{x_n}{c_n} = a \quad \Longleftrightarrow \quad \lim_{t\to\infty} \frac{\widehat{x}(t)}{\widehat{c}(t)} = a.$$

Proof of Lemma 7.90 The implication "$\Longleftarrow$" is obvious.

On the other hand, if $0 < a < \infty$, then the implication "$\Longrightarrow$" follows from the inequality

$$\left|\frac{\widehat{x}(t)}{\widehat{c}(t)} - a\right| \leq \left|\frac{x_{\lfloor t\rfloor}}{c_{\lfloor t\rfloor}} - a\right| + \left|\frac{x_{\lfloor t\rfloor+1}}{c_{\lfloor t\rfloor+1}} - a\right|,$$

while, if $a = 0$ or ∞, it follows from the inequalities

$$\min\left\{\frac{x_{\lfloor t\rfloor}}{c_{\lfloor t\rfloor}}, \frac{x_{\lfloor t\rfloor+1}}{c_{\lfloor t\rfloor+1}}\right\} \leq \frac{\widehat{x}(t)}{\widehat{c}(t)} \leq \max\left\{\frac{x_{\lfloor t\rfloor}}{c_{\lfloor t\rfloor}}, \frac{x_{\lfloor t\rfloor+1}}{c_{\lfloor t\rfloor+1}}\right\},$$

which completes the proof. □

7.8.3 Generalized Inverse Functions for Piecewise Linear Interpolations of Sequences

Let $\{x\} = \{x_n\}_{n\geq 0}$ be a sequence of real numbers such that $x_0 = 0$ and $\{x_n\}_{n\geq 0} \in \mathbb{S}^\infty$. When studying the asymptotic properties of the generalized inverse functions for a sequence $\{x_n\}_{n\geq 0}$, we also consider the generalized inverse function for its piecewise linear interpolation $\widehat{x}$. Among the generalized inverse functions for $\{x_n\}_{n\geq 0}$, the main attention is paid to the functions $\{x\}^{\leftarrowtail}$ and $\{x\}^{\rightarrowtail}$, and, among the generalized inverse functions for the piecewise linear interpolation $\widehat{x}$, the main focus is on $\hat{x}^{\leftarrowtail}$ and $\hat{x}^{\rightarrowtail}$. In the first place, this is explained by relations (7.69) and (7.70). Moreover, for all $s \geq 0$,

$$\{x\}^{\leftarrowtail}(s) - 1 \leq \hat{x}^{\leftarrowtail}(s) \leq \{x\}^{\leftarrowtail}(s) \tag{7.85}$$

and

$$\{x\}^{\rightarrowtail}(s) \leq \hat{x}^{\rightarrowtail}(s) \leq \{x\}^{\rightarrowtail}(s) + 1. \tag{7.86}$$

The following result describes a relationship between the generalized inverse functions for sequences and those for their piecewise linear interpolations.

Lemma 7.91 *Let $\{c_n\}_{n\geq 0} \in \mathbb{S}^\infty$. Then,*

1) $\{c\}^{\leftarrowtail} \sim \hat{c}^{\leftarrowtail}$ and $\{c\}^{\rightarrowtail} \sim \hat{c}^{\rightarrowtail}$, that is,

$$\lim_{s\to\infty} \frac{\{c\}^{\leftarrowtail}(s)}{\widehat{c}^{\leftarrowtail}(s)} = \lim_{s\to\infty} \frac{\{c\}^{\rightarrowtail}(s)}{\widehat{c}^{\rightarrowtail}(s)} = 1; \tag{7.87}$$

2) if the sequence $\{c_n\}_{n\geq 0}$ is increasing, then

$$\widehat{c}^{-1} = \widehat{c}^{\leftarrowtail} = \widehat{c}^{\leftarrow} = \widehat{c}^{\rightarrow} = \widehat{c}^{\rightarrowtail} \tag{7.88}$$

and

$$\{c\}^{\leftarrowtail} \sim \{c\}^{\leftarrow} \sim \{c\}^{\rightarrow} \sim \{c\}^{\rightarrowtail} \sim \widehat{c}^{-1}, \tag{7.89}$$

where $\widehat{c}^{-1}$ is the inverse function for $\widehat{c}$.

Proof of Lemma 7.91 The first statement follows directly from inequalities (7.85) and (7.86).

If the sequence $\{c_n\}_{n\geq 0}$ is increasing, then $\widehat{c} \in \mathbb{C}^\infty_{\text{inc}}$, whence (7.88) follows. In turn, (7.89) follows from the first statement of the lemma and from relations (7.88) and (7.72). □

7.8.4 The Limit Behavior of the Ratio of Asymptotically Quasi-inverse Functions for Sequences

Now we are in position to answer three of the questions above (see (7.79)–(7.81)). Throughout this section, sequences of real numbers $\{x_n\}_{n\ge 0}$ and $\{p_n\}_{n\ge 0}$ are such that $x_0 = p_0 = 0$. We start with conditions for (7.79).

Theorem 7.92 *Let $a \in (0, \infty)$ and*

$$\lim_{n\to\infty} \frac{x_n}{p_n} = a, \tag{7.90}$$

where $\{x_n\}_{n\ge 1}$ is a sequence of real numbers and $\{p_n\}_{n\ge 1}$ is an increasing sequence of positive numbers tending to infinity. Then,

1) if $\{p_n\}_{n\ge 1}$ is an SQI-sequence, then

$$\lim_{s\to\infty} \frac{\{x\}^{\sim}(s)}{\{p\}^{\sim}(s/a)} = 1 \tag{7.91}$$

for all functions

$$\{x\}^{\sim} \in \{\{x\}^{\mapsfrom}, \{x\}^{\leftarrow}, \{x\}^{\rightarrow}, \{x\}^{\mapsto}\}$$

and

$$\{p\}^{\sim} \in \{\{p\}^{\mapsfrom}, \{p\}^{\leftarrow}, \{p\}^{\rightarrow}, \{p\}^{\mapsto}, \widehat{p}^{-1}\},$$

where $\widehat{p}^{-1}$ is the inverse function of $\widehat{p}$, that is,

$$\widehat{p}^{-1}(s/a) = n + \frac{(s/a) - p_n}{p_{n+1} - p_n}, \qquad s \in [ap_n, ap_{n+1}),\ n \ge 0;$$

2) if

$$\liminf_{n\to\infty} \frac{n(p_{n+1} - p_n)}{p_n} > 0, \tag{7.92}$$

then (7.91) holds;

3) if $p_n = p(n)$, $n \ge 1$, where the function $p = (p(t), t \ge 1)$ is such that $p \in \mathbb{C}_{\text{inc}}$ and $p \in \mathcal{WSQI}$, then relation (7.91) holds with

$$\{p\}^{\sim} \in \{\{p\}^{\mapsfrom}, \{p\}^{\leftarrow}, \{p\}^{\rightarrow}, \{p\}^{\mapsto}, \widehat{p}^{-1}, p^{-1}\},$$

where p^{-1} is the inverse function of p.

Proof of Theorem 7.92 Let $\{p_n\}_{n\geq 0}$ be an SQI-sequence, that is, its piecewise linear interpolation $\widehat{p}$ is an SQI-function.

Relation (7.90) and Lemma 7.90 imply that

$$\lim_{t\to\infty} \frac{\widehat{x}(t)}{\widehat{p}(t)} = a,$$

where $\widehat{x}$ is the piecewise linear interpolation of the sequence $\{x_n\}_{n\geq 0}$. Since $\widehat{p}$ is an SQI-function, the first statement of Theorem 7.53 implies that

$$\lim_{s\to\infty} \frac{\widehat{x}^{\leftarrowtail}(s)}{\widehat{p}^{-1}(s/a)} = \frac{\widehat{x}^{\rightarrowtail}(s)}{\widehat{p}^{-1}(s/a)} = 1.$$

Now Lemma 7.91 yields

$$\lim_{s\to\infty} \frac{\{x\}^{\leftarrowtail}(s)}{\{p\}^{\sim}(s/a)} = \frac{\{x\}^{\rightarrowtail}(s)}{\{p\}^{\sim}(s/a)} = 1$$

for any $\{p\}^{\sim} \in \{\{p\}^{\leftarrowtail}, \{p\}^{\leftarrow}, \{p\}^{\rightarrow}, \{p\}^{\rightarrowtail}, \widehat{p}^{-1}\}$. It remains to note that the latter relation together with (7.69) imply (7.91). Thus, statement 1) is proved.

Statement 2) follows from statement 1) and statement 3) of Theorem 4.58. Finally, statement 3) follows from statement 1) and statement (c) of Lemma 7.88. Therefore, Theorem 7.92 is proved. □

Remark 7.93 Using Corollary 7.33 for statement 3), the assumption $p \in \mathcal{WSQI}$ can be exchanged with the equivalent one that $p^{-1} \in \mathcal{WPRV}$. Taking Remark 3.22 into account, we see that statement 3) is equivalent to Theorem 2.5. The proofs of statement 3) and Theorem 2.5 are different, but statement 3) exhibits the meaning of condition (2.23) which, by Remark 3.22 and Corollary 7.33, reduces to $a(\cdot) \in \mathcal{WSQI}$.

Remark 7.94 Theorem 7.92 contains sufficient conditions under which relation (7.90) implies (7.91). The conditions are not necessary for this implication, but it is hard to improve them without extra assumptions. For example, the sequence $p_n = \ln(n + 1)$, $n \geq 0$, does not satisfy the assumptions of any of the statements of Theorem 7.92 and thus this theorem is not helpful in proving the implication (7.90)$\Longrightarrow$(7.91). This is compatible with the construction in Sect. 2.2.2 showing that there exists a sequence $\{x_n\}_{n\geq 0}$ such that

$$\lim_{n\to\infty} \frac{x_n}{\ln n} = 1,$$

but the limit

$$\lim_{s\to\infty} \frac{\{x\}^{\sim}(s)}{e^s} \in (0, \infty)$$

does not exist for any of the generalized inverse functions for the sequence $\{x_n\}_{n\ge 0}$.

The following result contains some sufficient conditions for (7.80).

Theorem 7.95 *Let $\{x_n\}_{n\ge 1}$ be a sequence of real numbers tending to infinity and let $\{p_n\}_{n\ge 1}$ be an increasing sequence of positive numbers also tending to infinity. Let there exist a number $a \in (0, \infty)$ and a function $\{p\}^{\sim} \in \{\{p\}^{\leftarrowtail}, \{p\}^{\leftarrow}, \{p\}^{\rightarrow}, \{p\}^{\rightarrowtail}, \widehat{p}^{-1}\}$ such that*

$$\lim_{s\to\infty} \frac{\{x\}^{\leftarrowtail}(s)}{\{p\}^{\sim}(s/a)} = \lim_{s\to\infty} \frac{\{x\}^{\rightarrowtail}(s)}{\{p\}^{\sim}(s/a)} = 1. \tag{7.93}$$

1) If $\{p_n\}_{n\ge 0}$ is a PRV-sequence, then (7.90) holds, that is,

$$\lim_{n\to\infty} \frac{x_n}{p_n} = a.$$

2) If

$$\limsup_{n\to\infty} \frac{n(p_{n+1} - p_n)}{p_n} < \infty, \tag{7.94}$$

then (7.90) holds.

3) If $p_n = p(n)$, $n \ge 1$, where the function $p = (p(t), t \ge 1)$ is such that $p \in \mathbb{C}^{\infty}_{\text{inc}}$, $p \in \mathcal{WPRV}$, and if there exists a function

$$\{p\}^{\sim} \in \{\{p\}^{\leftarrowtail}, \{p\}^{\leftarrow}, \{p\}^{\rightarrow}, \{p\}^{\rightarrowtail}, \widehat{p}^{-1}, p^{-1}\}$$

such that (7.93) holds, then so does (7.90).

Proof of Theorem 7.95 Let $\{p_n\}_{n\ge 0}$ be a PRV-sequence, that is, its piecewise linear interpolation $\widehat{p}$ is a PRV-function.

By our assumptions, relation (7.93) holds for some function $\{p\}^{\sim}$ belonging to $\{\{p\}^{\leftarrowtail}, \{p\}^{\leftarrow}, \{p\}^{\rightarrow}, \{p\}^{\rightarrowtail}, \widehat{p}^{-1}\}$. Thus statement 2) of Lemma 7.91 implies relation (7.93), with $\{p\}^{\sim} = \widehat{p}^{-1}$. Taking statement 1) of Lemma 7.91 into account, we see that

$$\lim_{s\to\infty} \frac{\widehat{x}^{\leftarrowtail}(s)}{\widehat{p}^{-1}(s/a)} = \lim_{s\to\infty} \frac{\widehat{x}^{\rightarrowtail}(s)}{\widehat{p}^{-1}(s/a)} = 1. \tag{7.95}$$

Since $\widehat{p} \in \mathcal{PRV}$, relation (7.95) and statement 3) of Theorem 7.53 imply that

$$\lim_{s\to\infty} \frac{\widehat{x}(s)}{\widehat{p}(s)} = a.$$

This together with Lemma 7.90 yields (7.90). So, statement 1) is proved.

Statement 2) follows from statement 1) and statement 2) of Theorem 4.58. Finally, statement 3) follows from statement 1) and statements (a) and (b) of Lemma 7.88. Therefore Theorem 7.95 is proved. □

Theorems 7.92 and 7.95 provide sufficient conditions for (7.81). We collect these conditions in the following two corollaries.

Corollary 7.96 *Let $a \in (0, \infty)$ and $\{x_n\}_{n\ge1}$ be a sequence of real numbers tending to infinity. Let $\{p_n\}_{n\ge1}$ be an increasing sequence of positive numbers also tending to infinity. Let*

$$0 < \liminf_{n\to\infty} \frac{n(p_{n+1} - p_n)}{p_n} \quad \textit{and} \quad \limsup_{n\to\infty} \frac{n(p_{n+1} - p_n)}{p_n} < \infty. \tag{7.96}$$

1) If (7.90) holds, then (7.93) holds for all functions

$$\{x\}^{\sim} \in \{\{x\}^{\leftarrowtail}, \{x\}^{\leftarrow}, \{x\}^{\rightarrow}, \{x\}^{\rightarrowtail}\},$$

and

$$\{p\}^{\sim} \in \{\{p\}^{\leftarrowtail}, \{p\}^{\leftarrow}, \{p\}^{\rightarrow}, \{p\}^{\rightarrowtail}, \widehat{p}^{-1}\};$$

moreover, (7.91) holds for any

$$\{p\}^{\sim} \in \{\{p\}^{\leftarrowtail}, \{p\}^{\leftarrow}, \{p\}^{\rightarrow}, \{p\}^{\rightarrowtail}, \widehat{p}^{-1}\}.$$

2) If there exists a function $\{p\}^{\sim} \in \{\{p\}^{\leftarrowtail}, \{p\}^{\leftarrow}, \{p\}^{\rightarrow}, \{p\}^{\rightarrowtail}, \widehat{p}^{-1}\}$ satisfying (7.93), then (7.90) holds.

Corollary 7.97 *Let $a \in (0, \infty)$ and let $\{x_n\}_{n\ge1}$ be a sequence of real numbers tending to infinity. Further, let $p_n = p(n)$, $n \ge 1$ be such that the function $p = (p(t), t \ge 1)$ satisfies $p \in \mathbb{C}^{\infty}_{\mathrm{inc}}$ and $p \in \mathcal{WPOV}$. Then:*

1) If (7.90) holds, then (7.91) holds for all functions

$$\{x\}^{\sim} \in \{\{x\}^{\leftarrowtail}, \{x\}^{\leftarrow}, \{x\}^{\rightarrow}, \{x\}^{\rightarrowtail}\}$$

and

$$\{p\}^{\sim} \in \{\{p\}^{\leftarrowtail}, \{p\}^{\leftarrow}, \{p\}^{\rightarrow}, \{p\}^{\rightarrowtail}, \widehat{p}^{-1}, p^{-1}\};$$

moreover, (7.93) holds for any

$$\{p\}^{\sim} \in \{\{p\}^{\leftarrowtail}, \{p\}^{\leftarrow}, \{p\}^{\rightarrow}, \{p\}^{\rightarrowtail}, \widehat{p}^{-1}, p^{-1}\}.$$

2) *If there exists a function* $\{p\}^{\sim} \in \{\{p\}^{\leftarrowtail}, \{p\}^{\leftarrow}, \{p\}^{\rightarrow}, \{p\}^{\rightarrowtail}, \widehat{p}^{-1}, p^{-1}\}$ *satisfying (7.93), then (7.90) holds.*

Corollaries 7.96 and 7.97 can also be proved for the cases $a = 0$ and $a = \infty$ by using Theorem 7.63 and Lemma 7.90.

Corollary 7.98 *Let* $\{x_n\}_{n\geq 1}$ *be a sequence of real numbers tending to infinity and let* $\{p_n\}_{n\geq 1}$ *be an increasing sequence of positive numbers also tending to infinity. Assume that condition (7.96) holds. Then,*

$$\lim_{n\to\infty} \frac{x_n}{p_n} = 0 \quad \Longrightarrow \quad \lim_{s\to\infty} \frac{\{x\}^{\sim}(s)}{\{p\}^{\sim}(s)} = \infty \tag{7.97}$$

and

$$\lim_{n\to\infty} \frac{x_n}{p_n} = \infty \quad \Longrightarrow \quad \lim_{s\to\infty} \frac{\{x\}^{\sim}(s)}{\{p\}^{\sim}(s)} = 0 \tag{7.98}$$

for all functions

$$\{x\}^{\sim} \in \{\{x\}^{\leftarrowtail}, \{x\}^{\leftarrow}, \{x\}^{\rightarrow}, \{x\}^{\rightarrowtail}\},$$

and

$$\{p\}^{\sim} \in \{\{p\}^{\leftarrowtail}, \{p\}^{\leftarrow}, \{p\}^{\rightarrow}, \{p\}^{\rightarrowtail}, \widehat{p}^{-1}\}.$$

Corollary 7.99 *Let* $\{x_n\}_{n\geq 1}$ *be a sequence of real numbers tending to infinity and let* $p_n = p(n), n \geq 1$ *be such that the function* $p = (p(t), t \geq 1)$ *satisfies* $p \in \mathbb{C}^{\infty}_{\mathrm{inc}}$ *and* $p \in \mathcal{WPOV}$. *Then relations (7.97) and (7.98) hold for all functions*

$$\{x\}^{\sim} \in \{\{x\}^{\leftarrowtail}, \{x\}^{\leftarrow}, \{x\}^{\rightarrow}, \{x\}^{\rightarrowtail}\}$$

and

$$\{p\}^{\sim} \in \{\{p\}^{\leftarrowtail}, \{p\}^{\leftarrow}, \{p\}^{\rightarrow}, \{p\}^{\rightarrowtail}, \widehat{p}^{-1}, p^{-1}\}.$$

In view of the above results and Corollary 7.46, Corollaries 7.96–7.99 can be rewritten for RV-sequences $\{p_n\}_{n\geq 1}$ as follows.

Corollary 7.100 *Let* $a \in [0, \infty]$ *and let* $\{x_n\}_{n\geq 1}$ *be a sequence of real numbers tending to infinity. Let* $\{p_n\}_{n\geq 1}$ *be an increasing sequence of positive numbers also* *tending to infinity. If the limit*

$$\lim_{n\to\infty} \frac{n(p_{n+1} - p_n)}{p_n} = \rho \in (0, \infty) \tag{7.99}$$

exists, then

$$\lim_{n\to\infty} \frac{x_n}{p_n} = a \quad \Longrightarrow \quad \lim_{s\to\infty} \frac{\{x\}^{\sim}(s)}{\{p\}^{\sim}(s)} = \left(\frac{1}{a}\right)^{\frac{1}{\rho}} \tag{7.100}$$

for all functions

$$\{x\}^{\sim} \in \{\{x\}^{\leftarrowtail}, \{x\}^{\leftarrow}, \{x\}^{\rightarrow}, \{x\}^{\rightarrowtail}\},$$

and

$$\{p\}^{\sim} \in \{\{p\}^{\leftarrowtail}, \{p\}^{\leftarrow}, \{p\}^{\rightarrow}, \{p\}^{\rightarrowtail}, \widehat{p}^{-1}\},$$

where $1/\infty := 0$ *and* $1/0 := \infty$.

Corollary 7.101 *Let* $a \in [0, \infty]$ *and let* $\{x_n\}_{n\ge 1}$ *be a sequence of real numbers tending to infinity. Let* $p_n = p(n)$, $n \ge 1$ *be such that the function* $p = (p(t), t \ge 1)$ *satisfies* $p \in \mathbb{C}_{\mathrm{inc}}^{\infty}$ *and that* p *is an RV-function with positive index* ρ. *Then relation (7.100) holds for all functions*

$$\{x\}^{\sim} \in \{\{x\}^{\leftarrowtail}, \{x\}^{\leftarrow}, \{x\}^{\rightarrow}, \{x\}^{\rightarrowtail}\}$$

and

$$\{p\}^{\sim} \in \{\{p\}^{\leftarrowtail}, \{p\}^{\leftarrow}, \{p\}^{\rightarrow}, \{p\}^{\rightarrowtail}, \widehat{p}^{-1}, p^{-1}\}.$$

Corollary 7.102 *Let* $a \in (0, \infty)$ *and let* $\{x_n\}_{n\ge 1}$ *be a sequence of real numbers tending to infinity. Let* $\{p_n\}_{n\ge 1}$ *be an increasing sequence of positive numbers also tending to infinity. If the limit (7.99) exists, then*

$$\lim_{n\to\infty} \frac{x_n}{p_n} = a \quad \Longleftrightarrow \quad \lim_{s\to\infty} \frac{\{x\}^{\sim}(s)}{\{p\}^{\sim}(s)} = \left(\frac{1}{a}\right)^{\frac{1}{\rho}} \tag{7.101}$$

for all functions

$$\{x\}^{\sim} \in \{\{x\}^{\leftarrowtail}, \{x\}^{\leftarrow}, \{x\}^{\rightarrow}, \{x\}^{\rightarrowtail}\},$$

and

$$\{p\}^{\sim} \in \{\{p\}^{\leftarrowtail}, \{p\}^{\leftarrow}, \{p\}^{\rightarrow}, \{p\}^{\rightarrowtail}, \widehat{p}^{-1}\}.$$

Corollary 7.103 *Let $a \in (0, \infty)$ and let $\{x_n\}_{n \ge 1}$ be a sequence of real numbers tending to infinity. Let $p_n = p(n)$, $n \ge 1$ be such that the function $p = (p(t), t \ge 1)$ satisfies $p \in \mathbb{C}_{\text{inc}}^{\infty}$ and that p is an RV-function with positive index ρ. Then relation (7.101) holds for all functions*

$$\{x\}^{\sim} \in \{\{x\}^{\leftarrowtail}, \{x\}^{\leftarrow}, \{x\}^{\rightarrow}, \{x\}^{\rightarrowtail}\}$$

and

$$\{p\}^{\sim} \in \{\{p\}^{\leftarrowtail}, \{p\}^{\leftarrow}, \{p\}^{\rightarrow}, \{p\}^{\rightarrowtail}, \widehat{p}^{-1}, p^{-1}\}.$$

7.9 Stability of Solutions of Differential Equations

In this section, we study the problem of asymptotic stability with respect to the initial condition of a solution of the Cauchy problem for an ordinary differential equation. We shall see that the above results apply to this particular problem. Other applications will be considered in the following chapters.

7.9.1 Solutions of the Cauchy Problem for an Autonomous Differential Equation

Consider the Cauchy problem for an autonomous ordinary differential equation, i.e.

$$d\mu(t) = g(\mu(t))dt, \quad t \ge 0, \qquad \mu(0) = b > 0, \tag{7.102}$$

where $g = (g(u), u > 0)$ is a positive continuous function. Under this assumption, the problem (7.102) has a unique solution μ_b for all fixed $b > 0$.

In general, there is no simple representation for the solution μ_b in terms of the function g, but the inverse function is easy to evaluate. For doing so, we put

$$G_b(s) = \int_b^s \frac{du}{g(u)}, \qquad s \ge b, \tag{7.103}$$

and show that

$$\mu_b = G_b^{-1}, \tag{7.104}$$

that is, the solution of problem (7.102) is a continuously differentiable and increasing function which is the inverse of the function $G_b = (G_b(s), s \ge b)$.

Indeed, since $G_b(b) = 0$, we have $G_b^{-1}(0) = b$. Moreover, for $t > 0$,

$$\frac{dG_b^{-1}(t)}{dt} = \left(G_b^{-1}(t)\right)' = \frac{1}{(G_b)'\left(G_b^{-1}(t)\right)} = g\left(G_b^{-1}(t)\right),$$

which proves (7.104).

On the other hand, if μ_b is a solution of (7.102), then

$$t = \int_0^t d\tau = \int_0^t \frac{d\mu_b(\tau)}{g(\mu_b(\tau))} = \int_b^{\mu_b(t)} \frac{du}{g(u)} = G_b(\mu_b(t))$$

for all $t \geq 0$, that is, (7.104) holds. Thus, G_b^{-1} is the unique solution of (7.102).

The solution μ_b is a positive increasing function tending to infinity. Two cases may occur. The first is that an "implosion" happens, more precisely, that μ_b reaches infinity in finite time T_∞. This occurs if

$$T_\infty = \lim_{s\to\infty} G_b(s) = \int_b^\infty \frac{du}{g(u)} < \infty. \tag{7.105}$$

In this case, $\lim_{t\to T_\infty} \mu_b(t) = \infty$.

Example 7.104 Condition (7.105) is satisfied if $g(u) = u^r$, $u > 0$, for $r > 1$. Some simple algebra gives

$$T_\infty = \frac{1}{(r-1)b^{r-1}}$$

and

$$\mu_b(t) = \left(\frac{1}{b^{1-r} - (r-1)t}\right)^{\frac{1}{r-1}}, \qquad t \in [0, T_\infty).$$

Case (7.105) is not considered in the sequel.

The contrary condition to (7.105) is

$$T_\infty = \lim_{s\to\infty} G_b(s) = \int_b^\infty \frac{du}{g(u)} = \infty. \tag{7.106}$$

Condition (7.106) excludes an "implosion", that is, μ_b cannot reach infinity in finite time. In this case, $\lim_{t\to\infty} \mu_b(t) = \infty$.

Example 7.105 Condition (7.106) is satisfied for $g(u) = u^r$, $u > 0$, where $r \leq 1$. If $r < 1$, then a simple calculation proves that

$$\mu_b(t) = \left(b^{1-r} + (1-r)t\right)^{\frac{1}{1-r}}, \qquad t \in [0, \infty).$$

Similarly, if $r = 1$,

$$\mu_b(t) = be^t, \qquad t \in [0, \infty).$$

Remark 7.106 Since g is a positive and continuous function, condition (7.106) holds for all $b > 0$. Therefore, without loss of generality, one may restrict the consideration to the case of $b = 1$, for example. In this case, we put $G = G_1$.

7.9.2 Asymptotic Stability of a Solution of the Cauchy Problem with Respect to the Initial Condition for an Autonomous Differential Equation

A solution of the Cauchy problem (7.102) is called *asymptotically stable with respect to the initial condition* if

$$\lim_{t\to\infty} \frac{\mu_{b_1}(t)}{\mu_{b_2}(t)} = 1, \tag{7.107}$$

for all initial conditions $b_1 > 0$ and $b_2 > 0$.

Example 7.105 shows that a solution of problem (7.102) is asymptotically stable with respect to the initial condition if $g(u) = u^r$, $u > 0$, where $r < 1$. On the other hand, the solution is not asymptotically stable with respect to the initial condition, if $g(u) = u$, $u > 0$.

So, the natural question arises about sufficient conditions under which a solution of (7.102) is asymptotically stable with respect to the initial condition. An answer to this question is related to the conditions

$$\int_1^\infty \frac{du}{g(u)} = \infty \tag{7.108}$$

and

$$\liminf_{t\to\infty} \int_t^{ct} \frac{du}{g(u)G(u)} > 0, \quad \text{for all} \quad c > 1, \tag{7.109}$$

where the function G is as defined in Remark 7.106.

Condition (7.108) shows that

$$\lim_{s\to\infty} G(s) = \infty,$$

whence

$$G \in \mathbb{C}^\infty_{\text{inc}}, \tag{7.110}$$

and $G_b \in \mathbb{C}^\infty_{\text{inc}}$, for all $b > 0$ (see Remark 7.106).

In turn, since g is a positive, continuous function, statement 5) of Theorem 4.9, with $f = G$, $\theta = f' = 1/g$ and $f'_{\ln} = f'/f = 1/(gG)$, implies that condition (7.109) is equivalent to

$$G \in \mathcal{SQI} \tag{7.111}$$

(see Definition 3.28), that is,

$$G_*(c) > 1, \quad \text{for all} \quad c > 1,$$

where $G_*(c) = \liminf_{s\to\infty} G(cs)/G(s)$ is the lower limit function for G (see (3.1)).

According to condition (7.108),

$$\begin{aligned} G_*(c) &= \liminf_{s\to\infty} \frac{G(cs)}{G(s)} = \liminf_{s\to\infty} \left(\int_1^{cs} \frac{du}{g(u)} \Big/ \int_1^{s} \frac{du}{g(u)} \right) \\ &= \liminf_{s\to\infty} \left(\int_b^{cs} \frac{du}{g(u)} \Big/ \int_b^{s} \frac{du}{g(u)} \right) = \liminf_{s\to\infty} \frac{G_b(cs)}{G_b(s)} = (G_b)_*(c), \end{aligned}$$

for all $b > 0$ and $c > 0$, that is,

$$G_* = (G_b)_*,$$

for all $b > 0$. Therefore, conditions (7.108) and (7.109) imply that $G_b \in \mathcal{WSQI}$. Since the function G_b is continuous, $G_b \in \mathcal{SQI}$, for all $b > 0$.

Remark 7.107 Condition (7.108) follows from (7.109). Indeed, if condition (7.108) does not hold, then

$$\lim_{s\to\infty} G(s) = \sup_{s\geq 1} G(s) = C < \infty,$$

whence

$$\liminf_{t\to\infty} \int_t^{ct} \frac{du}{g(u)G(u)} = \frac{1}{C} \liminf_{t\to\infty} \int_t^{ct} \frac{du}{g(u)} = \frac{1}{C} \liminf_{t\to\infty}(G(ct) - G(t)) = 0,$$

for all $c > 1$, which contradicts (7.109).

Moreover, the implication (7.109) $\Longrightarrow$ (7.108) holds according to Corollary 3.64, since G is an SQI-function under (7.109).

The following result contains sufficient conditions for a solution of the Cauchy problem (7.102) to be asymptotically stable with respect to the initial condition.

Theorem 7.108 *Let condition (7.109) hold. Then a solution of the Cauchy problem (7.102) is asymptotically stable with respect to the initial condition.*

Proof of Theorem 7.108 Remark 7.107 and relation (7.109) imply (7.108). Since

$$\left|\frac{G_b(s)}{G(s)} - 1\right| \le \frac{\left|\int_1^b \frac{du}{g(u)}\right|}{G(s)}$$

for all $s > 0$ and $b > 0$, condition (7.108) yields

$$\lim_{s\to\infty} \frac{G_b(s)}{G(s)} = 1$$

for all $b > 0$. In turn, conditions (7.108) and (7.109) imply relations (7.110) and (7.111). Thus

$$G \in \mathbb{C}_{\mathrm{inc}}^{\infty} \cap \mathcal{SQI}$$

and this allows us to apply Corollary 7.52 to get

$$\lim_{s\to\infty} \frac{G_b^{-1}(s)}{G^{-1}(s)} = 1,$$

for all $b > 0$. Recalling equality (7.104) and Remark 7.106, we obtain that

$$\lim_{t\to\infty} \frac{\mu_b(t)}{\mu_1(t)} = 1,$$

for all $b > 0$, and this is equivalent to the fact that a solution of the Cauchy problem (7.102) is asymptotically stable with respect to the initial condition. □

Condition (7.109) holds, if $g(u) = u^r$, $u > 0$, in case of $r < 1$, and does not hold, if $r = 1$. These conclusions are compatible with Example 7.105.

Next, we consider some sufficient conditions for (7.109). Sometimes these conditions are more convenient in applications compared to (7.109).

Proposition 7.109 *Let g be a positive, continuous function, for which at least one of the following conditions holds:*

(i) $\limsup_{u\to\infty} (g(u)G(u)/u) < \infty$;
(ii) *g is an almost decreasing function (recall (4.34) and (4.36)), that is,*

$$\lim_{v\to\infty} \inf_{u\ge v} \frac{g(v)}{g(u)} > 0;$$

(iii) g is a nonincreasing function for sufficiently large arguments;
(iv) there exists an $\alpha < 1$ *such that*

$$0 < \liminf_{u\to\infty} u^{-\alpha} g(u) \qquad \textit{and} \qquad \limsup_{u\to\infty} u^{-\alpha} g(u) < \infty;$$

(v) $g^*(c) < c$ *for all* $c > 1$, *where* $g^*(c) = \limsup_{u\to\infty} g(cu)/g(u)$ *is the upper limit function for* g;

(vi) g *is an RV-function with index* $\rho < 1$;

(vii) g *is given by*

$$g = \frac{1}{\frac{1}{d_1 g_1} + \cdots + \frac{1}{d_n g_n}},$$

where g_k, $k = 1, \ldots, n$, *are positive, continuous functions, for which condition (7.109) holds, and where* d_k, $k = 1, \ldots, n$, *are almost constant functions (recall Definition 4.11).*

Then condition (7.109) is satisfied for the function g.

Proof of Proposition 7.109 Relation (7.109) follows from

a) condition (i) according to statement 1 of Proposition 4.50, with $f = G$ and $\theta = 1/g$;
b) condition (ii) according to statement 4) of Proposition 4.50, with $f = G$ and $\theta = 1/g$, by noting that the density $\theta = 1/g$ is almost increasing (see (4.35) and (4.37)) if and only if the function g is almost decreasing;
c) condition (iii), since (ii) follows from (iii);
d) condition (iv) according to statement 2) of Proposition 4.50, with $f = G$ and $\theta = 1/g$;
e) condition (v) according to Corollary 4.52, with $f = G$ and $\theta = 1/g$, by noting that $(1/g)_* = 1/g^*$;
f) condition (vi), since $g^*(c) = c^\rho$ and thus (iv) follows from (v);
g) condition (vii) according to Proposition 4.14, with $\theta = 1/g$ and $\theta_k = 1/g_k$, $k = 1, \ldots, n$. □

Remark 7.110 According to Corollary 4.51, if the function gG is almost increasing (see (4.35) and (4.37)), that is, if

$$\lim_{v\to\infty} \inf_{u\geq v} \frac{g(u)G(u)}{g(v)G(v)} > 0, \tag{7.112}$$

then condition (i) of Proposition 7.109 is equivalent to condition (7.109). In particular, condition (7.112) holds, if the function g is almost increasing, that is, if

$$\lim_{v\to\infty} \inf_{u\geq v} \frac{g(u)}{g(v)} > 0.$$

Remark 7.111 Condition (i) of Proposition 7.109 does not hold for RV-functions g with index $\rho = 1$, that is, for functions $g(t) = t\ell(t)$, where $\ell(\cdot)$ is an SV-function.

This follows from Parameswaran [290], where the equality

$$\lim_{t\to\infty} \ell(t) \int_1^t \frac{ds}{s\ell(s)} = \infty$$

is shown.

Remark 7.112 Corollary 7.33 proves that, under condition (7.108) and for an arbitrary initial condition $b > 0$, the solution μ_b of the Cauchy problem (7.102) is a PRV-function (and thus μ_b preserves the asymptotic equivalence of functions and sequences) if and only if condition (7.109) holds.

7.9.3 Asymptotic Stability of Solutions of the Cauchy Problem with Respect to the Initial Condition for a Non-autonomous Ordinary Differential Equation

Consider the Cauchy problem for the more general equation

$$d\tilde{\mu}(t) = g(\tilde{\mu}(t))\, \psi(t)\, dt, \qquad t \ge 0, \qquad \tilde{\mu}(0) = b > 0. \tag{7.113}$$

Here $g = (g(u), u > 0)$ is a positive, continuous function satisfying condition (7.108) and $\psi = (\psi(t), t \ge 0)$ is continuous and, in general, attains values of both signs. Assume also that

$$\Psi(t) = \int_0^t \psi(u)\, du > 0, \qquad t > 0. \tag{7.114}$$

Under these assumptions, the Cauchy problem (7.113) has a unique, continuously differentiable solution $\tilde{\mu}_b$, whatever $b > 0$ may be. Taking (7.104) into account, the solution can be represented as

$$\tilde{\mu}_b(t) = \mu_b\left(\Psi(t)\right) = G_b^{-1}\left(\Psi(t)\right), \qquad t \ge 0, \tag{7.115}$$

where μ_b is the solution of the Cauchy problem (7.102) and G_b^{-1} is the inverse function of G_b (see (7.103)). Indeed, for all $t > 0$,

$$d\mu_b\left(\Psi(t)\right) = (\mu_b)'\left(\Psi(t)\right)\left(\Psi(t)\right)'\, dt = g(\mu_b\left(\Psi(t)\right))\psi(t)\, dt$$

and, in addition, $\mu_b\left(\Psi(0)\right) = \mu_b(0) = b$. Thus $\mu_b \circ \Psi$ is a solution of the Cauchy problem (7.113). On the other hand, if $\tilde{\mu}_b$ is a solution of this problem, then

$$G\left(\tilde{\mu}_b(t)\right) = \int_b^{\tilde{\mu}_b(t)} \frac{du}{g(u)} = \int_0^t \frac{d\tilde{\mu}_b(u))}{g(\tilde{\mu}_b(u))} = \int_0^t \psi(u) = \Psi(t),$$

for all $t \ge 0$, that is, $G_b \circ \tilde{\mu}_b = \Psi$. Thus, $\tilde{\mu}_b = G_b^{-1} \circ \Psi$.

Therefore, relation (7.115) determines the unique solution $\tilde{\mu}_b$ of the Cauchy problem (7.113) under condition (7.114). If

$$\lim_{t\to\infty} \Psi(t) = \infty, \tag{7.116}$$

then condition (7.108) implies that $\tilde{\mu}_b$ is a positive function, tending to infinity as $t \to \infty$, which does not reach infinity in finite time. In contrast to the solution μ_b of the Cauchy problem (7.102), the function $\tilde{\mu}_b$ is not necessarily increasing at every point.

In Theorem 7.113, we mention some conditions under which a solution of the Cauchy problem (7.113) is asymptotically stable with respect to the initial condition.

Theorem 7.113 *Let conditions (7.109), (7.114), and (7.116) hold. Then a solution of the Cauchy problem (7.113) is asymptotically stable with respect to the initial condition.*

Theorem 7.113 follows immediately from Theorem 7.108 combined with relation (7.115).

7.10 Comments

This chapter is based on the works of Buldygin et al. [57, 59, 61, 62, 65, 66].

The idea of deriving some property for inverse objects from the corresponding property for the original ones can be found in the literature. Here we only mention number theory (Manstavicius [256]), random sets (Molchanov [283]), and limit theorems of probability theory (Saulis and Statulevicius [319]).

Section 7.1 Several other analogs of inverse functions are known in the literature. For example, Klement et al. [222, 223] introduce and study the so-called *quasi-inverse* and *pseudo-inverse* functions. Various notions of multivariate inverses in the context of quantile functions have also been introduced (see, e.g., Serfling [325], Chakak and Ihlami [80], Fraiman and López [133]).

Section 7.2 All four functions (GIF1)–(GIF4) are studied in Bingham et al. [41, Section 2.4.4] (see also Feng et al. [130] in relation to quantile transformations). Klement et al. [222, 223] introduce quasi-inverse functions in a different way as compared to Definition 7.1. Balkema et al. [21] consider inversely asymptotic functions which, in some cases are asymptotically inverse.

Section 7.5 Asymptotically inverse functions for the classes $\mathcal{RV}_\rho$, $\rho > 0$, have been considered in Bingham et al. [41]. Relations (7.5)–(7.7) are studied in Djurčić and Torgašev [108] if f_{inv} and g_{inv} are inverse or generalized inverse functions. However, Example 7.49 shows that the generalized inverse functions are not always the best choice for inverting the asymptotic equivalence. Some results like (7.5) for multivalued functions are discussed by Molchanov [277], where the implication (7.5) is called *inversion*.

Section 7.7 The proof of relation (7.2) can be found in Bingham et al. [41, Theorem 1.5.12]. Equivalence (7.55), in the particular case of increasing continuous functions f, is proved by Djurčić and Torgašev [108] (see also Buldygin et al. [59]). Some cases of Theorem 7.64 are considered by Djurčić et al. [106].

Some results complementing those presented in this section, but for the generalized inverse function $f^{\leftarrow}$ instead of $f^{\sim}$, are also known in the literature. We mention some of the related papers here: Theorems 7.74 and 7.75 are proved in Djurčić et al. [107] and Djurčić et al. [110], respectively; relation (7.54) is obtained by Djurčić and Torgašev [109].

Section 7.8 Some aspects of the theory of regularly varying sequences appeared in Polya [294] and Schmidt [321]. According to Karamata's [204] definition, a sequence $\{c_n\}_{n\geq 1}$ is called regularly varying if, for some $\rho > -1$,

$$\lim_{n\to\infty} \frac{1}{nc_n} \sum_{k=1}^{n} c_k = \frac{1}{\rho + 1}.$$

This definition recalls the main results of Karamata's theorem for functions (see Theorem 6.1), but in Karamata [205] he adopted a new definition which we use nowadays, namely

$$\lim_{n\to\infty} \frac{c_{[\lambda n]}}{c_n} = \lambda^{\rho}, \qquad \lambda > 0. \tag{7.117}$$

Bojanic and Seneta [46] define a regularly varying function by requiring that

$$\text{the limit} \quad \lim_{n\to\infty} \frac{c_{[\lambda n]}}{c_n} \quad \text{exists for all} \quad \lambda > 0 \tag{7.118}$$

and prove that this limit equals λ^{ρ}, for some $\rho \in \mathbf{R}$. The crucial condition $a_{n+1}/a_n \to 1$ (see the comments to Sect. 4.11) follows from (7.118). A simpler and shorter proof of the latter result is found by Weissman [365] (see also Higgins [182]). Galambos and Seneta [145] present an example showing that, if the limit in (7.118) is assumed to exist only for $\lambda \in \mathbf{N}$, then it is possible that $a_{n+1}/a_n \not\to 1$. However, if $\{c_n\}_{n\geq 1}$ is monotone, then (7.118) with $\lambda \in \mathbf{N}$ instead of $\lambda > 0$ implies that $a_{n+1}/a_n \to 1$ (see de Haan [164]).

Definition 7.87 extends "regularity" of sequences to the classes $\mathcal{PRV}$, $\mathcal{SQI}$, $\mathcal{ORV}$, and $\mathcal{POV}$. A unified theory of sequences in the class $\mathcal{RV}$ is presented by Bojanic and Seneta [46] (see also Galambos and Seneta [145]). Sequences in the class $\mathcal{ORV}$ are discussed in Božin and Djurčić [50]. Lemma 7.88 could also be proved for the class $\mathcal{ORV}$.

An application of regularly varying sequences to the study of the asymptotic behavior of decreasing solutions of half-linear difference equations is discussed by Matucci and Řehák [265]. Regularly varying sequences play an important role in the limit theory for functions of multivariate Markov chains (see Mikosch and

Wintenberger [275]). The asymptotic behavior of Fourier series can be determined if their coefficients belong to the class $\mathcal{ORV}$ (see Chen and Chen [82]).

Section 7.9 The question we addressed in this chapter can also be described as follows. Assume that particles are distributed along the y-axis at an initial time, after which they start evolving according to the same law. The question is whether or not all particles move along asymptotically equivalent paths (see Chicone [83] for historial comments and a less formal discussion of solutions of ordinary differential equations).

Two positive solutions μ_{b_1} and μ_{b_2} corresponding to initial conditions b_1 and b_2, respectively, are asymptotically equivalent in the sense of relation (7.107) if and only if $\log \mu_{b_1}(t) - \log \mu_{b_2}(t) \to 0$, $t \to \infty$. If the initial equation is $y' = g(y)$, then the log-solution $\log y$ corresponds to the log-equation $z' = zg(z)$ being again autonomous. If the above property holds for all pairs of initial conditions, then one may refer to this property as *asymptotic uniform Lyapunov stability* of the log-equation. If one accepts this terminology, then we are dealing with the asymptotic uniform Lyapunov stability for log-equations.

A related property, the *stability with respect ro initial time difference*, has been studied for ordinary differential equations (see Mishchenko and Rozov [276], Shaw and Yakar [332]), fuzzy differential equations (cf. Yakar et al. [369]), and fractional differential equations (Agarwal et al. [6]).

Other results on stability are given by Samoĭlenko and Stanzhytskyi [314]. Some further connections of regularly varying functions and ordinary differential equations are given by Marić [261] (see also Řehák [299]).

Chapter 8
Generalized Renewal Processes

8.1 Introduction

If the trajectories of a stochastic process $f = f(x)$ are continuous, then we can deal with its renewal process $x = x(t)$ as in Fig. 8.1.

But there may exist various possibilities for definitions (see Fig. 8.2, where x_1, x_2, and x_3, as functions of the argument t, represent (so-called) generalized renewal processes under some mild conditions, see Sect. 7.2).

In Fig. 8.2, $x_1(t)$, $x_2(t)$, and $x_3(t)$ are the three consecutive moments when f crosses the level t. All three moments could be generalized renewal processes under suitable assumptions on f. For obvious reasons, the process $x_1(t)$ is called the first exit time (from the interval $[0, t)$), while $x_3(t)$ is the last exit time (from $[0, t)$). Moreover, $x_1(t) + x_3(t) - x_2(t)$, the total length of dashed intervals, is the sojourn time (in the interval $[0, t)$), which could also be a representative of the class of generalized renewal processes.

For monotone, but discontinuous functions, the generalized inverse is an example of a generalized renewal process (see Fig. 8.3).

Again, if a function is not monotone, then there is a variety of "reasonable" definitions of renewal processes. In Fig. 8.4, $x_1(t)$, $x_2(t)$, $x_3(t)$ as well as the sojourn time, being equal to the total length of dashed intervals, are four versions of generalized renewal processes.

The asymptotic properties of generalized renewal processes constructed from stochastic processes in discrete or in continuous time is one of the main topics of this book. We are mainly concerned with asymptotic properties which hold almost surely (a.s.) In the first place, we study dualities between the strong law of large numbers (SLLN) for stochastic processes (in discrete or in continuous time) and their corresponding renewal or generalized renewal processes. Such questions for sequences of sums of independent, identically distributed random variables and the corresponding renewal processes have already been discussed in Chaps. 1 and 2.

V. V. Buldygin et al., *Pseudo-Regularly Varying Functions and Generalized Renewal Processes*, Probability Theory and Stochastic Modelling 91, https://doi.org/10.1007/978-3-319-99537-3_8

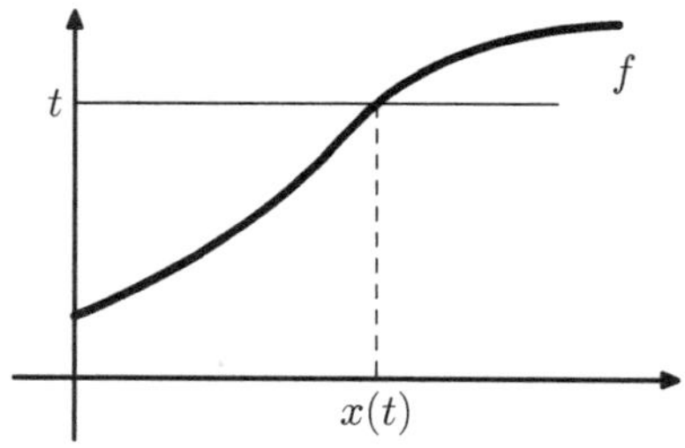

Fig. 8.1 Renewal process $x(t)$ constructed from an increasing continuous path of a stochastic process f. Equivalent definitions are given below: 1) $x(t)$ is the first time when f crosses the level t; 2) $x(t)$ is the last time when f crosses the level t; 3) $x(t)$ is the total time f spent below the level t

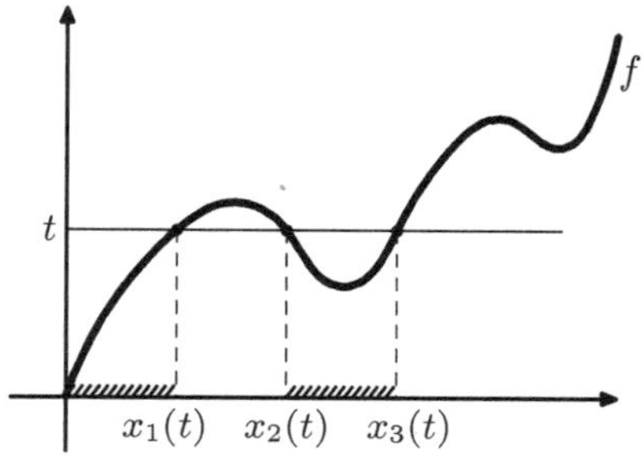

Fig. 8.2 Three generalized renewal processes $x_1(t)$, $x_2(t)$, and $x_1(t) + x_3(t) - x_2(t)$ constructed from a continuous, non-monotone path of a stochastic process f. Now the first time when f crosses the level t, the last time when f crosses the level t, and the total time f spent below the level t are different

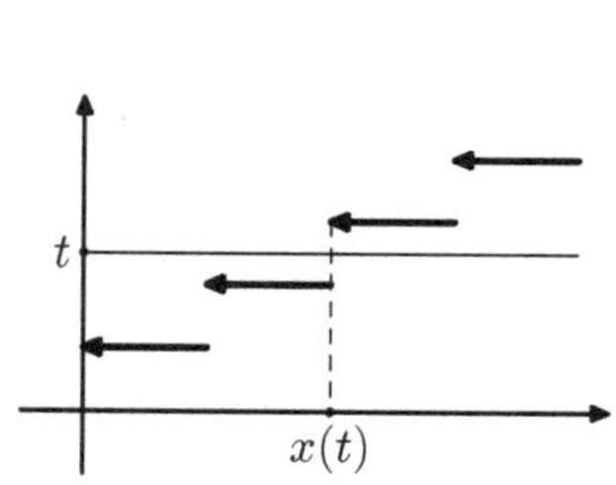

Fig. 8.3 Renewal process $x(t)$ constructed from a discontinuous path of a stochastic process f. In the theory of functions, this relates to the *generalized inverse function*

As pointed out in Chap. 2, much less is known about the behavior of renewal processes, if the underlying random variables are independent, but not identically distributed, or if they attain negative values with positive probability.

In Chap. 2, a general approach was considered to study the asymptotic properties of generalized renewal processes constructed from random sequences. The methods developed there allow us to prove the SLLN in Theorems 2.3 and 2.5. Lemma 2.2 precedes the proof of Theorems 2.3 and shows that, if condition (2.18) holds, then

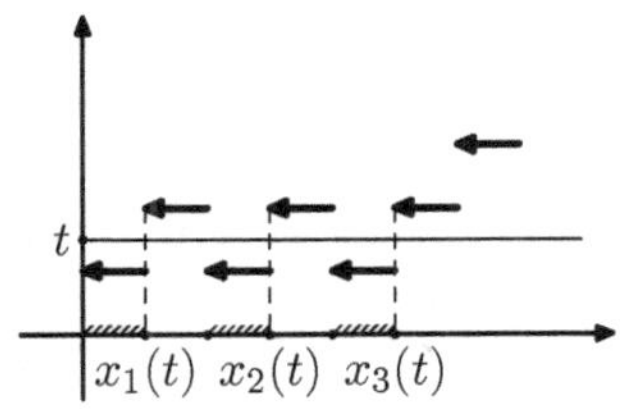

Fig. 8.4 Generalized renewal processes $x_1(t)$, $x_2(t)$, and $x_3(t)$ constructed from a discontinuous, non-monotone path of a stochastic process f. The total time f spent below the level t could also be considered as a generalized renewal process

the renewal processes M and L are asymptotically quasi-inverse for the normalizing sequence $\{a_n\}_{n\geq 1}$. This conclusion together with condition (2.23) is the basis of the proof of Theorem 2.3 which, in turn, is used to prove several other results. However, condition (2.18) in Theorem 2.3 is too restrictive. If this condition is omitted, then there are basically two ideas to investigate the relationship between an SLLN for the underlying stochastic process (or random sequence) and an SLLN for the corresponding renewal process. The first idea is demonstrated in the proof of Theorem 2.5 for the generalized renewal processes T, L and M, which are particular cases with an appropriate choice of the sequence $\{Z_n\}$. Condition (2.18) is dropped in Theorem 2.5, but another, rather technical, condition (2.23) is used there. On the other hand, the construction in Sect. 2.2.2 shows that one cannot neglect condition (2.23) in general.

Below we make use of a second idea, which already occurred in Chap. 7 for the first time. This idea is used there to investigate asymptotically quasi-inverse functions and to study conditions under which generalized inverse functions are asymptotically quasi-inverse functions as well. Another result of Chap. 7, related to that idea, is the duality between the asymptotic behavior of the ratio of functions (or sequences) and the asymptotic behavior of the ratio of their asymptotically quasi-inverse functions. Developing this idea further, we extend Theorem 2.5 and clarify the meaning of condition (2.23) which, from the point of view of the theory of pseudo-regularly varying functions developed in Chaps. 3 and 4, means that the normalizing function belongs to the class $\mathcal{SQI}$ of sufficiently quickly increasing functions. It is worth mentioning that the method used throughout this chapter allows us to study both stochastic processes and random sequences from a unified point of view.

Chapter 8 is organized as follows. In Sect. 8.2, generalized renewal processes constructed from random sequences are studied and a number of results are proved on the relationship between the SLLN for the underlying random sequence and the SLLN for the corresponding renewal process. These results explain the nature of the SQI-, PRV-, and POV-properties of the normalizing sequences under rather simple conditions, for which these properties are satisfied.

Generalized renewal processes constructed from stochastic processes are studied in Sect. 8.4. The results of this section also explain the role of the SQI-, PRV-, and

POV-properties of the normalizing sequences in the SLLN for renewal processes. In addition, we make clear why the requirement of continuity of the paths of the underlying process is important.

The results of Sects. 8.2 and 8.4 are used in Sects. 8.3 and 8.5, respectively. These sections contain a discussion of several particular questions. Among these questions is the investigation of the first exit time of a sequence of sums of random variables (not necessarily independent) from a domain with a curvilinear boundary (see Sect. 8.3) or the SLLN for generalized renewal processes constructed from nonhomogeneous compound Poisson processes (see Sect. 8.5).

8.2 Generalized Renewal Processes Constructed from Random Sequences

Let $\{\Omega, \mathfrak{F}, \mathsf{P}\}$ be a complete probability space. Generalized renewal processes constructed from random sequences have already been considered in Chaps. 1 and 2 and we adopt the notation introduced in Chap. 2.

Let $\{Z_n\} = \{Z_n\}_{n\geq 0}$ be a sequence of real-valued random variables with $Z_0 = 0$. In what follows, we mainly consider sequences $\{Z_n\}$ such that

$$\mathsf{P}\{Z_n \to \infty\} = \mathsf{P}\left\{\lim_{n\to\infty} Z_n = \infty\right\} = 1. \tag{8.1}$$

Given a random sequence $\{Z_n\}$, consider the following three *generalized renewal processes*:

$$M^+(s) = M^+_{\{Z_n\}}(s) = \min\{n \geq 0 : Z_n \geq s\}, \qquad s \geq 0, \tag{8.2}$$

$$L(s) = L_{\{Z_n\}}(s) = \max\{n \geq 0 : Z_n \leq s\}, \qquad s \geq 0, \tag{8.3}$$

$$T(s) = T_{\{Z_n\}}(s) = \sum_{n=1}^{\infty} \mathbb{I}\{Z_n \leq s\}, \qquad s \geq 0. \tag{8.4}$$

For the sake of brevity, we sometimes call them *renewal processes.*

The values of $M^+(s)$, $L(s)$, or $T(s)$ may be equal to infinity in the case of a general sequence $\{Z_n\}$. However, under condition (8.1), M^+, L, and T are almost surely real-valued stochastic processes.

Generalized renewal processes constructed from a random sequence have a clear interpretation. The random variable $M^+(s)$ is obviously the "*first hitting time*" of the set $[s, \infty)$ by the sequence $\{Z_n\}$ or the "*first exit time*" from the set $(-\infty, s)$. In turn, the random variables $L(s)$ and $T(s)$ can be viewed as the "*last exit time*" from the set $(-\infty, s]$ for the sequence $\{Z_n\}$ and the "*sojourn time*" in the set $(-\infty, s]$, respectively.

One can also consider other versions of generalized renewal processes. One of these versions is, for example, the renewal process $M(s)$, $s \geq 0$, from (2.8). Although we pay main attention to the above three processes we sometimes work with a general definition of a renewal process $R_{\{Z_n\}}$ (see below).

When considering the almost sure (a.s.) convergence of generalized renewal processes, our main aim is to establish a relationship between the almost sure convergence of the normalized sequences of the underlying random variables and the almost sure convergence of the normalized generalized renewal processes constructed from this sequence. Nevertheless, following tradition, we refer to results of this kind as the strong law of large numbers, even in cases where the underlying random sequences are not sums of independent random variables.

The duality between the strong law of large numbers for normalized sequences of random variables and for the corresponding generalized renewal processes has already been studied in Chaps. 1 and 2. Taking up this duality, we follow the general idea that the paths (or sample trajectories) of some generalized renewal processes are the generalized inverse functions for the corresponding realizations of the sequence $\{Z_n\}$ (see Sect. 7.8).

Indeed, for any $\omega \in \Omega$,

$$M^+(\omega, s) = \min\{n \geq 0 : Z_n(\omega) \geq s\} = \{Z_n(\omega)\}^{\leftarrow}(s), \qquad s \geq 0, \tag{8.5}$$

$$L(\omega, s) = \max\{n \geq 0 : Z_n(\omega) \leq s\} = \{Z_n(\omega)\}^{\rightarrow}(s), \qquad s \geq 0, \tag{8.6}$$

(see (7.68)) and, for any $\omega \in \Omega$ and all $s \geq 0$,

$$M^+(\omega, s) - 1 \leq T(\omega, s) \leq L(\omega, s). \tag{8.7}$$

The relations (8.5)–(8.7) allow us to apply the results of Sect. 7.8, where we established some properties of asymptotically quasi-inverse functions for sequences and the asymptotic behavior of the ratio of asymptotically quasi-inverse functions for these sequences.

8.2.1 The SLLN for Random Sequences and the Corresponding Renewal Processes

Throughout this section, normalizing sequences of real numbers $\{p_n\}_{n\geq 0}$ are such that $p_0 = 0$. Recall that (see Definition 7.87) $\{p_n\}$ is a PRV-, SQI-, POV-, or RV-sequence, if the piecewise linear interpolation $\widehat{p}$ of this sequence is a PRV-, SQI-, POV-, or RV-function.

We start with general conditions under which the strong law of large numbers for normalized sequences of random variables implies the strong law of large numbers for the corresponding generalized renewal processes. The following result is an appropriate version of Theorem 7.92.

Theorem 8.1 *Let*

$$\lim_{n\to\infty} \frac{Z_n}{p_n} = a \qquad a.s., \tag{8.8}$$

where $a \in (0, \infty)$, $\{Z_n\}_{n\geq 1}$ is a sequence of real-valued random variables and $\{p\} = \{p_n\}_{n\geq 1}$ is a nonrandom, increasing sequence of positive numbers tending to infinity. Then:

1) If $\{p\}$ is an SQI-sequence, then

$$\lim_{s\to\infty} \frac{M^{+}_{\{Z_n\}}(s)}{p^{\sim}(s/a)} = \lim_{s\to\infty} \frac{T_{\{Z_n\}}(s)}{p^{\sim}(s/a)} = \lim_{s\to\infty} \frac{L_{\{Z_n\}}(s)}{p^{\sim}(s/a)} = 1 \qquad a.s. \tag{8.9}$$

for any function $\{p\}^{\sim} \in \{\{p\}^{\leftarrow}, \{p\}^{\rightarrow}, \widehat{p}^{\,-1}\}$, where $\{p\}^{\leftarrow}$ and $\{p\}^{\rightarrow}$ are the generalized inverse functions for the sequence $\{p\}$ (see (7.68)), $\widehat{p}$ is the piecewise linear interpolation of $\{p\}$ (see (7.82)), and $\widehat{p}^{\,-1}$ is the inverse function of $\widehat{p}$, that is,

$$\widehat{p}^{\,-1}(s/a) = n + \frac{(s/a) - p_n}{p_{n+1} - p_n}, \qquad s \in [ap_n, ap_{n+1}),\ n \geq 0.$$

2) If

$$\liminf_{n\to\infty} \frac{n(p_{n+1} - p_n)}{p_n} > 0, \tag{8.10}$$

then (8.9) holds.

3) If $p_n = p(n)$, $n \geq 1$, where the function $p = (p(t), t \geq 1)$ is such that $p \in \mathbb{C}_{\text{inc}}$ and $p \in \mathcal{SQI}$, then relation (8.9) holds for any $\{p\}^{\sim} \in \{\{p\}^{\leftarrow}, \{p\}^{\rightarrow}, \widehat{p}^{\,-1}, p^{-1}\}$, where p^{-1} is the inverse function of p.

The meaning of Theorem 8.1 is that the strong law of large numbers (8.8) implies the strong law of large numbers (8.9), if the normalizing sequence $\{p_n\}_{n\geq 1}$ is an SQI-sequence. The second statement of this theorem contains the simple sufficient condition (8.10), under which $\{p_n\}_{n\geq 1}$ is an SQI-sequence. Theorem 4.63 shows that, under some extra restrictions, i.e., under (4.73) and (4.74), condition (8.10) is equivalent to the SQI-property of the sequence $\{p_n\}_{n\geq 1}$. Among such increasing sequences $\{p_n\}_{n\geq 1}$, for which condition (8.10) is equivalent to the SQI-property of the sequence $\{p_n\}_{n\geq 1}$, are $p_n = n^b$, $n \geq 1$, with $b > 0$, and $p_n = c^n$, $n \geq 1$, with $c > 0$.

The assumption $p \in \mathcal{SQI}$ in the third statement of Theorem 8.1 can be exchanged with the equivalent condition $p^{-1} \in \mathcal{PRV}$ (see Corollary 7.33). Taking Remark 3.22 into account, we see that statement 3) is equivalent to Theorem 2.5. The proof of statement 3) differs from that of Theorem 2.5, however, statement 3) explains condition (2.23) which, according to Remark 3.22 and Corollary 7.33, means that $a(\cdot) \in \mathcal{SQI}$. Recall that Chap. 4 contains various sufficient conditions under which an absolutely continuous function belongs to $\mathcal{PRV}$ or $\mathcal{SQI}$.

The sequence $p_n = \log(n+1)$, $n \geq 0$, does not satisfy the conditions of any of the statements of Theorem 8.1, and thus one cannot derive the strong law of large numbers (8.9) from (8.8) under the above normalization. Moreover, an example of a sequence $\{Z_n\}_{n\geq 1}$ for which the SLLN (8.8) holds, but the SLLN (8.9) does not, is given in Sect. 2.2.2. This example shows that the SQI-condition imposed on the sequence $\{p_n\}_{n\geq 1}$ in Theorem 8.1 cannot be dropped.

The following result is an appropriate version of Theorem 7.95 containing conditions under which (8.9) implies (8.8).

Theorem 8.2 *Let $\{Z_n\}_{n\geq 1}$ be a sequence of real-valued random variables, almost surely tending to infinity, and let $\{p\} = \{p_n\}_{n\geq 1}$ be a nonrandom, increasing sequence of positive numbers, also tending to infinity. Assume there are a number $a \in (0, \infty)$ and a function $\{p\}^{\sim} \in \{\{p\}^{\leftarrow}, \{p\}^{\rightarrow}, \widehat{p}^{-1}\}$ such that (8.9) holds. Then:*

1) If $\{p_n\}_{n\geq 0}$ is a PRV-sequence, then the SLLN (8.8) holds.
2) If

$$\limsup_{n\to\infty} \frac{n(p_{n+1} - p_n)}{p_n} < \infty, \tag{8.11}$$

then the SLLN (8.8) holds.
3) If $p_n = p(n)$, $n \geq 1$, where the function $p = (p(t), t \geq 1)$ is such that $p \in \mathbb{C}^{\infty}_{\mathrm{inc}}$, $p \in \mathcal{PRV}$, and if the SLLN (8.9) holds for some function $\{p\}^{\sim} \in \{\{p\}^{\leftarrow}, \{p\}^{\rightarrow}, \widehat{p}^{-1}, p^{-1}\}$, then the SLLN (8.8) also follows.

Theorem 8.2 is a converse statement to Theorem 8.1 and shows that the SLLN (8.9) implies the SLLN (8.8), if $\{p_n\}_{n\geq 1}$ is a PRV-sequence. Statement 2) of this theorem contains the simple condition (8.11) yielding that $\{p_n\}_{n\geq 1}$ is a PRV-sequence. Theorem 4.63 proves that, under some extra assumptions, i.e. under conditions (4.73) and (4.74), condition (8.11) is equivalent to the PRV-property of the sequence $\{p_n\}_{n\geq 1}$.

The sequences $p_n = \log(n+1)$, $n \geq 1$, and $p_n = n^b$, $n \geq 1$, with $b > 0$, serve as examples of increasing sequences $\{p_n\}_{n\geq 1}$, for which condition (8.11) is equivalent to the PRV-property of the sequence $\{p_n\}_{n\geq 1}$. In contrast, the sequence $p_n = c^n$, $n \geq 1$, with $c > 0$, does not satisfy the assumptions of any of the statements of Theorem 8.2, and thus one cannot derive the SLLN (8.8) from the SLLN (8.9) for this normalizing sequence. An example of a random sequence $\{Z_n\}_{n\geq 1}$ for which the SLLN (8.9) holds, but not the SLLN (8.8), is given in Remark 2.6. This example shows that one cannot drop the PRV-condition imposed on the sequence $\{p_n\}_{n\geq 1}$ in Theorem 8.2 in general.

Theorems 8.1 and 8.2 imply conditions under which the strong laws of large numbers (8.8) and (8.9) are equivalent. As in Sect. 7.8, we present these conditions in the following two results.

Theorem 8.3 *Let $a \in (0, \infty)$, $\{Z_n\}_{n\ge1}$ be a sequence of real-valued random variables, almost surely tending to infinity, and let $\{p_n\}_{n\ge1}$ be a nonrandom, increasing sequence of positive numbers, also tending to infinity. Let*

$$0 < \liminf_{n\to\infty} \frac{n(p_{n+1}-p_n)}{p_n} \quad \text{and} \quad \limsup_{n\to\infty} \frac{n(p_{n+1}-p_n)}{p_n} < \infty. \tag{8.12}$$

1) *If the SLLN (8.8) holds, then, for any function $\{p\}^{\sim} \in \{\{p\}^{\leftarrowtail}, \{p\}^{\rightarrowtail}, \widehat{p}^{-1}\}$, the SLLN (8.9) follows.*
2) *If there exists a function $\{p\}^{\sim} \in \{\{p\}^{\leftarrowtail}, \{p\}^{\rightarrowtail}, \widehat{p}^{-1}\}$ for which the SLLN (8.9) holds, then the SLLN (8.8) follows.*

Theorem 8.4 *Let $a \in (0, \infty)$, $\{Z_n\}_{n\ge1}$ be a sequence of real-valued random variables, almost surely tending to infinity, and let $p_n = p(n), n \ge 1$, where the function $p = (p(t), t \ge 1)$ is such that $p \in \mathbb{C}^{\infty}_{\mathrm{inc}}$ and $p \in \mathcal{POV}$.*

1) *If the SLLN (8.8) holds, then the SLLN (8.9) follows for any function $\{p\}^{\sim} \in \{\{p\}^{\leftarrowtail}, \{p\}^{\rightarrowtail}, \widehat{p}^{-1}, p^{-1}\}$.*
2) *If the SLLN (8.9) holds for some function $\{p\}^{\sim} \in \{\{p\}^{\leftarrowtail}, \{p\}^{\rightarrowtail}, \widehat{p}^{-1}, p^{-1}\}$, then the SLLN (8.8) also follows.*

Theorems 8.3 and 8.4 show that the SLLNs (8.8) and (8.9) are equivalent if $\{p_n\}_{n\ge1}$ is a POV-sequence.

Taking Corollaries 7.98 and 7.99 into account, we complete Theorems 8.3 and 8.4 with the cases $a = 0$ and $a = \infty$.

Theorem 8.5 *Let $\{Z_n\}_{n\ge1}$ be a sequence of real-valued random variables, almost surely tending to infinity, and let $\{p_n\}_{n\ge1}$ be an increasing sequence of positive numbers, also tending to infinity. Let condition (8.12) hold. Then, for any function $\{p\}^{\sim} \in \{\{p\}^{\leftarrowtail}, \{p\}^{\rightarrowtail}, \widehat{p}^{-1}\}$,*

1) *if*

$$\lim_{n\to\infty} \frac{Z_n}{p_n} = 0 \quad \text{a.s.}, \tag{8.13}$$

then

$$\lim_{s\to\infty} \frac{M^{+}_{\{Z_n\}}(s)}{p^{\sim}(s)} = \lim_{s\to\infty} \frac{T_{\{Z_n\}}(s)}{p^{\sim}(s)} = \lim_{s\to\infty} \frac{L_{\{Z_n\}}(s)}{p^{\sim}(s)} = \infty \quad \text{a.s.}; \tag{8.14}$$

2) if

$$\lim_{n\to\infty} \frac{Z_n}{p_n} = \infty \qquad a.s., \tag{8.15}$$

then

$$\lim_{s\to\infty} \frac{M^+_{\{Z_n\}}(s)}{p^\sim(s)} = \lim_{s\to\infty} \frac{T_{\{Z_n\}}(s)}{p^\sim(s)} = \lim_{s\to\infty} \frac{L_{\{Z_n\}}(s)}{p^\sim(s)} = 0 \qquad a.s. \tag{8.16}$$

Theorem 8.6 *Let $\{Z_n\}_{n\ge1}$ be a sequence of real-valued random variables, almost surely tending to infinity, and let $p_n = p(n)$, $n \ge 1$, where $p = (p(t), t \ge 1)$ is such that $p \in \mathbb{C}^\infty_{\text{inc}}$ and $p \in \mathcal{WPOV}$. Then Theorem 8.5 holds for all functions $\{p\}^\sim \in \{\{p\}^{\leftarrow}, \{p\}^{\rightarrow}, \widehat{p}^{-1}, p^{-1}\}$.*

For RV-sequences $\{p_n\}_{n\ge1}$, Theorems 8.3–8.6 take the following form in view of Corollaries 7.100–7.103.

Theorem 8.7 *Let $a \in [0,\infty]$, $\{Z_n\}_{n\ge1}$ be a sequence of real-valued random variables, almost surely tending to infinity, and let $\{p_n\}_{n\ge1}$ be an increasing sequence of positive numbers, also tending to infinity and for which the limit*

$$\lim_{n\to\infty} \frac{n(p_{n+1} - p_n)}{p_n} = \rho \in (0,\infty) \tag{8.17}$$

exists. If (8.8) holds, that is, if

$$\lim_{n\to\infty} \frac{Z_n}{p_n} = a \qquad a.s.,$$

then, for any function $\{p\}^\sim \in \{\{p\}^{\leftarrow}, \{p\}^{\rightarrow}, \widehat{p}^{-1}\}$,

$$\lim_{s\to\infty} \frac{M^+_{\{Z_n\}}(s)}{p^\sim(s)} = \lim_{s\to\infty} \frac{T_{\{Z_n\}}(s)}{p^\sim(s)} = \lim_{s\to\infty} \frac{L_{\{Z_n\}}(s)}{p^\sim(s)} = \left(\frac{1}{a}\right)^{\frac{1}{\rho}}, \tag{8.18}$$

where we set $1/\infty := 0$ and $1/0 := \infty$.

Theorem 8.8 *Let $a \in [0,\infty]$, $\{Z_n\}_{n\ge1}$ be a sequence of real-valued random variables, almost surely tending to infinity, and let $p_n = p(n)$, $n \ge 1$, where the function $p = (p(t), t \ge 1)$ is such that $p \in \mathbb{C}^\infty_{\text{inc}}$ and $p \in \mathcal{RV}$. If (8.8) holds, then (8.18) follows for all functions $\{p\}^\sim \in \{\{p\}^{\leftarrow}, \{p\}^{\rightarrow}, \widehat{p}^{-1}, p^{-1}\}$.*

Theorem 8.9 *Let $a \in (0,\infty)$, $\{Z_n\}_{n\ge1}$ be a sequence of real-valued random variables, almost surely tending to infinity, and let $\{p_n\}_{n\ge1}$ be an increasing sequence of positive numbers, also tending to infinity and for which the limit (8.17) exists. If (8.8) holds, then (8.18) follows for all functions $\{p\}^\sim \in \{\{p\}^{\leftarrow}, \{p\}^{\rightarrow}, \widehat{p}^{-1}\}$.*

Conversely, if (8.18) holds for some function $\{p\}^{\sim} \in \{\{p\}^{\leftarrowtail}, \{p\}^{\rightarrowtail}, \widehat{p}^{-1}\}$, *then (8.8) also follows.*

Theorem 8.10 *Let* $a \in (0, \infty)$, $\{Z_n\}_{n\ge 1}$ *be a sequence of real-valued random variables, almost surely tending to infinity, and let* $p_n = p(n)$, $n \ge 1$, *where the function* $p = (p(t), t \ge 1)$ *is such that* $p \in \mathbb{C}^{\infty}_{\mathrm{inc}}$ *and* $p \in \mathcal{RV}$. *If (8.8) holds, then (8.18) follows for any function* $\{p\}^{\sim} \in \{\{p\}^{\leftarrowtail}, \{p\}^{\rightarrowtail}, \widehat{p}^{-1}, p^{-1}\}$. *Conversely, if (8.18) holds for some function* $\{p\}^{\sim} \in \{\{p\}^{\leftarrowtail}, \{p\}^{\rightarrowtail}, \widehat{p}^{-1}, p^{-1}\}$, *then (8.8) also follows.*

Remark 8.11 The nonrandom constant a in Theorems 8.1–8.10 can be exchanged with a random variable $a(\omega), \omega \in \Omega$. If $a(\omega) \in (0, \infty)$ almost surely, then the statements of Theorems 8.1–8.10 remain valid. Otherwise, one should restrict the corresponding statements to the random events $\{a(\omega) \in (0, \infty)\}$ or $\{a(\omega) = 0\}$ or $\{a(\omega) = \infty\}$, respectively.

8.3 Examples

In this section, some examples of the asymptotic behavior of generalized renewal processes are considered for various sequences of random variables. A number of "classical" results, namely those related to partial sums of independent, identically distributed random variables, have been presented in Chaps. 1 and 2. So, we present some new examples below.

8.3.1 Generalized Renewal Processes Constructed from Stationary Sequences

Let $\{\xi_k\} = \{\xi_k, k \in \mathbf{Z}\}$ be a strictly stationary sequence of real-valued random variables such that

$$\mathsf{E}|\xi_0| < \infty \qquad \text{and} \qquad \mathsf{E}(\xi_0|\mathscr{I}_\infty) = \eta \qquad \text{a.s.,} \tag{8.19}$$

where $\mathscr{I}_\infty$ is the σ-algebra of random events which are invariant with respect to shifts.

Note that condition (8.19) yields that

$$\mathsf{E}|\eta| \le \mathsf{E}|\xi_0| < \infty,$$

whence $\mathsf{P}\{|\eta| = \infty\} = 0$. In addition, $\eta = \mathsf{E}\xi_0$ by the ergodicity of the sequence $\{\xi_k\}$. In particular, this is the case if the $\{\xi_k\}$ are independent, identically distributed random variables with mean $\mu = \mathsf{E}\xi_0$.

According to the ergodic theorem,

$$\lim_{n\to\infty} \frac{Z_n}{n} = \eta \qquad \text{a.s.,} \tag{8.20}$$

where

$$Z_n = \sum_{k=1}^{n} \xi_k, \qquad n \ge 1. \tag{8.21}$$

Relation (8.20) together with Theorem 8.7 and Remark 8.11 implies the following result.

Proposition 8.12 *Let $\{\xi_k\}$ be a strictly stationary sequence of real-valued random variables satisfying condition (8.19). Let $\{Z_n\}$ be as defined in (8.21). Then:*

1) *for the generalized renewal processes*

$$R_{\{Z_n\}} = M^+_{\{Z_n\}},\ L_{\{Z_n\}},\ T_{\{Z_n\}},$$

we have

$$\lim_{s\to\infty} \frac{R_{\{Z_n\}}(s)}{s} = \frac{1}{\eta} \qquad \textit{a.s.} \qquad \textit{on the event } \{\eta > 0\};$$

in particular, if $\mathsf{P}\{\eta > 0\} > 0$, then

$$\lim_{s\to\infty} \mathsf{P}\left\{ \frac{R_{\{Z_n\}}(s)}{s} < u \,\Big|\, \eta > 0 \right\} = \frac{\mathsf{P}\{\eta > (1/u)\}}{\mathsf{P}\{\eta > 0\}}$$

for all $u > 0$ such that $1/u$ is a continuity point of the distribution function of the random variable η;

2) *for the generalized renewal processes*

$$R_{\{|Z_n|\}} = M^+_{\{|Z_n|\}},\ L_{\{|Z_n|\}},\ T_{\{|Z_n|\}},$$

we have

$$\lim_{s\to\infty} \frac{R_{\{|Z_n|\}}(s)}{s} = \frac{1}{|\eta|} \qquad \textit{a.s.} \qquad \textit{on } \{|\eta| > 0\}; \tag{8.22}$$

in particular, if $\mathsf{P}\{|\eta| > 0\} > 0$, then

$$\lim_{s\to\infty} \mathsf{P}\left\{ \frac{R_{\{|Z_n|\}}(s)}{s} < u \,\Big|\, |\eta| > 0 \right\} = \frac{\mathsf{P}\{|\eta| > (1/u)\}}{\mathsf{P}\{|\eta| > 0\}}$$

for all $u > 0$ such that $1/u$ is a continuity point of the distribution function of the random variable $|\eta|$;

3) if

$$\limsup_{n\to\infty} |Z_n| = \infty \qquad a.s. \qquad on\ \{\eta = 0\},$$

then

$$\lim_{s\to\infty} \frac{R_{\{|Z_n|\}}(s)}{s} = \infty \qquad a.s. \qquad on\ \{\eta = 0\}.$$

Example 8.13 Let $\{\xi_k\}$ be a stationary Gaussian sequence such that $\mathsf{E}\xi_0 = \mu \in (-\infty, \infty)$ and let its spectral function $(G(\lambda), \lambda \in [-\pi, \pi])$ be continuous for all $\lambda \neq 0$. Assume that

$$\sigma_0^2 = \lim_{\lambda\downarrow 0}(G(\lambda) - G(-\lambda)) > 0.$$

Then $\eta = \mathsf{E}(\xi_0 | S_\infty)$ is a Gaussian random variable with parameters

$$\mathsf{E}\eta = \mu \qquad \text{and} \qquad \mathsf{var}\,\eta = \sigma_0^2.$$

Thus $1/|\eta|$ is an almost surely positive random variable with density

$$q(u) = \frac{1}{\sqrt{2\pi}\sigma_0 u^2}\left(\exp\left\{\frac{-(1-u\mu)^2}{2\sigma_0^2 u^2}\right\} + \exp\left\{\frac{-(1+u\mu)^2}{2\sigma_0^2 u^2}\right\}\right), \qquad u > 0.$$

Proposition 8.12 implies that, for the generalized renewal processes $R_{\{|Z_n|\}}$, relation (8.22) holds and

$$\lim_{s\to\infty} \mathsf{P}\left\{\frac{R_{\{|Z_n|\}}(s)}{s} < u\right\} = \mathsf{P}\left\{|\eta| > \frac{1}{u}\right\} = \int_{(1/u)}^{\infty} q(u)\,du$$

for all $u > 0$.

8.3.2 The Asymptotic Behavior of the First Exit Time from a Domain with a Curvilinear Boundary

Let $\{\xi_k\}$ be a strictly stationary sequence of real-valued random variables for which condition (8.19) holds, and let $\{\alpha_n\}_{n\geq 1}$ be a sequence of positive numbers such that

$$p_n = \frac{n}{\alpha_n} \uparrow \infty, \qquad n \uparrow \infty. \tag{8.23}$$

Along with the sequence $\{\xi_k\}$, we consider

$$S_n = \sum_{k=1}^{n} \xi_k, \qquad n \geq 1.$$

Let

$$B_\varphi(s) = \{(n, s\alpha_n), n \geq 1\}$$

be a curvilinear boundary in the set $\mathbf{N} \times \mathbf{R}_0$, where $s > 0$ is a positive parameter.

The random variable

$$M_{\{S_n\}}(B_\varphi(s)) = \min\{n : S_n > s\alpha_n\}$$

is called the *the first exit time* of the sequence $\{S_n\}_{n\geq 1}$, $S_0 = 0$, through the boundary $B_\varphi(s)$. The random variable $M_{\{S_n\}}(B_\varphi(s))$ plays an important role in nonlinear renewal theory and its asymptotic behavior is also of significant interest. Recall that a similar problem for sums of independent, identically distributed random variables $\{\xi_k\}_{k\geq 1}$, with $\mathsf{E}\xi_1 = a > 0$, has already been considered in Example 2.14 of Chap. 2. There we assumed that $\alpha_n = \alpha(n)$, where $(\alpha(t), t \geq 1)$ is a continuous positive function such that

$$p(t) = \frac{t}{\alpha(t)} \uparrow \infty, \qquad t \uparrow \infty.$$

To establish the asymptotic behavior of $M_{\{S_n\}}(B_\varphi(s))$ (as $s \to \infty$) we consider the auxiliary sequence

$$Z_n = \frac{S_n}{\alpha_n}, \qquad n \geq 1. \tag{8.24}$$

Note that

$$M^+_{\{Z_n\}}(s) \leq M_{\{S_n\}}(B_\varphi(s)) \leq 1 + M^+_{\{Z_n\}}(s) \tag{8.25}$$

for all $s > 0$. Further, according to the ergodic theorem,

$$\lim_{n\to\infty} \frac{Z_n}{p_n} = \eta \qquad \text{a.s.} \tag{8.26}$$

Relations (8.24)–(8.26) together with statement 2) of Theorem 8.1 and Remark 8.11 imply the following result.

Proposition 8.14 *Assume that condition (8.23) holds. If*

$$\liminf_{n\to\infty}\left((n+1)\left(\frac{\alpha_n}{\alpha_{n+1}}\right)-n\right)>0,$$

then

$$\lim_{s\to\infty}\frac{M_{\{S_n\}}(B_\varphi(s))}{\{p\}^{\sim}(s/\eta)}=1 \qquad a.s. \qquad \text{on the event } \{\eta>0\},$$

where

$$\{p\}^{\sim}(s/\eta)=\min\left\{n:\frac{n}{\alpha_n}\ge\frac{s}{\eta}\right\}.$$

8.3.3 Generalized Renewal Processes Constructed from Markovian Gaussian Sequences Under Strong Dependence

Let $\{Z_n\}_{n\ge1}$ be a centered Markovian Gaussian sequence such that $\sigma_n^2=\mathsf{E}Z_n^2>0$, $n\ge1$, and let $r_{n,n+1}\ne0$, $n\ge1$, where

$$r_{k,n}=\mathsf{E}(Z_kZ_n/\sigma_k\sigma_n)$$

is the covariance between Z_k and Z_n. Now, every centered Markovian Gaussian sequence is characterized by the relation

$$r_{k,n}=r_{k,m}r_{m,n}, \qquad 1\le k\le m\le n$$

(see, for example, Feller [129]), which implies that

$$r_{1,n}=\prod_{m=1}^{n-1}r_{m,m+1}\ne0, \qquad n\ge1,$$

where the limit $|r|_{1,\infty}=\lim_{n\to\infty}|r_{1,n}|$ exists and satisfies $|r|_{1,\infty}\in[0,1]$.

The following recurrence relations hold for every Markovian Gaussian sequence $\{Z_n\}_{n\ge1}$:

$$Z_1=\sigma_1w_1, \qquad Z_n=a_nZ_{n-1}+b_nw_n, \qquad n\ge2,$$

where $a_n=(\sigma_n/\sigma_{n-1})r_{n-1,n}$, $b_n^2=\sigma_n^2(1-r_{n-1,n}^2)$, $n\ge2$, and $\{w_n\}_{n\ge1}$ is a sequence of independent, standard normal random variables.

Consider the sequence $\{Y_n\}_{n\ge 1}$ defined by

$$Y_1 = Z_1, \qquad Y_n = \frac{Z_n}{a_2 \dots a_n} = \frac{\sigma_1 Z_n}{\sigma_n r_{1,n}}, \qquad n \ge 2.$$

Then, for $n \ge 2$,

$$\mathsf{E}\{Y_n \big| w_1, \dots, w_{n-1}\} = \frac{a_n}{a_2 \dots a_n} \mathsf{E}\{Z_{n-1} \big| w_1, \dots, w_{n-1}\} = \frac{Z_{n-1}}{a_2 \dots a_{n-1}} = Y_{n-1},$$

and thus the sequence $\{Y_n\}_{n\ge 1}$ is a Gaussian martingale.

Assume that

$$|r|_{1,\infty} > 0, \tag{8.27}$$

that is, $\{Z_n\}_{n\ge 1}$ is a sequence of strongly dependent Gaussian random variables. Since

$$\sup_{n\ge 2} \mathsf{E}|Y_n|^2 = \frac{\sigma_1^2}{|r|_{1,\infty}^2} < \infty,$$

the martingale Y_n converges almost surely by Doob's theorem and hence

$$\lim_{n\to\infty} \frac{Z_n}{\sigma_n} = \eta \qquad \text{a.s.},$$

where η is a standard normal random variable. This implies that

$$\lim_{n\to\infty} \frac{|Z_n|}{\sigma_n} = |\eta| \qquad \text{a.s.} \tag{8.28}$$

Now relation (8.28) together with statement 2) of Theorem 8.1 and Remark 8.11 yields the following result.

Proposition 8.15 *Let $\{Z_n\}_{n\ge 1}$ be a centered Markovian Gaussian sequence such that $\{\sigma_n\}_{n\ge 1}$ is a positive increasing sequence tending to infinity. If condition (8.27) holds and*

$$\liminf_{n\to\infty} \frac{n(\sigma_{n+1} - \sigma_n)}{\sigma_n} > 0,$$

then, for any of the generalized renewal processes

$$R_{\{|Z_n|\}} = M^+_{\{|Z_n|\}},\ L_{\{|Z_n|\}},\ T_{\{|Z_n|\}},$$

we have

$$\lim_{s\to\infty} \frac{R_{\{|Z_n|\}}(s)}{\{\sigma\}^{\sim}(s/\eta)} = 1 \qquad a.s.,$$

where

$$\{\sigma\}^{\sim}(s/\eta) = \min\left\{n : \sigma(n) \ge \frac{s}{\eta}\right\}.$$

8.3.4 Generalized Renewal Processes Constructed from Sums of Independent Random Variables

Let $\{\xi_k\}_{k\ge 1}$ be a sequence of independent, identically distributed, real-valued random variables with zero mean and finite second moment $\sigma^2 = \mathsf{E}\xi_1^2 < \infty$.

Along with $\{\xi_k\}_{k\ge 1}$, let $\{\mu_k\}_{k\ge 1}$ be a sequence of positive numbers such that

$$\lim_{n\to\infty} \frac{\sum_{k=1}^n \mu_k}{\sqrt{n\log\log n}} = \infty. \tag{8.29}$$

Consider the sequence of partial sums

$$Z_n = \sum_{k=1}^n (\xi_k + \mu_k), \qquad n \ge 1.$$

Since

$$\left|\frac{Z_n}{\sum_{k=1}^n \mu_k} - 1\right| = \frac{\left|\sum_{k=1}^n \xi_k\right|}{\sum_{k=1}^n \mu_k} = \frac{\sqrt{n\log\log n}}{\sum_{k=1}^n \mu_k} \cdot \frac{\left|\sum_{k=1}^n \xi_k\right|}{\sqrt{n\log\log n}},$$

the law of the iterated logarithm implies that

$$\lim_{n\to\infty} \frac{Z_n}{\sum_{k=1}^n \mu_k} = 1 \qquad \text{a.s.} \tag{8.30}$$

by condition (8.29).

From the SLLN (8.30) and statement 2) of Theorem 8.1, we conclude the SLLN for the generalized renewal processes constructed from the sequence $\{Z_n\}_{n\ge 1}$.

Proposition 8.16 *If condition (8.29) holds and*

$$\liminf_{n\to\infty} \frac{n\mu_{n+1}}{\sum_{k=1}^n \mu_k} > 0,$$

then

$$\lim_{s\to\infty} \frac{R_{\{Z_n\}}(s)}{\mathfrak{M}(s)} = 1 \qquad a.s.,$$

for any of the generalized renewal processes

$$R_{\{Z_n\}} = M^+_{\{Z_n\}},\ L_{\{Z_n\}},\ T_{\{Z_n\}},$$

where

$$\mathfrak{M}(s) = \min\left\{n : \sum_{k=1}^{n} \mu_k \ge s\right\}.$$

8.4 Renewal Processes Constructed from Stochastic Processes

We continue the investigation of asymptotic properties of generalized renewal processes. Next we treat the case of generalized renewal processes constructed from stochastic processes in continuous time.

Let $X(t) = (X(\omega, t),\ \omega \in \Omega), t \ge 0$, be a real-valued stochastic process defined on a complete probability space $\{\Omega, \mathfrak{F}, \mathsf{P}\}$. Put $X(\omega) = (X(\omega, t),\ t \ge 0), \omega \in \Omega$, and recall that, given any fixed $\omega \in \Omega$, the function $X(\omega)$ is called a *trajectory* or *sample path* of the process $X = (X(t), t \ge 0)$.

Consider the following three *generalized renewal processes*, which, for the sake of brevity, will also be called *renewal processes*. The first one is defined as

$$M^+_X(s) = M^+_X(\omega, s) = \inf\{t \ge 0 : X(\omega, t) \ge s\}, \tag{8.31}$$

$$\omega \in \{\omega : X(\omega) \in \mathbb{F}^{(\infty)}\},$$

for $s \ge \max\{0, X(0)\}$, and we put $M^+_X(s) = 0$, for $0 \le s < X(0)$. We also assume that $M^+_X(\omega, s) \equiv \infty$ for $\omega \in \{X \notin \mathbb{F}^{(\infty)}\}$.

If $\{X \in \mathbb{F}^{(\infty)}\} \in \mathfrak{F}$ (which holds if X is separable), then the restriction of the process M^+_X to the event $\{X \in \mathbb{F}^{(\infty)}\}$ is a stochastic process defined on $\{X \in \mathbb{F}^{(\infty)}\}$ and it possesses nonnegative values. If $\mathsf{P}\{X \in \mathbb{F}^{(\infty)}\} = 1$, then M^+_X is a stochastic process defined on the space $\{\Omega, \mathfrak{F}, \mathsf{P}\}$.

The process M^+_X is well-defined on the set $\{X \in \mathbb{F}^{\infty}\}$. Along with M^+_X, we define on this set the generalized renewal process L_X as

$$L_X(s) = L_X(\omega, s) = \sup\{t \ge 0 : X(\omega, t) \le s\}, \tag{8.32}$$

$$\omega \in \{\omega : X(\omega) \in \mathbb{F}^{\infty}\},$$

for $s \geq \max\{0, X(0)\}$, and we put $L_X(s) = 0$, for $0 \leq s < X(0)$. We also assume that $L_X(\omega, s) \equiv \infty$, if $\omega \in \{X \notin \mathbb{F}^\infty\}$.

If $\{X \in \mathbb{F}^\infty\} \in \mathfrak{F}$ (which holds if X is separable), then the restriction of the process L_X to the set $\{X \in \mathbb{F}^\infty\}$ is a stochastic process defined on $\{X \in \mathbb{F}^\infty\}$ and it possesses nonnegative values. If $\mathsf{P}\{X \in \mathbb{F}^\infty\} = 1$, then L_X is also a stochastic process defined on $\{\Omega, \mathfrak{F}, \mathsf{P}\}$.

If the stochastic process X is measurable, that is, if its sample paths $X(\omega)$ are measurable functions almost surely, then, along with M_X^+ and L_X, one can define a third generalized renewal process T_X on $\{X \in \mathbb{F}^\infty\}$, i.e.

$$\begin{aligned} T_X(s) = T_X(\omega, s) &= \operatorname{meas}(\{t \geq 0 : X(\omega, t) \leq s\}) \\ &= \int_0^\infty \mathbb{I}\{X(\omega, t) \leq s\}\, dt, \qquad \omega \in \{\omega : X(\omega) \in \mathbb{F}^\infty\}, \end{aligned} \tag{8.33}$$

for $s \geq \max\{0, X(0)\}$, and we put $T_X(s) = 0$, for $0 \leq s < X(0)$. We also assume that $T_X(\omega, s) \equiv \infty$, if $\omega \in \{X \notin \mathbb{F}^\infty\}$. If $\{X \in \mathbb{F}^\infty\} \in \mathfrak{F}$ (which holds if X is separable) then the restriction of T_X to the set $\{X \in \mathbb{F}^\infty\}$ is a stochastic process defined on $\{X \in \mathbb{F}^\infty\}$ and it possesses nonnegative values. If $\mathsf{P}\{X \in \mathbb{F}^\infty\} = 1$, then T_X is again a stochastic process defined on $\{\Omega, \mathfrak{F}, \mathsf{P}\}$.

Thus, if X is a measurable, separable stochastic process X, then the three generalized renewal processes M_X^+, L_X, and T_X are well-defined on the set $\{X \in \mathbb{F}^\infty\}$. Under the condition $\mathsf{P}\{X \in \mathbb{F}^\infty\} = 1$, M_X^+, L_X, and T_X are stochastic processes defined on the space $\{\Omega, \mathfrak{F}, \mathsf{P}\}$ and possess nonnegative values.

All three generalized renewal processes are nondecreasing and one has

$$M_X^+(s) \leq T_X(s) \leq L_X(s) \tag{8.34}$$

for $s \geq 0$.

There are also other possible versions of generalized renewal processes, but we concentrate on the above processes and sometimes use the general notation R_X for any of these processes.

The generalized renewal processes M_X^+, L_X, and T_X are similar to the corresponding renewal processes constructed from random sequences and have a similar meaning. For $s \geq \max\{0, X(0)\}$, the random variable $M_X^+(s)$ is the "*first hitting time*" of the set $[s, \infty)$ or the "*first exit time*" from $(-\infty, s)$ for the process X. Analogously, the random variables $L_X(s)$ and $T_X(s)$ are the "*last exit time*" from $(-\infty, s]$ and the "*sojourn time*" in $(-\infty, s]$, respectively.

Note that

$$M_X^+(s) = \min\{t \geq 0 : X(t) = s\}; \qquad L_X(s) = \max\{t \geq 0 : X(t) = s\} \tag{8.35}$$

if the process X has continuous trajectories.

As before in this chapter, the main aim of this section is to obtain general results on the duality between the SLLN for the underlying stochastic processes and the SLLN for the corresponding renewal processes.

When establishing the duality for stochastic processes we use the same property as in the case of random sequences (see Sect. 8.2), i.e., that the trajectories of the generalized renewal processes are generalized inverse functions for the corresponding paths of the process X. Indeed, for any $\omega \in \Omega$

$$M_X^+(\omega, s) = (X(\omega))^{\hookleftarrow}(s), \qquad s \geq 0, \tag{8.36}$$

where $(X(\omega))^{\hookleftarrow}$ is one of the generalized inverse functions for the sample path $X(\omega)$, and

$$L_X(\omega, s) = (X(\omega))^{\hookrightarrow}(s), \qquad s \geq 0, \tag{8.37}$$

where $(X(\omega))^{\hookrightarrow}$ is another generalized inverse function for the path $X(\omega)$ (recall the definitions in Sect. 7.2).

Relations (8.34)–(8.37) allow us to use the techniques developed in Chap. 7 for asymptotically quasi-inverse functions and, in particular, to apply the results of Sect. 7.5 concerning the duality between the asymptotic behavior of the ratio of functions and that of the ratio of their asymptotically quasi-inverse functions.

8.4.1 *The SLLN for Stochastic Processes and Their Corresponding Renewal Processes*

Throughout this section, we assume that an SLLN holds for the underlying process, i.e.

$$\lim_{t\to\infty} \frac{X(t)}{f(t)} = \eta \qquad \text{a.s.}, \tag{8.38}$$

where η is a random variable.

We start the discussion on the asymptotic behavior of generalized renewal processes with the case where the sample paths of X are continuous functions. An important feature of such processes is that

$$X(\omega, M_X^+(\omega, s)) = s$$

for all $s \geq \max\{X(\omega), 0\}$ and for

$$\omega \in \{\limsup_{t\to\infty} X(t) = \infty\} = \left\{\omega : \limsup_{t\to\infty} X(\omega; t) = \infty\right\},$$

which follows from relation (8.35). The above property implies that

$$\lim_{s\to\infty} \frac{X(M_X^+(s))}{s} = 1 \qquad \text{on the set} \qquad \left\{\limsup_{t\to\infty} X(t) = \infty\right\}.$$

Correspondingly,

$$\lim_{s\to\infty} \frac{X(M_X^+(s))}{s} = \frac{X(L_X(s))}{s} = 1 \qquad \text{on} \qquad \left\{\lim_{t\to\infty} X(t) = \infty\right\}.$$

The above relation is basically the result stated in the following lemma, which is similar to Lemma 2.2. The only difference is that Lemma 8.17 does not contain a condition similar to (2.18) from Lemma 2.2.

Lemma 8.17 *Let a stochastic process X, a nonrandom, positive function f, and a random variable $\eta \in (0, \infty)$ be given such that the SLLN (8.38) holds.*

1) If $X \in \mathbb{C}^{(\infty)}$ a.s., then

$$\lim_{s\to\infty} \frac{f(M_X^+(s))}{s} = \frac{1}{\eta} \qquad a.s. \tag{8.39}$$

2) If $X \in \mathbb{C}^{\infty}$ a.s., then

$$\lim_{s\to\infty} \frac{f(M_X^+)}{s} = \lim_{s\to\infty} \frac{f(L_X(s))}{s} = \frac{1}{\eta} \qquad a.s. \tag{8.40}$$

3) If $X \in \mathbb{C}^{\infty}$ a.s. and the function f is either nondecreasing or asymptotically equivalent to a nondecreasing function, then

$$\lim_{s\to\infty} \frac{f(M_X^+(s))}{s} = \lim_{s\to\infty} \frac{f(L_X(s))}{s} = \lim_{s\to\infty} \frac{f(T_X(s))}{s} = \frac{1}{\eta} \qquad a.s. \tag{8.41}$$

Proof of Lemma 8.17. The first two statements follow from relations (8.35)–(8.38) and from statements 3) and 4) of Theorem 7.16. In turn, the third statement is a corollary of the second statement and relation (8.34). □

Relations (8.39)–(8.41) give us some information on the asymptotic almost sure behavior of the processes $f \circ M_X^+$, $f \circ L_X$, and $f \circ T_X$. To pass to the renewal processes M_X^+, L_X, and T_X themselves, one needs to impose some extra assumptions, the main one of which is that $f \in \mathcal{WSQI}$ (see condition (3.14)).

Theorem 8.18 *Let $X \in \mathbb{C}^{\infty}$ a.s., $f \in \mathbb{F}^{\infty}$, $f \sim f_0 \in \mathbb{F}_{\text{ndec}}$, $\mathbb{AI}(f_0) \neq \varnothing$, $\psi_0 \in \mathbb{AI}(f_0)$, and $\eta \in (0, \infty)$ a.s.*

1) *If the SLLN (8.38) holds and* $f \in \mathcal{WSQI}$, *then*

$$\lim_{s\to\infty} \frac{M_X^+(s)}{\psi_0(s/\eta)} = \lim_{s\to\infty} \frac{L_X(s)}{\psi_0(s/\eta)} = \lim_{s\to\infty} \frac{T_X(s)}{\psi_0(s/\eta)} = 1 \quad a.s. \tag{8.42}$$

and

$$\lim_{s\to\infty} \frac{M_X^+(s)}{f^\sim(s/\eta)} = \lim_{s\to\infty} \frac{L_X(s)}{f^\sim(s/\eta)} = \lim_{s\to\infty} \frac{T_X(s)}{f^\sim(s/\eta)} = 1 \quad a.s. \tag{8.43}$$

for every $f^\sim \in \mathbb{AQI}(f)$.

2) *If the SLLN (8.38) holds,* $f \in \mathcal{WSQI}$, *and* $f \in \mathbb{C}^\infty$, *then*

$$\lim_{s\to\infty} \frac{M_X^+(s)}{f^\sim(s/\eta)} = \lim_{s\to\infty} \frac{L_X(s)}{f^\sim(s/\eta)} = \lim_{s\to\infty} \frac{T_X(s)}{f^\sim(s/\eta)} = 1 \quad a.s.,$$

where $f^\sim \in \{f^{\leftarrow}, f^{\rightarrow}\}$.

3) *If* $f \sim f_0 \in \mathbb{C}^\infty_{\mathrm{inc}}$, *then statements 1) and 2) are satisfied, if* $f^\sim = \psi_0 = f_0^{-1}$, *where* f_0^{-1} *is the inverse function of* f_0.

Proof of Theorem 8.18. This follows from relations (8.35)–(8.38), Lemma 8.17, and Theorems 7.50–7.53. □

8.4.2 *The SLLN for Stochastic Processes and Their Corresponding Renewal Processes Under WPRV Normalizations*

If a stochastic process X is not continuous, then a result similar to Lemma 8.17 holds under the condition $f \in \mathcal{WPRV}$.

Lemma 8.19 *Let a measurable separable stochastic process* X, *a nonrandom positive function* f, *and a random variable* $\eta \in (0, \infty)$ *be such that* $f \in \mathcal{WPRV}$ *and the SLLN (8.38) holds a.s.*

1) *If* $X \in \mathbb{F}^{(\infty)}$ *a.s., then the SLLN (8.39) holds.*
2) *If* $X \in \mathbb{F}^\infty$ *a.s., then the SLLN (8.40) holds.*
3) *If* $X \in \mathbb{F}^\infty$ *a.s. and the function* f *is either nondecreasing or asymptotically equivalent to a nondecreasing function, then the SLLN (8.41) holds.*

Proof of Lemma 8.19. Since $f \in \mathcal{WPRV}$ and $\eta \in (0, \infty)$ a.s., the SLLN (8.38) implies that $X \in \mathcal{WPRV}$ a.s., whence we conclude that $X \in \mathcal{PRV}$ a.s., since the process X is measurable. Therefore, the first and second statement follow from relations (8.35)–(8.38) combined with statements 1) and 2) of Theorem 7.16. In turn, the third statement is a corollary of the second statement and relation (8.34), which completes the proof. □

The following result is a converse to Theorem 8.18. It follows from Theorem 7.53 and shows that the SLLN (8.42) implies the SLLN (8.38) under the assumption $f \in \mathcal{WPRV}$ (see condition (3.10))

Theorem 8.20 *Let $X \in \mathbb{F}^\infty$ a.s., $f \in \mathbb{F}^\infty$, $f \sim f_0 \in \mathbb{F}_{\text{ndec}}$, $\mathbb{AI}(f_0) \neq \varnothing$, $\psi_0 \in \mathbb{AI}(f_0)$, and $\eta \in (0, \infty)$ a.s. If $f \in \mathcal{WPRV}$ and*

$$\lim_{s\to\infty} \frac{M_X^+(s)}{\psi_0(s/\eta)} = \lim_{s\to\infty} \frac{L_X(s)}{\psi_0(s/\eta)} = 1 \qquad a.s., \tag{8.44}$$

then the SLLN (8.38) holds.

Note that the continuity of the process X is not assumed in Theorem 8.20.

Corollary 8.21 *Let $X \in \mathbb{C}^\infty$ a.s., $f \in \mathbb{F}^\infty$, $f \sim f_0 \in \mathbb{F}_{\text{ndec}}$, $\mathbb{AI}(f_0) \neq \varnothing$, $\psi_0 \in \mathbb{AI}(f_0)$ and $\eta \in (0, \infty)$ a.s. If $f \in \mathcal{WPOV}$, then the SLLNs (8.38) and (8.44) are equivalent.*

Proof of Corollary 8.21. This follows from Theorems 8.18 and 8.20. □

8.4.3 *The SLLN for Stochastic Processes and Their Corresponding Renewal Processes Under POV Normalizations*

If the normalizing function f belongs to $\mathcal{POV}$, then Corollary 8.21 can be extended by applying Theorem 7.57.

Theorem 8.22 *Let $f \in \mathcal{POV}$, $f^\sim \in \mathbb{AQI}(f)$, $X \in \mathbb{F}^{(\infty)}$ a.s. and $\eta \in (0, \infty)$ a.s.*

1) *If X is measurable, then the SLLNs (8.38) and (8.43) are equivalent.*
2) *If $X \in \mathbb{F}^\infty_{\text{inc}}$ a.s., then*

$$\lim_{t\to\infty} \frac{X(t)}{f(t)} = \eta \quad a.s. \qquad \Longleftrightarrow \qquad \lim_{s\to\infty} \frac{M_X^+(s)}{f^\sim(s/\eta)} = 1 \quad a.s. \tag{8.45}$$

3) *If $X \in \mathbb{C}^\infty$ a.s., then (8.45) holds.*
4) *Statements 2) and 3) hold for*

$$f^\sim \in \{f^{\leftarrowtail}, f^{\leftarrow}, f^{\rightarrow}, f^{\rightarrowtail}\},$$

where $f^{\leftarrowtail}, f^{\leftarrow}, f^{\rightarrow}, f^{\rightarrowtail}$ are the generalized inverse functions of f.

5) *If* $f \sim f_0 \in \mathbb{C}^{\infty}_{\mathrm{inc}}$*, then statements 1)–3) hold for* $f^{\sim} = f_0^{-1}$*, where* f_0^{-1} *is the inverse function of* f_0.

Relation (8.42) transforms as follows, if the random variable η in (8.38) attains values 0 or ∞.

Theorem 8.23 *Let* X *be an a.s. continuous stochastic process and* $f \in \mathcal{POV}$ *be a nonrandom function.*

(a) If

$$\lim_{t\to\infty} \frac{X(t)}{f(t)} = \infty \qquad a.s.,$$

then

$$\lim_{s\to\infty} \frac{M_X^+(s)}{f^{\sim}(s)} = \lim_{s\to\infty} \frac{L_X(s)}{f^{\sim}(s)} = \lim_{s\to\infty} \frac{T_X(s)}{f^{\sim}(s)} = 0 \qquad a.s.,$$

where $f^{\sim}$ *is an arbitrary asymptotically quasi-inverse function for* f.

(b) If $X \in \mathbb{C}^{\infty}$ *a.s. and*

$$\lim_{t\to\infty} \frac{X(t)}{f(t)} = 0 \qquad a.s.,$$

then

$$\lim_{s\to\infty} \frac{M_X^+(s)}{f^{\sim}(s)} = \lim_{s\to\infty} \frac{L_X(s)}{f^{\sim}(s)} = \lim_{s\to\infty} \frac{T_X(s)}{f^{\sim}(s)} = \infty \qquad a.s.,$$

where $f^{\sim}$ *is an arbitrary asymptotically quasi-inverse function for* f.

(c) Statements (a) and (b) hold for

$$f^{\sim} \in \{f^{\leftarrowtail}, f^{\leftarrow}, f^{\rightarrow}, f^{\rightarrowtail}\},$$

where $f^{\leftarrowtail}, f^{\leftarrow}, f^{\rightarrow}, f^{\rightarrowtail}$ *are the generalized inverse functions for* f.

Proof of Theorem 8.23. This follows from Theorem 7.63. □

8.4.4 The SLLN for Stochastic Processes and Their Corresponding Renewal Processes Under RV Normalizations

Corollary 7.46 implies that Theorem 8.22 can be stated more precisely if the normalizations are RV-functions with positive indices.

Theorem 8.24 *Let f be an RV-function with positive index ρ, $f^{\sim} \in \mathbb{AQI}(f)$, $X \in \mathbb{F}^{(\infty)}$ a.s., and $\eta \in (0, \infty)$ a.s.*

1) If the process X is measurable, then

$$\lim_{t\to\infty} \frac{X(t)}{f(t)} = \eta \qquad a.s.$$

if and only if

$$\lim_{s\to\infty} \frac{M_X^+(s)}{f^{\sim}(s)} = \lim_{s\to\infty} \frac{L_X(s)}{f^{\sim}(s)} = \lim_{s\to\infty} \frac{T_X(s)}{f^{\sim}(s)} = \left(\frac{1}{\eta}\right)^{\frac{1}{\rho}} \qquad a.s.$$

2) If $X \in \mathbb{F}_{\mathrm{inc}}^{\infty}$ a.s., then

$$\lim_{t\to\infty} \frac{X(t)}{f(t)} = \eta \quad a.s. \qquad \Longleftrightarrow \qquad \lim_{s\to\infty} \frac{M_X^+(s)}{f^{\sim}(s)} = \left(\frac{1}{\eta}\right)^{\frac{1}{\rho}} \quad a.s. \tag{8.46}$$

3) If $X \in \mathbb{C}^{\infty}$ a.s., then (8.46) holds.
4) Statements 1)–3) hold for

$$f^{\sim} \in \{f^{\leftarrowtail}, f^{\leftarrow}, f^{\rightarrow}, f^{\rightarrowtail}\},$$

where $f^{\leftarrowtail}, f^{\leftarrow}, f^{\rightarrow}, f^{\rightarrowtail}$ are generalized inverse functions for f.
5) If $f \sim f_0 \in \mathbb{C}_{\mathrm{inc}}^{\infty}$, then statements 1)–3) hold for $f^{\sim} = f_0^{-1}$, where f_0^{-1} is the inverse function of f_0.

8.5 Further Examples

Below we consider more examples of applications of the results obtained in the preceding section. The first of these examples is related to renewal processes constructed from stochastic processes with independent and stationary increments.

Example 8.25 Let $(X(t), t \geq 0)$ be a measurable, separable, real-valued stochastic process with independent and stationary increments and let $\mathsf{E}X(1) = a > 0$. Then

$$\frac{X(t)}{t} \to a \qquad \text{a.s.},$$

that is, condition (8.45) holds with $f(t) = t$ and $\eta = a > 0$. This together with Theorem 8.24 implies that

$$\lim_{s\to\infty} \frac{M_X^+(s)}{f^\sim(s)} = \lim_{s\to\infty} \frac{L_X(s)}{f^\sim(s)} = \lim_{s\to\infty} \frac{T_X(s)}{f^\sim(s)} = \frac{1}{a} \qquad \text{a.s.}$$

In the following two examples, the asymptotic behavior of integrals of stationary processes is considered.

Example 8.26 Let $(Y(t), t \in \mathbf{R})$ be a measurable, real-valued, strictly stationary stochastic process such that

$$\mathsf{E}|Y(0)|^{1+\varepsilon} < \infty \tag{8.47}$$

for some $\varepsilon > 0$. Put

$$X(t) = \int_0^t Y(u)du, \qquad t \ge 0.$$

Note that the process X is continuous almost surely under (8.47). In addition, let

$$\mathsf{E}\left[Y(0)|\mathscr{S}_\infty\right] = \eta > 0 \qquad \text{a.s.},$$

where $\mathscr{S}_\infty$ is the σ-algebra of random events invariant with respect to shifts. Then condition (8.45) follows from the ergodic theorem and thus Theorem 8.24 implies that

$$\lim_{s\to\infty} \frac{M_X^+(s)}{f^\sim(s)} = \lim_{s\to\infty} \frac{L_X(s)}{f^\sim(s)} = \lim_{s\to\infty} \frac{T_X(s)}{f^\sim(s)} = \frac{1}{a} \qquad \text{a.s.}$$

Note that $\mathsf{E}|\eta| \le \mathsf{E}|Y(0)| < \infty$, and hence $\mathsf{P}\{|\eta| = \infty\} = 0$. In addition, $\eta = \mathsf{E}Y(0)$ by the ergodicity of the process Y.

Example 8.27 Let $(Y(t), t \in \mathbf{R})$ be a measurable, real-valued, second order stationary stochastic process and let $\mathsf{E}Y(0) = a > 0$. Assume that the covariance function B of the process is such that

$$\text{the integral} \qquad \int_3^\infty \frac{B(t)}{t\log t} \log\log t \, dt \qquad \text{converges.} \tag{8.48}$$

As in Example 8.25, consider the process

$$X(t) = \int_0^t Y(u)du, \qquad t \ge 0,$$

which is continuous almost surely. It is shown in Gaposhkin [146] that (8.48) implies the SLLN

$$X(t)/t \to a \qquad \text{a.s.}$$

Thus condition (8.45) holds with $f(t) = t$ and $\eta = a > 0$, whence we conclude, by Theorem 8.24, that

$$\lim_{s\to\infty} \frac{M_X^+(s)}{f^\sim(s)} = \lim_{s\to\infty} \frac{L_X(s)}{f^\sim(s)} = \lim_{s\to\infty} \frac{T_X(s)}{f^\sim(s)} = \frac{1}{a} \qquad \text{a.s.}$$

8.5.1 *Generalized Renewal Processes Constructed from Compound Counting Processes*

Let $(\nu(t), t \geq 0)$ be a random counting process taking values in the set of nonnegative integers and having almost surely right-continuous, nondecreasing trajectories. Assume that

$$\nu(0) = 0, \qquad \lim_{t\to\infty} \nu(t) = \infty \qquad \text{a.s.}$$

and

$$\lim_{t\to\infty} \frac{\nu(t)}{f(t)} = 1 \qquad \text{a.s.}, \tag{8.49}$$

where $f \in \mathbb{F}^\infty$.

In addition, let $\{\xi_k\} = \{\xi_k\}_{k\geq 1}$ be a sequence of independent, identically distributed random variables with mean

$$\mu = \mathsf{E}\xi_1 > 0. \tag{8.50}$$

We assume that the sequence $\{\xi_k\}_{k\geq 1}$ is independent of the process $\nu(\cdot)$.

Making use of $\nu(\cdot)$ and $\{\xi_k\}$, we construct the *compound counting process*

$$X(t) = \sum_{k=1}^{\nu(t)} \xi_k, \qquad t \geq 0, \tag{8.51}$$

where we set $X(t) = 0$ if $\nu(t) = 0$.

Relation (8.49) and Kolmogorov's strong law of large numbers imply that

$$\lim_{t\to\infty} \frac{X(t)}{f(t)} = \lim_{t\to\infty} \frac{X(t)}{\nu(t)} \cdot \lim_{t\to\infty} \frac{\nu(t)}{f(t)} = \mu \qquad \text{a.s.} \tag{8.52}$$

Note that the trajectories of the process X are almost surely right-continuous step-functions. Thus the trajectories are measurable and hence X is a measurable, separable, real-valued process. In addition, relations (8.50) and (8.52) imply that $X \in \mathbb{F}^\infty$ almost surely.

Consider the three generalized renewal processes M_X^+, L_X, and T_X as above. For the case where the function f in (8.49) and (8.52) belongs to $\mathcal{WPRV}$, Lemma 8.19 can be rewritten as follows.

Proposition 8.28 *Let X be the stochastic process as defined in (8.51). Assume that (8.49) holds and $f \in \mathcal{WPRV}$. Then:*

1) The SLLN holds, i.e.

$$\lim_{s\to\infty} \frac{f(M_X^+(s))}{s} = \lim_{s\to\infty} \frac{f(L_X(s))}{s} = \frac{1}{\mu} \qquad a.s.$$

2) If f is either a nondecreasing function or asymptotically equivalent to a nondecreasing function, then

$$\lim_{s\to\infty} \frac{f(M_X^+(s))}{s} = \lim_{s\to\infty} \frac{f(L_X(s))}{s} = \lim_{s\to\infty} \frac{f(T_X(s))}{s} = \frac{1}{\mu} \qquad a.s.$$

If the function f in the SLLNs (8.49) and (8.52) belongs to $\mathcal{POV}$, then, in view of Theorem 8.22, the SLLN (8.52) is equivalent to

$$\lim_{s\to\infty} \frac{M_X^+(s)}{f^\sim(s/\mu)} = \lim_{s\to\infty} \frac{L_X(s)}{f^\sim(s/\mu)} = \lim_{s\to\infty} \frac{T_X(s)}{f^\sim(s/\mu)} = 1 \qquad \text{a.s.,} \tag{8.53}$$

where $f^\sim$ is an asymptotically quasi-inverse function for f. As a result, relations (8.50)–(8.52) and Theorems 8.22 and 8.24 yield the following proposition.

Proposition 8.29 *Let X be the stochastic process as defined in (8.51) and assume $f \in \mathcal{POV}$. Then:*

1) The SLLN (8.53) holds for any asymptotically quasi-inverse function $f^\sim$ for f.
2) If f is an RV-function with positive index ρ, then

$$\lim_{s\to\infty} \frac{M_X^+(s)}{f^\sim(s)} = \lim_{s\to\infty} \frac{L_X(s)}{f^\sim(s)} = \lim_{s\to\infty} \frac{T_X(s)}{f^\sim(s)} = \left(\frac{1}{\mu}\right)^{\frac{1}{\rho}} \qquad a.s.$$

for any asymptotically quasi-inverse function $f^\sim$ for f.
3) Statements 1) and 3) hold for

$$f^\sim \in \{f^{\leftarrowtail}, f^{\leftarrow}, f^{\rightarrow}, f^{\rightarrowtail}\},$$

where $f^{\leftarrowtail}$, $f^{\leftarrow}$, $f^{\rightarrow}$, $f^{\rightarrowtail}$ are the generalized inverse functions for f.

4) *If* $f \sim f_0 \in \mathbb{C}^{\infty}_{\text{inc}}$, *then statements 1) and 2) hold for* $f^{\sim} = f_0^{-1}$, *where* f_0^{-1} *is the inverse function of* f_0.

8.5.2 Generalized Renewal Processes Constructed from Compound Poisson Processes

Let $(\nu(t), t \geq 0)$ be a (nonhomogeneous) Poisson process with nonnegative and measurable intensity function $(\lambda(t), t \geq 0)$ satisfying

$$\int_0^t \lambda(s)\, ds < \infty, \qquad \text{for all } t > 0, \qquad \text{and} \qquad \int_0^\infty \lambda(s)\, ds = \infty. \tag{8.54}$$

Put

$$\Lambda(t) = \int_0^t \lambda(s) ds, \qquad t > 0, \tag{8.55}$$

and note that $(\nu(t), t \geq 0)$ is a stochastic process with independent increments such that the random variable $\nu(t_2) - \nu(t_1)$ is Poissonian with parameter $\Lambda(t_2) - \Lambda(t_1)$ for $0 \leq t_1 < t_2$. The values of this process are nonnegative integers and its sample paths are almost surely right-continuous, nondecreasing step-functions. If $\lambda(s) > 0$, $s \geq 0$, then $\Lambda \in \mathbb{C}^{\infty}_{\text{inc}}$.

Let $\{\tau_n\}_{n \geq 0}$ be a sequence such that $\tau_0 = 0$ and $\Lambda(\tau_n) = n$, $n \geq 1$. If $n(t) = \max\{n : \tau_n \leq t\}$, $t > 0$, then the Kolmogorov strong law of large numbers implies

$$\lim_{t\to\infty} \frac{\nu(\tau_{n(t)})}{\Lambda(\tau_{n(t)})} = \lim_{t\to\infty} \frac{\nu(\tau_{n(t)})}{n(t)} = 1 \qquad \text{a.s.}$$

Since

$$n(t) = \Lambda(\tau_{n(t)}) \leq \Lambda(t) \leq \Lambda(\tau_{n(t)+1}) = n(t) + 1,$$

whence

$$\frac{n(t)}{n(t)+1} \cdot \frac{\nu(\tau_{n(t)})}{n(t)} \leq \frac{\nu(t)}{\Lambda(t)} \leq \frac{n(t)+1}{n(t)} \cdot \frac{\nu(\tau_{n(t)+1})}{n(t)+1},$$

for all $t > 0$, and $\lim_{t\to\infty} n(t) = \infty$, we obtain

$$\lim_{t\to\infty} \frac{\nu(t)}{\Lambda(t)} = 1 \qquad \text{a.s.}, \tag{8.56}$$

that is, condition (8.49) holds with $f = \Lambda$. Applying Propositions 8.28 and 8.29 together with relation (8.56), we obtain the following result.

Corollary 8.30 *Let X be a compound process as in (8.51), where $(\nu(t), t \geq 0)$ is a Poisson process with nonnegative, measurable intensity function $(\lambda(s), s \geq 0)$ satisfying condition (8.54). Then:*

1) If $\Lambda \in \mathcal{PRV}$, then

$$\lim_{s\to\infty} \frac{\Lambda(M_X^+(s))}{s} = \lim_{s\to\infty} \frac{\Lambda(L_X(s))}{s} = \lim_{s\to\infty} \frac{\Lambda(T_X(s))}{s} = \frac{1}{\mu} \qquad a.s.$$

2) If $\Lambda \in \mathcal{POV}$, then

$$\lim_{s\to\infty} \frac{M_X^+(s)}{\Lambda^\sim(s/\mu)} = \lim_{s\to\infty} \frac{L_X(s)}{\Lambda^\sim(s/\mu)} = \lim_{s\to\infty} \frac{T_X(s)}{\Lambda^\sim(s/\mu)} = 1 \qquad a.s.,$$

where $\Lambda^\sim$ is an arbitrary asymptotically quasi-inverse function for Λ.

3) If $\lambda(s) > 0$, $s \geq 0$, and $\Lambda \in \mathcal{POV}$, then

$$\lim_{s\to\infty} \frac{M_X^+(s)}{\Lambda^{-1}(s/\mu)} = \lim_{s\to\infty} \frac{L_X(s)}{\Lambda^{-1}(s/\mu)} = \lim_{s\to\infty} \frac{T_X(s)}{\Lambda^{-1}(s/\mu)} = 1 \qquad a.s.,$$

where Λ^{-1} is the inverse function of Λ.

4) If Λ is an RV-function with positive index ρ, then

$$\lim_{s\to\infty} \frac{M_X^+(s)}{\Lambda^\sim(s)} = \lim_{s\to\infty} \frac{L_X(s)}{\Lambda^\sim(s)} = \lim_{s\to\infty} \frac{T_X(s)}{\Lambda^\sim(s)} = \left(\frac{1}{\mu}\right)^{\frac{1}{\rho}} \qquad a.s.,$$

where $\Lambda^\sim$ is an arbitrary asymptotically quasi-inverse function for Λ.

5) If $\lambda(s) > 0$, $s \geq 0$, and Λ is an RV-function with positive index ρ, then

$$\lim_{s\to\infty} \frac{M_X^+(s)}{\Lambda^{-1}(s)} = \lim_{s\to\infty} \frac{L_X(s)}{\Lambda^{-1}(s)} = \lim_{s\to\infty} \frac{T_X(s)}{\Lambda^{-1}(s)} = \left(\frac{1}{\mu}\right)^{\frac{1}{\rho}} \qquad a.s.,$$

where Λ^{-1} is the inverse function of Λ.

6) Statements 2) and 4) hold for any $\Lambda^\sim \in \{M_\Lambda^+, L_\Lambda, T_\Lambda\}$.

Remark 8.31 The function Λ is absolutely continuous with density λ. Recall that conditions for absolutely continuous functions to belong to certain classes of functions, in particular to $\mathcal{POV}$, have been studied in Chap. 4. In view of Theorem 4.9 and under condition (8.54), $\Lambda \in \mathcal{POV}$ if and only if

$$\limsup_{c\to 1} \limsup_{t\to\infty} \int_t^{ct} \frac{\lambda(u)}{\Lambda(u)} du = 0$$

and

$$\liminf_{t\to\infty} \int_t^{ct} \frac{\lambda(u)}{\Lambda(u)} du > 0$$

for all $c > 1$.

Below some sufficient conditions are presented under which $\Lambda \in \mathcal{POV}$ (recall Proposition 4.54, Corollary 4.55, and Proposition 4.14).

Proposition 8.32 *Let the intensity function λ be measurable, nonnegative, and let condition (8.54) hold. Additionally, let at least one of the following conditions be satisfied:*

1) $0 < \liminf_{t\to\infty} \frac{t\lambda(t)}{\Lambda(t)}$ *and* $\limsup_{t\to\infty} \frac{t\lambda(t)}{\Lambda(t)} < \infty$;
2) there exists an $\alpha > -1$ such that

$$0 < \liminf_{t\to\infty} \frac{\lambda(t)}{t^\alpha} \qquad \textit{and} \qquad \limsup_{t\to\infty} \frac{\lambda(t)}{t^\alpha} < \infty;$$

3) $\liminf_{c\downarrow 1} \lambda^*(c) \le 1$ *and* $c\lambda_*(c) > 1$ *for all* $c > 1$;
4) λ is an RV-function with index $\rho \in (-1, \infty)$;
5) the function λ satisfies condition (8.54) and is of the form

$$\lambda = d_1 \cdot \lambda_1 + \cdots + d_n \cdot \lambda_n,$$

where λ_k, $k = 1, \ldots, n$, are nonnegative, measurable POV-functions and where d_k, $k = 1, \ldots, n$, are almost constant functions (see Definition 4.11).

Then $\Lambda \in \mathcal{POV}$ and statements 2), 3), and 6) of Corollary 8.30 hold.

8.5.3 Generalized Renewal Processes Constructed from Compound Poisson Processes with Piecewise Constant Intensity Functions

Let $(\nu(t), t \ge 0)$ be a Poisson process with positive, *piecewise constant intensity function* $(\lambda(t), t \ge 0)$, that is,

$$\lambda(t) = \sum_{n=0}^{\infty} \lambda_{n+1} \mathbb{I}\{t \in [t_n, t_{n+1})\}, \qquad t \ge 0, \tag{8.57}$$

where

$$\lambda_n > 0, \ n \geq 1; \qquad t_0 = 0; \ \Delta t_n = t_n - t_{n-1} > 0, \ n \geq 0; \qquad \lim_{n\to\infty} t_n = \infty, \tag{8.58}$$

and

$$\sum_{n=0}^{\infty} \lambda_{n+1} \Delta t_n = \infty. \tag{8.59}$$

Relations (8.57)–(8.59) imply that Λ (see (8.55)) is a positive, increasing, piecewise linear function, i.e.

$$\Lambda(t) = \Lambda_n + \lambda_{n+1}(t - t_n), \qquad t \in [t_n, t_{n+1}), \qquad n \geq 0, \tag{8.60}$$

where

$$\Lambda_0 = 0 \qquad \text{and} \qquad \Lambda_{n+1} = \sum_{k=0}^{n} \lambda_{k+1} \Delta t_k, \qquad n \geq 0,$$

and that condition (8.54) is satisfied.

Using the properties of piecewise linear functions established in Sect. 4.11, we obtain conditions for the sequences $\{\lambda_n\}$ and $\{t_n\}$ under which $\Lambda \in \mathcal{POV}$.

Example 8.33 If

$$\lim_{n\to\infty} \frac{t_{n+1}}{t_n} = 1$$

and

$$0 < \liminf_{n\to\infty} \frac{t_n \lambda_{n+1}}{\Lambda_n}, \qquad \limsup_{n\to\infty} \frac{t_n \lambda_{n+1}}{\Lambda_n} < \infty,$$

then Theorem 4.67 and Proposition 8.32 implies that $\Lambda \in \mathcal{POV}$ and the third statement of Corollary 8.30 holds.

8.6 Comments

This chapter is based on Buldygin et al. [57, 59, 62, 65, 70]. Note that the notion of "generalized renewal processes" is attributed in Borovkov and Borovkov [48] exclusively to the case of (8.4), while (8.2) and (8.3) are not studied there in the framework of renewal theory.

Section 8.1 The soujourn times and first exit times are treated as representatives of generalized renewal processes in this chapter. These functionals have been studied by other methods for many classes of stochastic processes. Below are some examples. Sojourn times is a traditional topic in the theory of Gaussian stochastic processes (see, for example, Berman [30]). Occupation times (which is another name for sojourn times) for Markov processes are studied by Bingham [37]. Another well-studied topic for Gaussian stochastic processes is the limit behavior of extreme values (see, e.g., the survey by Leadbetter and Rootzen [248]). For us, these two functionals are asymptotically quasi-inverse functions to each other. Berman [31] is the first to relate a sojourn time of a Gaussian stochastic process $\{X(t)\}$ and its extreme values $\{\max_{s\le t} X(s)\}$ (see Berman [33]).

Rubino and Sericola [310] discuss sojourn times for Markov chains and their applications. Novikov and Kordzakhia [286] consider the first exit time for a first-order autogressive sequence.

The quality of a stochastic network (computer and telecommunications network, internet, logistics and supply-chain network, parallel simulation and distributed processing system etc.) is determined by the duration of travel and sojourn times in it, such as the time for a customer to travel from one server to another or the amount of time it takes for a customer to visit a certain set of nodes. Issues concerning soujourn times become very important (even critical) for many situations in practice. More about this topic can be found in Serfozo [326].

Sacerdote and Smith [311] establish almost sure convergence of the first exit times for diffusion processes.

Section 8.2 Iglehart and Whitt [188] write that "... an associated counting process ... is essentially an inverse to the partial sum process ... for nonnegative random variables, not necessarily independent or identically distributed". Using this idea, they prove the equivalence between functional central limit theorems for both objects. Gut et al. [160] establish the equivalence of the strong law of large numbers, rate of convergence, and law of the iterated logarithm for independent, identically distributed random variables (see also Frolov et al. [138] for related results). Glynn and Whitt [150] study the strong law of large numbers for counting processes constructed from nondecreasing sequences of random variables. Glynn and Whitt [150] clearly state that these two objects are inverses each to other. In general, the idea of duality is widely used in Gut [159].

An application of the method of dual objects is used by Doukhan et al. [112, 113] to relate the almost sure asymptotic behavior of the number of records until time n and the moment of the nth record if the underlying sequence of random variables is an F^{α}-scheme.

Section 8.4 Condition (8.49) is the starting point of our considerations in Sect. 8.5.1. In turn, one may ask about sufficient conditions for (8.49). This question can be answered by using the same ideas as those presented in Chap. 8. Subordinators, i.e., increasing processes possessing independent and homogeneous increments, are the simplest examples for $\nu(t)$ in Sect. 8.5.1 (see, e.g., Bertoin [35]). In the case of subordinators, one may get a sharper result for compound renewal processes, like the law of the iterated logarithm (see, for example, Bertoin and Caballero [36]). Fristedt and Pruitt [137] address the problem on the duality between subordinators and their corresponding first passage times.

General Lévy processes, whose trajectories tend to infinity almost surely, are also reasonable models for $\nu(t)$, for which one could easily provide tractable conditions for (8.49) (see, e.g., Sato [318, Chapter 7]).

Some other models of interest for compound Poisson processes are given by *doubly stochastic Poisson processes* (see Grandell [152]) and *compound Poisson process with a Poisson subordinator* (see Di Crescenzo et al. [102]).

Chapter 9
Asymptotic Behavior of Solutions of Stochastic Differential Equations

9.1 Introduction

It is common knowledge that the value $X(t)$ at time $t > 0$ of a bank account earning interests at a fixed rate $r > 0$ can be written in differential form as

$$dX(t) = rX(t)\,dt, \qquad t > 0. \tag{9.1}$$

However, the financial world is more complex and allows interest rates to be functions of time. Moreover, more realistic models for interest rates involve random rates $r = r(t)$ rather than constant r as in (9.1).

Vašíček [358] introduced the so-called *one-factor short rate model of interest rates* that can be written as an Itô stochastic differential equation

$$dX(t) = a(b - X(t))\,dt + \sigma\,dW(t),$$

where W is a Wiener process, while a, b, and σ have a financial meaning as "*speed of reversion*", "*long term mean level*", and "*volatility*", respectively. (Here we do not aim at a full explanation of the financial meaning of these words; an interested reader may consult, e.g., the introductory text on stochastic interest rate models by Privault [295].) A drawback of the Vašíček model is that it allows for negative values of $X(t)$. The Cox–Ingersoll–Ross [95] equation solves the problem of positivity of $X(t)$ encountered with the Vašíček model by the use of the following nonlinear equation

$$dX(t) = a(b - X(t))\,dt + \sigma X^{1/2}(t)\,dW(t).$$

There are plenty of stochastic models for interest rates discussed in the literature. We mention only one of them, the *affine model*. A particular case of affine models,

V. V. Buldygin et al., *Pseudo-Regularly Varying Functions and Generalized Renewal Processes*, Probability Theory and Stochastic Modelling 91, https://doi.org/10.1007/978-3-319-99537-3_9

important to motivate the results of this chapter, is given by

$$dX(t) = a(t)X(t)\,dt + \sigma(t)X^{1/2}(t)\,dW(t). \tag{9.2}$$

Another class of *constant elasticity of variance models* is concerned with the studies of stochastic differential equations like

$$dX(t) = aX^{\alpha}(t)\,dt + bX^{\beta}(t)\,dW(t).$$

Vašíček, Cox–Ingersoll–Ross, and constant elasticity of variance equations are particular cases of the following stochastic differential equation (SDE)

$$dX(t) = g(X(t))\,dt + \sigma(X(t))\,dW(t), \qquad t \geq 0, \qquad X(0) \equiv b > 0, \tag{9.3}$$

called an *autonomous* SDE, where W is a standard Wiener process and g, σ are positive, continuous functions such that there exists a unique continuous solution of equation (9.3). In turn, the affine equation (9.2) is a particular case of equation (9.16) below.

For all stochastic models of interest rates, the behavior of their solutions is quite irregular, which perhaps reflects the random nature of a solution. However, one may want to have a simpler (possibly nonrandom) approximation of the solution that explains the main trend of the fluctuations. There can be many meanings of the word "approximation" in this situation. For example, one could try to find a deterministic function ϕ for which

$$|X(t) - \phi(t)| \quad \text{is small} \tag{9.4}$$

or

$$\frac{X(t)}{\phi(t)} \quad \text{is close to 1.} \tag{9.5}$$

It is not only finance that motivates the study of approximations of solutions of the SDEs described above. In agronomy, for example, when modelling growth of animals, the so-called growth curves play a crucial role. They are described by solutions of the SDE

$$dX(t) = ae^{-bt}X(t)\,dt + \sigma X(t)\,dW(t), \tag{9.6}$$

being a representative of equation (9.16) studied in this chapter. (The deterministic analog of (9.6) is called the *Gompertz equation*.)

Many other examples of stochastic differential equations can be found in engineering, physics, chemistry, and biology as well as in mathematical problems of stochastic filtration, optimal stopping, control etc.

This chapter aims at finding nonrandom approximations as in (9.5) (a precise definition is given below) of solutions of a general class of stochastic differential equations that includes all equations mentioned above. We follow the setting by Gihman and Skorohod [149], however the results below are more general.

9.1.1 Gihman and Skorohod's Setting and Approach

Consider an autonomous stochastic differential equation as in (9.3). It is further assumed that

$$\mathsf{P}\left\{\lim_{t\to\infty} X(t) = \infty\right\} > 0$$

and that X does not reach infinity in finite time, i.e., that no explosion occurs. Throughout this chapter, a continuous solution means that the paths of the process X are continuous.

Along with the SDE (9.3) consider the corresponding autonomous ordinary differential equation (ODE)

$$d\mu(t) = g(\mu(t))\,dt, \qquad t \geq 0, \qquad \mu(0) = b > 0, \tag{9.7}$$

which coincides with the SDE (9.3), if $\sigma(u) \equiv 0$. Suppose that a unique solution μ of equation (9.7) exists and tends to ∞ as $t \to \infty$. Moreover, we assume that μ also does not reach infinity in finite time. We have shown in Sect. 7.9 that these conditions are satisfied for the positive, continuous function g, if

$$\int_b^\infty \frac{du}{g(u)} = \infty. \tag{9.8}$$

Gihman and Skorohod [149] found sufficient conditions for the almost sure asymptotic equivalence of solutions of the SDE (9.3) and the ODE (9.7) on the set $\{\lim_{t\to\infty} X(t) = \infty\}$, that is, for

$$\lim_{t\to\infty} \frac{X(t)}{\mu(t)} = 1 \qquad \text{a.s. on the set} \qquad \left\{\lim_{t\to\infty} X(t) = \infty\right\}. \tag{9.9}$$

If

$$\lim_{t\to\infty} X(t) = \infty \qquad \text{a.s.,} \tag{9.10}$$

then (9.9) can be rewritten as

$$\lim_{t\to\infty} \frac{X(t)}{\mu(t)} = 1 \qquad \text{a.s.} \tag{9.11}$$

Since (9.9) and (9.11) are satisfied on the set $\{\lim_{t\to\infty} X(t) = \infty\}$, they describe the rate of growth to infinity of a solution of the SDE (9.3).

The method of proof of (9.9) and (9.11) proposed in Gihman and Skorohod [149] consists of two steps. The first step is to consider the stochastic process

$$Y(t) = G_b(X(t)), \qquad t \geq 0,$$

where

$$G_b(s) = \int_b^s \frac{du}{g(u)}, \qquad s \geq b,$$

and to prove

$$\lim_{t\to\infty} \frac{G_b(X(t))}{t} = 1 \quad \text{a.s. on the set} \quad \left\{\lim_{t\to\infty} X(t) = \infty\right\} \tag{9.12}$$

under certain assumptions.

The function G_b has already appeared in Sect. 7.9 (see (7.103)). If g is continuous and positive, then G_b is the inverse function of μ, that is,

$$\mu(G_b(s)) = s, \qquad s \geq b. \tag{9.13}$$

The second step is to prove that relation (9.12) implies relation (9.9) under some additional conditions. In particular, it is assumed in Gihman and Skorohod [149] that

$$\lim_{\varepsilon\to 0} \sup_{z>C} \sup_{|\frac{z}{u}-1|\leq\varepsilon} \left|\frac{\mu(z)}{\mu(u)} - 1\right| = 0 \tag{9.14}$$

for some $C > 0$.

Remark 3.22 explains that the function μ belongs to the class $\mathcal{PRV}$ *of pseudo-regularly varying functions* if equality (9.14) holds. Thus, by Theorem 3.42 (see Remark 3.44), μ preserves the asymptotic equivalence of functions tending to infinity. On using (9.13) together with (9.12), we get

$$\lim_{t\to\infty} \frac{X(t)}{\mu(t)} = \lim_{t\to\infty} \frac{\mu(G_b(X(t)))}{\mu(t)} = 1 \quad \text{a.s. on the set} \quad \left\{\lim_{t\to\infty} X(t) = \infty\right\},$$

and hence (9.9) holds.

The above reasoning differs from the proof in Gihman and Skorohod [149]. It shows that the implication (9.12) $\Longrightarrow$ (9.9) is based on the property that $\mu \in \mathcal{PRV}$.

9.1.2 Organization of the Current Chapter

Condition (9.14) is given in terms of the function μ, that is, in terms of a solution of the ODE (9.7). There are also various cases where conditions expressed in terms of the underlying functions g and G_b look more suitable. The corresponding conditions can easily be obtained with the help of Corollary 7.33, which asserts that μ is a PRV-function if and only if G_b is an SQI-function, that is, if condition (3.14) holds.

Properties of functions belonging to the classes $\mathcal{PRV}$ and $\mathcal{SQI}$ have been studied in Chaps. 3 and 4. The results obtained there can effectively be used to find assumptions for (9.9) and (9.11) to hold, that is, for the asymptotic equivalence of solutions of an autonomous SDE and the corresponding ODE.

The problem of studying the asymptotic equivalence of solutions of SDEs and ODEs discussed above can be generalized in various ways. One possibility is to study the so-called $\varphi_{1,2}$-asymptotic equivalence, that is,

$$\lim_{t\to\infty} \frac{\varphi_1(X(t))}{\varphi_2(\mu(t))} = 1 \qquad \text{a.s. on the set} \qquad \left\{\lim_{t\to\infty} X(t) = \infty\right\}, \tag{9.15}$$

where φ_1 and φ_2 are some nonrandom, positive, continuous functions.

It is also reasonable to study the asymptotic equivalence (or φ-asymptotic equivalence) for solutions of two different ODEs or SDEs. The asymptotic equivalence of two stochastic differential equations is of special interest if one of them is a numerical approximation of the other. This kind of result may apply to various approximation schemes.

A more general stochastic differential equation than (9.3) is written as follows

$$\begin{gathered} dX(t) = g(X(t))\,\psi(t)\,dt + \sigma(X(t))\,\theta(t)\,dW(t), \qquad t \ge 0, \\ X(0) \equiv b > 0, \end{gathered} \tag{9.16}$$

having a corresponding ODE with separable variables, i.e.

$$d\mu(t) = g(\mu(t))\,\psi(t)\,dt, \qquad t \ge 0, \qquad \mu(0) = b > 0.$$

These equations differ from their autonomous analogs in view of the presence of time dependent factors, ψ and θ, in both terms on the right-hand sides of the equations. We obtain conditions for the asymptotic equivalence for such type of equations, as well.

All these problems are investigated in this chapter. When studying the asymptotic equivalence of solutions of ordinary differential equations and stochastic differential equations, we apply the results about PRV-, SQI-, and POV-functions and about their asymptotically quasi-inverse functions obtained in the previous chapters.

Chapter 9 is organized as follows. The asymptotic equivalence of solutions of an SDE and that of its corresponding ODE is studied in Sect. 9.2.

The results of Sect. 9.2 are applied in Sect. 9.3 to study the asymptotic behavior of generalized renewal processes constructed from solutions of SDEs.

Some corollaries of the results of Sect. 9.2 for autonomous SDEs are separately discussed in Sect. 9.4.

In Sect. 9.5, we consider the φ-asymptotic equivalence of solutions of autonomous SDEs and ODEs.

Section 9.6 is devoted to the $\varphi_{1,2}$-asymptotic equivalence of solutions of two autonomous ODEs, while Sect. 9.7 deals with the $\varphi_{1,2}$-asymptotic equivalence of solutions of autonomous SDEs and ODEs.

Conditions for the $\varphi_{1,2}$-asymptotic equivalence of solutions of two autonomous SDEs are given in Sect. 9.8.

9.2 Order of Growth of Solutions of Stochastic Differential Equations

In this section, the asymptotic behavior as $t \to \infty$ of a solution $\tilde{X} = (\tilde{X}(t), t \geq 0)$ of the Cauchy problem for the SDE

$$d\tilde{X}(t) = g(\tilde{X}(t))\, \psi(t)\, dt + \sigma(\tilde{X}(t))\, \theta(t)\, dW(t), \qquad t \geq 0, \tag{9.17}$$
$$\tilde{X}(0) \equiv b > 0,$$

is studied, where W is a standard Wiener process, defined on a complete probability space $\{\Omega, \mathfrak{F}, \mathsf{P}\}$, and b is a nonrandom, positive constant (initial condition). The coefficients of the drift and diffusion in equation (9.17) depend on both the phase and time variable. This equation coincides with the SDE (9.3) if $\psi(t) \equiv \theta(t) \equiv 1$. To indicate the dependence of a solution $\tilde{X}$ on the initial condition, we sometimes write $\tilde{X}_b$ instead of $\tilde{X}$.

We also assume that the real-valued functions

$$g = (g(u),\ u \in (-\infty, \infty)), \qquad \sigma = (\sigma(u),\ u \in (-\infty, \infty)),$$

$$\psi = (\psi(t),\ t \geq 0), \qquad \theta = (\theta(t),\ t \geq 0)$$

are continuous and, in addition, that g, σ, and θ are positive. Note that, in general, the function ψ may attain values of both signs. Further, we assume that all these functions are such that, on $\{\Omega, \mathfrak{F}, \mathsf{P}\}$, there exists a continuous solution $\tilde{X}$ of the SDE (9.17) with

$$\mathsf{P}\left\{\lim_{t\to\infty} \tilde{X}(t) = \infty\right\} > 0,$$

and that $\tilde{X}$ does not reach infinity in finite time.

Along with the Cauchy problem for the SDE (9.17) the Cauchy problem for the ODE with the same initial condition is studied, i.e.

$$d\tilde{\mu}(t) = g(\tilde{\mu}(t))\,\psi(t)\,dt, \qquad t \geq 0, \qquad \tilde{\mu}(0) = b > 0. \tag{9.18}$$

The ODE (9.18) coincides with the SDE (9.17), if $\sigma(x) \equiv 0$. We also consider the Cauchy problem for the autonomous ODE

$$d\mu(t) = g(\mu(t))\,dt, \qquad t \geq 0, \qquad \mu(0) \equiv b > 0, \tag{9.19}$$

coinciding with (9.18), if $\psi(t) \equiv 1$.

To indicate the dependence of solutions $\tilde{\mu}$ and μ on the initial condition we sometimes denote them by $\tilde{\mu}_b$ and μ_b, respectively.

Recall that problems (9.19) and (9.18) appeared in Sect. 7.9, where the asymptotic stability of solutions of these problems with respect to the initial condition has been studied (see Theorem 7.108 and 7.113).

The aim of this section is to find conditions under which a solution of the SDE (9.17) is asymptotically equivalent to a solution of the ODE (9.18) on the set $\left\{\lim_{t\to\infty} \tilde{X}(t) = \infty\right\}$, that is,

$$\lim_{t\to\infty} \frac{\tilde{X}(t)}{\tilde{\mu}(t)} = 1 \qquad \text{a.s. on the set} \qquad \left\{\lim_{t\to\infty} \tilde{X}(t) = \infty\right\}. \tag{9.20}$$

9.2.1 *Auxiliary Functions and Their Properties*

An important role in our studies of the asymptotic behavior of the function $\tilde{\mu}$ and the process $\tilde{X}$ is played by the functions G_a, G, Ψ, Ψ_1, and Θ, where

$$G_a(v) = \int_a^v \frac{du}{g(u)}, \qquad v \geq a \geq 0, \tag{9.21}$$

$$G(v) = G_1(v) = \int_1^v \frac{du}{g(u)}, \qquad v \geq 1, \tag{9.22}$$

$$\Psi(t) = \int_0^t \psi(s)ds, \qquad t \geq 0, \tag{9.23}$$

$$\Psi_1(t) = \int_0^t |\psi(s)|ds, \qquad t \geq 0, \tag{9.24}$$

$$\Theta(t) = \int_0^t \theta^2(s)ds, \qquad t \geq 0. \tag{9.25}$$

The following conditions imposed on these functions will be used throughout this chapter:

$$\lim_{v\to\infty} G(v) = \int_1^\infty \frac{du}{g(u)} = \infty; \tag{9.26}$$

$$\liminf_{v\to\infty} \int_v^{cv} \frac{du}{g(u)G(u)} > 0 \qquad \text{for all} \qquad c > 1; \tag{9.27}$$

$$\lim_{c\downarrow 1} \limsup_{v\to\infty} \int_v^{cv} \frac{du}{g(u)G(u)} = 0; \tag{9.28}$$

$$\Psi(t) > 0, \qquad t > 0; \tag{9.29}$$

$$\lim_{t\to\infty} \Psi(t) = \int_0^\infty \psi(s)ds = \infty; \tag{9.30}$$

$$\limsup_{t\to\infty} \frac{\Psi_1(t)}{\Psi(t)} < \infty; \tag{9.31}$$

$$\liminf_{t\to\infty} \int_t^{ct} \frac{\psi(s)}{\Psi(s)} ds > 0 \qquad \text{for all} \qquad c > 1; \tag{9.32}$$

$$\sum_{n=0}^{\infty} \frac{\Theta(2^{n+1}) - \Theta(2^n)}{\Psi_1^2(2^n)} < \infty. \tag{9.33}$$

As known from Sect. 7.9, condition (9.26) implies that there exists a unique continuously differentiable solution μ_b of the Cauchy problem (9.19) with initial condition $b > 0$. Moreover, this solution tends to infinity as $t \to \infty$ and admits the representation

$$\mu_b(t) = G_b^{-1}(t), \qquad t \geq 0,$$

where G_b^{-1} is the inverse function of G_b, that is, $G_b = \mu_b^{-1}$.

Moreover (see Sect. 7.9), conditions (9.26), (9.29), and (9.30) imply that the Cauchy problem (9.18) with initial condition $b > 0$ has a unique continuously differentiable solution $\tilde{\mu}_b$ tending to infinity as $t \to \infty$ and admitting the representation

$$\tilde{\mu}_b(t) = \mu_b(\Psi(t)) = G_b^{-1}(\Psi(t)), \qquad t \geq 0. \tag{9.34}$$

Condition (9.27) appeared in Sect. 7.9 for the first time (see (7.109)). This important property means that $G \in \mathcal{SQI}$ (recall Definition 3.28) by statement 5) of Theorem 4.9, with $f = G$, $\theta = 1/g$, and $f'_{\ln} = 1/(gG)$, since g is positive and continuous, that is, $G_*(c) > 1$ for all $c > 1$, where $G_*(c) = \liminf_{v\to\infty} G(cv)/G(v)$ is the lower limit function for G (see (3.1)).

Recall that condition (9.26) follows from (9.27) (see Remark 7.107). If (9.26) holds, then

$$\lim_{v\to\infty} \frac{G_{a_1}(v)}{G_{a_2}(v)} = 1 \tag{9.35}$$

for all $a_1 \geq 0$ and $a_2 \geq 0$, whence we conclude that, that for all $a \geq 0$,

$$(G_a)_* = G_*. \tag{9.36}$$

Thus condition (9.26) implies that $G_a \in \mathcal{SQI}$ for all $a \geq 0$, if $G \in \mathcal{SQI}$.

In turn, Corollary 7.33 shows that $G_b^{-1} \in \mathcal{PRV}$, and then $\mu_b \in \mathcal{PRV}$ if and only if $G_b \in \mathcal{SQI}$. Summarizing this, condition (9.27) is equivalent to $G_b^{-1} \in \mathcal{PRV}$ (or to $\mu_b \in \mathcal{PRV}$) for all $b > 0$.

Theorem 3.42 together with Remark 3.22 implies the following result.

Proposition 9.1 *If, for some $b > 0$, the function G_b^{-1} (as well as the function μ_b) preserves the asymptotic equivalence of functions tending to infinity or, equivalently, if (9.14) holds, then condition (9.27) follows. Conversely, if condition (9.27) holds, then the function G_b^{-1} (as well as the function μ_b), for all $b > 0$, preserves the asymptotic equivalence of functions tending to infinity and (9.14) follows.*

Some sufficient conditions for (9.27) have been given in Proposition 7.109.

Condition (9.28) is equally important. Since g is positive and continuous, statement 3) of Theorem 4.9, with $f = G$, $\theta = 1/g$, and $f'_{\ln} = 1/(gG)$, asserts that (9.28) is equivalent to $G \in \mathcal{PRV}$ or to $\mu_b^{-1} \in \mathcal{PRV}$. Now Theorem 3.42 and relation (9.36), which follows from (9.26), imply the next result.

Proposition 9.2 *Let condition (9.26) hold.*

1) *If, for some $b > 0$, the function G_b as well as the function μ_b^{-1} preserves the asymptotic equivalence of functions tending to infinity, then condition (9.28) holds.*
2) *Conversely, if condition (9.28) holds, then, for all $b > 0$, the function G_b^{-1} as well as the function μ_b preserves the asymptotic equivalence of functions tending to infinity.*

Let us consider some sufficient conditions for (9.28), which are sometimes easier to check than (9.28) itself.

Proposition 9.3 *Let g be a positive continuous function and let at least one of the following conditions hold:*

(i) $\liminf_{u\to\infty} g(u)G(u)/u > 0$;
(ii) *g is an almost increasing function (see (4.35), (4.37)), that is,*

$$\lim_{v\to\infty} \inf_{u\geq v} \frac{g(u)}{g(v)} > 0;$$

(iii) *g is a nondecreasing function for sufficiently large arguments;*
(iv) *there exists an $\alpha < 1$ such that*

$$0 < \liminf_{u\to\infty} u^{-\alpha} g(u) \qquad \textit{and} \qquad \limsup_{u\to\infty} u^{-\alpha} g(u) < \infty;$$

(v) *the Lebesgue measure of the set $\{c \in (0,1] : g^*(c) < \infty\}$ is positive, where*

$$g^*(c) = \limsup_{t\to\infty} \frac{g(ct)}{g(t)}$$

is the upper limit function for g;
(vi) *g is an ORV-function;*
(vii) *the representation*

$$\frac{1}{g} = \frac{d_1}{g_1} + \cdots + \frac{d_n}{g_n}$$

holds, where g_k, $k = 1, \ldots, n$, are nonnegative, continuous functions satisfying conditions (9.26) and (9.28), and where d_k, $k = 1, \ldots, n$, are almost constant functions (see Definition 4.11).

Then condition (9.28) is satisfied

Proof of Proposition 9.3. Condition (9.28) follows from

- (i) by statement 1) of Proposition 4.17, with $f = G$ and $\theta = 1/g$;
- (ii) by statement 1) of Corollary 4.40, with $f = G$ and $\theta = 1/g$, since the density $\theta = 1/g$ is almost decreasing (see (4.34)) if and only if the function g is almost increasing;
- (iii), since (iii) implies (ii);
- (iv) by statement 3) of Proposition 4.44, with $f = G$ and $\theta = 1/g$;
- (v) by statement 1) Corollary 4.47, with $f = G$ and $\theta = 1/g$, since $(1/g)_* = 1/g^*$ (see statement (x) of Lemma 3.1) and g^* is measurable as the upper limit function of a continuous function g;
- (vi) by statement 2) of Corollary 4.47, with $f = G$ and $\theta = 1/g$, since $(1/g) \in \mathcal{ORV}$ if and only if $g \in \mathcal{ORV}$ (see statement (x) of Lemma 3.1) and g^* is measurable as the upper limit function of a continuous function g. □

Condition (9.32) for the function Ψ is similar to (9.27) and, by statement 5) of Theorem 4.9, it is equivalent to $\Psi \in \mathcal{SQI}$.

Note that condition (9.32) implies (9.30). Moreover, the function Ψ is continuous. Thus, $\Psi \in \mathbb{C}^\infty$. Taking Proposition 7.27 into account, we get the following result.

Proposition 9.4 *Let condition (Ψ4) hold and let Ψ be either nondecreasing or asymptotically equivalent to a nondecreasing function. Then any asymptotically* *quasi-inverse function* $\Psi^{\sim}$ *(in particular,* $\Psi^{\sim} = M_{\Psi}^{+}, L_{\Psi}$*) preserves the asymptotic equivalence of functions.*

9.2.2 Main Results

The following Theorems 9.5 and 9.9 are the main results of this section.

Theorem 9.5 *Let* ψ *be a continuous function and* g, σ, θ *be positive, continuous functions such that there exists a continuous solution* $\tilde{X}_{b_1}$ *of the Cauchy problem for the SDE (9.17) with an arbitrary initial condition* $b_1 > 0$. *Additionally to conditions (9.27),* $(\Psi 1)$–$(\Psi 3)$, *and (9.33), we assume that*

(T1) the function σ/g *is bounded, that is,*

$$\sup_{-\infty<u<\infty} \frac{\sigma(u)}{g(u)} < \infty;$$

(T2) the function g *is continuously differentiable on* $(-\infty, \infty)$ *and its derivative* $(g'(u), u \in (-\infty, \infty))$ *satisfies*

$$\lim_{t\to\infty} \frac{\int_0^t \left| g'(\tilde{X}_{b_1}(s))\theta^2(s) \right| ds}{\Psi_1(t)} = 0 \qquad \text{a.s. on the set} \qquad \left\{ \lim_{t\to\infty} \tilde{X}_{b_1}(t) = \infty \right\}. \tag{9.37}$$

Then, for all $b_1 > 0$ *and* $b_2 > 0$,

$$\lim_{t\to\infty} \frac{\tilde{X}_{b_1}(t)}{\tilde{\mu}_{b_2}(t)} = 1 \qquad \text{a.s. on the set} \qquad \left\{ \lim_{t\to\infty} \tilde{X}_{b_1}(t) = \infty \right\}, \tag{9.38}$$

where $\tilde{\mu}_{b_2}$ *is a solution of the Cauchy problem (9.18) with initial condition* $b_2 > 0$. *In particular, if* $b_1 = b_2 = b$, *then relation (9.20) is satisfied.*

Remark 9.6 A unique continuous solution $\tilde{X}_b$ of the Cauchy problem for the SDE (9.17) exists for an arbitrary initial condition $b > 0$ and continuous coefficients

$$a(t,x) = \psi(t)g(x), \qquad b(t,x) = \theta(t)\sigma(x),$$

if

a) for any $T \in (0, \infty)$, there exists a number $K = K(T)$ such that, for all $t \in [0, T]$ and $x \in (-\infty, \infty)$,

$$|a(t,x)| + |b(t,x)| \le K(1 + |x|);$$

b) for all $C \in (0, \infty)$ and $T \in (0, \infty)$, there exists a number $L = L(C, T)$ such that

$$|a(t, x) - a(t, y)| + |b(t, x) - b(t, y)| \le L|x - y|$$

for $|x| \le C$, $|y| \le C$ and $t \in [0, T]$ (see Gihman and Skorohod [149]).

Remark 9.7 In view of (9.34), relation (9.38) can be rewritten as

$$\lim_{t\to\infty} \frac{\tilde{X}_{b_1}(t)}{\mu_{b_2}(\Psi(t))} = 1 \qquad \text{a.s. on the set} \qquad \left\{\lim_{t\to\infty} \tilde{X}_{b_1}(t) = \infty\right\}, \tag{9.39}$$

where μ_{b_2} is a solution of the Cauchy problem for the ODE (9.19) with an initial condition b_2.

Remark 9.8 Relation (9.37) holds if

$$\lim_{u\to\infty} g'(u) = 0 \qquad \text{and} \qquad \limsup_{t\to\infty} \frac{\Theta(t)}{\Psi_1(t)} < \infty, \tag{9.40}$$

or, if

$$\limsup_{u\to\infty} \left|g'(u)\right| < \infty \qquad \text{and} \qquad \lim_{t\to\infty} \frac{\Theta(t)}{\Psi_1(t)} = 0. \tag{9.41}$$

9.2.3 The Exact Order of Growth of the Process $G_a \circ \tilde{X}_b$

The proof of Theorem 9.5 follows the general approach described in Sect. 9.1. First we establish the exact order of growth as $t \to \infty$ of the process

$$G_a \circ \tilde{X}_b = (G_a(\tilde{X}_b(t)),\ t \ge 0).$$

Theorem 9.9 *Let ψ be a continuous function and g, σ, θ be positive, continuous functions such that there exists a continuous solution $\tilde{X}_b$ of the Cauchy problem for the SDE (9.17) with an arbitrary initial condition $b > 0$. Along with conditions (9.26), (Ψ1)–(Ψ3) and (9.33) assume (T1) and (T2). Then, for all $a \ge 0$,*

$$\lim_{t\to\infty} \frac{G_a(\tilde{X}_b(t))}{\Psi(t)} = 1 \qquad \textit{a.s. on the set} \qquad \left\{\lim_{t\to\infty} \tilde{X}(t) = \infty\right\}. \tag{9.42}$$

9.2.4 Some Auxiliary Results

To prove Theorem 9.9, we consider the Cauchy problem for the auxiliary SDE

$$dZ(t) = \Big(g_1(Z(t))\psi(t) + g_2(Z(t))\,\theta^2(t)\Big)\,dt + \sigma_1(Z(t))\,\theta(t)\,dw(t), \qquad t \ge 0,$$
$$Z(0) \equiv d > 0, \tag{9.43}$$

where W is a standard Wiener process, defined on a complete probability space $\{\Omega, \mathfrak{F}, \mathsf{P}\}$, and where d is a nonrandom, positive constant. Assume that the continuous function ψ and the positive, continuous functions g_1, g_2, σ_1, θ are such that there exists a continuous solution $Z = (Z(t), t \ge 0)$ of the Cauchy problem for the SDE (9.43).

The following result describes the exact order of growth of the process Z.

Lemma 9.10 *Along with conditions* $(\Psi 1)$*–*$(\Psi 3)$ *and (9.33) assume that*

(A1) $\lim_{u\to\infty} g_1(u) = \kappa \in (0, \infty)$,
(A2) the function σ_1 *is bounded,*
(A3)

$$\lim_{t\to\infty} \frac{\int_0^t g_2(Z(s))\theta^2(s)ds}{\Psi_1(t)} = 0 \qquad \textit{a.s. on the set} \qquad \left\{\lim_{t\to\infty} Z(t) = \infty\right\}.$$

Then

$$\lim_{t\to\infty} \frac{Z(t)}{\Psi(t)} = \kappa \qquad \textit{a.s. on the set} \qquad \left\{\lim_{t\to\infty} Z(t) = \infty\right\}. \tag{9.44}$$

Proof of Lemma 9.10. Since

$$Z(t) = d + \int_0^t g_1(Z(s))\,\psi(s)\,ds + \int_0^t g_2(Z(s))\,\theta^2(s)\,ds + \int_0^t \sigma_1(Z(s))\,\theta(s)\,dW(s),$$

conditions $(\Psi 1)$–$(\Psi 3)$ and (A3) imply that Lemma 9.10 follows from

$$\lim_{t\to\infty} \frac{1}{\Psi(t)} \int_0^t g_1(Z(s))\,\psi(s)\,ds = \kappa \qquad \text{a.s. on the set} \qquad \left\{\lim_{t\to\infty} Z(t) = \infty\right\} \tag{9.45}$$

and

$$\lim_{t\to\infty} \frac{1}{\Psi_1(t)} \int_0^t \sigma_1(Z(s))\,\theta(s)\,dW(s) = 0 \qquad \text{a.s.} \tag{9.46}$$

Condition (A1) yields that, for all $\omega \in \{\lim_{t\to\infty} Z(t) = \infty\}$ and $\varepsilon > 0$, there exists an $s_\varepsilon = s_\varepsilon(\omega) > 0$ such that $|g_1(Z(s)) - \kappa| \le \varepsilon$ for $s \ge s_\varepsilon$. Thus, for all $t \ge s_\varepsilon$,

$$\frac{\left|\int_{s_\varepsilon}^t (g_1(Z(s)) - \kappa)\,\psi(s)\,ds\right|}{\Psi(t)} \le \frac{\varepsilon \int_{s_\varepsilon}^t |\psi(s)|\,ds}{\Psi(t)} \le \varepsilon\left(\frac{\Psi_1(t)}{\Psi(t)}\right).$$

In view of condition (Ψ2), we see that

$$\limsup_{t\to\infty} \frac{\left|\int_0^t (g_1(Z(s)) - \kappa)\,\psi(s)\,ds\right|}{\Psi(t)} = \limsup_{t\to\infty} \frac{\left|\int_{s_\varepsilon}^t (g_1(Z(s)) - \kappa)\,\psi(s)\,ds\right|}{\Psi(t)}$$
$$\le \varepsilon\left(\limsup_{t\to\infty} \frac{\Psi_1(t)}{\Psi(t)}\right).$$

This together with (Ψ3) implies (9.45).

The final step is to prove relation (9.46). Let

$$\xi(t) = \int_0^t \sigma_1(Z(s))\,\theta(s)\,dW(s), \qquad t \ge 0.$$

Let $E(t) = \mathsf{E}\xi^2(t)$ and note that E is an increasing function, since ξ, being a stochastic integral, is a martingale. Let G be an increasing function such that $E(t) \le G(t)$, $t \ge 0$. Further, let b be an arbitrary continuous function increasing to ∞ and such that

$$\int_0^\infty \frac{1}{b^2(t)}\,dG < \infty. \tag{9.47}$$

Let $t_0 = 0$ and $t_n = \sup\{t : b(t) \le 2^n\}$, $n \ge 1$. Then

$$\mathsf{E}\sup_{t\ge 0}\left(\frac{\xi(t)}{b(t)}\right)^2 \le \sum_{n=0}^\infty \mathsf{E}\sup_{t_n\le t\le t_{n+1}}\left(\frac{\xi(t)}{b(t)}\right)^2 \le \sum_{n=0}^\infty \frac{1}{b^2(t_n)}\mathsf{E}\sup_{t\le t_{n+1}}\xi^2(t).$$

By Doob's inequality for martingales,

$$\mathsf{E}\sup_{t\le t_{n+1}}\xi^2(t) \le 4E(t_{n+1}) \le 4G(t_{n+1}),$$

whence

$$\mathsf{E}\sup_{t\ge 0}\left(\frac{\xi(t)}{b(t)}\right)^2 \le 4\sum_{n=0}^\infty \frac{1}{b^2(t_n)}\sum_{k=0}^n \int_{t_k}^{t_{k+1}} dG = 4\sum_{k=0}^\infty \int_{t_k}^{t_{k+1}} dG \sum_{n=k}^\infty \frac{1}{b^2(t_n)}.$$

Since

$$\sum_{n=k}^{\infty}\frac{1}{b^2(t_n)}=4\sum_{n=k}^{\infty}\frac{1}{2^{2n}}=\frac{16}{3}\cdot\frac{1}{2^{2k}},$$

we conclude that

$$\begin{aligned}\mathsf{E}\sup_{t\ge 0}\left(\frac{\xi(t)}{b(t)}\right)^2&\le\frac{64}{3}\sum_{k=0}^{\infty}\frac{1}{2^{2k}}\int_{t_k}^{t_{k+1}}dG\le\frac{256}{3}\sum_{k=0}^{\infty}\int_{t_k}^{t_{k+1}}\frac{1}{b^2(t)}\,dG\\&=\frac{256}{3}\int_0^{\infty}\frac{1}{b^2(t)}\,dG.\end{aligned}$$

This means that

$$\mathsf{E}\sup_{t\ge 0}\left(\frac{\xi(t)}{b(t)}\right)^2<\infty$$

provided (9.47) holds. Moreover, if b satisfies (9.47), then

$$\limsup_{t\to\infty}\left|\frac{\xi(t)}{b(t)}\right|<\infty\qquad\text{almost surely.}\tag{9.48}$$

Now

$$E(t)=\int_0^t\mathsf{E}\sigma_1^2(Z(s))\theta^2(s)\,ds\le M\Theta(t),$$

where $M=\sup_u\sigma_1^2(u)<\infty$ by Assumption (A2). Put $G(t)=M\Theta(t)$, $t\ge 0$, and choose an increasing function b such that (9.47) holds and

$$\lim_{t\to\infty}\frac{b(t)}{\Psi_1(t)}=0.\tag{9.49}$$

Then, as shown above, (9.48) holds, whence we immediately get (9.46).

It remains to show that a function b exists possessing the corresponding properties. First, with G defined above,

$$\begin{aligned}\frac{1}{M}\int_1^{\infty}\frac{1}{\Psi_1^2(s)}\,dG&=\int_1^{\infty}\frac{\theta^2(s)}{\Psi_1^2(s)}\,ds=\sum_{n=0}^{\infty}\int_{2^n}^{2^{n+1}}\frac{\theta^2(s)}{\Psi_1^2(s)}\,ds\\&\le\sum_{n=0}^{\infty}\frac{\Theta(2^{n+1})-\Theta(2^n)}{\Psi_1^2(2^n)}<\infty\end{aligned}$$

by (9.33), that is

$$\int_1^\infty \frac{1}{\Psi_1^2(s)}\, dG < \infty.$$

According to Lemma 9.11 below, the latter property allows us to choose a function b for which both relations (9.47) and (9.49) hold. This completes the proof of Lemma 9.10. □

Lemma 9.11 *Let G be an increasing continuous function. Let f be a positive decreasing continuous function such that $I \stackrel{def}{=} \int_1^\infty f(s)\,dG < \infty$. Then there exists a positive continuous function g increasing to ∞ such that*

$$\int_1^\infty f(s)g(s)\,dG < \infty. \tag{9.50}$$

The proof of Lemma 9.11 is based on the following construction. Let $s_1 = 1$ and let, for $n > 1$, s_n be an arbitrary number such that $s_n > s_{n-1}$ and $\int_{s_{n-1}}^{s_n} f(s)\,dG \le I \cdot 2^{-(n-1)}$. If $g(s_n) = n$ for any $n \ge 1$ and $g(s)$ is a linear function in each interval $[s_n, s_{n+1}]$, then g is a positive continuous function increasing to ∞. Moreover,

$$\int_{s_n}^{s_{n+1}} f(s)g(s)\,dG \le (n+1)\int_{s_n}^{s_{n+1}} f(s)\,dG \le I \cdot \frac{n+1}{2^n}, \qquad n \ge 1.$$

Summing up these inequalities we prove (9.50).

The proof of Theorem 9.9 is based on the following auxiliary result.

Lemma 9.12 *Let ψ be a continuous function and g, σ, θ be positive, continuous functions such that conditions $(\Psi 1)$–$(\Psi 3)$ and (9.33) are satisfied. Assume there exists a continuous solution $\tilde{X}$ of the Cauchy problem for the SDE (9.17) with initial condition $b > 0$. Further, let there exist an increasing and twice continuously differentiable function $f = (f(u),\ u \in (-\infty, \infty))$ such that*

(B1) $\lim_{u\to\infty} f(u) = \infty$,
(B2) $\lim_{u\to\infty} f'(u)g(u) = C \in (0, \infty)$,
(B3) the function $f'\sigma$ is bounded,
(B4)

$$\lim_{t\to\infty} \frac{\int_0^t f''(\tilde{X}(s))\sigma^2(\tilde{X}(s))\theta^2(s)ds}{\Psi_1(t)} = 0 \quad \text{a.s. on the set} \quad \left\{\lim_{t\to\infty} \tilde{X}(t) = \infty\right\}.$$

Then

$$\lim_{t\to\infty} \frac{f(\tilde{X}(t))}{\Psi(t)} = C \quad \text{a.s. on the set} \quad \left\{\lim_{t\to\infty} \tilde{X}(t) = \infty\right\}. \tag{9.51}$$

Proof of Lemma 9.12. Put

$$Z(t) = f(\tilde{X}(t)), \qquad t > 0.$$

Then

$$\tilde{X}(t) = f^{-1}(Z(t)), \qquad t > 0.$$

Condition (B1) implies that

$$\left\{\lim_{t\to\infty} \tilde{X}(t) = \infty\right\} = \left\{\lim_{t\to\infty} Z(t) = \infty\right\}. \tag{9.52}$$

Applying Itô's formula to the process $f(\tilde{X}(\cdot))$, we see that

$$\begin{aligned} dZ(t) &= \left[f'(\tilde{X}(t))\, g(\tilde{X}(t))\psi(t) + \frac{1}{2} f''(\tilde{X}(t))\, \sigma^2(\tilde{X}(t))\, \theta^2(t) \right] dt \\ &\quad + f'(\tilde{X}(t))\sigma(\tilde{X}(t))\theta(t)\, dW(t) \\ &= \left[f'(f^{-1}(Z(t))g\left(f^{-1}(Z(t))\right)\psi(t) + \frac{1}{2} f''(f^{-1}(Z(t)))\sigma^2(f^{-1}(Z(t)))\theta^2(t)) \right] dt \\ &\quad + f'(f^{-1}(Z(t))\sigma\left(f^{-1}(Z(t))\right)\theta(t)\, dW(t). \end{aligned}$$

Thus, the process Z is a continuous solution of the SDE

$$dZ(t) = \left(g_1(Z(t))\psi(t) + g_2(Z(t))\theta^2(t)\right) dt + \sigma_1(Z(t))\theta(t)\, dW(t),$$

where

$$g_1(u) = f'(f^{-1}(u))g(f^{-1}(u)), \qquad g_2(u) = \frac{1}{2} f''(f^{-1}(u))\sigma^2(f^{-1}(u))$$

and

$$\sigma_1(u) = f'(f^{-1}(u))\sigma(f^{-1}(u)).$$

Moreover,

$$Z(0) = f(\tilde{X}(0)) = f(b) > 0.$$

The above SDE is similar to the SDE (9.43), allowing us to use Lemma 9.10, for which we check assumptions (A1)–(A3).

Conditions (B1) and (B2) imply that

$$\lim_{u\to\infty} g_1(u) = \lim_{u\to\infty} f'(f^{-1}(u))g(f^{-1}(u)) = \lim_{u\to\infty} f'(u)g(u) = C,$$

and thus condition (A1) holds with $\kappa = C$. Condition (A2) immediately follows from (B1) and (B3).

Since

$$\lim_{t\to\infty} \frac{\int_0^t g_2(Z(s))\theta^2(s)ds}{\Psi_1(t)} = \frac{1}{2} \lim_{t\to\infty} \frac{\int_0^t f''(\tilde{X}(s))\sigma^2(\tilde{X}(s))\theta^2(s)ds}{\Psi_1(t)},$$

condition (B4) and relation (9.52) imply condition (A3).

Now Lemma 9.10 yields

$$\lim_{t\to\infty} \frac{f(\tilde{X}(t))}{\Psi(t)} = \lim_{t\to\infty} \frac{Z(t)}{\Psi(t)} = C \qquad \text{a.s. on the set} \qquad \left\{\lim_{t\to\infty} Z(t) = \infty\right\},$$

whence we conclude (9.51) from (9.52), which completes the proof. □

9.2.5 *Proof of the Main Results*

The above auxiliary results allow us to prove Theorems 9.5 and 9.9. We start with Theorem 9.9.

Proof of Theorem 9.9. Let

$$f(u) = \begin{cases} G_0(u) = \displaystyle\int_0^u \frac{dv}{g(v)}, & u \ge 0, \\ -G_0(|u|) = -\displaystyle\int_u^0 \frac{dv}{g(v)}, & u < 0. \end{cases}$$

We check the assumptions of Lemma 9.12 for the function f.

Indeed, by the assumptions of Theorem 9.9, the function g is positive and continuously differentiable. Thus, f is twice continuously differentiable with

$$f' = 1/g > 0 \qquad \text{and} \qquad f'' = -g'/g^2. \tag{9.53}$$

This together with (9.26) implies that f is increasing and that condition (B1) holds.

Further,

$$\lim_{u\to\infty} f'(u)g(u) = \lim_{u\to\infty} \frac{g(u)}{g(u)} = 1,$$

that is, condition (B2) holds with $C = 1$. Moreover,

$$f'\sigma = \frac{\sigma}{g},$$

whence condition (B3) follows from condition (T1) of Theorem 9.9.

In turn, conditions (T1) and (T2) of Theorem 9.9 together with relation (9.53) imply that

$$\begin{aligned}\lim_{t\to\infty} \frac{\left|\int_0^t f''(\tilde{X}(s))\sigma^2(\tilde{X}(s))\theta^2(s)\,ds\right|}{\Psi_1(t)} &= \lim_{t\to\infty} \frac{\left|\int_0^t \frac{\sigma^2(\tilde{X}(s))}{g^2(\tilde{X}(s))} \cdot g'(\tilde{X}(s))\theta^2(s)\,ds\right|}{\Psi_1(t)} \\ &\le \lim_{t\to\infty} \frac{L\int_0^t \left|g'(\tilde{X}(s))\theta^2(s)\right|\,ds}{\Psi_1(t)} = 0\end{aligned}$$

a.s. on the set $\left\{\lim_{t\to\infty} \tilde{X}(t) = \infty\right\}$, where

$$L = \sup_u \frac{\sigma^2(u)}{g^2(u)} < \infty.$$

Hence condition (B4) holds as well.

So, all assumptions of Lemma 9.12 are satisfied for $C = 1$, and we apply this lemma to show that

$$\lim_{t\to\infty} \frac{G_0(\tilde{X}(t))}{\Psi(t)} = \lim_{t\to\infty} \frac{f(\tilde{X}(t))}{\Psi(t)} = 1 \qquad \text{a.s. on the set} \qquad \left\{\lim_{t\to\infty} \tilde{X}(t) = \infty\right\}.$$

Taking (9.26) and relation (9.35) into account, we obtain

$$\lim_{t\to\infty} \frac{G_a(\tilde{X}(t))}{\Psi(t)} = \lim_{t\to\infty} \frac{G_0(\tilde{X}(t))}{\Psi(t)} \cdot \lim_{t\to\infty} \frac{G_a(\tilde{X}(t))}{G_0(\tilde{X}(t))} = 1$$

for all $a \ge 0$ a.s. on the set $\left\{\lim_{t\to\infty} \tilde{X}(t) = \infty\right\}$, that is, relation (9.42) holds. This proves Theorem 9.9. □

Now we turn to the proof of Theorem 9.5.

Proof of Theorem 9.5. Since condition (9.26) follows from relation (9.27) (see Remark 7.107), all assumptions of Theorem 9.5 are stronger than the corresponding assumptions of Theorem 9.9. Thus

$$\lim_{t\to\infty} \frac{G_{b_2}(\tilde{X}_{b_1}(t))}{\Psi(t)} = 1 \qquad \text{a.s. on the set} \qquad \left\{\lim_{t\to\infty} \tilde{X}_{b_1}(t) = \infty\right\}$$

for all $b_1 > 0$ and $b_2 > 0$ (see relation (9.42)). By Proposition 9.1, condition (9.27) yields that the function $G_{b_2}^{-1}$ preserves the asymptotic equivalence of functions tending to infinity. Thus

$$\lim_{t\to\infty} \frac{\tilde{X}_{b_1}(t)}{G_{b_2}^{-1}(\Psi(t))} = \lim_{t\to\infty} \frac{G_{b_2}^{-1}(G_{b_2}(\tilde{X}_{b_1}(t)))}{G_{b_2}^{-1}(\Psi(t))} = \frac{G_{b_2}(\tilde{X}_{b_1}(t))}{\Psi(t)} = 1$$

a.s. on the set $\left\{\lim_{t\to\infty} \tilde{X}_{b_1}(t) = \infty\right\}$. This implies relation (9.38), since

$$G_{b_2}^{-1}(\Psi(t)) = \tilde{\mu}_{b_2}(t), \qquad t \ge 0,$$

(see (9.34)), which completes the proof of Theorem 9.5. □

9.2.6 A Discussion of the Main Results

Each of Theorems 9.5 and 9.9 provides a certain piece of information about the asymptotic behavior of the solution $\tilde{X}_b$ of the SDE (9.17), namely, Theorem 9.5 provides the asymptotic (9.38), while Theorem 9.9 results in (9.42). So, the natural question arises about a relationship between these two asymptotics.

Analyzing the second step of the proof of Theorem 9.5 (i.e., the part after Theorem 9.9) we see that, in fact, the following assertion from "function analysis" is proved there: *if a nonrandom, positive function* $(x(t), t \ge 0)$ *is such that* $\lim_{t\to\infty} x(t) = \infty$, *then condition* (9.27) *yields the implication*

$$\lim_{t\to\infty} \frac{G_a(x(t))}{\Psi(t)} = 1 \quad \exists a > 0 \qquad \Longrightarrow \qquad \frac{x(t)}{\mu_c(\Psi(t))} = 1 \quad \forall c > 0. \tag{9.54}$$

The implication (9.54) follows from Proposition 9.1. In turn, Proposition 9.2 shows that *conditions* (9.26) *and* (9.28) *imply the converse implication*

$$\lim_{t\to\infty} \frac{G_a(x(t))}{\Psi(t)} = 1 \quad \forall a > 0 \qquad \Longleftarrow \qquad \frac{x(t)}{\mu_c(\Psi(t))} = 1 \quad \exists c > 0. \tag{9.55}$$

So, the two implications together mean that *conditions* (9.27) *and* (9.28) *for all* $a > 0$ *and* $c > 0$ *yield the equivalence*

$$\lim_{t\to\infty} \frac{G_a(x(t))}{\Psi(t)} = 1 \qquad \Longleftrightarrow \qquad \frac{x(t)}{\mu_c(\Psi(t)} = 1. \tag{9.56}$$

The implications (9.54)–(9.56) allow us to conclude that, under the assumptions of Theorem 9.9 (including condition (9.26)), we have

(A) if condition (9.27) is satisfied and if there are $a > 0$ and $b > 0$ such that (9.42) holds, then (9.38) follows for all $b_1 > 0$ and $b_2 > 0$;
(B) if condition (9.28) is satisfied and there are $b_1 > 0$ and $b_2 > 0$ such that (9.38) holds, then (9.42) follows for all $a > 0$ and $b > 0$;
(C) if conditions (9.27) and (9.28) hold, then (9.38) and (9.42) are equivalent for all $a > 0$, $b > 0$, $b_1 > 0$, and $b_2 > 0$.

9.2.7 Asymptotic Stability of Solutions of Stochastic Differential Equations with Respect to Initial Conditions

Assume there exists a continuous solution $\tilde{X}_b$ of the SDE (9.17) with an arbitrary initial condition $b > 0$. This solution is said to be *asymptotically stable at infinity with respect to initial conditions* if

$$\lim_{t\to\infty} \frac{\tilde{X}_{b_1}(t)}{\tilde{X}_{b_2}(t)} = 1 \tag{9.57}$$

for all $b_1 > 0$ and $b_2 > 0$ a.s. on the set $\left\{\lim_{t\to\infty} \tilde{X}_{b_1}(t) = \infty\right\} \cap \left\{\lim_{t\to\infty} \tilde{X}_{b_2}(t) = \infty\right\}$.

The asymptotic stability with respect to initial conditions has been studied in Theorem 7.108 for solutions of the ODE (9.18) (see also (7.113)). Theorem 7.108 shows that, under conditions (9.27), (9.29), and (9.30), a solution $\tilde{\mu}_b$ of the ODE (9.18) is asymptotically stable with respect to initial conditions. Now, in addition, Theorem 9.5 provides conditions for the asymptotic stability at infinity with respect to initial conditions for the SDE (9.17).

Corollary 9.13 *Assume that the assumptions of Theorem 9.5 hold. Then the solution $\tilde{X}_b$ of the SDE (9.17) is asymptotically stable at infinity with respect to initial conditions.*

Proof of Corollary 9.13. According to Theorem 9.5, for all $b_1 > 0$ and $b_2 > 0$,

$$\lim_{t\to\infty} \frac{\tilde{X}_{b_1}(t)}{\tilde{\mu}_{b_2}(t)} = 1 \quad \text{a.s. on the set} \quad \left\{\lim_{t\to\infty} \tilde{X}_{b_1}(t) = \infty\right\}$$

and

$$\lim_{t\to\infty} \frac{\tilde{X}_{b_2}(t)}{\tilde{\mu}_{b_1}(t)} = 1 \quad \text{a.s. on the set} \quad \left\{\lim_{t\to\infty} \tilde{X}_{b_2}(t) = \infty\right\}.$$

This implies (9.57). □

If condition (9.27) is dropped, then Theorem 9.9 provides conditions for the G-asymptotic stability at infinity with respect to initial conditions for the SDE (9.17).

Corollary 9.14 *Let the assumptions of Theorem 9.9 hold. Then, for all $b_1 > 0$ and $b_2 > 0$,*

$$\lim_{t\to\infty} \frac{G(\tilde{X}_{b_1}(t))}{G(\tilde{X}_{b_2}(t))} = 1$$

a.s. on the set $\left\{\lim_{t\to\infty} \tilde{X}_{b_1}(t) = \infty\right\} \cap \left\{\lim_{t\to\infty} \tilde{X}_{b_2}(t) = \infty\right\}$.

9.3 Renewal Processes Constructed from Solutions of Stochastic Differential Equations

Generalized renewal processes constructed from stochastic processes in continuous time and their asymptotic properties have been considered in Sects. 8.4 and 8.5. In the present section, we study generalized renewal processes constructed from a solution $\tilde{X}_b$ of the SDE (9.17). Recall that solutions of these equations have continuous paths.

For the process $\tilde{X}_b$, we consider the three *generalized renewal processes*

$$(M^+_{\tilde{X}_b}(s), s \ge b), \qquad (L_{\tilde{X}_b}(s), s \ge b), \qquad (T_{\tilde{X}_b}(s), s \ge b),$$

where $M^+_{\tilde{X}_b}(s)$ is the *first hitting time of the level s by the process $\tilde{X}_b$*, $L_{\tilde{X}_b}(s)$ is the *last time when the process $\tilde{X}_b$ crosses the level s*, and $T_{\tilde{X}_b}(s)$ is the *total sojourn time in the set $(-\infty, s]$ for the process $\tilde{X}_b$*.

Since the process $\tilde{X}_b$ has continuous paths,

$$M^+_{\tilde{X}_b}(s) = \inf\left\{t \ge 0 : \tilde{X}_b(t) \ge s\right\} = \min\left\{t \ge 0 : \tilde{X}_b(t) = s\right\},$$

$$L_{\tilde{X}_b}(s) = \sup\left\{t \ge 0 : \tilde{X}_b(t) \le s\right\} = \max\left\{t \ge 0 : \tilde{X}_b(t) = s\right\},$$

$$T_{\tilde{X}_b}(s) = \operatorname{meas}\left(\left\{t \ge 0 : \tilde{X}_b(t) \le s\right\}\right) = \int_0^\infty \mathbb{I}\left\{\tilde{X}_b(t) \le s\right\} dt$$

for $s \ge b$ (see relations (8.31)–(8.33) and (8.35)).

The asymptotic behavior of the above generalized renewal processes is considered in the following result.

Theorem 9.15 *Assume that the assumptions of Theorem 9.9 hold. Then:*

1) for all $b > 0$,

$$\lim_{s\to\infty} \frac{\Psi(M^{+}_{\tilde{X}_b}(s))}{G(s)} = \lim_{s\to\infty} \frac{\Psi(L_{\tilde{X}_b}(s))}{G(s)} = 1 \tag{9.58}$$

a.s. on the set $\left\{\lim_{t\to\infty} \tilde{X}_b(t) = \infty\right\}$, and, equivalently,

$$\lim_{s\to\infty} \frac{\Psi(M^{+}_{\tilde{X}_b}(G^{-1}(s)))}{s} = \lim_{s\to\infty} \frac{\Psi(L_{\tilde{X}_b}(G^{-1}(s)))}{s} = 1 \tag{9.59}$$

a.s. on the set $\left\{\lim_{t\to\infty} \tilde{X}_b(t) = \infty\right\}$, where G and Ψ are the functions defined in (9.22) and (9.23) and where G^{-1} is the inverse function of G;

2) if the function Ψ is either nondecreasing or asymptotically equivalent to a nondecreasing function, then, in addition to (9.58) and (9.59),

$$\lim_{s\to\infty} \frac{\Psi(T_{\tilde{X}_b}(s))}{G(s)} = 1 \quad \text{a.s. on the set} \quad \left\{\lim_{t\to\infty} \tilde{X}_b(t) = \infty\right\}, \tag{9.60}$$

for all $b > 0$, and, equivalently,

$$\lim_{s\to\infty} \frac{\Psi(T_{\tilde{X}_b}(G^{-1}(s)))}{s} = 1 \quad \text{a.s. on the set} \quad \left\{\lim_{t\to\infty} \tilde{X}_b(t) = \infty\right\}; \tag{9.61}$$

3) if condition (9.27) holds, then

$$\lim_{s\to\infty} \frac{G^{-1}(\Psi(M^{+}_{\tilde{X}_b}(s)))}{s} = \lim_{s\to\infty} \frac{G^{-1}(\Psi(L_{\tilde{X}_b}(s)))}{s} = 1, \tag{9.62}$$

for all $b > 0$, a.s. on the set $\left\{\lim_{t\to\infty} \tilde{X}_b(t) = \infty\right\}$;

4) if condition (9.27) holds and if the function Ψ is either nondecreasing or asymptotically equivalent to a nondecreasing function, then, in addition to (9.62),

$$\lim_{s\to\infty} \frac{G^{-1}(\Psi(T_{\tilde{X}_b}(s)))}{s} = 1 \quad \text{a.s. on the set} \quad \left\{\lim_{t\to\infty} \tilde{X}_b(t) = \infty\right\}, \tag{9.63}$$

for all $b > 0$;

5) *if condition (Ψ4) holds and if the function Ψ is either nondecreasing or asymptotically equivalent to a nondecreasing function, then*

$$\lim_{s\to\infty}\frac{M^{+}_{\tilde{X}_b}(s)}{\Psi^{\sim}(G(s))}=\lim_{s\to\infty}\frac{L_{\tilde{X}_b}(s)}{\Psi^{\sim}(G(s))}=\lim_{s\to\infty}\frac{T_{\tilde{X}_b}(s)}{\Psi^{\sim}(G(s))}=1, \tag{9.64}$$

for all $b>0$ and any asymptotic quasi-inverse function $\Psi^{\sim}$, in particular, for $\Psi^{\sim}=M^{+}_{\Psi}, L_{\Psi}, T_{\Psi}$, a.s. on the set $\left\{\lim_{t\to\infty}\tilde{X}_b(t)=\infty\right\}$ or, equivalently,

$$\lim_{s\to\infty}\frac{M^{+}_{\tilde{X}_b}(s)}{\Psi^{\sim}(G_a(s))}=\lim_{s\to\infty}\frac{L_{\tilde{X}_b}(s)}{\Psi^{\sim}(G_a(s))}=\lim_{s\to\infty}\frac{T_{\tilde{X}_b}(s)}{\Psi^{\sim}(G_a(s))}=1, \tag{9.65}$$

for all $a>0$ and $b>0$, a.s. on the set $\left\{\lim_{t\to\infty}\tilde{X}_b(t)=\infty\right\}$;

6) *if the function Ψ is increasing and condition (Ψ4) holds, then*

$$\lim_{s\to\infty}\frac{M^{+}_{\tilde{X}_b}(s)}{\Psi^{-1}(G(s))}=\lim_{s\to\infty}\frac{L_{\tilde{X}_b}(s)}{\Psi^{-1}(G(s))}=\lim_{s\to\infty}\frac{T_{\tilde{X}_b}(s)}{\Psi^{-1}(G(s))}=1, \tag{9.66}$$

for all $b>0$, a.s. on the set $\left\{\lim_{t\to\infty}\tilde{X}_b(t)=\infty\right\}$ or, equivalently,

$$\lim_{s\to\infty}\frac{M^{+}_{\tilde{X}_b}(s)}{\Psi^{-1}(G_a(s))}=\lim_{s\to\infty}\frac{L_{\tilde{X}_b}(s)}{\Psi^{-1}(G_a(s))}=\lim_{s\to\infty}\frac{T_{\tilde{X}_b}(s)}{\Psi^{-1}(G_a(s))}=1, \tag{9.67}$$

for all $a>0$ and $b>0$, a.s. on the set $\left\{\lim_{t\to\infty}\tilde{X}_b(t)=\infty\right\}$, where Ψ^{-1} is the inverse function of Ψ;

7) *if the function Ψ is increasing and condition (Ψ4) holds, then*

$$\lim_{s\to\infty}\frac{M^{+}_{\tilde{X}_b}(s)}{\tilde{\mu}_a^{-1}(s)}=\lim_{s\to\infty}\frac{L_{\tilde{X}_b}(s)}{\tilde{\mu}_a^{-1}(s)}=\lim_{s\to\infty}\frac{T_{\tilde{X}_b}(s)}{\tilde{\mu}_a^{-1}(s)}=1, \tag{9.68}$$

for all $a>0$ and $b>0$, a.s. on the set $\left\{\lim_{t\to\infty}\tilde{X}_b(t)=\infty\right\}$, where $\tilde{\mu}_a^{-1}$ is the inverse function of $\tilde{\mu}_a$ and where $\tilde{\mu}_a$ is the solution of the ODE (9.18) with initial condition $b=a$.

Proof of Theorem 9.15. In view of Theorem 9.9,

$$\lim_{t\to\infty}\frac{G(\tilde{X}_b(t))}{\Psi(t)}=1 \qquad \text{a.s. on the set} \qquad \left\{\lim_{t\to\infty}\tilde{X}_b(t)=\infty\right\}$$

for all $b > 0$. This implies that

$$\lim_{s\to\infty} \frac{G(\tilde{X}_b(M^+_{\tilde{X}_b}(s)))}{\Psi(M^+_{\tilde{X}_b}(s))} = 1 \quad \text{a.s. on the set} \quad \left\{\lim_{t\to\infty} \tilde{X}_b(t) = \infty\right\}.$$

Since the process $\tilde{X}_b$ has continuous paths,

$$\tilde{X}_b(M^+_{\tilde{X}_b}(s)) = s$$

for all $s > b$. Hence,

$$\lim_{s\to\infty} \frac{\Psi(M^+_{\tilde{X}_b}(s))}{G(s)} = 1 \quad \text{a.s. on the set} \quad \left\{\lim_{t\to\infty} \tilde{X}_b(t) = \infty\right\}.$$

The same arguments yield

$$\lim_{s\to\infty} \frac{\Psi(L_{\tilde{X}_b}(s))}{G(s)} = 1 \quad \text{a.s. on the set} \quad \left\{\lim_{t\to\infty} \tilde{X}_b(t) = \infty\right\}.$$

Therefore, (9.58) is proved. Since $G^{-1} \in \mathbb{C}^\infty_{\text{inc}}$, (9.58) and (9.59) are equivalent, which proves statement 1).

Taking relation (8.34) into account, statement 2) follows from statement 1).

If condition (9.27) holds, then, by (9.17), the function G^{-1} preserves the asymptotic equivalence of functions tending to infinity. Thus relations (9.62) and (9.63) follow from (9.58) and (9.60), respectively. So, statements 3) and 4) are proved.

By Proposition 9.4 and relation (9.35), statement 5) is a consequence of statements 1)–2). In turn, statement 6) follows from statement 5). Statement 7) is the same as statement 6), since $\tilde{\mu}_a^{-1} = \Psi^{-1} \circ G_a$ (see (9.34)). □

9.4 Order of Growth of Solutions of Autonomous Stochastic Differential Equations

In this section, we study the asymptotic behavior as $t \to \infty$ of a solution $X = (X(t), t \ge 0)$ of the Cauchy problem for the *autonomous SDE* (9.3), that is, of the Cauchy problem for the SDE

$$dX(t) = g(X(t))\,dt + \sigma(X(t))\,dW(t), \qquad t \ge 0, \qquad X(0) \equiv b > 0,$$

where W is a standard Wiener process, defined on a complete probability space $\{\Omega, \mathfrak{F}, \mathsf{P}\}$, and where b is a nonrandom, positive constant (initial condition). To indicate the dependence of a solution X on the initial condition we sometimes denote it by X_b.

The coefficients

$$g = (g(u), u \in (-\infty, \infty)) \qquad \text{and} \qquad \sigma = (\sigma(u), u \in (-\infty, \infty))$$

in (9.3) depend on the phase variable. Throughout this section we assume that g and σ are positive, continuous functions such that there exists a continuous solution X_b, for all $b > 0$. Moreover, we assume that the coefficients g and σ are such that

$$\mathsf{P}\left\{\lim_{t\to\infty} X_b(t) = \infty\right\} = 1, \tag{9.69}$$

for $b > 0$, and that X_b does not reach infinity in finite time.

Remark 9.16 The Cauchy problem for the SDE (9.3) has a unique continuous solution X_b, for which condition (9.69) holds and that does not reach infinity in finite time, if

a) there exists a positive number K such that

$$|g(u)| + |\sigma(u)| \le K(1 + |u|)$$

for all $u \in (-\infty, \infty)$;

b) given an arbitrary positive number $C > 0$, there exists a number $L_C > 0$ such that

$$|g(u) - g(v)| + |\sigma(u) - \sigma(v)| \le L_C|u - v|$$

for $|u| \le C$ and $|v| \le C$;

c)

$$\int_{-\infty}^{x} \exp\left\{-\int_0^z \frac{2g(u)}{\sigma^2(u)}du\right\} dz = \infty \quad \text{and} \quad \int_x^{\infty} \exp\left\{-\int_0^z \frac{2g(u)}{\sigma^2(u)}du\right\} dz < \infty$$

for all $x \in (-\infty, \infty)$ (see Gihman and Skorohod [149], §§15, 16).

Example 9.17 If $\inf \sigma(u) > 0$, then condition c) of Remark 9.16 holds only if

$$\int_0^{\infty} g(u)du = \infty. \tag{9.70}$$

Note that (9.70) is not sufficient for condition c). Indeed, if

$$\sigma^2(u) \equiv 2 \qquad \text{and} \qquad g(u) = \frac{1}{1+u}, \qquad u \ge 0,$$

then condition (9.70) is satisfied. But,

$$\int_0^\infty \exp\left\{-\int_0^z g(u)du\right\} dz = \int_0^\infty \exp\{-\ln(1+z)\}\, dz$$
$$= \int_0^\infty \frac{dz}{1+z} = \infty,$$

that is, condition c) does not hold. On the other hand, if

$$\sigma^2(u) \equiv 2 \quad \text{and} \quad g(u) = 1 + |u|^\varepsilon, \qquad u \in (-\infty, \infty),$$

with $\varepsilon \geq 0$, then condition c) holds.

Along with the SDE (9.3) we consider the Cauchy problem for the *autonomous ODE* (9.7), that is, the Cauchy problem for the ODE

$$d\mu(t) = g(\mu(t))\, dt, \qquad t \geq 0, \qquad \mu(0) = b > 0,$$

corresponding to the SDE (9.3) for $\sigma(u) \equiv 0$. Since g is a positive, continuous function, the latter equation, with an initial condition $b > 0$, has a unique solution $\mu_b = (\mu_b(t), t \geq 0)$ tending to ∞ as $t \to \infty$ (see Sect. 7.9). Moreover, we have shown in Sect. 7.9 that the solution μ_b does not reach infinity in finite time if condition (9.26) holds, that is, if

$$\int_b^\infty \frac{du}{g(u)} = \infty.$$

For example, condition (9.26) is satisfied if condition a) in Remark 9.16 holds.

The aim of this subsection is to find conditions under which (9.11) holds, that is, under which the solution of the autonomous SDE (9.3) is asymptotically equivalent to the solution of the corresponding autonomous ODE (9.7), i.e.

$$\lim_{t\to\infty} \frac{X_b(t)}{\mu_b(t)} = 1 \qquad \text{a.s.} \tag{9.71}$$

Therefore we are interested in the exact order of growth of the solution of an autonomous SDE.

Note that the SDE (9.3) is a particular case of (9.17) if

$$\psi(t) \equiv \theta(t) \equiv 1. \tag{9.72}$$

Similarly, the ODE (9.7) is a particular case of (9.18) if $\psi(t) \equiv 1$. This allows us to apply Theorems 9.5 and 9.9.

If (9.72) holds, then (9.23)–(9.25) imply that

$$\Psi(t) = \Psi_1(t) = \Theta(t) = t, \qquad t \geq 0.$$

Thus conditions (9.29)–(9.32) as well as $(\Theta\Psi_1)$ hold.

Hence, Theorem 9.9 can be rewritten as follows.

Theorem 9.18 *Let g and σ be positive, continuous functions such that the SDE (9.3) has a continuous solution X_b, for any initial condition $b > 0$, satisfying condition (9.69). Further, let conditions (9.26), (T1), and (T2) hold. Then*

$$\lim_{t\to\infty} \frac{G_a(X_b(t))}{t} = 1 \qquad a.s. \tag{9.73}$$

for all $a \geq 0$.

Recalling Remark 9.8, we see that Theorem 9.18 implies the following result (see Gihman and Skorohod [149], §16).

Corollary 9.19 *Let g and σ be positive, continuous functions such that the SDE (9.3) has a continuous solution X_b, for any initial condition $b > 0$, satisfying condition (9.69). Further, suppose that condition (9.26) holds, the function σ/g is bounded, the function g is continuously differentiable, and*

$$\lim_{u\to\infty} g'(u) = 0. \tag{9.74}$$

Then (9.73) follows for all $a \geq 0$.

In turn, Theorem 9.5 takes the following form.

Theorem 9.20 *Let g and σ be positive, continuous functions such that the SDE (9.3) has a continuous solution X_b, for every initial condition $b > 0$, satisfying condition (9.69). Further, suppose that conditions (9.27), (T1), and (T2) hold. Then*

$$\lim_{t\to\infty} \frac{X_{b_1}(t)}{\mu_{b_2}(t)} = 1 \qquad a.s. \tag{9.75}$$

for all $b_1 > 0$ and $b_2 > 0$. In particular, if $b_1 = b_2 = b$, then relation (9.71) follows.

In view of Remark 9.8, Theorem 9.20 implies the following result.

Corollary 9.21 *Let g and σ be positive, continuous functions such that the SDE (9.3) has a continuous solution X_b, for any initial condition $b > 0$, satisfying condition (9.69). Further, let the function σ/g be bounded, the function g be continuously differentiable, and let relations (9.27) and (9.74) hold. Then relation (9.75) follows for all $b_1 > 0$ and $b_2 > 0$. In particular, if $b_1 = b_2 = b$, then relation (9.71) holds.*

Corollary 9.20 and Proposition 9.1 yield the following result (see Gihman and Skorohod [149], §16).

Corollary 9.22 *Let g and σ be positive, continuous functions such that the SDE (9.3) has a continuous solution X_b, for every initial condition $b > 0$, satisfying condition (9.69). Further, let the function σ/g be bounded, the function g be continuously differentiable, and let relations (9.14) and (9.74) be satisfied. Then relation (9.75) holds for all $b_1 > 0$ and $b_2 > 0$. In particular, if $b_1 = b_2 = b$, then relation (9.71) follows.*

9.4.1 Renewal Processes Constructed from Solutions of Autonomous Stochastic Differential Equations

In this subsection, we study the asymptotic behavior of generalized renewal processes constructed from solutions X_b of autonomous SDEs as given in (9.3). The following result immediately follows from Theorem 9.15, provided condition (9.72) holds.

Theorem 9.23 *Let the assumptions of Theorem 9.18 (or, of Corollary 9.19) hold. Then*

$$\lim_{s\to\infty} \frac{M^+_{X_{b_1}}(s)}{G_{b_2}(s)} = \lim_{s\to\infty} \frac{L_{X_{b_1}}(s)}{G_{b_2}(s)} = \lim_{s\to\infty} \frac{T_{X_{b_1}}(s)}{G_{b_2}(s)} = 1 \qquad a.s.$$

for all $b_1 > 0$ and $b_2 > 0$, and, equivalently,

$$\lim_{s\to\infty} \frac{M^+_{X_{b_1}}(s)}{\mu_{b_2}^{-1}(s)} = \lim_{s\to\infty} \frac{L_{X_{b_1}}(s)}{\mu_{b_2}^{-1}(s)} = \lim_{s\to\infty} \frac{T_{X_{b_1}}(s)}{\mu_{b_2}^{-1}(s)} = 1 \qquad a.s.,$$

where $\mu_{b_2}^{-1}$ is the inverse function of μ_{b_2}.

9.5 φ-Order of Growth of Solutions of Autonomous Stochastic Differential Equations

The asymptotic equivalence for solutions of SDEs and the corresponding ODEs has been studied in Sects. 9.2 and 9.4. The results of these sections can be extended to the so-called *φ-asymptotic equivalence* of solutions of SDEs and ODEs. For the sake of simplicity, we study autonomous SDEs and ODEs, that is, we are going to generalize the results of Sect. 9.4.

The main aim of this section is to find conditions under which

$$\lim_{t\to\infty} \frac{\varphi(X_b(t))}{\varphi(\mu_b(t))} = 1 \qquad \text{a.s.,} \tag{9.76}$$

where, as in Sect. 9.4, $X_b = (X_b(t), t \geq 0)$ is the continuous solution of the Cauchy problem for the autonomous SDE (9.3), that is, of the Cauchy problem for

$$dX(t) = g(X(t))\,dt + \sigma(X(t))\,dW(t), \qquad t \geq 0, \qquad X(0) \equiv b > 0,$$

and where $\mu_b = (\mu_b(t), t \geq 0)$ is the positive, continuous solution of the Cauchy problem for the autonomous ODE (9.7), that is, of the Cauchy problem for

$$d\mu(t) = g(\mu(t))\,dt, \qquad t \geq 0, \quad \mu(0) = b > 0,$$

corresponding to the SDE (9.3). We suppose that these solutions tend to ∞ as $t \to \infty$ and that $\varphi = (\varphi(u), u \in \mathbf{R})$ for $u \geq u_0 \geq 0$ is a positive, increasing, continuously differentiable function tending to ∞ as $u \to \infty$.

Relation (9.76) generalizes relation (9.73), corresponding to $a = b$, as well as relation (9.75), corresponding to $b_1 = b_2 = b$. Indeed, if $\varphi(u) = u$ for $u > 0$, then (9.76) coincides with (9.75), and, if $\varphi = \mu_b^{-1} = G_b$, then (9.76) coincides with (9.73).

For $a > 0$, put

$$G_a^{(\varphi)} = G_a \circ \varphi^{-1}, \qquad g^{(\varphi)} = (g \circ \varphi^{-1})(\varphi' \circ \varphi^{-1}), \tag{9.77}$$

where G_a is as defined in (9.21), φ^{-1} is the inverse function of φ, and φ' is the derivative of φ.

For example, if $\varphi(u) = \ln(u)$, then $G^{(\ln)} = G \circ \exp$ and $g^{(\ln)} = (g \circ \exp)/\exp$, while, if $\varphi(u) = u$, then $G^{(\varphi)} = G$ and $g^{(\varphi)} = g$.

The definitions of $g^{(\varphi)}$ and $G_a^{(\varphi)}$ imply that

$$\frac{d}{du} G_a^{(\varphi)}(u) = \frac{1}{g^{(\varphi)}(u)} \tag{9.78}$$

for $u > \varphi(a)$. Since $G_b = \mu_b^{-1}$,

$$G_b^{(\varphi)} = (\varphi \circ \mu_b)^{-1}, \tag{9.79}$$

that is, $(G_b(\varphi^{-1}(s)), s \geq \varphi(b))$, is the inverse function of $(\varphi(\mu_b(t)), t \geq 0)$.

Along with (9.76) consider

$$\lim_{t\to\infty} \frac{G(X_b(t))}{t} = 1 \qquad \text{a.s.,} \tag{9.80}$$

where $G = G_1$. It is clear that (9.80) is equivalent to

$$\lim_{t\to\infty} \frac{G_b(X_b(t))}{t} = 1 \qquad \text{a.s.} \tag{9.81}$$

We are going to derive a relationship between (9.76) and (9.80). Consider the two conditions

$$\liminf_{v\to\infty} \int_v^{cv} \frac{du}{g^{(\varphi)}(u)G^{(\varphi)}(u)} = \liminf_{v\to\infty} \int_{\varphi^{-1}(v)}^{\varphi^{-1}(cv)} \frac{du}{g(u)G(u)} > 0, \qquad \text{for all} \quad c > 1, \tag{9.82}$$

and

$$\lim_{c\downarrow 1} \limsup_{v\to\infty} \int_v^{cv} \frac{du}{g^{(\varphi)}(u)G^{(\varphi)}(u)} = \lim_{c\downarrow 1} \limsup_{t\to\infty} \int_{\varphi^{-1}(v)}^{\varphi^{-1}(cv)} \frac{du}{g(u)G(u)} = 0. \tag{9.83}$$

Note that condition (9.82) coincides with condition (9.27), with $G^{(\varphi)}$ instead of G (see Proposition 9.1). Similarly, condition (9.83) coincides with (9.28), with $G^{(\varphi)}$ instead of G (see Proposition 9.2). Moreover, condition (9.26) follows from condition (9.82) (see Remark 7.107).

Proposition 9.24 *Let g and σ be positive continuous functions such that the SDE (9.3) has a continuous solution X_b, for any initial condition $b > 0$, satisfying condition (9.69). Further, assume that $\varphi(u)$, $u \ge u_0 \ge 0$, is a positive, increasing, and continuously differentiable function tending to ∞ as $u \to \infty$. Then:*

1) if condition (9.82) holds, then relation (9.80) implies (9.76);
2) if conditions (9.26) and (9.83) hold, then relation (9.76) implies (9.80);
3) if conditions (9.82) and (9.83) hold, then (9.76) and (9.80) are equivalent.

Proof of Proposition 9.24. Since the function $g^{(\varphi)}$ is positive and continuous, relation (9.78) and statement 5) of Theorem 4.9, with $f = G^{(\varphi)}$, $\theta = 1/g^{(\varphi)}$ and $f'_{\ln} = 1/(g^{(\varphi)}G^{(\varphi)})$, imply that condition (9.82) is equivalent to $G^{(\varphi)} \in \mathcal{SQI}$ and $G_b^{(\varphi)} \in \mathcal{SQI}$, for all $b > 0$. This together with Corollary 7.33 results in $\varphi \circ \mu_b \in \mathcal{PRV}$, by relation (9.79), and thus the function $\varphi \circ \mu_b$ preserves the asymptotic equivalence of functions tending to infinity (see Definition 3.39). Hence (9.80) implies (9.81) and Theorem 3.42 yields

$$\lim_{t\to\infty} \frac{\varphi(X_b(t))}{\varphi(\mu_b(t))} = \lim_{t\to\infty} \frac{\varphi(\mu_b(G_b(X_b(t))))}{\varphi(\mu_b(t))} = \lim_{t\to\infty} \frac{G_b(X_b(t))}{t} = 1 \qquad \text{a.s.,}$$

since $\mu_b = G_b^{-1}$. So, statement 1) is proved.

Now, let condition (9.83) hold. Then (9.26), relation (9.78), and statement 3) of Theorem 4.9, with $f = G^{(\varphi)}, \theta = 1/g^{(\varphi)}$, and $f'_{\ln} = 1/(g^{(\varphi)}G^{(\varphi)})$, imply that (9.83) is equivalent to $G^{(\varphi)} \in \mathcal{PRV}$ and $G_b^{(\varphi)} \in \mathcal{PRV}$, for all $b > 0$. This together with Theorem 3.42 gives that the function $G_b^{(\varphi)} = G_b(\varphi^{-1})$ preserves the asymptotic equivalence of functions tending to infinity. Therefore (9.76) yields

$$\lim_{t\to\infty} \frac{G_b(X_b(t))}{t} = \lim_{t\to\infty} \frac{G_b(X_b(t))}{G_b(\mu_b(t))}$$

$$= \lim_{t\to\infty} \frac{G_b^{(\varphi)}(\varphi(X_b(t)))}{G_b^{(\varphi)}(\varphi(\mu_b(t)))} = \lim_{t\to\infty} \frac{\varphi(X_b(t))}{\varphi(\mu_b(t))} = 1 \qquad \text{a.s.},$$

since $G_b = \mu_b^{-1}$. So, statement 2) is also proved.

Since condition (9.26) follows from condition (9.82), statement 3) follows from 1) and 2). □

The following result contains conditions for (9.76).

Theorem 9.25 *Let g and σ be positive, continuous functions such that conditions (T1) and (T2) hold and the SDE (9.3) has a continuous solution X_b, for any initial condition $b > 0$, satisfying (9.69) (cf. Remark 9.16). Additionally, let $\varphi(u)$, $u \geq u_0 \geq 0$, be a positive, increasing, continuously differentiable function tending to ∞ as $u \to \infty$. If condition (9.82) holds, then relation (9.76) holds as well.*

Theorem 9.25 is a corollary of Theorem 9.18 and Proposition 9.24.

Considering Remark 9.8, we see that Theorem 9.25 implies the following result.

Corollary 9.26 *Let g and σ be positive, continuous functions such that the SDE (9.3) has a continuous solution X_b, for any initial condition $b > 0$, satisfying (9.69). Further, let the function σ/g be bounded, the function g be continuously differentiable, and condition (9.74) be satisfied. In addition, let $\varphi(u)$, $u \geq u_0 \geq 0$, be a positive, increasing, continuously differentiable function tending to ∞ as $u \to \infty$. If condition (9.82) holds, then relation (9.76) follows.*

In order to find rather simple conditions for (9.82) one can apply Proposition 7.109, with the functions $g^{(\varphi)}$ and $G^{(\varphi)}$ instead of g and G, respectively. This results in the following proposition.

Proposition 9.27 *Let g be a positive, continuous function and $\varphi(u)$, $u \geq u_0 \geq 0$, be a positive, increasing, continuously differentiable function tending to ∞ as $u \to \infty$. Assume that at least one of the following conditions holds:*

(i) $\limsup_{u\to\infty} g^{(\varphi)}(u)G^{(\varphi)}(u)/u = \limsup_{u\to\infty} g(u)G(u)\varphi'(u)/\varphi(u) < \infty$;
(ii) $g\varphi'$ *is an almost decreasing function (see (4.34), (4.36)), that is,*

$$\lim_{v\to\infty} \inf_{u\geq v} \frac{g(v)\varphi'(v)}{g(u)\varphi'(u)} > 0;$$

(iii) $g\varphi'$ *is a nonincreasing function for sufficiently large arguments;*
(iv) *there exists an* $\alpha < 1$ *such that*

$$0 < \liminf_{u\to\infty} u^{-\alpha} g^{(\varphi)}(u) \qquad \text{and} \qquad \limsup_{u\to\infty} u^{-\alpha} g^{(\varphi)}(u) < \infty;$$

(v) $(g^{(\varphi)})^*(c) < c$ *for all* $c > 1$, *where*

$$(g^{(\varphi)})^*(c) = \limsup_{u\to\infty} g^{(\varphi)}(cu)/g^{(\varphi)}(u)$$

is the upper limit function for $g^{(\varphi)}$*;*
(vi) $g^{(\varphi)}$ *is an RV-function with index* $\rho < 1$*;*
(vii) $g^{(\varphi)}$ *admits the representation*

$$g^{(\varphi)} = \frac{1}{\frac{1}{d_1 g_1^{(\varphi)}} + \cdots + \frac{1}{d_n g_n^{(\varphi)}}},$$

where $g_k^{(\varphi)}$, $k = 1, \ldots, n$, *are functions satisfying (9.82), and* d_k, $k = 1, \ldots, n$, *are almost constant functions (see Definition 4.11).*

Then condition (9.82) holds.

Now we are going to describe sufficient conditions for (9.83). To do so, we apply Proposition 9.3, with the functions $g^{(\varphi)}$ and $G^{(\varphi)}$ instead of g and G, respectively.

Proposition 9.28 *Let* g *be a positive, continuous function and let condition (9.26) hold. Assume that* $\varphi(u)$, $u \ge u_0 \ge 0$, *is a positive, increasing, and continuously differentiable function, tending to* ∞ *as* $u \to \infty$. *Let at least one of the following conditions hold:*

(i) $\liminf_{u\to\infty} g^{(\varphi)}(u)G^{(\varphi)}(u)/u = \liminf_{u\to\infty} g(u)G(u)\varphi'(u)/\varphi(u) > 0$;
(ii) $g\varphi'$ *is an almost increasing function (see (4.35) and (4.37)), that is,*

$$\lim_{v\to\infty} \inf_{u\ge v} \frac{g(u)\varphi'(u)}{g(v)\varphi'(v)} > 0;$$

(iii) $g\varphi'$ *is a nondecreasing function for sufficiently large arguments;*
(iv) *there exists an* $\alpha < 1$ *such that*

$$0 < \liminf_{u\to\infty} u^{-\alpha} g^{(\varphi)}(u) \qquad \text{and} \qquad \limsup_{u\to\infty} u^{-\alpha} g^{(\varphi)}(u) < \infty;$$

(v) *the Lebesgue measure of the set* $\left\{c \in (0,1] : (g^{(\varphi)})^*(c) < \infty\right\}$ *is positive, where* $(g^{(\varphi)})^*(c)$ *is the upper limit function for* $g^{(\varphi)}$*;*
(vi) $g^{(\varphi)}$ *is an ORV-function;*

(vii) $g^{(\varphi)}$ *admits the representation*

$$g^{(\varphi)} = \frac{1}{\frac{1}{d_1 g_1^{(\varphi)}} + \cdots + \frac{1}{d_n g_n^{(\varphi)}}},$$

where $g_k^{(\varphi)}$, $k = 1, \dots, n$, *are functions satisfying (9.83), and* d_k, $k = 1, \dots, n$, *are almost constant functions (see Definition 4.11).*

Then condition (9.83) holds.

Example 9.29 Let $g(u) = u$, $u > 0$, and $\varphi(u) = \ln(1+u)$, $u > 0$. Then

$$\lim_{u\to\infty} \frac{g^{(\varphi)}(u)G^{(\varphi)}(u)}{u} = \lim_{u\to\infty} \frac{g(u)G(u)\varphi'(u)}{\varphi(u)} = 1.$$

Thus, Propositions 9.27 and 9.28 imply (9.82) and (9.83).

9.6 $\varphi_{1,2}$-Equivalence of Solutions of Autonomous Ordinary Differential Equations

Consider two autonomous differential equations with positive initial conditions, i.e.

$$\begin{cases} d\mu_1(t) = g_1(\mu_1(t))\,dt, & t \ge 0, \quad \mu_1(0) = b_1 > 0, \\ d\mu_2(t) = g_2(\mu_2(t))\,dt, & t \ge 0, \quad \mu_2(0) = b_2 > 0, \end{cases} \tag{9.84}$$

where g_1 and g_2 are positive, continuous functions such that

$$\int_1^\infty \frac{du}{g_k(u)} = \infty, \qquad k = 1, 2. \tag{9.85}$$

Then equations (9.84) have unique solutions μ_1 and μ_2 tending to ∞ as $t \to \infty$, and both of them do not reach infinity in finite time (see Sect. 7.9). Note that

$$\begin{cases} \mu_1(t) = G_1^{-1}(t), & t \ge 0, \\ \mu_2(t) = G_2^{-1}(t), & t \ge 0, \end{cases} \tag{9.86}$$

where

$$G_k(v) = \int_{b_k}^{v} \frac{du}{g_k(u)}, \qquad v \ge b_k, \qquad k = 1, 2, \tag{9.87}$$

and G_k^{-1} is the inverse function of G_k, $k = 1, 2$.

Moreover, let two functions $\varphi_k = (\varphi_k(u), u \in \mathbf{R})$, $k = 1, 2$, be given, being positive, increasing, and continuously differentiable for $u \geq u_0 \geq 0$, tending to ∞ as $u \to \infty$.

The aim of this section is to find conditions for

$$\lim_{t\to\infty} \frac{\varphi_1(\mu_1(t))}{\varphi_2(\mu_2(t))} = 1. \tag{9.88}$$

If $\varphi_1(u) = \varphi_2(u) = u$, for $u \geq u_0 \geq 0$, then (9.88) takes the form

$$\lim_{t\to\infty} \frac{\mu_1(t)}{\mu_2(t)} = 1, \tag{9.89}$$

which can be considered as the asymptotic stability of the solution of the ODE

$$d\mu(t) = g(\mu(t))\,dt, \qquad t \geq 0, \qquad \mu(0) = b > 0,$$

with respect to perturbations of the initial condition b and the function g. Recall that conditions for the asymptotic stability of the solution with respect to the initial condition b have been studied in Sect. 7.9 (see Theorem 7.108).

To find conditions for (9.88) we use the functions which appeared in the previous section (see (9.77)). Let

$$G_k^{(\varphi_k)} = G_k \circ \varphi_k^{-1}, \qquad g_k^{(\varphi_k)} = (g_k \circ \varphi_k^{-1})(\varphi_k' \circ \varphi_k^{-1}), \qquad k = 1, 2, \tag{9.90}$$

where G_k, $k = 1, 2$, are the functions defined in (9.87), φ_k^{-1} is the inverse function of φ_k, and φ_k' is the derivative of φ_k, $k = 1, 2$.

The definitions of $g_k^{(\varphi_k)}$ and $G_k^{(\varphi_k)}$ immediately imply that

$$\frac{d}{du} G_k^{(\varphi_k)}(u) = \frac{1}{g_k^{(\varphi_k)}(u)}, \qquad k = 1, 2, \tag{9.91}$$

for $u > \varphi_k(b_k)$. Moreover, since $G_k = \mu_k^{-1}$ (see (9.86)),

$$G_k^{(\varphi_k)} = (\varphi_k \circ \mu_k)^{-1}, \qquad k = 1, 2, \tag{9.92}$$

that is, $(G_k(\varphi_k^{-1}(s)), s \geq \varphi_k(b_k))$ and $(\varphi_k(\mu_k(t)), t \geq 0)$, $k = 1, 2$, are inverse to each other.

Note also that (9.85) implies

$$\lim_{v\to\infty} G_k^{(\varphi_k)}(v) = \infty, \qquad k = 1, 2. \tag{9.93}$$

9.6.1 Conditions for the $\varphi_{1,2}$-Asymptotic Equivalence of Solutions of Ordinary Differential Equations in Terms of the Functions $G_k^{(\varphi_k)}$

In Theorem 9.31 below, we establish conditions for the implications

$$\lim_{v\to\infty}\frac{G_1^{(\varphi_1)}(v)}{G_2^{(\varphi_2)}(v)}=1 \quad\Longrightarrow\quad \lim_{t\to\infty}\frac{\varphi_1(\mu_1(t))}{\varphi_2(\mu_2(t))}=1, \tag{9.94}$$

$$\lim_{v\to\infty}\frac{G_1^{(\varphi_1)}(v)}{G_2^{(\varphi_2)}(v)}=1 \quad\Longleftarrow\quad \lim_{t\to\infty}\frac{\varphi_1(\mu_1(t))}{\varphi_2(\mu_2(t))}=1, \tag{9.95}$$

$$\lim_{v\to\infty}\frac{G_1^{(\varphi_1)}(v)}{G_2^{(\varphi_2)}(v)}=1 \quad\Longleftrightarrow\quad \lim_{t\to\infty}\frac{\varphi_1(\mu_1(t))}{\varphi_2(\mu_2(t))}=1. \tag{9.96}$$

Consider two conditions similar to (9.27) and (9.28) (see also (9.82) and (9.83)), i.e., for $k=1,2$,

$$\liminf_{v\to\infty}\int_v^{cv}\frac{du}{g_k^{(\varphi_k)}(u)G_k^{(\varphi_k)}(u)}$$
$$=\liminf_{v\to\infty}\int_{\varphi_k^{-1}(v)}^{\varphi_k^{-1}(cv)}\frac{du}{g_k(u)G_k(u)}>0 \qquad \text{for all} \qquad c>1, \tag{9.97}$$

and

$$\lim_{c\downarrow 1}\limsup_{v\to\infty}\int_v^{cv}\frac{du}{g_k^{(\varphi_k)}(u)G_k^{(\varphi_k)}(u)}$$
$$=\lim_{c\downarrow 1}\limsup_{v\to\infty}\int_{\varphi_k^{-1}(v)}^{\varphi_k^{-1}(cv)}\frac{du}{g_k(u)G_k(u)}=0. \tag{9.98}$$

Remark 9.30 If $g=g_k^{(\varphi_k)}$ and $G=G_k^{(\varphi_k)}$, $k=1,2$, then conditions (9.82) and (9.83) coincide with (9.97) and (9.98), respectively. Thus, sufficient conditions for (9.97) and (9.98) can be derived from Propositions 9.27 and 9.28.

Theorem 9.31 *Let g_k, $k=1,2$, be positive, continuous functions, and φ_k, $k=1,2$, be positive, increasing, and continuously differentiable functions, tending to ∞ as $u\to\infty$.*

1) *If condition (9.97) holds for at least one $k=1,2$, then relation (9.94) follows.*
2) *If conditions (9.85) and (9.98) hold for at least one $k=1,2$, then relation (9.95) follows.*

3) *If conditions (9.97) and (9.98) hold for at least one $k = 1, 2$, then relation (9.95) follows.*

Proof of Theorem 9.31. If condition (9.97) holds for at least one $k = 1, 2$, then at least one of the functions $G_k^{(\varphi_k)}$, $k = 1, 2$, belongs to the class $\mathcal{SQI}$ (see the proof of Proposition 9.24). Thus statement 1) follows from Corollary 7.52.

If conditions (9.85) and (9.98) hold for at least one $k = 1, 2$, then at least one of the functions $G_k^{(\varphi_k)}$, $k = 1, 2$, belongs to $\mathcal{PRV}$ (see the proof of Proposition 9.24). From this together with Corollary 7.33, and by relation (9.92), we conclude that at least one of the functions $\varphi_k \circ \mu_k$, $k = 1, 2$, belongs to $\mathcal{SQI}$. Thus, statement 2) follows from Corollary 7.52.

Therefore, given $k = 1, 2$, condition (9.85) follows from relation (9.97) (see Remark 7.107) and thus statement 3) follows from 1) and 2). □

Example 9.32 Let g_k, $k = 1, 2$, be positive, continuous functions, and φ_k, $k = 1, 2$, be positive, increasing, and continuously differentiable functions, tending to ∞ as $u \to \infty$. If at least one of the functions $g_k^{(\varphi_k)}$, $k = 1, 2$, belongs to $\mathcal{RV}_\rho$, with $\rho < 1$, then Propositions 9.27 and 9.28 and Theorem 9.31 yield relation (9.96).

9.6.2 Conditions for the $\varphi_{1,2}$-Asymptotic Equivalence of Solutions of Ordinary Differential Equations in Terms of the Functions $g_k^{(\varphi_k)}$

We are interested in finding conditions for the following implications and equivalence:

$$\lim_{v\to\infty} \frac{g_1^{(\varphi_1)}(v)}{g_2^{(\varphi_2)}(v)} = 1 \quad \Longrightarrow \quad \lim_{t\to\infty} \frac{\varphi_1(\mu_1(t))}{\varphi_2(\mu_2(t))} = 1, \tag{9.99}$$

$$\lim_{v\to\infty} \frac{g_1^{(\varphi_1)}(v)}{g_2^{(\varphi_2)}(v)} = 1 \quad \Longleftarrow \quad \lim_{t\to\infty} \frac{\varphi_1(\mu_1(t))}{\varphi_2(\mu_2(t))} = 1, \tag{9.100}$$

$$\lim_{v\to\infty} \frac{g_1^{(\varphi_1)}(v)}{g_2^{(\varphi_2)}(v)} = 1 \quad \Longleftrightarrow \quad \lim_{t\to\infty} \frac{\varphi_1(\mu_1(t))}{\varphi_2(\mu_2(t))} = 1. \tag{9.101}$$

First of all we consider an example showing that (9.99) may fail even for RV-functions g_1 and g_2.

Example 9.33 Let $\varphi_1(u) = \varphi_2(u) = u$,

$$g_1(u) = u, \qquad g_2(u) = u + \sqrt{u}, \qquad u > 0,$$

and $b_1 = b_2 = 1$. Then

$$\mu_1(t) = e^t, \qquad \mu_2(t) = \left(2e^{t/2} - 1\right)^2, \qquad t \geq 0.$$

Thus,

$$\lim_{t\to\infty} \frac{g_1(t)}{g_2(t)} = 1, \qquad \text{but} \qquad \lim_{t\to\infty} \frac{\mu_2(t)}{\mu_1(t)} = 4.$$

Note that both g_1 and g_2 belong to $\mathcal{RV}_1$.

9.6.3 An Application of Karamata's Theorem

In view of Theorem 9.31 and relations (9.94)–(9.96) the problem of determining sufficient conditions for (9.99)–(9.101) is close to the problem of finding a relationship between the asymptotic equivalence of absolutely continuous functions and that of their densities.

Let the functions $(f_1(u), u \geq u_0 \geq 0)$ and $(f_2(u), u \geq u_0 \geq 0)$ be positive and locally integrable (with respect to Lebesgue measure; see Sect. 6.2). Consider the Lebesgue integrals

$$F_k(v) = \int_{a_k}^{v} f_k(u)\,du, \qquad v \geq a_k, \qquad k = 1, 2,$$

where $a_1 > 0$ and $a_2 > 0$. We assume throughout this subsection that

$$\lim_{v\to\infty} F_k(v) = \infty, \qquad k = 1, 2. \tag{9.102}$$

A natural question is about sufficient conditions for the following implications and equivalence:

$$\lim_{u\to\infty} \frac{f_1(u)}{f_2(u)} = 1 \qquad \Longrightarrow \qquad \lim_{v\to\infty} \frac{F_1(v)}{F_2(v)} = 1, \tag{9.103}$$

$$\lim_{u\to\infty} \frac{f_1(u)}{f_2(u)} = 1 \qquad \Longleftarrow \qquad \lim_{v\to\infty} \frac{F_1(v)}{F_2(v)} = 1, \tag{9.104}$$

$$\lim_{u\to\infty} \frac{f_1(u)}{f_2(u)} = 1 \qquad \Longleftrightarrow \qquad \lim_{v\to\infty} \frac{F_1(v)}{F_2(v)} = 1. \tag{9.105}$$

Since

$$\liminf_{u\to\infty} \frac{f_1(u)}{f_2(u)} \le \liminf_{v\to\infty} \frac{F_1(v)}{F_2(v)} \quad \text{and} \quad \limsup_{v\to\infty} \frac{F_1(v)}{F_2(v)} \le \limsup_{u\to\infty} \frac{f_1(u)}{f_2(u)},$$

relation (9.103) follows from condition (9.102) without any additional assumption.

In contrast to (9.103), its converse relation (9.104) may fail.

Example 9.34 Let $f_1(u) = 2u$ and $f_2(u) = 2u(1 + \cos u^2)$, $u > 0$. Then $F_1(v) = v^2$, $F_2(v) = v^2 + \sin v^2$, $v > 0$, and

$$\lim_{v\to\infty} \frac{F_2(v)}{F_1(v)} = \lim_{v\to\infty} \left(1 + \frac{\sin v^2}{v^2}\right) = 1,$$

but

$$\frac{f_2(u)}{f_1(u)} = 1 + \cos u^2,$$

and the limit does not exist as $u \to \infty$. Note that $f_1 \in \mathcal{RV}_1$, but f_2 is not an RV-function.

Thus, relation (9.104) requires some additional conditions on the functions f_1 and f_2. In the case of RV-functions, the corresponding conditions are contained in the direct part of Karamata's theorem on the asymptotic behavior of integrals of RV-functions (see Sect. 6.2).

Lemma 9.35 *Let f_1 and f_2 be RV-functions with indices $\rho_1 > -1$ and $\rho_2 > -1$. Then relations (9.104) and (9.105) are satisfied.*

Proof of Lemma 9.35. Under the assumptions of the lemma, relation (9.102) as well as relation (9.103) hold, and hence one only needs to check that (9.104) is satisfied.

Let

$$\lim_{v\to\infty} \frac{F_2(v)}{F_1(v)} = 1. \tag{9.106}$$

According to Karamata's theorem (direct part),

$$\lim_{u\to\infty} \frac{t f_k(u)}{(1+\rho_k)F_k(u)} = 1, \qquad k = 1, 2, \tag{9.107}$$

since $\rho_k > -1$ (see relation (6.1)). This implies that F_k is an RV-function with index $1 + \rho_k > 0$, $k = 1, 2$. By (9.106), we see that $\rho_1 = \rho_2 = \rho$ and

$$\lim_{u\to\infty} \frac{f_1(u)}{f_2(u)} = \lim_{t\to\infty} \frac{u f_1(u)}{(1+\rho)F_1(u)} \cdot \lim_{t\to\infty} \frac{F_1(u)}{F_2(u)} \cdot \lim_{u\to\infty} \frac{(1+\rho)F_2(u)}{u f_2(u)} = 1.$$

Therefore relation (9.104) is proved. □

Now we turn to sufficient conditions for (9.99)–(9.101).

Theorem 9.36 *Let g_k, $k = 1, 2$, be positive, continuous functions, and φ_k, $k = 1, 2$, be positive, increasing, and continuously differentiable functions, tending to ∞ as $u \to \infty$. If condition (9.97) holds for at least one $k = 1, 2$, then relation (9.99) follows. In particular, if $\varphi_1 = \varphi_2 = \varphi$, then*

$$\lim_{u\to\infty} \frac{g_1(u)}{g_2(u)} = 1 \qquad \Longrightarrow \qquad \lim_{t\to\infty} \frac{\varphi(\mu_1(t))}{\varphi(\mu_2(t))} = 1. \tag{9.108}$$

Since relation (9.103) follows from (9.102) and since (9.85), corresponding to (9.102), follows from (9.97), Theorem 9.36 is a consequence of Theorem 9.31.

Theorem 9.36 and Example 9.32 imply the following result.

Corollary 9.37 *Let g_k, $k = 1, 2$, be positive, continuous functions, and φ_k, $k = 1, 2$, be positive, increasing, and continuously differentiable functions, tending to ∞ as $u \to \infty$. If at least one of the functions $g_k^{(\varphi_k)}$, $k = 1, 2$, belongs to $\mathcal{RV}_{\rho_k}$, with index $\rho_k < 1$, then relation (9.99) holds. In particular, if $\varphi_1 = \varphi_2 = \varphi$, then relation (9.108) follows.*

Example 9.33 shows that, in general, the condition $\rho_k < 1$ in Corollary 9.37 cannot be replaced by $\rho_k \le 1$.

Theorem 9.38 *Let g_k, $k = 1, 2$, be positive, continuous functions, and φ_k, $k = 1, 2$, be positive, increasing, and continuously differentiable functions, tending to ∞ as $u \to \infty$. If both functions $g_k^{(\varphi_k)}$, $k = 1, 2$, belong to $\mathcal{RV}_{\rho_k}$, with $\rho_k < 1$, then relation (9.101) holds. In particular, if $\varphi_1 = \varphi_2 = \varphi$, then*

$$\lim_{u\to\infty} \frac{g_1(u)}{g_2(u)} = 1 \qquad \Longleftrightarrow \qquad \lim_{t\to\infty} \frac{\varphi(\mu_1(t))}{\varphi(\mu_2(t))} = 1. \tag{9.109}$$

Theorem 9.38 follows from Lemma 9.35 and Corollary 9.37.

Corollary 9.39 *Let g_1 and g_2 be positive, continuous RV-functions, with indices $\rho_1 < 1$ and $\rho_2 < 1$. Then*

$$\lim_{u\to\infty} \frac{g_1(u)}{g_2(u)} = 1 \qquad \Longleftrightarrow \qquad \lim_{t\to\infty} \frac{\mu_1(t)}{\mu_2(t)} = 1. \tag{9.110}$$

9.7 $\varphi_{1,2}$-Order of Growth of Solutions of Autonomous Stochastic Differential Equations

In this section, the results of Sect. 9.5 are generalized, that is, we find sufficient conditions for

$$\lim_{t\to\infty} \frac{\varphi_1(X_1(t))}{\varphi_2(\mu_2(t))} = 1 \qquad \text{a.s.}, \tag{9.111}$$

where $X_1 = (X_1(t), t \geq 0)$ is the continuous solution of the Cauchy problem for the autonomous SDE

$$dX_1(t) = g_1(X_1(t))\,dt + \sigma_1(X_1(t))\,dW(t), \qquad t \geq 0, \qquad X_1(0) \equiv b_1 > 0, \tag{9.112}$$

and where $\mu_2 = (\mu_2(t), t \geq 0)$ is the continuous solution of the Cauchy problem for the autonomous ODE

$$d\mu_2(t) = g_2(\mu_2(t))\,dt, \quad t \geq 0, \qquad \mu_2(0) = b_2 > 0. \tag{9.113}$$

For the case of $\varphi_1 = \varphi_2 = \varphi$, $g_1 = g_2 = g$, this problem has been considered in Sect. 9.5.

Since

$$\lim_{t\to\infty} \frac{\varphi_1(X_1(t))}{\varphi_2(\mu_2(t))} = \lim_{t\to\infty} \frac{\varphi_1(X_1(t))}{\varphi_1(\mu_1(t))} \cdot \lim_{t\to\infty} \frac{\varphi_1(\mu_1(t))}{\varphi_2(\mu_2(t))}, \tag{9.114}$$

the conditions for relation (9.111) can immediately be derived from the results of Sects. 9.5 and 9.6. For example, Corollary 9.26 and Theorem 9.31 imply the following assertion.

Theorem 9.40 *Assume that*

(1) g_1 and σ_1 are positive, continuous functions such that the SDE (9.112) has a continuous solution X_1, for any initial condition $b_1 > 0$, a.s. tending to ∞ as $t \to \infty$;
(2) the function σ_1/g_1 is bounded;
(3) the function g_1 is continuously differentiable and

$$\lim_{u\to\infty} g_1'(u) = 0;$$

(4) g_2 is a positive, continuous function;
(5) φ_1 and φ_2, $u \geq u_0 \geq 0$, are positive, increasing, and continuously differentiable functions, tending to ∞ as $u \to \infty$;
(6) for at least one $k = 1, 2$,

$$\liminf_{v\to\infty} \int_{\varphi_k^{-1}(v)}^{\varphi_k^{-1}(cv)} \frac{du}{g_k(u)G_k(u)} > 0 \qquad \text{for all} \qquad c > 1,$$

with

$$G_k(v) = \int_{b_k}^{v} \frac{du}{g_k(u)}, \qquad v \geq b_k, \qquad k = 1, 2,$$

where φ_k^{-1} is the inverse function of φ_k, $k = 1, 2$.

Then

$$\lim_{v\to\infty}\frac{G_1(\varphi_1^{-1}(v))}{G_2(\varphi_2^{-1}(v))}=1 \implies \lim_{t\to\infty}\frac{\varphi_1(X_1(t))}{\varphi_2(\mu_2(t))}=1 \quad a.s. \tag{9.115}$$

and

$$\lim_{v\to\infty}\frac{g_1(\varphi_1^{-1}(v))\varphi_1'(\varphi_1^{-1}(v))}{g_2(\varphi_2^{-1}(v))\varphi_2'(\varphi_2^{-1}(v))}=1 \implies \lim_{t\to\infty}\frac{\varphi_1(X_1(t))}{\varphi_2(\mu_2(t))}=1 \quad a.s. \tag{9.116}$$

In particular, if $\varphi_1=\varphi_2=\varphi$, *then*

$$\lim_{v\to\infty}\frac{G_1(v)}{G_2(v)}=1 \implies \lim_{t\to\infty}\frac{\varphi(X_1(t))}{\varphi(\mu_2(t))}=1 \quad a.s. \tag{9.117}$$

and

$$\lim_{v\to\infty}\frac{g_1(v)}{g_2(v)}=1 \implies \lim_{t\to\infty}\frac{\varphi(X_1(t))}{\varphi(\mu_2(t))}=1 \quad a.s. \tag{9.118}$$

In turn, Corollaries 9.26 and 9.37 imply the following result.

Theorem 9.41 *Let assumptions* (1)–(5) *of Theorem 9.40 hold. If at least one of the functions* $g_k^{(\varphi_k)}$, $k=1,2$, *belongs to* $\mathcal{RV}_{\rho_k}$, *with* $\rho_k<1$, *then (9.115)–(9.118) hold.*

Below are some illustrative examples.

Example 9.42 (See Gihman and Skorohod [149], §16, Corollary 1) Let assumptions (1)–(3) of Theorem 9.40 hold. Furthermore, let $g_1(u)\sim Cu^\beta$ as $u\to\infty$, where $0\le\beta<1$ and $C>0$. If $g_2(x)=Cu^\beta$, $u>0$, and $\varphi_1(u)=\varphi_2(u)=u$, $u>0$, then, by Theorem 9.41,

$$\lim_{t\to\infty}\frac{X_1(t)}{(C(1-\beta)t)^{1/(1-\beta)}}=1 \quad \text{a.s.},$$

since $\mu_2(t)\sim(C(1-\beta)t)^{1/(1-\beta)}$ as $t\to\infty$.

Example 9.43 Let assumptions (1)–(3) of Theorem 9.40 hold. Furthermore, let $g_1(u)\sim Cu/(\ln u)^\gamma$ as $u\to\infty$, where $\gamma>0$ and $C>0$. If

$$\varphi_1(u)=\varphi_2(u)=(\ln(1+u))^{1+\gamma}, \qquad u>0,$$

and $g_2(x) = C(1+u)/(\ln(1+u))^\gamma$, $u > 0$, then $g_1^{(\varphi_1)}(u) \sim C(1+\gamma)$ as $u \to \infty$, and, by Theorem 9.41,

$$\lim_{t\to\infty} \frac{(\ln X_1(t))^{1+\gamma}}{C(1+\gamma)t} = 1 \qquad \text{a.s.,}$$

since $\varphi_2(\mu_2(t)) \sim (C(1+\gamma)t)$ as $t \to \infty$.

Example 9.44 Let assumptions (1)–(3) of Theorem 9.40 hold. Furthermore, let $g_1(u) \sim Cu\exp(-(\ln u)^r)$ as $u \to \infty$, where $0 < r < 1$ and $C > 0$. Note that $\exp((\ln u)^r)$, $u > 1$, is an SV-function and

$$\exp\left((\ln u)^r\right)/(\ln u)^\gamma \to \infty, \qquad u \to \infty,$$

for all $\gamma > 0$. If

$$\varphi_1(u) = \varphi_2(u) = \exp\left((\ln u)^r\right), \qquad u > 0,$$

and

$$g_2(u) = Cu\exp\left(-(\ln u)^r\right), \qquad u > 1,$$

then $g_1^{(\varphi_1)}(u) \sim r(\ln u)^{(r-1)/r}$ as $u \to \infty$, and, by Theorem 9.41,

$$\lim_{t\to\infty} \frac{\exp((\ln X_1(t))^r)}{\exp((\ln \mu_2(t))^r)} = 1 \qquad \text{a.s.}$$

9.8 $\varphi_{1,2}$-Equivalence of Solutions of Autonomous Stochastic Differential Equations

Consider two autonomous SDEs with positive initial conditions, i.e.

$$\begin{cases} dX_1(t) = g_1(X_1(t))\,dt + \sigma_1(X_1(t))\,dW_1(t), & t \ge 0, \\ dX_2(t) = g_2(X_2(t))\,dt + \sigma_2(X_2(t))\,dW_2(t), & t \ge 0, \end{cases} \tag{9.119}$$

$$\begin{cases} X_1(0) \equiv b_1 > 0, \\ X_2(0) \equiv b_2 > 0, \end{cases}$$

where W_1 and W_2 are standard Wiener processes, defined on a complete probability space $\{\Omega, \mathfrak{F}, \mathsf{P}\}$. For our considerations below the dependency between W_1 and W_2 does not play a role.

We assume that the coefficients of the equations, i.e.

$$g_1 = (g_1(u), u \in (-\infty, \infty)), \qquad g_2 = (g_2(u), u \in (-\infty, \infty))$$

and

$$\sigma_1 = (\sigma_1(u), u \in (-\infty, \infty)), \qquad \sigma_2 = (\sigma_2(u), u \in (-\infty, \infty))$$

are positive, continuous functions such that continuous solutions X_1 and X_2 exist, for all $b_1 > 0$ and $b_2 > 0$, that X_1 and X_2 a.s. tend to ∞ as $t \to \infty$, and that they do not reach infinity in finite time (see Remark 9.16).

Along with the SDEs (9.119) consider the corresponding ODEs

$$\begin{cases} d\mu_1(t) = g_1(\mu_1(t))\,dt, & t \geq 0, \qquad \mu_1(0) = b_1 > 0, \\ d\mu_2(t) = g_2(\mu_2(t))\,dt, & t \geq 0, \qquad \mu_2(0) = b_2 > 0. \end{cases} \tag{9.120}$$

If condition (9.85) holds, then these equations have unique solutions μ_1 and μ_2, tending to ∞ as $t \to \infty$, and both of them do not reach infinity in finite time (see Sect. 7.9).

Moreover, let $\varphi_k = (\varphi_k(u), u \in \mathbf{R})$, $k = 1, 2$, be two positive, increasing, and continuously differentiable functions, for $u \geq u_0 \geq 0$, tending to ∞ as $u \to \infty$.

The aim of this section is to find sufficient conditions for

$$\lim_{t\to\infty} \frac{\varphi_1(X_1(t))}{\varphi_2(X_2(t))} = 1 \qquad \text{a.s.} \tag{9.121}$$

Since

$$\lim_{t\to\infty} \frac{\varphi_1(X_1(t))}{\varphi_2(X_2(t))} = \lim_{t\to\infty} \frac{\varphi_1(X_1(t))}{\varphi_1(\mu_1(t))} \cdot \lim_{t\to\infty} \frac{\varphi_1(\mu_1(t))}{\varphi_2(\mu_2(t))} \cdot \lim_{t\to\infty} \frac{\varphi_2(\mu_2(t))}{\varphi_2(X_2(t))}, \tag{9.122}$$

the results from the previous sections can be applied. For example, Corollary 9.26 and Theorem 9.31 together with relation (9.122) imply the following result.

Theorem 9.45 *Assume that*

(1) g_k *and* σ_k, $k = 1, 2$, *are positive, continuous functions such that the SDEs (9.119) have continuous solutions* X_1 *and* X_2, *for all initial conditions* $b_1 > 0$ *and* $b_2 > 0$, *and that* X_1 *and* X_2 *a.s. tend to* ∞ *as* $t \to \infty$;

(2) the functions σ_1/g_1 *and* σ_2/g_2 *are bounded;*

(3) the functions g_1 *and* g_2 *are continuously differentiable with*

$$\lim_{u\to\infty} g_k'(u) = 0, \qquad k = 1, 2;$$

(4) φ_1 *and* φ_2 *are positive, increasing, and continuously differentiable functions, for* $u \geq u_0 \geq 0$, *tending to* ∞ *as* $u \to \infty$;

(5) for at least one $k = 1, 2$,

$$\liminf_{v\to\infty} \int_{\varphi_k^{-1}(v)}^{\varphi_k^{-1}(cv)} \frac{du}{g_k(u)G_k(u)} > 0 \qquad \textit{for all} \qquad c > 1,$$

with

$$G_k(v) = \int_{b_k}^{v} \frac{du}{g_k(u)}, \qquad v \geq b_k, \qquad k = 1, 2,$$

where φ_k^{-1} *is the inverse function of* φ_k, $k = 1, 2$.

Then

$$\lim_{v\to\infty} \frac{G_1(\varphi_1^{-1}(v))}{G_2(\varphi_2^{-1}(v))} = 1 \quad \Longrightarrow \quad \lim_{t\to\infty} \frac{\varphi_1(X_1(t))}{\varphi_2(X_2(t))} = 1 \qquad \text{a.s.} \tag{9.123}$$

and

$$\lim_{v\to\infty} \frac{g_1(\varphi_1^{-1}(v))\varphi_1'(\varphi_1^{-1}(v))}{g_2(\varphi_2^{-1}(v))\varphi_2'(\varphi_2^{-1}(v))} = 1 \quad \Longrightarrow \quad \lim_{t\to\infty} \frac{\varphi_1(X_1(t))}{\varphi_2(X_2(t))} = 1 \qquad \text{a.s.} \tag{9.124}$$

In particular, if $\varphi_1 = \varphi_2 = \varphi$, *then*

$$\lim_{v\to\infty} \frac{G_1(v)}{G_2(v)} = 1 \quad \Longrightarrow \quad \lim_{t\to\infty} \frac{\varphi(X_1(t))}{\varphi(X_2(t))} = 1 \qquad \text{a.s.} \tag{9.125}$$

and

$$\lim_{v\to\infty} \frac{g_1(v)}{g_2(v)} = 1 \quad \Longrightarrow \quad \lim_{t\to\infty} \frac{\varphi(X_1(t))}{\varphi(X_2(t))} = 1 \qquad \text{a.s.} \tag{9.126}$$

In turn, Corollaries 9.26 and 9.37 together with relation (9.122) yield the following result.

Theorem 9.46 *Let assumptions* (1)–(4) *of Theorem 9.45 hold and let at least one of the functions* $g_k^{(\varphi_k)}$, $k = 1, 2$, *belong to* $\mathcal{RV}_{\rho_k}$, *with* $\rho_k < 1$. *Then relations (9.123)–(9.126) hold.*

On using the equality

$$\lim_{t\to\infty} \frac{\varphi_1(\mu_1(t))}{\varphi_2(\mu_2(t))} = \lim_{t\to\infty} \frac{\varphi_1(\mu_1(t))}{\varphi_1(X_1(t))} \cdot \lim_{t\to\infty} \frac{\varphi_1(X_1(t))}{\varphi_2(X_2(t))} \cdot \lim_{t\to\infty} \frac{\varphi_2(X_2(t))}{\varphi_2(\mu_2(t))} \tag{9.127}$$

and applying Theorems 9.45 and 9.31, we obtain the following result complementing Theorem 9.45.

Theorem 9.47 *Let assumptions* (1)–(4) *of Theorem 9.45 be satisfied. Let*

(i) for $k = 1, 2$,

$$\liminf_{v\to\infty} \int_{\varphi_k^{-1}(v)}^{\varphi_k^{-1}(cv)} \frac{du}{g_k(u)G_k(u)} > 0 \quad \text{for all} \quad c > 1;$$

(ii) for $k = 1, 2$,

$$\lim_{c\downarrow 1} \limsup_{v\to\infty} \int_{\varphi_k^{-1}(v)}^{\varphi_k^{-1}(cv)} \frac{du}{g_k(u)G_k(u)} = 0.$$

Then

$$\lim_{v\to\infty} \frac{G_1(\varphi_1^{-1}(v))}{G_2(\varphi_2^{-1}(v))} = 1 \iff \lim_{t\to\infty} \frac{\varphi_1(X_1(t))}{\varphi_2(X_2(t))} = 1 \quad a.s. \tag{9.128}$$

In particular, if $\varphi_1 = \varphi_2 = \varphi$, *then*

$$\lim_{v\to\infty} \frac{G_1(v)}{G_2(v)} = 1 \iff \lim_{t\to\infty} \frac{\varphi(X_1(t))}{\varphi(X_2(t))} = 1 \quad a.s. \tag{9.129}$$

In turn, Theorems 9.47 and 9.38 together with relation (9.127) imply the next result complementing Theorem 9.46.

Theorem 9.48 *Let assumptions* (1)–(4) *of Theorem 9.45 hold and let the functions* $g_k^{(\varphi_k)}$ *belong to* $\mathcal{RV}_{\rho_k}$, *with* $\rho_k < 1$, $k = 1, 2$. *Then, along with (9.128)–(9.129),*

$$\lim_{v\to\infty} \frac{g_1(\varphi_1^{-1}(v))}{g_2(\varphi_2^{-1}(v))} = 1 \iff \lim_{t\to\infty} \frac{\varphi_1(X_1(t))}{\varphi_2(X_2(t))} = 1 \quad a.s.$$

In particular, if $\varphi_1 = \varphi_2 = \varphi$, *then*

$$\lim_{v\to\infty} \frac{g_1(v)}{g_2(v)} = 1 \iff \lim_{t\to\infty} \frac{\varphi(X_1(t))}{\varphi(X_2(t))} = 1 \quad a.s.$$

Corollary 9.49 *Let assumptions* (1)–(3) *of Theorem 9.45 hold and let both* g_1 *and* g_2 *be RV-functions with indices* $\rho_1 < 1$ *and* $\rho_2 < 1$. *Then*

$$\lim_{v\to\infty} \frac{g_1(v)}{g_2(v)} = 1 \iff \lim_{t\to\infty} \frac{X_1(t)}{X_2(t)} = 1 \quad a.s.$$

9.9 Comments

For an introductory text to stochastic interest models and SDEs see, e.g., Privault [295] (see also Brigo and Mercurio [52] and Carmona and Tehranchi [78]). Stochastic growth equations like (9.6) are widely discussed in the literature (see, for example, Ditlevsen and Samson [103]). A reader-friendly introduction to SDEs in science and engineering is Henderson and Plaschko [177]. Several applications of SDEs in physics and chemistry are described in Kampen [202]. Some results concerning the almost sure limit behavior of solutions of certain stochastic differential equations for quantum states are studied by Barchielli and Paganoni [25]. Growth models with stochastic differential equations in tumor immunology have been studied by Rosenkranz [307]. Stochastic filtration, optimal stopping, control and links to SDEs are discussed by Øksendal [287].

Section 9.1 This section is based on Buldygin et al. [63–67, 71]. Equation (9.16) has been studied by Buldygin and Timoshenko [74–76]. In the theory of ordinary differential equations, relations (9.4) and (9.5) mean that the equation $X'(t) = f(X)+g(X,t)$ is *asymptotically autonomous*, where $g(X,t)$ is called a perturbation (see, for example, Strauss and Yorke [351] and the references therein). In this language, we study conditions for the stochastic differential equation (9.3) to be asymptotically autonomous under the perturbation $\sigma(X)\,dW(t)$.

The study of the asymptotic behavior, as $t \to \infty$, of a solution of an autonomous stochastic differential equation traces back to Gikhman and Skorokhod [149]. This problem was also studied later by Keller et al. [211]. Kersting [214] investigates a similar problem for multi-dimensional SDEs. A converse to the property that no explosion of a solution of an SDE occurs has also been studied in the literature (see, for example, Chow and Khas'minskiĭ [85]).

Similar questions about the asymptotic behavior of solutions for stochastic *difference* equations have also been discussed in many papers starting with Kesten [216] (see, for example, Keller et al. [212], Klebaner [221], Appleby et al. [15], to mention a few).

The problem of asymptotic stability of solutions of stochastic linear systems has been studied by Khas'minskiĭ [217] (see also Khas'minskiĭ [218]). Some extensions have been obtained by Friedman and Pinsky [136].

Sufficient conditions for the crucial assumption, i.e. $X(t) \to \infty$ almost surely, are considered by Khas'minskiĭ [217] for linear systems of SDEs (see also Klesov and Timoshenko [231] for extensions to non-autonomous SDEs). A related problem, under $\limsup |X(t)| = \infty$ almost surely, is discussed in Appleby et al. [16] for autonomous SDEs. Berkolaiko and Rodkina [29] contains a discussion of unboundedness as well as that of the vanishing of solutions of stochastic difference equation. The unboundedness of solutions of linear stochastic systems has been studied by Friedman [134] (see also Friedman [135]).

The asymptotic closeness in the sense of relation (9.4) of solutions of stochastic and ordinary differential equations is studied in Samoilenko and Stanzhyts'kyi [314], Samoilenko et al. [315], and Krenevich [244, 245]. In a certain sense, the closeness means the same as asymptotic equivalence for a different equation, namely (9.4) means (9.5) with $e^{X(t)}$ instead of $X(t)$ and $e^{\phi(t)}$ instead of $\phi(t)$. (The converse implication also holds with $\log X(t)$ and $\log \phi(t)$ in place of $X(t)$ and $\phi(t)$ in (9.4).)

Section 9.2 This approach to studying the asymptotic behavior of solutions of SDEs has been developed in Buldygin et al. [63–65, 67, 71], and in Buldygin and Tymoshenko [75, 76]. Moreover, relations (9.9), (9.11), and the corresponding results of Chap. 8 can help to obtain limit theorems for generalized renewal processes constructed from solutions of the autonomous SDE (9.3) in the same way as in Buldygin et al. [63, 64].

Section 9.3 The first passage time is an example of a generalized renewal process. Keller et al. [213] discuss some limit distributions for the first passage times constructed from diffusion processes. Sciadas [322] studies first exit problems for solutions of SDEs. Imkeller et al. [190] consider the first exit times for Lévy driven diffusions. Lerche [249] carefully studies boundary crossings for a Brownian motion. A recent collection of papers (see Metzler et al. [269]) is entirely devoted to applications of first passage times.

Another generalized renewal process, the sojourn time, has been discussed by Ikeda and Watanabe [189] for solutions of SDEs.

Section 9.5 The limit $\lim_{t\to\infty}(\log|X(t)|)/t$ is called the *Lyapunov exponent* if $X(t)$ is a solution of a differential equation. A number of publications have been devoted to the estimation of a Lyapunov exponent for stochastic differential equations. The first paper we could find on this topic is Kozin [240], however many more publications have appeared since then (see, e.g., Mao [257] or, more recently, Mao [258], for a survey of results and further references). Guo et al. [157] discuss the same problem for stochastic differential delay equations. In our language, this is related to the problem of log-equivalence of the solution of an SDE and the solution of its corresponding ODE if this solution grows linearly.

We would also like to mention the so-called *rough large deviations*

$$\ln \mathsf{P}(S_n \geq x) \sim c_n = \ln e^{c_n}, \qquad n \to \infty,$$

which is the log-equivalence of $\mathsf{P}(S_n \geq x)$ and e^{c_n} in our language (see, for example, Dembo and Zeitouni [100]). Ventsel' [359] was the first to discuss rough large deviation theorems (for Markov processes in his case).

The results of this section have been obtained in Buldygin et al. [71] and in Buldygin and Tymoshenko [76].

Section 9.6 The strong law of large numbers and the law of the iterated logarithm for solutions of SDEs can be embedded into our framework. For example, let X_i, $i = 1, 2$, be solutions of SDEs with drift and diffusion coefficients a_i and σ_i. Further, let μ_i be the solutions of the ODEs obtained from the corresponding SDEs by deleting the diffusion terms. If $X_i(t) \sim \mu_i(t)$ almost surely as $t \to \infty$, $i = 1, 2$, and $\mu_1(t) = o(\mu_2(t))$, then $\lim_{t\to\infty}(X_1(t)/\mu_2(t)) = 0$ almost surely. The latter result is a strong law of large numbers for X_1 with nonrandom normalization μ_2. The classical case corresponds to $\mu_2(t) = t$. An SLLN for linear stochastic systems is proved by Palamarchuk [289].

An analogous reasoning applies to the law of the iterated logarithm, that is, to a relation like $\limsup_{t\to\infty}(X_1(t)/\mu_2(t)) = 1$ almost surely, or to the bounded law of the iterated logarithm, i.e., $\limsup_{t\to\infty}|X_1(t)/\mu_2(t)| < \infty$. Other methods for the law of the iterated logarithm can be found in Sundar [352], Kouritzin and Heunes [239], Wang [364], and Makhno [255]. Applications of the law of the iterated logarithm for solutions of SDEs to financial markets are discussed by Appleby and Wu [17].

Section 9.8 Relation (9.15) has been discussed in Buldygin et al. [67] and Buldygin et al. [71]. Conditions for the solutions of equations (9.119) to be asymptotically equivalent can be found in Buldygin and Tymoshenko [75, 76].

Chapter 10
Asymptotics for Renewal Processes Constructed from Multi-indexed Random Walks

In this chapter, we study the asymptotic behavior of the renewal function and that of the renewal process constructed from a random walk over a restricted domain of multidimensional time. In doing so, we essentially apply the $\mathcal{PRV}$ property studied in Sect. 3.3.1 (see Chap. 3).

10.1 Introduction

The importance of renewal processes is well-known; here we only mention *renewal risk models* as generalizations of the classical *Cramér–Lundberg model* in insurance (see also the other examples mentioned in the Preface). The renewal process defined in (1.1) can be viewed as an additive functional of a *marked point process* (see (2.9)), where the indicators play the role of marks. In a wide variety of scientific and engineering disciplines, a tremendous increase has taken place in the use of point process models to describe data from which images of objects of interest are formed. This explains the emphasis given to point processes on multidimensional spaces, especially to processes in two dimensions. However, the definition of the renewal process as in (1.1) does not work in this case, since the maximum of objects is not well-defined in the case of only partially ordered sets (like, e.g., the plane, space etc.). On the other hand, a definition as in (2.9) would still work perfectly.

The classical renewal process can be described as the last time until t when an event occurs (see the definition in (1.1)), that is, events may occur at an arbitrary point of the positive half-axis treated as an *occurrence time*. If events may occur at an arbitrary point of a plane, then the renewal process cannot uniquely be defined in this way. On the other hand, the interpretation of the classical renewal process as the *total number* of events which occurred until time t leads to a possible definition in the multi-dimensional case. In contrast to the process, the classical definition of its

V. V. Buldygin et al., *Pseudo-Regularly Varying Functions and Generalized Renewal Processes*, Probability Theory and Stochastic Modelling 91, https://doi.org/10.1007/978-3-319-99537-3_10

main characteristic, the *renewal function*, can be transferred from the classical case to the multi-dimensional one without any problem.

Another feature of our setting is that we allow an observer to detect the events in a specified domain rather than in the whole space. In the classical setting, this corresponds to the following definition. Let $\{S_n\}_{n\ge 1}$ be the sequence of cumulative sums of independent, identically distributed, positive random variables and $\{k_n\}_{n\ge 1}$ be a fixed subsequence. Then we may define the renewal process of an argument t as the total number of integers in the subsequence $\{k_n\}_{n\ge 1}$ such that $S_{k_n} \le t$. In other words, an observer can detect the occurrence of events only at moments belonging to the subsequence $\{k_n\}_{n\ge 1}$ (a side of the earth cannot be captured by a satellite happening to be at the opposite side; this will only be feasible after some time when the satellite will see the required side again). Having obtained observations from a *restricted* domain one nevertheless may want to make statistical inferences about the parameters of the underlying stochastic process.

In the multidimensional case, the choice of a domain for observations is much wider than in the classical case. The most simple domains can be described as the "viewing angle of a device". It turns out that a characteristic of the domain fully describes the asymptotic behavior of the renewal process and renewal function (see Theorems 10.3 and 10.4). The characteristic of the domain (see (10.15)) which describes the asymptotic behavior of the renewal process is different for different domains and differs from the classical one in most cases. Here, the $\mathcal{PRV}$ property (see Sect. 3.3.1) of this characteristic is needed, since otherwise the behavior of the process may be too irregular.

This chapter is organized as follows. Sections 10.2 and 10.3 recall some basic facts from classical renewal theory and from the theory of limit theorems for multi-indexed sums of independent, identically distributed random variables, respectively.

The definition of the renewal function and renewal process constructed from a random walk with d-dimensional time is discussed in Sects. 10.4 and 10.5. The latter plays a crucial role in describing the asymptotic behavior of a renewal process (like cumulative sums in the classical case; see Chap. 1). The process is defined by the representation (10.13), which also explains why (10.10) is called the "renewal function" in earlier works for $d > 1$.

The asymptotic behavior of the renewal function and renewal processes with restricted d-dimensional time is obtained in Sects. 10.6 and 10.7, respectively. The asymptotics depend on the dimension of indices and on the domain, where the process is defined. The assumption for (10.15), i.e., that the counting function for this domain belongs to the class $\mathcal{PRV}$, is of major importance.

Several examples of domains for which the latter assumption is valid are discussed in Sect. 10.8.

The asymptotic behavior of the renewal process and function is heavily related to the so-called *Dirichlet divisor problem* in number theory. The coefficient of the main term of the asymptotic is determined by the leading coefficient of the *Dirichlet divisor function*. To be self-contained we provide an elementary calculation of this coefficient in Sect. 10.9 for all $d \ge 1$.

An example of a domain for which the counting function does not belong to the class $\mathcal{PRV}$ is presented in Sect. 10.10.

10.2 Some Classical Results from Renewal Theory

Let $\{X_n\}_{n\ge1}$ be independent, identically distributed random variables and $\{S_n\}_{n\ge1}$ be their partial sums, i.e. $S_n = \sum_{k=1}^n X_k$. The sequence $S = \{S_n\}_{n\ge1}$ is also called a *random walk*. The renewal process $N = \{N(t)\}$ constructed from the random walk S has been defined in (1.1) (and denoted $L = \{L(t)\}$ there). The stochastic process N is well-defined almost surely if $\mathsf{P}(\Omega') = 1$, where

$$\Omega' = \left\{\omega \in \Omega : \lim_{n\to\infty} S_n(\omega) = \infty\right\}. \tag{10.1}$$

Then $N(t)$ is finite for all $\omega \in \Omega'$ and $t \ge 0$. For $\omega \notin \Omega'$, the values of $N(\cdot,\omega)$ do not matter; one may put (say) $N(\cdot,\omega) = 0$. Condition (10.1) holds if, e.g., the random variables X_n are nonnegative with positive expectation $\mu = \mathsf{E}\,X_1$. The latter assumption is of key importance in classical renewal theory, for which central topics are to investigate the asymptotic behavior, as $t \to \infty$, of the renewal process $\{N(t)\}$ and that of its expectation $\{U(t)\}$, where $U(t) = \mathsf{E}\,N(t)$, called the *renewal function*. It is not difficult to show that $U(t)$ is finite for all $t > 0$, if the random variables $\{X_n\}_{n\ge1}$ are nonnegative and nondegenerate.

The (so-called) *renewal theorem* asserts that, if the random variables $\{X_n\}$ are nonnegative and

$$0 < \mu = \mathsf{E}\,X_1 < \infty, \tag{10.2}$$

then

$$\lim_{t\to\infty} \frac{U(t)}{t} = \frac{1}{\mu}.$$

The dual result for renewal processes holds under the same assumption (10.2), that is,

$$\lim_{t\to\infty} \frac{N(t)}{t} = \frac{1}{\mu} \qquad \text{a.s.}$$

(see also Theorem 1.14 above).

There is another very useful representation of the renewal process. In the case of nonnegative random variables $\{X_n\}_{n\ge1}$, it is clear that

$$N(t) = \sum_{n=1}^{\infty} \mathbb{1}\{S_n \le t\}, \tag{10.3}$$

where $\mathbb{1}\{\cdot\}$ denotes the indicator function of a random event.

Passing to the expectation in (10.3), we get

$$U(t) = \sum_{n=1}^{\infty} \mathsf{P}(S_n \le t).$$

Along with the function U, its "density" u is also studied in renewal theory. If the random variables $\{X_n\}_{n\ge1}$ attain only integer values, then the function

$$u(t) = \sum_{n=1}^{\infty} \mathsf{P}(S_n = t)$$

is well-defined for positive integers t (and we set $u(t) = 0$ for other t). The well-known renewal theorem (see Erdős et al. [122]) asserts that

$$\lim_{t\to\infty} u(t) = \frac{1}{\mu},$$

provided the random variables $\{X_n\}_{n\ge1}$ are aperiodic and condition (10.2) holds.

It is clear that the results for u can be extended to U as well. However, it is worth mentioning that the conditions are not optimal if the limit properties for U are derived from the corresponding results for u. Moreover, noninteger-valued random variables cannot be treated this way.

10.3 Limit Theorems for Multiple Sums

There are many generalizations of the above renewal theorem. In this section, we consider the case where, instead of the usual partial sums, multi-indexed sums of random variables are used to define the renewal process N. First we briefly discuss some notation and results for multi-indexed sums.

Denote by $\mathbf{N}^d$ the space of vectors with d (≥ 1) positive integer coordinates. The elements of $\mathbf{N}^d$ will be denoted by $\mathbf{k}$, $\mathbf{n}$ etc. Consider a family $\{X(\mathbf{n}), \mathbf{n} \in \mathbf{N}^d\}$ of independent, identically distributed random variables and their multiple sums

$$S(\mathbf{n}) = \sum_{\mathbf{k}\preceq\mathbf{n}} X(\mathbf{k}),$$

where the symbol "$\preceq$" denotes the coordinate-wise partial ordering in $\mathbf{N}^d$, that is, $k_1 \le n_1, \dots, k_d \le n_d$ for $\mathbf{k} = (k_1, \dots, k_d)$ and $\mathbf{n} = (n_1, \dots, n_d)$. In view of the analogy with the case $d = 1$, the family $\{S(\mathbf{n}), \mathbf{n} \in \mathbf{N}^d\}$ is called a *random walk with multidimensional time* $\mathbf{n}$.

Some properties of random walks with multidimensional time follow immediately from the corresponding results for classical random walks. For example, if the expectation $\mu = \mathsf{E}\, X_1$ exists, then

$$\frac{S(\mathbf{n})}{|\mathbf{n}|} \xrightarrow{\mathsf{P}} \mu \qquad \text{as} \qquad |\mathbf{n}| \to \infty,$$

where $\mathbf{n} = (n_1, \ldots, n_d)$ and $|\mathbf{n}| = n_1 \cdots n_d$. The proof of this result is obvious and is based on Khintchine's classical law of large numbers for independent, identically distributed random variables.

But there are other properties of random walks with multidimensional time, which are no longer immediate from the corresponding classical results. In what follows we apply the *strong law of large numbers for multi-indexed sums* proved in Smythe [338]. For notational convenience, we introduce a random variable X, which is a copy of the random variables $X(\mathbf{k})$, and we set $\log^+ z = \log(1+z)$ for $z \geq 0$.

Theorem 10.1 *Let $\{X(\mathbf{n}), \mathbf{n} \in \mathbf{N}^d\}$ be independent, identically distributed random variables such that*

$$\mathsf{E}\, X = \mu \qquad \textit{exists} \tag{10.4}$$

and

$$\mathsf{E}\, |X| \left(\log^+ |X|\right)^{d-1} < \infty. \tag{10.5}$$

Then

$$\mathsf{P}\left(\left|\frac{S(\mathbf{n})}{|\mathbf{n}|} - \mu\right| \geq \varepsilon \ i.o.\right) = 0, \tag{10.6}$$

for all $\varepsilon > 0$, where here and in what follows "i.o." stands for "infinitely often".

In fact, (10.6) and (10.4)–(10.5) are equivalent (see Smythe [338]).

Various generalizations of Theorem 10.1 are known in the literature. The following *strong law of large numbers for subsets* has been obtained in Indlekofer and Klesov [196].

Theorem 10.2 *Let $\{X(\mathbf{n}), \mathbf{n} \in \mathbf{N}^d\}$ be independent, identically distributed random variables and $D \subseteq \mathbf{N}^d$. The strong law of large numbers*

$$\mathsf{P}\left(\left|\frac{S(\mathbf{n})}{|\mathbf{n}|} - \mu\right| \geq \varepsilon \ i.o. \, for \, \mathbf{n} \in D\right) = 0 \tag{10.7}$$

holds if and only if (10.4) is true together with

$$\sum_{\mathbf{n} \in D} \mathsf{P}(|X| \geq |\mathbf{n}|) < \infty. \tag{10.8}$$

We apply Theorem 10.2 below in this chapter.

Theorems 10.1 and 10.2 coincide if $D = \mathbf{N}^d$, since relation (10.8) is equivalent to (10.5) in this case.

Theorem 10.2 has been proved in Indlekofer and Klesov [196] only for the case $d = 2$ and for subsets D of a special form described below. Indeed, Theorem 10.2 may not be true in the case of a general subset D. We nevertheless believe that this result is valid for all $d \ge 2$ and for a rather general class of subsets D.

10.4 The Renewal Function for Multi-indexed Random Walks

The function

$$u_d(t) = \sum_{\mathbf{n}\in\mathbf{N}^d} \mathsf{P}(S(\mathbf{n}) = t) \tag{10.9}$$

has been studied in Ney and Wainger [283] for the case $d = 2$ and positive integer t, if the random variables $X(\mathbf{k})$ are integer-valued. In [283], $u_d(t)$ is called the *renewal sequence constructed from a random walk with multidimensional time*. This name for u_d clearly results from the analogy with the case $d = 1$, however, the "renewal process" itself has not been defined in [283] for $d > 1$. On the other hand, for $d > 1$, the problems considered in [283] are very similar to their counterparts in the case $d = 1$. Namely, the main problem is to find the asymptotic behavior of the function $u_d(t)$ as $t \to \infty$ (for integer-valued t).

It has been realized by Ney and Wainger [283] that the behavior of u_d in the case $d > 1$ differs from that in the case $d = 1$. For example, u_d is not bounded, if $d > 1$, in contrast to the case $d = 1$. Moreover, the classical method of making use of a difference equation for $u(t)$ (called the renewal equation, in the general non-lattice case) fails if $d > 1$, since "*There does not appear to be a natural analog of this equation in dimension two, mainly because lattice points of the plane are not linearly ordered under the natural order*" (cf. Ney and Wainger [283]). A similar observation holds true for higher dimensions as well. Despite these circumstances, the asymptotic behavior of u_d has been found in Ney and Wainger [283] by using direct Fourier analysis. The result for u_d obtained in [283] also allows us to study the asymptotic behavior of

$$U_d(t) = \sum_{\mathbf{n}\in\mathbf{N}^d} \mathsf{P}(S(\mathbf{n}) < t), \tag{10.10}$$

if $d = 2$. If $\tau_k = \operatorname{card}\{\mathbf{n} : |\mathbf{n}| = k\}$, $k \ge 1$, then the renewal function (10.10) can be rewritten as

$$U_d(t) = \sum_{k=1}^{\infty} \tau_k \, \mathsf{P}(S_k < t), \tag{10.11}$$

where $S_k = S(\mathbf{n})$ if $k = |\mathbf{n}|$. The sequence $\{\tau_k\}$ is called the *Dirichlet divisor function*. Finding the asymptotic behavior of $\tau_1 + \cdots + \tau_k$ as $k \to \infty$ is one of the classical problems in number theory.

Some of the assumptions in Ney and Wainger [283] have been weakened by Mori [254], also for the case of a general d. However, the assumptions in [254] are only optimal for $d = 2$ and $d = 3$. Nevertheless, as mentioned in [254], "*For the present our results are valid only for $d = 2$ or $d = 3$. However, it might be true for $d \geq 4$ if an order estimate in the divisor problem is improved for such d*". The method of proof in Mori [254] is the same as in Ney and Wainger [283] except that a non-uniform local limit theorem is used in [254] instead of a uniform local limit theorem (for details we refer to [254]).

The investigation of this topic has been continued in Galambos and Kátai [143, 144], where a finer estimate in the local central limit theorem is used to improve the results of Mori [254]. Nevertheless, it is explicitly stated in [143, 144] that "*...lack of knowledge in number theory imposes limitations on our results* [about $u_d(t)$]".

In all papers cited above, the asymptotic behavior of u_d is described by

$$\lim_{t\to\infty} \frac{u_d(t)}{(\log t)^{d-1}} = \frac{1}{\mu(d-1)!}$$

(for integer-valued t). Note, however, that this result can be proved only for $d \leq 3$ at present. We should like to stress once again that the case of $d \geq 4$ could also be treated if a "conjectured" rate of decay of the error of approximation in the generalized Dirichlet divisors problem holds true.

Clearly, the behavior of u_d depends on the distribution function F of the random variables $X(\mathbf{k})$. This dependence, however, disappears asymptotically if the distribution belongs to a certain (wide enough) class of distributions. To indicate the dependence of u_d on the distribution function F, we sometimes write $u_{d,F}$ instead of u_d.

A successful attempt has been made in Galambos et al. [142] to describe the behavior of $u_{d,F}$ in terms of $u_{d,\Phi}$, where Φ is the standard normal distribution function, i.e.

$$u_{d,F}(t) = u_{d,\Phi}(t) + o\left((\log t)^{d-1}\right), \qquad t \to \infty$$

(for integer-valued t). The idea of approximating by the Gaussian distribution is widely accepted in probability theory. In the framework of the renewal theorem for random walks with multidimensional time, this also allows us to improve the moment conditions, by reducing them to a finite second moment, and to involve the case of $d \geq 3$ as well. It is worth mentioning that the asymptotic behavior of $u_{d,\Phi}$ depends, in turn, on the rate of approximation in the generalized Dirichlet divisors problem.

In contrast to u_d, the asymptotic behavior of U_d can be treated for all $d \geq 1$. It is shown in Klesov [224] that

$$\lim_{t\to\infty} \frac{U_d(t)}{(\log t)^{d-1}} = \frac{1}{\mu(d-1)!},$$

if the expectation $\mathsf{E}\, X = \mu$ exists and is positive. Note that, for proving the latter asymptotic, direct probabilistic methods have been developed in Klesov [224]. A further result in [224] asserts that *there exists a polynomial* $\mathscr{P}$ *of order* $d-1$ *such that*

$$\lim_{t\to\infty} \left[\frac{U_d(t)}{t} - \frac{1}{\mu} \mathscr{P}\left(\log \frac{t}{\mu}\right) \right] = 0. \tag{10.12}$$

The polynomial $\mathscr{P}$ is strongly related to the polynomial in the approximation of the number of divisors in the Dirichlet problem. For example, the leading coefficient of the polynomial $\mathscr{P}$ equals $\frac{1}{(d-1)!}$. The condition imposed on the $X(\mathbf{k})$'s in Klesov [224] requires that

$$t(\log t)^{2(d-1)}\, \mathsf{P}(X \geq t) \to 0, \qquad t \to \infty,$$

which is much weaker than the finiteness of the second moment. Note also that this condition is even weaker than (10.2), if $d = 1$, and thus can be dropped in this case.

The method of proof in Klesov [224] differs from those in the earlier papers. It avoids the framework of analytical methods and makes use of a direct probabilistic reasoning instead. As a result, the problem can be reduced to the law of large numbers for multi-indexed sums $S(\mathbf{n})$.

10.5 Renewal Processes Constructed from Multi-indexed Random Walks

In the previous sections, the main attention has been paid to the renewal function U_d and its density u_d, while nothing is said about the renewal process itself. The reason is that relation (1.1) is meaningless if $d > 1$, since the space $\mathbf{N}^d$ is not linearly ordered with respect to the natural (coordinate-wise) ordering. This means that the definition of the renewal process is not obvious for $d > 1$. Relation (10.3) helps to solve this problem, that is, the representation of N as a sum of indicator functions may serve as a definition for all $d \geq 1$, i.e.

$$N_d(t) = \operatorname{card}\{\mathbf{n} \colon S(\mathbf{n}) \leq t\} = \sum_{\mathbf{n}\in\mathbf{N}^d} \mathbb{1}\{S(\mathbf{n}) \leq t\}. \tag{10.13}$$

The asymptotic behavior of the process N_d defined in this way has been studied in Klesov and Steinebach [229]. The idea is to reduce the problem of the asymptotic behavior of the renewal process to known results from probability theory, namely to the strong law of large numbers for multi-indexed sums (see Theorem 10.1). Expansions of type (10.12) have also been obtained in Klesov and Steinebach [229], for $d = 2$ and $d = 3$, while for $d > 3$ they only hold if the hypothesis about the rate of approximations in the generalized Dirichlet divisors problem is valid.

Below we consider the problem for the renewal process introduced in Indlekofer and Klesov [194]. Namely, let $D \subseteq \mathbf{N}^d$ and put

$$N_D(t) = \operatorname{card}\{\mathbf{n} \in D\colon S(\mathbf{n}) \le t\} = \sum_{\mathbf{n}\in D} \mathbb{1}\{S(\mathbf{n}) \le t\}. \tag{10.14}$$

For notational convenience we drop the subscript D and write $N(t)$ instead of $N_D(t)$. The result below holds for all $d \ge 1$ and all sets D, for which the function $A_D(t) \stackrel{\text{def}}{=} A(0, t)$ defined by (10.15) is pseudo-regularly varying (recall Definition 3.16). We omit the subscript D for A as well.

10.6 Asymptotics for Renewal Functions Constructed from Random Walks with Multidimensional Time

Throughout this section we assume that the random variables $X(\mathbf{k})$, $\mathbf{k} \in \mathbf{N}^d$, are non-negative, non-degenerate and such that condition (10.4) holds with $\mu > 0$. Moreover, let X denote a copy of the $X(\mathbf{k})$'s.

For a subset D of the space $\mathbf{N}^d$, consider the counting process

$$N(t) = \sum_{\mathbf{n}\in D} \mathbb{1}\{S(\mathbf{n}) \le t\}, \quad t > 0,$$

and its expectation

$$U(t) = \sum_{\mathbf{n}\in D} \mathsf{P}(S(\mathbf{n}) \le t), \quad t > 0.$$

For $x < y$, put

$$C(x, y) = \{\mathbf{n} \in D\colon x \le |\mathbf{n}| < y\},$$

and

$$A(x, y) = \operatorname{card}\{C(x, y)\}. \tag{10.15}$$

We assume that $A(0,t)$ is an *Avakumović–Karamata function*, that is, $A(0,t) \to \infty$, as $t \to \infty$, and

$$\lim_{c \to 1} \limsup_{t \to \infty} \frac{A(0, ct)}{A(0, t)} = 1$$

(see relation (3.10)). The Avakumović–Karamata functions have been studied in detail in Sect. 3.3. Recall that an important subclass of Avakumović–Karamata functions is $\mathcal{PRV}$, the class of pseudo-regularly varying functions.

Theorem 10.3 *Assume that* $X(\mathbf{k})$, $\mathbf{k} \in \mathbf{N}^d$, *are independent, identically distributed random variables with* $0 < \mathsf{E}\, X = \mu < \infty$. *If* D *is a subset of the space* $\mathbf{N}^d$ *such that* $A(0,t)$ *is an Avakumović–Karamata function, then*

$$\lim_{t \to \infty} \frac{U(t)}{A(0, t/\mu)} = 1.$$

Proof of Theorem 10.3 For all $0 < \varepsilon < 1$, we have

$$U(t) = \sum_{\mathbf{n} \in C_1(\varepsilon,t)} \mathsf{P}(S(\mathbf{n}) \le t) + \sum_{\mathbf{n} \in C_2(\varepsilon,t)} \mathsf{P}(S(\mathbf{n}) \le t) + \sum_{\mathbf{n} \in C_3(\varepsilon,t)} \mathsf{P}(S(\mathbf{n}) \le t),$$

where

$$C_1(\varepsilon, t) = C(0, (1-\varepsilon)t/\mu), \quad C_2(\varepsilon, t) = C((1-\varepsilon)t/\mu, (1+\varepsilon)t/\mu),$$
$$C_3(\varepsilon, t) = C((1+\varepsilon)t/\mu, \infty).$$

We consider each term separately.

The term $C_3(\varepsilon, t)$. Let $r > 0$ be such that

$$\mu^r = \int_{[0,r)} x \, dF(x) > \frac{\mu}{1+\varepsilon},$$

where F is the joint distribution function of the $X(\mathbf{k})$'s, $\mathbf{k} \in \mathbf{N}^d$. Consider their truncated versions at the level r, i.e. $X^r(\mathbf{k}) = X(\mathbf{k}) \mathbb{1}\{X(\mathbf{k}) < r\}$, which are again independent, identically distributed random variables. Put $S^r(\mathbf{n}) = \sum_{\mathbf{k} \le \mathbf{n}} X^r(\mathbf{k})$. The random variables $X^r(\mathbf{k})$ are almost surely bounded from above by r and $\mathsf{E}\, X^r(\mathbf{k}) = \mu^r$. Moreover,

$$\sum_{\mathbf{n} \in C_3(\varepsilon,t)} \mathsf{P}(S(\mathbf{n}) \le t) \le \sum_{\mathbf{n} \in C_3(\varepsilon,t)} \mathsf{P}(S^r(\mathbf{n}) \le t).$$

Since $t - \mu^r |\mathbf{n}| < 0$ for $\mathbf{n} \in C_3(\varepsilon, t)$, we have

$$\sum_{\mathbf{n} \in C_3(\varepsilon,t)} \mathsf{P}(S(\mathbf{n}) \le t) \le \sum_{\mathbf{n} \in \mathbf{N}^d} \mathsf{P}\left(|S^r(\mathbf{n}) - |\mathbf{n}|\mu^r| \ge \delta |\mathbf{n}|\right),$$

where $\delta = \mu^r - \frac{\mu}{1+\varepsilon} > 0$. The series on the right-hand side converges, for all $\delta > 0$, according to Smythe's [339] generalization of the Hsu–Robbins–Erdős theorem on the complete convergence of sums (see Hsu and Robbins [185], Erdős [120]). The necessary and sufficient moment condition for the convergence of this series is that

$$\mathsf{E}\, X^r(\mathbf{n}) = \mu^r, \qquad \mathsf{E}\left(X^r(\mathbf{n})\right)^2 \mathrm{Log}^{d-1}(|X^r(\mathbf{n})|) < \infty,$$

which obviously holds in the case under consideration, since the random variables are bounded a.s.. Here and in what follows we write $\mathrm{Log}(z) = \log(1 + |z|)$ for real numbers z.

The term $C_2(\varepsilon, t)$. Clearly,

$$\sum_{\mathbf{n} \in C_2(\varepsilon,t)} \mathsf{P}(S(\mathbf{n}) \le t) \le A(0, (1+\varepsilon)t/\mu) - A(0, (1-\varepsilon)t/\mu).$$

The term $C_1(\varepsilon, t)$. For $\mathbf{n} \in C_1(\varepsilon, t)$, we have

$$1 - \mathsf{P}(S(\mathbf{n}) \le t) \le \mathsf{P}(|S(\mathbf{n}) - |\mathbf{n}|\mu| \ge t - |\mathbf{n}|\mu) \le \mathsf{P}(|S(\mathbf{n}) - |\mathbf{n}|\mu| \ge |\mathbf{n}|\varepsilon/(1-\varepsilon)).$$

Since $\mathsf{P}(|S(\mathbf{n}) - |\mathbf{n}|\mu| \ge |\mathbf{n}|\varepsilon/(1-\varepsilon)) \to 0$ as $|\mathbf{n}| \to \infty$, the law of large numbers implies that

$$\lim_{t\to\infty} \frac{1}{A(0, (1-\varepsilon)t/\mu)} \sum_{\mathbf{n} \in C(0,(1-\varepsilon)t/\mu)} \mathsf{P}(S(\mathbf{n}) \le t) = 0.$$

On the other hand, let $n_0 = n_0(\varepsilon)$ be such that $\mathsf{P}(|S(\mathbf{n}) - |\mathbf{n}|\mu| < |\mathbf{n}|\varepsilon) \ge 1 - \varepsilon$ for all $\mathbf{n}$ with $|\mathbf{n}| \ge n_0$. Put $C_0(\varepsilon, t) = C(n_0, (1+\varepsilon)t/\mu)$. Then

$$U(t) \ge (1-\varepsilon) A(n_0, (1+\varepsilon)t/\mu).$$

A combination of the above relations completes the proof of Theorem 10.3. □

10.7 Asymptotics for Renewal Processes Constructed from Random Walks with Multidimensional Time

Now we are in a position to state the main result on the strong law of large numbers for renewal processes constructed from random walks with multidimensional time.

Theorem 10.4 *Assume that* $\{X(\mathbf{n}), \mathbf{n} \in \mathbf{N}^d\}$ *are nonnegative and independent, identically distributed random variables satisfying relations (10.4) and (10.5). Let*

$\mu > 0$ and let $D \subseteq \mathbf{N}^d$ be an infinite subset of the space $\mathbf{N}^d$ such that

$$\lim_{c \downarrow 1} \limsup_{t \to \infty} \frac{A(ct)}{A(t)} = 1, \tag{10.16}$$

where the function $A(t) \stackrel{def}{=} A(0, t)$ is as defined in (10.15) (depending on D). Then

$$\lim_{t \to \infty} \frac{N(t)}{A(t/\mu)} = 1 \qquad a.s.,$$

where the process N is given in (10.14) (and the subscript D is dropped here as well as in (10.16)).

Recall that a function satisfying condition (10.16) is called pseudo-regularly varying (see Sect. 3.3).

Corollary 10.5 *Assume that $\{X(\mathbf{n}), \mathbf{n} \in \mathbf{N}^d\}$ are nonnegative and independent, identically distributed random variables satisfying conditions (10.4) and (10.5). Let $\mu > 0$. If $D = \mathbf{N}^d$, then*

$$\lim_{t \to \infty} \frac{N(t)}{t(\log t)^{d-1}} = \frac{1}{\mu(d-1)!} \qquad a.s.$$

Corollary 10.5 follows immediately from Theorem 10.4, since

$$A(t) \sim \frac{1}{(d-1)!} t (\log t)^{d-1}, \qquad t \to \infty.$$

The above asymptotic behavior of the function A results from a "rough" estimate in the Dirichlet divisors problem. Indeed, in this case $A(t)$ is the number of solutions of the inequality $|\mathbf{n}| \le t$. Many proofs of this result are known (for an elementary one see, e.g., Klesov and Steinebach [230]).

Proof of Theorem 10.4 First we consider the special case of $\mu = 1$. Fix $0 < \varepsilon < 1$ and introduce the three processes

$$N_1(t) = \sum_{\mathbf{n} \in D:\ |\mathbf{n}| < (1-\varepsilon)t} \mathbb{1}\{S(\mathbf{n}) \le t\},$$

$$N_2(t) = \sum_{\mathbf{n} \in D:\ (1-\varepsilon)t \le |\mathbf{n}| < (1+\varepsilon)t} \mathbb{1}\{S(\mathbf{n}) \le t\},$$

$$N_3(t) = \sum_{\mathbf{n} \in D:\ |\mathbf{n}| \ge (1+\varepsilon)t} \mathbb{1}\{S(\mathbf{n}) \le t\}.$$

Next we separately derive bounds for the processes N_1, N_2, and N_3.

The process N_1. Since $\mu = 1$, we get

$$|N_1(t) - A((1-\varepsilon)t)| = \sum_{\mathbf{n} \in D\colon |\mathbf{n}| < (1-\varepsilon)t} \mathbb{1}\{S(\mathbf{n}) > t\}$$
$$\leq \sum_{\mathbf{n} \in D\colon |\mathbf{n}| < (1-\varepsilon)t} \mathbb{1}\left\{\frac{S(\mathbf{n})}{|\mathbf{n}|} - \mu \geq \varepsilon_1\right\},$$

where $\varepsilon_1 = \frac{\varepsilon}{1-\varepsilon}$. Let

$$\xi(\varepsilon, \omega) = \operatorname{card}\left\{\mathbf{n} \in D\colon \left|\frac{S(\mathbf{n}, \omega)}{|\mathbf{n}|} - \mu\right| \geq \varepsilon\right\}$$

for $\varepsilon > 0$. Finally,

$$|N_1(t) - A((1-\varepsilon)t)| \leq \xi(\varepsilon_1). \tag{10.17}$$

The process N_2. Since the indicator function does not exceed 1,

$$N_2(t) \leq A((1+\varepsilon)t) - A((1-\varepsilon)t). \tag{10.18}$$

The process N_3. Put

$$\varepsilon_2 = \frac{\varepsilon}{1+\varepsilon}.$$

Then

$$N_3(t) \geq \sum_{\mathbf{n} \in D\colon |\mathbf{n}| \geq (1+\varepsilon)t} \mathbb{1}\left\{\frac{S(\mathbf{n})}{|\mathbf{n}|} - \mu \leq -\varepsilon_2\right\} \leq \xi(\varepsilon_2). \tag{10.19}$$

On combining estimates (10.17)–(10.19), we obtain

$$N(t) \leq |N_1(t) - A((1-\varepsilon)t)| + A((1-\varepsilon)t) + N_2(t) + N_3(t)$$
$$\leq \xi(\varepsilon_1) + A((1+\varepsilon)t) + \xi(\varepsilon_2).$$

According to Theorem 10.1, there are two random events Ω_1 and Ω_2 such that $\mathsf{P}(\Omega_1) = \mathsf{P}(\Omega_2) = 1$ and $\xi(\varepsilon_1, \omega)$ is finite for every $\omega \in \Omega_1$, while $\xi(\varepsilon_2, \omega)$ is finite for every $\omega \in \Omega_2$. Thus, for $\omega \in \Omega_1 \cap \Omega_2$,

$$\limsup_{t\to\infty} \frac{N(t, \omega)}{A(t/\mu)} \leq \limsup_{t\to\infty} \frac{A((1+\varepsilon)t/\mu)}{A(t/\mu)},$$

whence

$$\limsup_{t\to\infty} \frac{N(t,\omega)}{A(t/\mu)} \le \lim_{\varepsilon\downarrow 0} \limsup_{t\to\infty} \frac{A((1+\varepsilon)t/\mu)}{A(t/\mu)} = 1, \tag{10.20}$$

for every $\omega \in \Omega_1 \cap \Omega_2$ by condition (10.16).

A lower bound can be obtained in a similar way, i.e.

$$\begin{aligned} N(t) &\ge -|N_1(t) - A((1-\varepsilon)t)| + A((1-\varepsilon)t) + N_2(t) + N_3(t) \\ &\ge -\xi(\varepsilon_1) + A((1-\varepsilon)t), \end{aligned}$$

and, for every $\omega \in \Omega_1$,

$$\liminf_{t\to\infty} \frac{N(t,\omega)}{A(t/\mu)} \ge \lim_{\varepsilon\downarrow 0} \liminf_{t\to\infty} \frac{A((1-\varepsilon)t/\mu)}{A(t/\mu)} = 1$$

by condition (10.16). The latter bound together with (10.20) completes the proof of the theorem in the case $\mu = 1$.

Now, if $\mu \ne 1$, we introduce the random variables $X_1(\mathbf{n}) = X(\mathbf{n})/\mu$, their sums $S_1(\mathbf{n})$, and the corresponding renewal process $N_1(t)$. Clearly, $\mathsf{E}\, X_1(\mathbf{n}) = 1$ and $N_1(t) = N(t\mu)$, and an application of the above result for unit expectation gives

$$\lim_{t\to\infty} \frac{N(t\mu)}{A(t)} = 1 \qquad \text{a.s.},$$

which is equivalent to what has to be proved. □

Remark 10.6 We used condition (10.5) in the proof of Theorem 10.4. It can easily be seen from the proof that this condition is optimal if $D = \mathbf{N}^d$, that is, for the proof of Corollary 10.5. For some special cases of D, however, this condition is too restrictive, but its optimality can be proved in many "regular" cases (see Indlekofer and Klesov [196]).

10.8 Examples

Example 10.7 Let $D = \mathbf{N}^d$. Then $A(0,t) \sim t\mathcal{P}(\log t)$ as $t \to \infty$, where $\mathcal{P}$ denotes a polynomial of order $d-1$ with leading coefficient $\frac{1}{(d-1)!}$. Although this fact is well-known, we give an elementary proof in Sect. 10.9. In this case, Theorem 10.3 implies that

$$\lim_{t\to\infty} \frac{U(t)}{t\log^{d-1} t} = \frac{1}{\mu(d-1)!}.$$

Example 10.8 Let $d = 2$ and $\theta \geq 1$. Consider the sector

$$D = \left\{(m, n) \colon \theta^{-1} m \leq n \leq \theta m\right\}.$$

We distinguish between the two cases $\theta = 1$ and $\theta > 1$.

If $\theta = 1$, then $A(0, t) \sim \sqrt{t}$, and Theorem 10.3 implies that

$$\lim_{t \to \infty} \frac{U(t)}{\sqrt{t}} = \frac{1}{\sqrt{\mu}}. \tag{10.21}$$

If $\theta > 1$, then $A(0, t) \sim t \log \theta$ and

$$\lim_{t \to \infty} \frac{U(t)}{t} = \frac{\log \theta}{\mu}. \tag{10.22}$$

Example 10.7 provides a kind of boundary case for the asymptotic behavior of $U(t)$ given in (10.22), where $\theta \to \infty$, that is,

$$\lim_{t \to \infty} \frac{U(t)}{t \log t} = \frac{1}{\mu}.$$

Example 10.9 Consider the case $d = 2$. Let f be an increasing function such that

$$f(x) \leq x \qquad \text{for all } x \geq 0,$$

and define the subset

$$D = \{(m, n) \colon f(m) \leq n \leq m\}.$$

Under certain conditions on f, $A(0, t)$ is an Avakumović–Karamata function. In what follows we assume that f is continuous, whence the inverse function of $x f(x)$ exists. Suppose that

$$f(x) = cx + o(x), \qquad x \to \infty, \tag{10.23}$$

for some $0 < c < 1$.

It is clear that $A(t)$ is the number of points (m, n) in the set D whose coordinates are such that $mn \leq t$. Thus

$$\begin{aligned} A(t) &= \sum_{m \leq m_1} (m - [f(m)] + O(1)) + \sum_{m_1 < m \leq m_2} \left(\left[\frac{t}{m}\right] - [f(m)] + O(1)\right) \\ &= \sum_{m \leq m_1} (m - f(m)) + \sum_{m_1 < m \leq m_2} \left(\frac{t}{m} - f(m)\right) + O(m_2), \end{aligned}$$

where $m_1 = [x_1]$, $m_2 = [x_2]$ and $x_1 = \sqrt{t}$, $x_2 f(x_2) = t$. From the asymptotic expansion of the harmonic series we get

$$A(t) = \frac{m_1(m_1+1)}{2} + t\log\frac{m_2}{m_1} - \sum_{m\le m_2} f(m) + O(x_2).$$

Now, (10.23) implies that

$$A(t) = \frac{m_1(m_1+1)}{2} + t\log\frac{m_2}{m_1} - \frac{cm_2(m_2+1)}{2} + o\left(x_2^2\right)$$

or

$$A(t) = \frac{x_1^2 - cx_2^2}{2} + t\log\frac{x_2}{x_1} + o\left(x_2^2\right) = \frac{t}{2}\left(1 - c\frac{x_2^2}{x_1^2}\right) + t\log\frac{x_2}{x_1} + o\left(x_2^2\right).$$

Therefore the asymptotic behavior of $A(t)$ is determined by that of the ratio x_2/x_1 and the remainder term $O(x_2)$. Writing the expression for x_2 explicitly in terms of t, we obtain $x_2(t) = g^{-1}(t)$, where $g(x) = xf(x)$. Since, by assumption (10.23), $g(t)/t^2 \to c$ as $t \to \infty$, we can apply the first statement of Theorem 7.57 with the numerator $g(t)$ and denominator t^2, which is a POV-function. Obviously, not only quasi-inverse functions, but also inverse functions exist in the case under investigation, and therefore,

$$\frac{1}{\sqrt{c}} = \lim_{t\to\infty}\frac{g^{-1}(t)}{\sqrt{t}} = \lim_{t\to\infty}\frac{x_2}{x_1}.$$

Hence $x_2 = O\left(\sqrt{t}\right)$ and

$$A(0,t) = -\frac{t}{2}\log c + o(t), \qquad t\to\infty,$$

if $0 < c < 1$, that is, the renewal theorem (10.22) holds for $\theta = c^{-1/2}$.

The case $c = 0$, that is,

$$f(x) = o(x), \qquad x\to\infty,$$

can be considered similarly.

Let m_x be the unique integer solution of the inequalities

$$m_x[f(m_x)] \le x < (m_x+1)[f(m_x+1)],$$

where $[z]$, as usual, denotes the integer part of a number z. Let

$$\delta(x) = \frac{1}{x} f^2(m_x).$$

Then

$$A(0,t) = t \log \frac{1}{\delta(t)} + O(t), \qquad t \to \infty,$$

whence

$$\lim_{t\to\infty} \frac{U(t)}{t \log \delta^{-1}(t/\mu)} = \frac{1}{\mu}.$$

For example, if $f(x) = x/\log(x)$, then $m_x \asymp \sqrt{x \log(x)}$ and $\delta^{-1}(x) \asymp \log(x)$ as $x \to \infty$. Thus

$$\lim_{t\to\infty} \frac{U(t)}{t \log\log t} = \frac{1}{\mu}.$$

Note that the asymptotic behavior of $A(0,t)$ is more complicated in the case $c = 1$, which is not treated here.

10.9 The Leading Coefficient in the Dirichlet Divisor Problem

Let $|\mathbf{n}| = n_1 n_2 \dots n_d$ for $\mathbf{n} = (n_1, n_2, \dots, n_d) \in \mathbf{N}^d$ and $T^{(d)}(x) = \operatorname{card}\{\mathbf{n} : |\mathbf{n}| \le x\}$ for $x \ge 1$. The aim of this section is to provide an elementary proof of the relation

$$T^{(d)}(x) = \frac{1}{(d-1)!} x \log^{d-1} x + g_d(x) x \log^{d-2} x, \tag{10.24}$$

for $d \ge 2$ and $x \ge 1$, where the function g_d is such that $\sup_{x\ge1} |g_d(x)| < \infty$.

We need two auxiliary results.

Lemma 10.10 *Let $\alpha \ge 0$. Then*

$$\sum_{i=1}^{[x]} \frac{1}{i} \log^\alpha i = \frac{1}{\alpha+1} \log^{\alpha+1} x + h_\alpha(x), \qquad x \to \infty,$$

where

$$h_\alpha(x) = \psi(x) + \int_1^x \rho(t) \frac{\log^\alpha t - \alpha \log^{\alpha-1} t}{t^2}\, dt$$

and

$$\rho(x) = [x] - x - \frac{1}{2}, \qquad \psi(x) = \begin{cases} \frac{\log^\alpha [x]}{2[x]}, & \textit{for } \alpha > 0, \\ \frac{1}{2}\left(\frac{1}{[x]} + 1\right), & \textit{for } \alpha = 0. \end{cases} \tag{10.25}$$

Note that

$$\sup_{x \ge 1} |h_\alpha(x)| < \infty, \quad \text{for all} \quad \alpha \ge 0, \quad \text{and}$$

$$h_\alpha(x) \to \int_1^\infty \rho(t) \frac{\log^\alpha t - \alpha \log^{\alpha-1} t}{t^2}\, dt, \qquad x \to \infty.$$

Lemma 10.11 *Let C_n^m denote the binomial coefficient $\binom{n}{m}$. Then*

$$\sum_{l=0}^{d-2} C_{d-2}^{l} (-1)^{d-2-l} \frac{1}{d-1-l} = \frac{1}{d-1}.$$

Proof of (10.24) Put $|\mathbf{n}|_{2,\dots,d} = n_2 \cdots n_d$. Then

$$T^{(d)}(x) = \text{card}\left\{ \mathbf{n} : n_1 \le x, |\mathbf{n}|_{2,\dots,d} \le \frac{x}{n_1} \right\}.$$

Thus

$$T^{(d)}(x) = \sum_{n_1=1}^{[x]} T^{(d-1)}(x/n_1). \tag{10.26}$$

We use (10.26) to prove (10.24) by induction. Note that, for $d = 1$,

$$T^{(1)}(x) = \sum_{k_1=1}^{[x]} \mathbb{1}_{k_1} = [x] = x + g_1(x)$$

where $g_1(x) = [x] - x$.

Next we consider the case $d = 2$. Lemma 10.10 implies that

$$T^{(2)}(x) = \sum_{n_1=1}^{[x]} T^{(1)}\left(\frac{x}{n_1}\right) = \sum_{n_1=1}^{[x]} \left(\frac{x}{n_1} + g_1\left(\frac{x}{n_1}\right)\right) = x \log x + g_2(x) x$$

as $\alpha = 0$, where

$$g_2(x) = h_0(x) + \frac{1}{x} \sum_{n_1=1}^{[x]} g_1\left(\frac{x}{n_1}\right).$$

Now, assume that (10.24) is valid for all dimensions less than d, and we shall then prove it for dimension d as well. According to (10.26) and by the induction assumption,

$$T^{(d)}(x) = \frac{1}{(d-2)!} \sum_{i=1}^{[x]} \frac{x}{i} \log^{d-2} \frac{x}{i} + g_d^{(1)}(x) x \log^{d-2} x,$$

where

$$g_d^{(1)}(x) = \frac{1}{\log^{d-2} x} \sum_{i=1}^{[x]} g_{d-1}\left(\frac{x}{i}\right) \frac{1}{i} \log^{d-3} \frac{x}{i}.$$

Now

$$\begin{aligned} \sum_{i=1}^{[x]} \frac{1}{i} \log^{d-2} \frac{x}{i} &= \sum_{i=1}^{[x]} \frac{1}{i} (\log x - \log i)^{d-2} \\ &= \sum_{l=0}^{d-2} C_{d-2}^{l} (-1)^{d-2-l} \log^{l} x \sum_{i=1}^{[x]} \frac{1}{i} \log^{d-2-l} i. \end{aligned}$$

Using Lemma 10.10 once again, with $\alpha = d-2-l$, we get

$$\sum_{i=1}^{[x]} \frac{1}{i} \log^{d-2-l} i = \frac{1}{d-1-l} \log^{d-1-l} x + h_{d-2-l}(x).$$

Lemma 10.11 implies that

$$\frac{1}{(d-2)!} \sum_{i=1}^{[x]} \frac{1}{i} \log^{d-2} \frac{x}{i} = \frac{1}{(d-1)!} \log^{d-1} x + g_d^{(2)}(x) x \log^{d-2} x,$$

where

$$g_d^{(2)}(x) = \frac{1}{x \log^{d-2} x} \frac{1}{(d-2)!} \sum_{l=0}^{d-2} C_{d-2}^{l} (-1)^{d-2-l} \left(\log^{l} x\right) h_{d-2-l}(x).$$

This proves (10.24) with $g_d(x) = g_d^{(1)}(x) + g_d^{(2)}(x)$. □

Proof of Lemma 10.10 We apply the Euler–Maclaurin summation formula for the function $f(x) = (1/x)\log^\alpha x$ and with $\ell_1 = 1$ and $\ell_2 = [x]$. Then $f'(x) = x^{-2}\left(\alpha\log^{\alpha-1} x - \log^\alpha x\right)$ and thus

$$\sum_{i=1}^{[x]} \frac{\log^\alpha i}{i} = \int_1^{[x]} \frac{\log^\alpha t}{t}\,dt + \psi(x) + \int_1^{[x]} \rho(t)\frac{\alpha\log^{\alpha-1} t - \log^\alpha t}{t^2}\,dt, \tag{10.27}$$

where ψ and ρ are as defined in (10.25). This proves Lemma 10.10, since $|\rho(t)| \le \frac{1}{2}$ and the integral $\int_1^\infty \frac{\log^\alpha t - \alpha\log^{\alpha-1} t}{t^2}\,dt$ converges absolutely. □

Proof of Lemma 10.11 It is clear that

$$\begin{aligned}
\sum_{l=0}^{d-2} C_{d-2}^l (-1)^{d-2-l}\frac{1}{d-1-l} &= \sum_{k=1}^{d-1} C_{d-2}^{d-1-k}(-1)^{k-1}\frac{1}{k}\\
&= \frac{1}{d-1}\sum_{k=1}^{d-1} C_{d-1}^k(-1)^{k-1}\\
&= \frac{1}{d-1}\left(1 - \sum_{k=0}^{d-1} C_{d-1}^k(-1)^k\right)\\
&= \frac{1}{d-1}\left(1-(1-1)^{d-1}\right) = \frac{1}{d-1}.
\end{aligned}$$

This completes the proof. □

10.10 An Example of Kátai

Let f and g be two real functions such that

$$f(x) \le x \le g(x), \qquad x \ge 1.$$

Put $D = \{(i,j)\colon f(i) \le j \le g(i)\}$. Condition (10.16) holds for a rather wide class of pairs of functions f and g. The question is whether or not it is true in general.

Kátai [208] gave a negative answer to this question by constructing a counter-example. Moreover, the function A in his construction does not belong to the class $\mathcal{ORV}$ (recall Definition 3.7). Kátai's construction is based on two integer sequences $\{x_n\}$ and $\{y_n\}$ satisfying $y_{n-1} < x_n < y_n$ for all $n > 1$ and $1 = x_1 < y_1$. The precise expression for these sequences is given below. In the intervals $(y_{n-1}, y_n]$, $n \ge 1$, the functions f and g are then given by

$$f(x) = g(x) = x \qquad \text{for } y_{n-1} < x \le x_n,$$
$$f(x) = x_n \quad \text{and} \quad g(x) = y_n \qquad \text{for } x_n < x \le y_n.$$

Note that, if the sequences $\{x_n\}$ and $\{y_n\}$ are such that

$$\lim_{n\to\infty} \frac{y_n}{x_n} = 1 \qquad \text{and} \qquad \lim_{n\to\infty} \frac{A\left(y_n^2\right)}{A\left(x_n^2\right)} = \infty, \tag{10.28}$$

then

$$\limsup_{t\to\infty} \frac{A(ct)}{A(t)} \ge \limsup_{n\to\infty} \frac{A\left(cx_n^2\right)}{A\left(x_n^2\right)} \ge \limsup_{n\to\infty} \frac{A\left(y_n^2\right)}{A\left(x_n^2\right)} = \infty$$

for all $c > 1$, since A increases. This implies that $A \notin \mathcal{POV}$.

Now we define the sequences $\{x_n\}$ and $\{y_n\}$ recursively in such a way that (10.28) is satisfied. Note that

$$A\left(x_n^2\right) = A\left(y_{n-1}^2\right) + x_n - y_{n-1} \quad \text{and} \quad A\left(y_n^2\right) = A\left(x_n^2\right) + (y_n - x_n)^2 \tag{10.29}$$

for arbitrary sequences $\{x_n\}$ and $\{y_n\}$. Once a member x_n of the sequence $\{x_n\}$ is defined, the member y_n of the sequence $\{y_n\}$ is chosen as

$$y_n = x_n + \sqrt{nA\left(x_n^2\right)}. \tag{10.30}$$

Relation (10.29) implies that

$$\frac{A\left(y_n^2\right)}{A\left(x_n^2\right)} = 1 + \frac{(y_n - x_n)^2}{A\left(x_n^2\right)} = 1 + n,$$

whence the second relation in (10.28) holds. Using (10.30) together with (10.29), we get

$$\frac{y_n}{x_n} = 1 + \frac{\sqrt{nA\left(x_n^2\right)}}{x_n} = 1 + \frac{\sqrt{n\left(A\left(y_{n-1}^2\right) + x_n - y_{n-1}\right)}}{x_n}.$$

If a member y_{n-1} of the sequence $\{y_n\}$ is defined, then the next member x_n of the sequence $\{x_n\}$ is chosen such that

$$\frac{x_n}{n} \ge \sqrt{\left(A\left(y_{n-1}^2\right) + x_n - y_{n-1}\right)},$$

whence

$$1 \le \frac{y_n}{x_n} \le 1 + \frac{1}{\sqrt{n}}.$$

This also proves the first relation in (10.28) and completes the example.

10.11 Comments

This chapter is based on the works of Indlekofer and Klesov [194], Klesov and Steinebach [230], and Buldygin et al. [56].

As far as the classical one-dimensional case is concerned, see Feller [125]. Lotka [253] contains a bibliography of 74 earlier papers on the subject.

Section 10.3 A rather complete treatment of limit theorems for sums of multi-indexed random variables is given by Klesov [225].

Section 10.4 The renewal function constructed from a random walk with two-dimensional time is considered for the first time in Ney and Wainger [283]. In fact, a purely analytical problem is studied there, which is that of the asymptotic behavior of a certain function represented by a series. In fact, analytical approaches dominate the first developments of renewal theory for dimension $d > 1$ (see, e.g. Maejima and Mori [254], Galambos and Kátai [143, 144], Galambos et al. [142], Hagwood [169], Abay [1], and Indlekofer et al. [191]). These authors do not study the renewal process, their primary interest was to find analytically asymptotics of the function (10.9) or of (10.10).

Section 10.5 The probabilistic approach in multidimensional time first defines the renewal process and then introduces the renewal function as its expectation (see, e.g., Klesov [224], Klesov and Steinebach [229]). We have followed this approach in Chap. 10. A different definition of the renewal process has been introduced by Ivanoff and Merzbach [197] for $d > 1$. However the asymptotic behavior of the processes defined in this way has not been studied yet.

Section 10.6 The function $U_d(t)$ given in (10.11) is an example of a so-called *weighted renewal function* corresponding to the weights $\{\tau_k\}$. Some results for weighted renewal functions are discussed by Smith [337]. More recently, Omey and Teugels [288] deal with $\mathcal{RV}_\alpha$, $\mathcal{ORV}$, and more general weights. The renewal function with weights $1/k$, $k \ge 1$, is called the *harmonic renewal function* (see Greenwood at al. [153] or extensions by Stam [346]; Borovkov and Borovkov [49] is a more recent reference). Alsmeyer [12] provides some relationships between harmonic renewal measures and a representative of the family of generalized renewal process, the first passage time. The proof presented in Sect. 10.6 is simpler than the one in Smith [337] for the weights $\{\tau_k\}$, even in the case of $D = \mathbf{N}^d$.

The function $A(x, y)$ defined by (10.15) describes the number of integer points in the domain $C(x, y)$. Evaluation of the number of integer points in subsets of $\mathbf{N}^d$ and determining the asymptotic behavior of the number of points in growing sets is a traditional topic in number theory (see, e.g., the Gauss circle problem) and in combinatorics. For almost all of our purposes, "rough" bounds are sufficient (see the monographs by Gelfond and Linnik [147] and Krätzel [243] for finer results).

Section 10.7 Corollary 10.5 is the main result in Klesov and Steinebach [229]. A relationship between the Riemann ζ function conjecture and some improved asymptotics for renewal functions for multi-indices of higher dimensions are also mentioned in [229].

Section 10.8 Example 10.7 describes the main result of Klesov [224]. A sector domain, as studied in Example 10.9, has been considered quite often in the case of multiple sums (see, for example, Gabriel [139], Gut [158], or Indlekofer and Klesov [192]).

Section 10.9 The proof of (10.24) is presented here just for the sake of completeness. Much more precise results are known in number theory (see, for example, Kolesnik [233] or Heath-Brown [176]). For our purposes, however, we only need a "rough" representation like (10.24) for the Dirichlet function.

Chapter 11
Spitzer Series and Regularly Varying Functions

11.1 Introduction

Let X, $\{X_n\}_{n\ge1}$ be independent, identically distributed random variables with distribution function F and let $\{S_n\}_{n\ge1}$ be the sequence of their partial sums.

Let w and φ be two positive functions. Put $w_k = w(k)$ and $\varphi_k = \varphi(k)$. We study the convergence and the asymptotic behavior with respect to small parameters ε of the series

$$Q(\varepsilon) = \sum_{k=1}^{\infty} w_k \, \mathsf{P}(|S_k| \ge \varepsilon\varphi_k), \qquad \varepsilon > 0. \tag{11.1}$$

We assume that both functions w and φ are regularly varying (see Definition 3.2). Recall that the notation $f \in \mathcal{RV}_a$ means that f is a measurable, regularly varying function of index a. Consider the case where the series (11.1) is constructed from large deviation probabilities, meaning that $S_k/\varphi_k \to 0$ in probability as $k \to \infty$. One can also study the case of (so-called) *small* or *moderate* deviation probabilities. Perhaps, the most known series of this type

$$\sum_{n=1}^{\infty} \mathsf{P}(|S_n| \ge \varepsilon n) \tag{11.2}$$

corresponds to the case $w_n = 1$ and $\varphi_n = 1$ for all $n \ge 1$.

Series like that in (11.1) often appear in various applications of limit theorems in probability theory. We only mention below two particular cases related to the main topic of this chapter.

V. V. Buldygin et al., *Pseudo-Regularly Varying Functions and Generalized Renewal Processes*, Probability Theory and Stochastic Modelling 91, https://doi.org/10.1007/978-3-319-99537-3_11

Strong Law of Large Numbers Let $\mathfrak{P}_2 = \{1, 2, 4, \dots, 2^m, \dots\}$ and

$$\varphi_n = n, \quad n \geq 1, \qquad w_n = \begin{cases} 1, & n \in \mathfrak{P}_2, \\ 0, & n \notin \mathfrak{P}_2. \end{cases}$$

Then the series (11.1) reduces to

$$\sum_{m=0}^{\infty} \mathsf{P}(|S_{2^m}| \geq \varepsilon 2^m) = \sum_{m=0}^{\infty} \mathsf{P}(|S_{2^{m+1}} - S_{2^m}| \geq \varepsilon 2^m).$$

Thus the convergence of series (11.1) is equivalent in this case to

$$\lim_{m\to\infty} \frac{S_{2^m}}{2^m} = 0 \qquad \text{almost surely}$$

by the Borel–Cantelli lemma. As is known, the latter relation is equivalent to the Kolmogorov strong law of large numbers

$$\lim_{n\to\infty} \frac{S_n}{n} = 0 \qquad \text{almost surely.}$$

Similar arguments lead to reducing the Marcinkiewicz–Zygmund strong law of large numbers (a rate of convergence result for the Kolmogorov strong law of large numbers) to the question about the convergence of a series of the form (11.1) for all $\varepsilon > 0$.

Note also that the convergence for all $\varepsilon > 0$ of the above series

$$\sum_{m=0}^{\infty} \mathsf{P}(|S_{2^m}| \geq \varepsilon 2^m) \tag{11.3}$$

is equivalent to that of the series

$$\sum_{n=1}^{\infty} \frac{1}{n} \mathsf{P}(|S_n| \geq \varepsilon n), \tag{11.4}$$

which we call the Spitzer series (the name is explained below). To outline the proof for the case of symmetrically distributed random variables X_n, denote for a moment the series in (11.3) by $Q_1(\varepsilon)$ and that in (11.4) by $Q_2(\varepsilon)$. Then, by the

Lévy inequality,

$$Q_2(\varepsilon) = \sum_{m=0}^{\infty} \sum_{n=2^m}^{2^{m+1}-1} \frac{1}{n} \mathsf{P}(|S_n| \ge \varepsilon n) \le 2 \sum_{m=0}^{\infty} \mathsf{P}(|S_{2^{m+1}}| \ge \varepsilon 2^m) \sum_{n=2^m}^{2^{m+1}-1} \frac{1}{n}$$
$$\le 2 \sum_{m=0}^{\infty} \mathsf{P}(|S_{2^{m+1}}| \ge \varepsilon 2^m) \le 2 Q_1(\varepsilon/2)$$

and similarly $Q_2(\varepsilon) \ge \frac{1}{2} Q_1(2\varepsilon)$ which proves the desired equivalence.

Complete Convergence According to the definition, a sequence S_n/n converges completely to zero if

$$\sum_{k=1}^{\infty} \mathsf{P}\left(\left|\frac{S_k}{k}\right| \ge \varepsilon\right) < \infty \quad \text{for all} \quad \varepsilon > 0.$$

Clearly, series (11.2) coincides with $Q(\varepsilon)$ from (11.1) if $w_k = 1$ and $\varphi_k = k$, $k \ge 1$. Hsu and Robbins [185] proved that $Q(\varepsilon) < \infty$ for all $\varepsilon > 0$ if

$$\mathsf{E}\, X = 0, \qquad \mathsf{E}\, X^2 < \infty. \tag{11.5}$$

The converse was proved by Erdős [120].

The results of Hsu and Robbins [185] and Erdős [120] were later generalized for series of the form (11.1) with a variety of functions w and φ. Below is an (incomplete) list of cases which have been studied in detail:

(c_1) $w_k = 1$, $\varphi_k = k$ (cf. Hsu and Robbins [185], Erdős [120]);
(c_2) $w_k = 1/k$, $\varphi_k = k$ (cf. Spitzer [343]);
(c_3) $w_k = k^r$, $\varphi_k = k^{1/p}$ (cf. Katz [209], Baum and Katz [28]);
(c_4) $w \in \mathcal{RV}_r$, where $w(t)/t^r$ increases, $\varphi_k = k^{1/p}$ (cf. Heyde and Rohatgi [181]).

According to the cases (c_1), (c_2), (c_3), and (c_4), we attribute the names Hsu–Robbins, Baum–Katz, Spitzer, and Heyde–Rohatgi to the corresponding series $Q(\varepsilon)$.

The Spitzer series under (c_2) is of special interest, since it corresponds to the "boundary" case in the range of power sequences $w_k = k^r$. Indeed, the series $\sum w_k$ converges, if $r < -1$, and diverges otherwise.

A necessary and sufficient condition for convergence of the Spitzer series under (c_2), for an arbitrary $\varepsilon > 0$, is

$$\mathsf{E}\, X = 0 \tag{11.6}$$

(see Spitzer [343]).

It is clear that $Q(\varepsilon)$ increases as ε decreases and moreover $Q(\varepsilon) \to \infty$ as $\varepsilon \to 0$. The asymptotic behavior of $Q(\varepsilon)$ as $\varepsilon \downarrow 0$ has been studied in Heyde [180], where it is proved that *if* (11.5) *holds, then*

$$\lim_{\varepsilon \downarrow 0} \varepsilon^2 Q(\varepsilon) = \sigma^2 \tag{11.7}$$

for the Hsu–Robbins series under (c_1), *with* $\sigma^2 = \operatorname{var} X$.

The Spitzer series corresponding to $r = -1$ exhibits a different behavior as $\varepsilon \downarrow 0$ as compared to the Baum–Katz case of $r > -1$, which can be viewed as an analogue of the corresponding series of real numbers, namely that, as $n \to \infty$,

$$\sum_{k=1}^{n} k^r \sim \begin{cases} \dfrac{n^{r+1}}{r+1}, & r > -1, \\ \log(n), & r = -1. \end{cases}$$

In other words, $\sum_{k=1}^{n} k^r$ does not have a power asymptotic behavior in the case $r = -1$.

The asymptotic behavior of the Spitzer series in case (c_2), as $\varepsilon \downarrow 0$, is given in Chow and Lai [87]. There it is shown that

$$\lim_{\varepsilon \downarrow 0} \frac{Q(\varepsilon)}{\ln(1/\varepsilon)} = 2, \tag{11.8}$$

provided condition (11.5) holds. Recall that the finiteness of $Q(\varepsilon)$, for all $\varepsilon > 0$, follows from condition (11.6). This indicates that condition (11.5) may be too restrictive to obtain the asymptotic in (11.8).

The main aim of this chapter is to determine the asymptotic behavior of the series (11.1) for the case when $w \in \mathcal{RV}_r$, $r \geq -1$, and $\varphi \in \mathcal{RV}_{1/p}$, $0 < p < \alpha$. Note that this case is even more general than (c_4).

A Statistical Motivation A simple motivation to study the asymptotics as $\varepsilon \downarrow 0$ of series like (11.1) is the following. Let $w_n = 1$ and $\varphi_n = 1$ for all $n \geq 1$, that is, we deal with the Hsu–Robbins series. Having (11.7) in mind, one may view the empirical version of $Q(\varepsilon)$, i.e.,

$$\hat{Q}(\varepsilon) = \sum_{n=1}^{\infty} \mathbb{1}_{\{|S_n| \geq \varepsilon n\}},$$

as a statistical estimator of the variance. This estimator $\hat{Q}(\varepsilon)$ is easy to calculate (which is particularly important for big data). To do so, one only needs to detect how many members of the sequence S_n, $n \geq 1$, fall in the region bounded by two straight lines $y = -\varepsilon x$ and $y = \varepsilon x$. This binary decision leads to the binary data in the estimator. Of course, there are also drawbacks of this approach, but we do not go into details here.

This chapter is organized as follows. In Sect. 11.2, we state the main result, i.e. Theorem 11.1, together with some corollaries, which exhibit possible asymptotics for simple choices of the functions $w(\cdot)$ and $\varphi(\cdot)$ in (11.1), like e.g. $w(x) = x^r \ln^q(n)$, or $\varphi(x) = x^{1/p} \ln^q(x)$, or a combination of the latter for various choices of r, q, and p. We also give a negative answer to a conjecture of Gut and Spătaru [161] claiming that the asymptotic behavior in the case of normal attraction is the same as that for nonnormal attraction, if $r > -1$.

In Sect. 11.3, the proof of Theorem 11.1 is given.

11.2 The Asymptotic Behavior of Large Deviation Probabilities

Another assumption is used in Spătaru [342] to study the asymptotic behavior of the Spitzer series under (c_2), namely,

$$\begin{gathered} F \text{ belongs to the domain of attraction of a } \textit{nondegenerate} \\ \alpha\text{-stable law with index } 1 < \alpha \le 2. \end{gathered} \tag{11.9}$$

Under assumptions (11.6) and (11.9), it is proved in Spătaru [342] that

$$\lim_{\varepsilon \downarrow 0} \frac{Q(\varepsilon)}{\ln(1/\varepsilon)} = \frac{\alpha}{\alpha - 1}. \tag{11.10}$$

Condition (11.5) implies that the distribution function F belongs to the domain of attraction of the standard Gaussian law, where (11.9) is satisfied with $\alpha = 2$. In this case, (11.8) and (11.10) coincide. Note also that (11.9) implies $\mathsf{E}\,|X|^\eta < \infty$, for all $1 < \eta < \alpha$, and the latter condition is weaker than (11.5), but stronger than (11.6). Nevertheless condition (11.9) seems to be too restrictive for the problem under consideration. Note that in Spătaru [342] it is not assumed that the limit law has to be nondegenerate, but the result may fail for a degenerate limit law. Indeed, if the limit law is, e.g., concentrated at zero, then the random variable $X = 0$ is attracted to the law $\mathbb{1}_{[0,\infty)}(x)$. In this case, $\mathsf{P}(|S_k| \ge \varepsilon k) = 0$ for all $\varepsilon > 0$ and (11.10) fails.

The Baum–Katz series under (c_3) (including the case of $r = -1$) has been studied in Gut and Spătaru [161] under assumption (11.9); Gut and Steinebach [162] have shown that the asymptotic behavior of $Q(\varepsilon)$ can also be obtained under (11.9) with $0 < \alpha < 1$. More general results concerning the asymptotics of the series (11.1), with $\alpha \ne 2$, are given in Rozovskiĭ [308] under assumption (11.9).

A less restrictive condition (compared to (11.9)) for the Baum–Katz series under (c_3) is that

$$\begin{gathered} F \text{ belongs to the domain of attraction} \\ \text{of a } \textit{semistable} \text{ law of index } \alpha > 0. \end{gathered} \tag{11.11}$$

Further results on the asymptotic behavior of $Q(\varepsilon)$, as $\varepsilon \downarrow 0$, have been obtained by Chen [81] for the Baum–Katz series under (c_3), with $r \geq 0$, including Heyde's result (11.7) as well. Note that the moment condition used in [81] is the same as the sufficient condition for complete convergence.

The proof in Chow and Lai [87] has much in common with traditional techniques of proof of limit theorems in probability, that is, it uses truncation, symmetrization, the Berry–Esseen inequality, and desymmetrization. The proof in Chen [81] essentially follows the method of Heyde [180] consisting of the following two steps: first, the theorem is proved for Gaussian random variables, and then the latter result is used for approximating the general case.

Heyde's [180] method is applied in Spătaru [342] in the case of attraction to a stable law. In Scheffler [320], some properties of distributions attracted to semistable laws are used to prove a similar result. Another method, used by Rozovskiĭ [308], is based on large deviations bounds in the case of attraction to a stable law. (In fact, these bounds were obtained earlier in Heyde [178], but Rosovskiĭ [308] refers to some other work.)

Note that the proof in Spătaru [342] is not complete in a certain sense. Indeed, relation (11.33) below is proved in [342], which means that the limit is determined only for the case of normal attraction to the limit law. We shall fill this gap by evaluating the limit in an even more general setting (see Remark 11.15).

There is a big difference between the series under (c_2) and (c_3) with $r > -1$. Namely, Heyde's result (11.7) asserts that the behavior of the Hsu–Robbins series under (c_1) (as a "representative" of the series under (c_3)) is "almost" independent of the distribution function F, that is, the asymptotic behavior depends on F only through the value of the variance σ^2. Roughly speaking, Heyde's result is invariant with respect to multiplication of the underlying random variables by a constant. The situation changes in the case of the Spitzer series under (c_2), whose behavior depends on F via the index α of the stable law to which F is attracted. Other properties of F do not affect the limit result.

In a certain sense, a distribution function F attracted to an α-stable law can "uniquely" be described in terms of the normalizing sequence, which is needed for attraction to the stable law. More precisely, condition (11.9) implies that there are two sequences of real numbers $\{a_n\}_{n\geq 1}$ and $\{c_n\}_{n\geq 1}$ such that the distribution of the random variables $S_n/c_n - a_n$ converges weakly to a stable law of index $\alpha > 1$. If one assumes (11.6), then one can restrict consideration to the case $a_n = 0$, $n \geq 1$ (see Lemma A.16). On the other hand, even if assumption (11.6) holds, the sequence $\{c_n\}_{n\geq 1}$ is still involved in the weak convergence. Thus one may expect that some properties of $\{c_n\}_{n\geq 1}$ are somehow reflected in (11.10). It is known that $c_n = n^\alpha h(n)$, with some slowly varying function h. The index α is involved in (11.10), however the function h is not. In other words, relation (11.10) is the same for both, normal attraction, where $h(x) \equiv \text{const}$, and general nonnormal attraction, where $h(x) \not\equiv \text{const}$.

The latter observation leads to the following natural questions. Fix some function $w \in \mathcal{RV}_{-1}$.

- Is there any difference in the behavior of $Q(\varepsilon)$ in the case of normal attraction (for which $c_n = cn^{1/\alpha}$, $c > 0$) and in the case of general attraction (for which $c_n = n^\alpha h(n)$, where h is a certain slowly varying the function)?
- If the above phenomenon is *not* a universal law, then why doesn't it occur in the case $w_k = 1/k$, $k \geq 1$?

We shall answer these questions in Remark 11.7 by applying Theorem 11.1.

Assume that condition (11.9) holds. We shall show that the asymptotic behavior can be studied not only for power functions, but also for the more general case of regularly varying functions.

The proof of the main result consists of two standard steps: first, we obtain the behavior for the stable limit law, and then we approximate the general case by the special one for the stable law. Although both steps sound standard, we propose new approaches for each of them. The standard way in the first step is to apply the Euler–Maclaurin formula, while our approach is based on an Abel type theorem for slowly varying functions (see Aljančić et al. [9]). This approach allows us to study not only stable laws but also much more general distribution functions.

For the second step, we apply new estimates for the large deviations of the tails of distribution functions belonging to the domain of attraction of certain stable limit distributions. In doing so, the cases $\alpha \neq 2$ and $\alpha = 2$ are different and each of them requires its own consideration. The bounds allow us to involve the case $\alpha = 1$ as well, which shows the additional feature that the corresponding bounds also contain the centering constants in the attraction to the limit law.

Note that the classical large deviation results deal with sufficiently large arguments of the tails, while our proof requires a result valid for *all* positive arguments, since a small parameter ε tending to 0 leads to arguments for the tail probabilities on the whole positive axis. An auxiliary result needed in the proof is considered in the Appendix (see Sect. A.3).

The proof of the large deviations bounds extends the method of Heyde [178]. We also obtain estimates for all $x \geq 0$, while the bounds in [178] are only valid for $x \geq x_n$, where $\{x_n\}$ is a sequence of real numbers such that $x_n \to \infty$ as $n \to \infty$. The price to pay for this is that we only get upper bounds, which means that we cannot obtain the exact asymptotics of the large deviations as in Heyde [178]. However, the upper bounds are sufficient for our purposes. An extra gain we get from extending the domain, where the upper bounds hold, is a simplification of the proof of the main result.

The proof in Rozovskiĭ [308] is also based on an asymptotic result like the large deviation bounds in Heyde [178]. The series in (11.1) are studied in [308] for the case $\alpha \neq 2$. On the other hand, the class of admissible sequences $\{w_n\}$ and $\{\varphi_k\}$ considered in [308] is even larger than in this chapter. Nevertheless, the class we study allows us to obtain the asymptotic behavior explicitly in contrast to [308].

The asymptotic behavior of the series $Q(\varepsilon)$ is expressed in terms of

$$W(t) = W(n), \ n \le t < n+1, \quad \text{where} \quad W(n) = \sum_{k=1}^{n} w_k, \tag{11.12}$$

together with the normalizing sequence involved in the attraction to the stable limit law. As a consequence, the asymptotic behavior in the case of normal attraction (for which the normalizing sequence is $cn^{1/\alpha}$, $c > 0$) differs from the behavior in the case of nonnormal attraction (where the normalizing sequence is $n^{1/\alpha}h(n)$, with a slowly varying function h). Therefore, the asymptotic behavior of the series $Q(\varepsilon)$ in the case of nonnormal attraction depends on the function h. On the other hand, if $w_n = 1/n$, for all $n \ge 1$, then the formula for the asymptotics does not involve h (see Remark 11.7).

It is known from the theory of regularly varying functions that $W \in \mathcal{RV}_{r+1}$, if $w \in \mathcal{RV}_r$ and $r > -1$ (this result is proved in Karamata [206] for integrals instead of sums). Moreover, $tw(t) \sim (r+1)W(t)$ in this case and thus W replaces w in the asymptotics of the series (11.1) if $r > -1$. This is not the case if $r = -1$. It is shown in Parameswaran [290] that $W \in \mathcal{RV}_0$ if $r = -1$, that is, W is a slowly varying function. Moreover, in this case, $tw(t) = o(W(t))$ as $t \to \infty$. This means that the asymptotic behavior of the series $Q(\varepsilon)$ cannot be expressed only in terms of the function w if $r = -1$.

Throughout this section, we assume that X, $\{X_n\}_{n\ge1}$ are independent, identically distributed random variables. As always, S_n, $n \ge 1$, denotes the nth partial sum of the random variables X_k, $1 \le k \le n$.

We write $X \in \mathcal{DA}(\alpha, \{c_n\}, \{a_n\})$ if there exists a *nondegenerate*, α-stable random variable Z_α to which the sums S_n are attracted, that is, if

$$\frac{X_1 + \cdots + X_n}{c_n} - a_n \xrightarrow{d} Z_\alpha, \tag{11.13}$$

where the symbol $\xrightarrow{d}$ denotes weak convergence and where $\{X_n\}_{n\ge1}$ are assumed to be independent, identically distributed copies of the random variable X. In this case, we also say that X belongs to the domain of attraction of a nondegenerate random variable Z_α with normalizing sequence $\{c_n\}$ and centering sequence $\{a_n\}$. We write $X \in \mathcal{DA}(\alpha, \{c_n\}, \{0_n\})$ if $a_n = 0$, $n \ge 1$.

Let the weights in (11.1) be such that

$$w_k = w(k), \quad \text{with} \quad w \in \mathcal{RV}_r, \tag{11.14}$$

and let the normalizing sequences for the large deviations probabilities in (11.1) be of the form

$$\varphi_k = \varphi(k), \quad \text{with} \quad \varphi \in \mathcal{RV}_{1/p}. \tag{11.15}$$

If $X \in \mathcal{DA}(\alpha, \{c_n\}, \{a_n\})$ and $p < \alpha$, then there exists a sequence $\{b_n\}$ such that $X \in \mathcal{DA}(\alpha, \{b_n\}, \{a_n\})$ and properties (i)–(iii) of Lemma A.15 hold. This, for example, means that $b \in \mathcal{RV}_{1/\alpha}$, that is,

$$b_k = b(k), \quad \text{where} \quad b(x) = x^{1/\alpha} h(x) \tag{11.16}$$

with a slowly varying function h such that $\varphi(x)/b(x)$ is continuous and increasing. In what follows we assume that the sequence $\{b_n\}_{n\geq 1}$ is chosen according to Lemma A.15, that is, all inverse functions appearing in the proof exist.

Put

$$\psi(x) = \varphi(x)/b(x) \quad \text{and} \quad \psi_k = \psi(k). \tag{11.17}$$

Note that the inverse function ψ^{-1} exists and, moreover, $\psi^{-1} \in \mathcal{RV}_{\alpha p/(\alpha-p)}$ (see, e.g., Seneta [324], Section 1.5, statement 5°). For W defined by (11.12), we put

$$U(x) = W\left(\psi^{-1}(x)\right). \tag{11.18}$$

It is easy to see that $U \in \mathcal{RV}_{(r+1)\alpha p/(\alpha-p)}$ (see Lemma A.20).

Throughout the sequel, we assume that at least one of the following conditions on the attraction to a stable distribution holds:

$$X \in \mathcal{DA}(\alpha, \{b_n\}, \{a_n\}) \tag{11.19}$$

or

$$X \in \mathcal{DA}(\alpha, \{b_n\}, \{0_n\}). \tag{11.20}$$

Recall that we agreed on assuming that each of the latter conditions means that the limit distribution is nondegenerate.

Now we are ready to state the main result.

Theorem 11.1 *Let X, $\{X_n\}_{n\geq 1}$ be independent, identically distributed random variables. Assume that condition (11.19) holds. Moreover, let condition (11.20) be satisfied, if $\alpha = 1$, and let condition (11.6) hold, if $\alpha > 1$. Let Z_α be an α-stable random variable to which X is attracted. Also let conditions (11.14) and (11.15) hold for some $r \geq -1$ and $0 < p < \alpha$. Finally, let the functions U and Q be as defined in (11.1) and (11.18), respectively. If the series $\sum w_k$ diverges and $\alpha > p(r+2)$, then*

$$\lim_{\varepsilon\downarrow 0} \frac{1}{U(1/\varepsilon)} \sum_{k=1}^{\infty} w_k \, \mathsf{P}(|S_k| \geq \varepsilon\varphi_k) = \mathsf{E}\,|Z_\alpha|^{(r+1)\alpha p/(\alpha-p)}. \tag{11.21}$$

Remark 11.2 The assumptions of Theorem 11.1 seem to be different for the cases $\alpha = 1$ and $\alpha \neq 1$, that is, condition (11.20) is assumed, if $\alpha = 1$, whereas (11.19) should hold, if $\alpha \neq 1$. Condition (11.20) is easy to check for $\alpha < 1$; it also follows from (11.6), if $\alpha > 1$ (see Lemma A.16). Therefore, we assume that (11.20) holds for all $0 < \alpha \le 2$, in fact. This actually means that there is no difference in the assumptions for the cases $\alpha = 1$ and $\alpha \neq 1$. If condition (11.20) is dropped for $\alpha = 1$, but (11.19) still holds, then Theorem 11.1 remains true with the modification that $\sum w_k \,\mathsf{P}(|S_k| \ge \varepsilon\varphi_k)$ has to be replaced by $\sum w_k \,\mathsf{P}(|S_k - a_k b_k| \ge \varepsilon\varphi_k)$. The proof of this result is similar to the proof of Theorem 11.1, if Corollary A.5 is used instead of Proposition A.4. We only deal with one of these series, namely with $\sum w_k \,\mathsf{P}(|S_k| \ge \varepsilon\varphi_k)$, and thus assume that condition (11.20) holds, if $\alpha = 1$. Note that, if $\mathsf{E}\, X = \mu \neq 0$ for $\alpha > 1$, then the asymptotic holds for the series $\sum w_k \,\mathsf{P}(|S_k - k\mu| \ge \varepsilon\varphi_k)$.

Remark 11.3 If $\alpha > p(r+2)$, then the moment of Z_α on the right-hand side of (11.21) is finite, since $(r+1)\alpha p/(\alpha - p) < \alpha$ (see (A.24)). Otherwise, that is, if $\alpha \le p(r+2)$, the right-hand side of (11.21) is infinite, if $\alpha \neq 2$, which means that the asymptotic behavior of $U(1/\varepsilon)$ is different from that of $\sum w_k \,\mathsf{P}(|S_k| \ge \varepsilon\varphi_k)$ in this case.

Remark 11.4 If $\alpha < p(r+2)$ and $\alpha \neq 2$, then not only the right-hand side of (11.21) is infinite, but also the series on the left-hand side of (11.21) diverges for all $\varepsilon > 0$, that is, the problem we are concerned with disappears in this case. The divergence of the series on the left-hand side of (11.21) can rather easily be proved for $\alpha < \min\{2, p(r+2)\}$. Indeed, fix $\varepsilon > 0$ and put $x_k = \varphi_k / b_k$. Since $p < \alpha$, we have $x_k \to \infty$ as $k \to \infty$ and one can use the large deviation result of Heyde [180], i.e.

$$\mathsf{P}(|S_k| \ge \varepsilon\varphi_k) \sim k\,\mathsf{P}(|X_1| \ge \varepsilon\varphi_k), \qquad k \to \infty.$$

This implies that both series $\sum w_k \,\mathsf{P}(|S_k| \ge \varepsilon\varphi_k)$ and $\sum k w_k \,\mathsf{P}(|X_1| \ge \varepsilon\varphi_k)$ converge or diverge simultaneously. In view of (A.26), the latter series converges if and only if

$$\sum_{k=1}^{\infty} \frac{k w_k g(\varepsilon\varphi_k)}{\varphi_k^\alpha} < \infty.$$

This condition fails if $\alpha < p(r+2)$, since g is a regularly varying function. In the boundary case of $\alpha = p(r+2)$, both convergence and divergence of the series on the left-hand side of (11.21) may happen (see Rozovskiĭ [308]).

Remark 11.5 The convergence of the series in (11.1), for $\alpha > p(r+2)$ and for all $\varepsilon > 0$, can be proved by using Heyde's principle above. To avoid repetition and to involve the case $\alpha = 2$ in addition to $\alpha \neq 2$, we apply another method based on the Baum and Katz [28] theorem. Indeed, choose $r' > r$ and $p' > p$ such that $\alpha > p'(r'+2)$. Then, in view of Lemma A.9, $w_k \le \text{const}\; k^{r'}$ and $\varphi_k \ge \text{const}\; k^{1/p'}$ for all $k \ge 1$. Thus the convergence of the series in (11.1), for all $\varepsilon > 0$, follows from the convergence of the series

$$\sum_{k=1}^{\infty} k^{r'} \mathsf{P}\left(|S_k| \ge \varepsilon k^{1/p'}\right),$$

for all $\varepsilon > 0$. The latter series converges, for all $\varepsilon > 0$, if and only if $\mathsf{E}\,|X|^{p'(r'+2)} < \infty$ (see Baum and Katz [28]). This moment is finite in the case under consideration, which follows from (A.27)). The above reasoning shows that the moment condition in Theorem 11.1 is more restrictive than the condition needed for the convergence of the series in (11.1). It is also true that the result of Theorem 11.1 is stronger than the convergence result itself.

Remark 11.6 We should like to add that the condition of $\alpha > p(r+2)$ coincides with the condition in Theorem 1 of Gut and Spătaru [161], which can easily be seen after a change of notation.

Remark 11.7 Now we are in a position to answer both questions posed in Sect. 11.1 above. The answers are clear in view of Theorem 11.1. Consider, for example, the case $p = 1$. Then Theorem 11.1, with weight function $w_k = 1/k$, $k \ge 1$, implies that the series $Q(\varepsilon)$ is asymptotically equivalent to $\ln\left(\psi^{-1}(1/\varepsilon)\right)$ as $\varepsilon \downarrow 0$, where ψ^{-1} is the inverse of the function defined in (11.17). The function h in (11.16) describes the kind of attraction to the limit distribution (if $h(t) \equiv c$, then the attraction is *normal*).

It is known from the theory of regularly varying functions that $\ln(f(t)) \sim \beta \ln(t)$, as $t \to \infty$, if $f \in \mathcal{RV}_\beta$ (see, for example, statement 2° on p. 18 in Seneta [324]). Since the logarithm "eliminates" any slowly varying function asymptotically and $\psi^{-1} \in \mathcal{RV}_{\alpha/(\alpha-1)}$, the function h disappears from the asymptotics in (11.10), so that only the part corresponding to *normal* attraction is left. This reasoning answers the second question.

To answer the first question, consider the simplest case of $\varphi(t) = t$. Put $\beta = \alpha/(\alpha-1)$ and introduce a function w such that $w(t) = 1$, for $0 \le t < 2$, and

$$w(t) = \frac{e^{\ln^{1/2}(t)}}{t \ln^{1/2}(t)}, \qquad \text{for} \quad t \ge 2.$$

Then $W(t) \sim 2e^{\ln^{1/2}(t)}$ as $t \to \infty$, where W is as defined in (11.12). Note that $w \in \mathcal{RV}_{-1}$. Thus, in view of Lemma A.21, $W \in \mathcal{RV}_0$. In case of normal attraction, we have $b(t) = ct^{1/\alpha}$, for some $c > 0$, whence $\psi^{-1}(t) = (ct)^\beta$, that is, the series $Q(\varepsilon)$ is equivalent to $U(1/\varepsilon) \sim 2e^{\beta^{1/2}\ln^{1/2}(1/\varepsilon)}$, as $\varepsilon \downarrow 0$, according to Theorem 11.1.

Now we consider a special case of nonnormal attraction, with $\psi^{-1}(t) = t^\beta e^{\ln^a(t)}$, $0 < a < 1$. The case of $a = 0$ formally corresponds to normal attraction. Then $\psi^{-1} \in \mathcal{RV}_\beta$ and thus the normalizing sequence b can easily be derived from ψ^{-1}. Theorem 11.1 implies that the series $Q(\varepsilon)$ is asymptotically equivalent to $2e^{(\beta\ln(1/\varepsilon)+\ln^a(1/\varepsilon))^{1/2}}$, as $\varepsilon \downarrow 0$. To distinguish the functions U for normal and nonnormal attraction, we write U_0 and U_a, respectively. It is easy to see that

$$\frac{U_a(x)}{U_0(x)} \sim \exp\left\{\frac{\ln^a(x)}{\sqrt{\beta\ln(x) + \ln^a(x)} + \sqrt{\beta\ln(x)}}\right\}, \qquad x \to \infty,$$

and thus $U_0 = o(U_a)$, as $a > 1/2$, while $U_a \sim U_0$, as $0 < a < 1/2$, and $U_a \sim e^{1/2\sqrt{\beta}}U_0$, as $a = 1/2$. The asymptotic behavior for normal and nonnormal attraction is different, if $a > 1/2$. This answers the first question.

Obviously U_a "dominates" U_0 in this example. Moreover, one may even get the impression that this is a universal law. Nevertheless, if $\psi^{-1}(t) = t^\beta e^{-\ln^a(t)}$, $\frac{1}{2} < a < 1$, then $U_a = o(U_0)$, so that this impression is wrong.

Below are some applications of Theorem 11.1 for various specific cases. First we consider the case when $r > -1$.

Corollary 11.8 *Let X, $\{X_n\}_{n\ge1}$ be independent, identically distributed random variables. Assume that condition (11.19) is satisfied, with $b_n = cn^{1/\alpha}$, $c > 0$ (this means that we are dealing with normal attraction). Moreover, we assume that condition (11.20) holds, if $\alpha = 1$; otherwise we assume that condition (11.6) holds. Let Z_α be an α-stable random variable to which X is attracted, and let $r > -1$, $0 < p < \alpha$, $\alpha > p(r+2)$. Put $\nu = (r+1)\alpha p/(\alpha - p)$. Then*

$$\sum_{k=1}^{\infty} k^r\,\mathsf{P}\left(|S_k| \ge \varepsilon k^{1/p}\right) \sim \left(\frac{1}{\varepsilon}\right)^{\nu} \frac{c^\nu}{r+1}\mathsf{E}\,|Z_\alpha|^\nu, \qquad \varepsilon \downarrow 0.$$

Corollary 11.8 has been proved in Gut and Spătaru [161]. Just the constant c^ν is missing in the statement of Theorem 1 in [161]. The proof of Corollary 11.8 is rather simple. Note that, in this case, $\psi(t) = t^{(\alpha-p)/\alpha p}/c$, whence $\psi^{-1}(t) = (ct)^{\alpha p/(\alpha-p)}$. Since $W(t) \sim t^{r+1}/(r+1)$, as $t \to \infty$ (cf. Lemma A.20), Corollary 11.8 immediately follows from Theorem 11.1.

Now one can give a negative solution to the conjecture, stated in Gut and Spătaru [161], that the asymptotic behavior in the case of nonnormal attraction coincides with that obtained in Corollary 11.8 for normal attraction. In fact, any slowly varying function can appear in the asymptotics of the series of large deviation probabilities. Below we provide the corresponding result for the function $\ln^q(t)$.

Corollary 11.9 *Let X, $\{X_n\}_{n\ge1}$ be independent, identically distributed random variables. Assume that condition (11.19) is satisfied, with $b(t) = t^{1/\alpha}\ln^q(t)$. Moreover, let condition (11.20) hold, if $\alpha = 1$, and let condition (11.6) hold, if $\alpha > 1$. Let Z_α be an α-stable random variable to which X is attracted, and let $r > -1$, $0 < p < \alpha$, $\alpha > p(r+2)$. Put $\nu = (r+1)\alpha p/(\alpha - p)$. Then*

$$\sum_{k=1}^{\infty} k^r\,\mathsf{P}\left(|S_k| \ge \varepsilon k^{1/p}\right) \sim \left(\frac{\ln^q(1/\varepsilon)}{\varepsilon}\right)^{\nu} \left(\frac{\alpha p}{\alpha - p}\right)^{\nu q} \frac{1}{r+1}\mathsf{E}\,|Z_\alpha|^\nu, \qquad \varepsilon \downarrow 0.$$

Note that the normalizing function $b(t) = t^{1/\alpha} \ln^q(t)$ is involved in the theorem on the attraction to an α-stable limit distribution and thus

$$\mathsf{P}(|X| \ge x) \sim \text{const} \left(\frac{\ln^q(x)}{x} \right)^{\alpha}, \qquad x \to \infty,$$

(see condition (A.28)).

Corollary 11.9 follows from Theorem 11.1, since $\psi(t) = t^{(\alpha-p)/\alpha p} \ln^{-q}(t)$ in this case, whence

$$\psi^{-1}(t) \sim \left(\frac{\alpha p}{\alpha - p} \right)^{q\alpha p/(\alpha-p)} t^{\alpha p/(\alpha-p)} \ln^{q\alpha p/(\alpha-p)}(t).$$

The behavior of the function W has been given in the proof of Corollary 11.8. This completes the proof of Corollary 11.9.

Now we consider the case $r = -1$.

Corollary 11.10 *Let X, $\{X_n\}_{n\ge1}$ be independent, identically distributed random variables. Assume that condition (11.19) is satisfied. Moreover, let condition (11.20) hold, if $\alpha = 1$, and let condition (11.6) hold, if $\alpha > 1$. Let $0 < p < \alpha$. Then*

$$\sum_{k=1}^{\infty} \frac{1}{k} \mathsf{P}(|S_k| \ge \varepsilon k^{1/p}) \sim \ln(1/\varepsilon) \frac{\alpha p}{\alpha - p}, \qquad \varepsilon \downarrow 0.$$

Corollary 11.10 has been proved in Gut and Spătaru [161]. For a more general case, in which condition (11.11) substitutes condition (11.19), Corollary 11.10 has been proved in Scheffler [320].

Corollary 11.10 immediately follows from Theorem 11.1 together with the well-known representation of the harmonic series (see (A.40)). Indeed, in this case $W(t) \sim \ln(t)$ and $\psi^{-1} \in \mathcal{RV}_{\alpha p/(\alpha-p)}$. Thus $U(x) \sim \frac{\alpha p}{\alpha - p} \ln(x)$ (see statement 2° on p. 18 in Seneta [324]).

We provide two more results for the case $r = -1$.

Corollary 11.11 *Let X, $\{X_n\}_{n\ge1}$ be independent, identically distributed random variables. Assume that condition (11.19) is satisfied. Moreover, let condition (11.20) hold, if $\alpha = 1$, and let condition (11.6) hold, if $\alpha > 1$. Let $0 < p < \alpha$ and $q > -1$. Then*

$$\sum_{k=2}^{\infty} \frac{\ln^q(k)}{k} \mathsf{P}\left(|S_k| \ge \varepsilon k^{1/p} \right) \sim \ln^{q+1}(1/\varepsilon) \frac{1}{q+1} \left(\frac{\alpha p}{\alpha - p} \right)^{q+1}, \qquad \varepsilon \downarrow 0.$$

Indeed, in this case, $W(t) \sim \frac{1}{q+1} \ln^{q+1}(t)$ and $\psi \in \mathcal{RV}_{(\alpha-p)/\alpha p}$. This implies $\psi^{-1} \in \mathcal{RV}_{\alpha p/(\alpha-p)}$, whence Corollary 11.11 follows.

The first part of Corollary 11.11 is meaningless if $q = -1$. It turns out that the true asymptotic of the series is completely different in this case.

Corollary 11.12 *Let X, $\{X_n\}_{n\ge 1}$ be independent, identically distributed random variables. Assume that condition (11.19) is satisfied. Moreover, let condition (11.20) hold, if $\alpha = 1$, and let condition(11.6) hold, if $\alpha > 1$. Let $0 < p < \alpha$. Then*

$$\sum_{k=2}^{\infty} \frac{1}{k\ln(k)} \mathsf{P}\left(|S_k| \ge \varepsilon k^{1/p}\right) \sim \ln\ln(1/\varepsilon), \qquad \varepsilon \downarrow 0.$$

Indeed, the asymptotic in this case is $W(t) \sim \ln\ln(t)$, from which Corollary 11.12 follows.

The case of $q < -1$ in Corollary 11.12 is special, since then the series $\sum \ln^q(k)/k$ converges and thus, by Lebesgue's dominated convergence theorem,

$$\lim_{\varepsilon\downarrow 0} \sum_{k=2}^{\infty} \frac{\ln^q(k)}{k} \mathsf{P}(|S_k| \ge \varepsilon k^{1/p}) = \sum_{k=2}^{\infty} \frac{\ln^q(k)}{k} \mathsf{P}(S_k \ne 0).$$

Up to now we have been concerned with the case of $\varphi(x) = x^{1/p}$. Now we show how the asymptotic of the series $Q(\varepsilon)$ changes if the function $x^{1/p}$ is replaced by other functions φ.

Corollary 11.13 *Let X, $\{X_n\}_{n\ge 1}$ be independent, identically distributed random variables. Assume that condition (11.19) is satisfied, with $b_n = cn^{1/\alpha}$ for some $c > 0$ (that is, we are dealing with normal attraction).*

Moreover, let condition (11.20) hold, if $\alpha = 1$, and let condition (11.6) hold, if $\alpha > 1$. Put $\nu = (r+1)\alpha p/(\alpha - p)$. If $r > -1$, $p > 0$ and $\alpha > p(r+2)$, then

$$\sum_{k=1}^{\infty} k^r \mathsf{P}\left(|S_k| \ge \varepsilon k^{1/p}\ln^q(k)\right) \sim \left(\frac{1}{\varepsilon\ln^q(1/\varepsilon)}\right)^{\nu} \left[c\left(\frac{\alpha-p}{\alpha p}\right)^q\right]^{\nu} \frac{1}{r+1} \mathsf{E}\,|Z_\alpha|^{\nu}$$

as $\varepsilon \downarrow 0$.

Indeed, in this case, $W(t) \sim \frac{1}{r+1}t^{r+1}$ and $\psi(t) = \frac{1}{c}x^{(\alpha-p)/\alpha p}\ln^q(t)$. Thus

$$\psi^{-1}(t) \sim c' t^{\alpha p/(\alpha-p)} \ln^{-q\alpha p/(\alpha-p)}(t),$$

where $c' = \left[c\left(\frac{\alpha-p}{\alpha p}\right)^q\right]^{\alpha p/(\alpha-p)}$. The latter relation proves Corollary 11.13.

Corollary 11.14 *Let X, $\{X_n\}_{n\ge 1}$ be independent, identically distributed random variables. Assume that condition (11.19) is satisfied. Moreover, let condition (11.20) hold, if $\alpha = 1$ and let condition (11.6) hold, if $\alpha > 1$. If $0 < p < \alpha$, then*

$$\sum_{k=2}^{\infty} \frac{1}{k} \mathsf{P}\left(|S_k| \ge \varepsilon k^{1/p}\ln^q(k)\right) \sim \ln(1/\varepsilon) \cdot \frac{\alpha p}{\alpha - p}, \qquad \varepsilon \downarrow 0.$$

Indeed, in this case, $W(t) \sim \ln(t)$ and thus Corollary 11.14 is a consequence of Theorem 11.1.

The asymptotic behavior of the series in (11.1) changes essentially if $p = \alpha$. Some related result has been obtained by Scheffler [320].

11.3 Proof of Theorem 11.1

Fix $0 < \theta < 1$ and put $t = \psi^{-1}\left(\frac{1}{\theta\varepsilon}\right)$. Then $\frac{1}{\varepsilon} = \theta\psi(t)$. Recall that $U \circ \psi = W$, $\psi \in \mathcal{RV}_{\frac{\alpha-p}{\alpha p}}$, $\psi^{-1} \in \mathcal{RV}_{\frac{\alpha p}{\alpha-p}}$ (see statement 5° in Section 1.5 of Seneta [324]) and $U \in \mathcal{RV}_{(r+1)\frac{\alpha p}{\alpha-p}}$. This means that

$$\lim_{t\to\infty} \frac{U(ct)}{U(t)} = c^{\beta} \tag{11.22}$$

for all $c > 0$, where $\beta = (r+1)\frac{\alpha p}{\alpha-p}$. Note that $\alpha > \beta$, since $\alpha > p(r+2)$.

Let the sequence $\{b_n\}$ and the random variable Z_α be as defined in (A.32). The distribution function of Z_α is denoted by G. The further proof consists of five steps.

Step 1 Put

$$\Delta_n = \sup_x |\,\mathsf{P}(|S_n| \ge b_n x) - \mathsf{P}(|Z_\alpha| \ge x)|.$$

Since G is a continuous function (see (A.22) below), $\Delta_n \to 0$ as $n \to \infty$. Thus

$$\lim_{m\to\infty} \frac{1}{W(m)} \sum_{k=1}^{m} w_k \Delta_k = 0,$$

whence, by (11.22),

$$\lim_{\varepsilon\downarrow 0} \frac{1}{U(1/\varepsilon)} \sum_{k\le\psi^{-1}\left(\frac{1}{\theta\varepsilon}\right)} w_k \Delta_k = 0. \tag{11.23}$$

Step 2 We show that

$$\limsup_{\varepsilon\downarrow 0} \frac{1}{U(1/\varepsilon)} \sum_{k>\psi^{-1}\left(\frac{1}{\theta\varepsilon}\right)} w_k\, \mathsf{P}\left(|Z_\alpha| \ge \varepsilon\psi_k\right) \le C\theta^{\alpha-\beta} \tag{11.24}$$

for some constant $C > 0$ not depending on θ. The latter inequality can be rewritten in an equivalent form as

$$\limsup_{t\to\infty} \frac{1}{U(\theta\psi(t))} \sum_{k>t} w_k \, \mathsf{P}\left(|Z_\alpha| \geq \frac{\psi_k}{\theta\psi(t)}\right) \leq C\theta^{\alpha-\beta}. \tag{11.25}$$

According to the asymptotic behavior of a stable distribution (see (A.23)), the latter inequality follows from the bound

$$\limsup_{t\to\infty} \frac{1}{U(\theta\psi(t))} \sum_{k>t} w_k \left(\frac{\theta\psi(t)}{\psi_k}\right)^\alpha \leq C\theta^{\alpha-\beta}. \tag{11.26}$$

In turn, this bound follows from the general result on the asymptotic of the tail of a series with a slowly varying function (see Lemma A.14, where $\lambda = \frac{\alpha-p}{p} - r$). Then

$$\sum_{k>t} \frac{w_k}{\psi_k^\alpha} \leq D \frac{tw(t)}{\psi^\alpha(t)}.$$

Therefore,

$$\limsup_{t\to\infty} \frac{1}{U(\theta\psi(t))} \sum_{k>t} w_k \left(\frac{\theta\psi(t)}{\psi_k}\right)^\alpha \leq D\theta^\alpha \limsup_{t\to\infty} \frac{U(\psi(t))}{U(\theta\psi(t))} \cdot \frac{tw(t)}{W(t)}.$$

Hence inequalities (11.26) and (11.24) follow, since $\frac{U(\psi(t)}{U(\theta\psi(t))} \to \theta^{-\beta}$, as $t \to \infty$, by (11.22) and since $\frac{tw(t)}{W(t)} \to r + 1$, as $t \to \infty$, by Lemma A.20.

Step 3 We show that

$$\lim_{\varepsilon\downarrow 0} \frac{1}{U(1/\varepsilon)} \sum_{k>\psi^{-1}\left(\frac{1}{\theta\varepsilon}\right)} kw_k \, \mathsf{P}(|X| \geq \varepsilon\varphi_k) \leq C\theta^{\alpha-\beta} \tag{11.27}$$

for some constant $C > 0$ not depending on θ. With $t = \psi^{-1}\left(\frac{1}{\theta\varepsilon}\right)$ as above, this inequality can be rewritten in an equivalent form as

$$\lim_{t\to\infty} \frac{1}{U(\theta\psi(t))} \sum_{k>t} kw_k \, \mathsf{P}\left(|X| \geq \frac{\varphi_k}{\theta\psi(t)}\right) \leq C\theta^{\alpha-\beta}. \tag{11.28}$$

Since the tail of a stable distribution is a regularly varying function (see (A.26)), the latter bound follows from

$$\lim_{t\to\infty} \frac{1}{U(\theta\psi(t))} \sum_{k>t} kw_k \frac{g(\varphi_k/\theta\psi(t))}{(\varphi_k/\theta\psi(t))^\alpha} \leq C\theta^{\alpha-\beta}.$$

Recall that g is a slowly varying function, that is,

$$\sum_{k>t} kw_k \frac{g(v\varphi_k)}{\varphi_k^\alpha} \le Dt^2 w(t) \frac{g(v\varphi(t))}{\varphi^\alpha(t)},$$

by Lemma A.14, with $\lambda = \frac{\alpha}{p} - 1 - r$, where $v = 1/\theta\psi(t)$ and where the constant D does not depend on t and v. Therefore,

$$\frac{1}{U(\theta\psi(t))} \sum_{k>t} kw_k \frac{g(\varphi_k/\theta\psi(t))}{(\varphi_k/\theta\psi(t))^\alpha} \le D\cdot\theta^\alpha \cdot \frac{g(b(t)/\theta)}{g(b(t))} \cdot \frac{tw(t)}{W(t)} \cdot \frac{U(\psi(t))}{U(\theta\psi(t))} \cdot \frac{tg(b(t))}{b^\alpha(t)}.$$

The term $\frac{g(b(t)/\theta)}{g(b(t))}$ tends to 1, since g is a slowly varying function. The term $\frac{tw(t)}{W(t)}$ tends to $r+1$, by Lemma A.20, while the term $\frac{U(\psi(t))}{U(\theta\psi(t))}$ tends to $\theta^{-\beta}$ by (11.22). Finally, $\frac{tg(b(t))}{b^\alpha(t)}$ is bounded. To prove this boundedness, note that Lemma A.15 (iii) together with (A.26) imply that $\lim_{t\to\infty} f(t)$ is finite, where

$$f(t) = \frac{tg\,(b(t))}{b^\alpha(t)}.$$

Since g is a slowly varying function and b is a regularly varying function, the limit $\lim_{t\to\infty} f(t)$ is finite. Therefore, relation (11.28) as well as relation (11.27) are proved.

Step 4.1 We prove that, for $\alpha < 2$,

$$\lim_{\varepsilon\downarrow 0} \frac{1}{U(1/\varepsilon)} \sum_{k>\psi^{-1}\left(\frac{1}{\theta\varepsilon}\right)} w_k \,\mathsf{P}(|S_k| \ge \varepsilon\varphi_k) \le C\theta^{\alpha-\beta}, \tag{11.29}$$

for some constant $C > 0$ not depending on θ. Again, introducing the variable t as above, the relation (11.29) is equivalent to

$$\lim_{t\to\infty} \frac{1}{U(\theta\psi(t))} \sum_{k>t} w_k \,\mathsf{P}(|S_k| \ge xb_k) \le C\theta^{\alpha-\beta}, \tag{11.30}$$

where x is defined by the equation $\varphi(k)/\theta\psi(t) = xb(k)$. Now Proposition A.4 and (11.28) easily imply that (11.30) holds, whence relation (11.29) follows.

Step 4.2 Consider the case $\alpha = 2$. Fix a real number η such that $\beta < \eta < 2$. We show that

$$\lim_{\varepsilon\downarrow 0} \frac{1}{U(1/\varepsilon)} \sum_{k>\psi^{-1}\left(\frac{1}{\theta\varepsilon}\right)} w_k \,\mathsf{P}(|S_k| \ge \varepsilon\varphi_k) \le C\theta^{\eta-\beta}, \tag{11.31}$$

for some constant $C > 0$ not depending on θ. This relation is equivalent to

$$\lim_{t\to\infty} \frac{1}{U(\theta\psi(t))} \sum_{k>t} w_k \,\mathsf{P}\left(|S_k| \geq \frac{\varphi_k}{\theta\psi(t)}\right) \leq C\theta^{\eta-\beta}. \tag{11.32}$$

The Markov inequality implies that

$$\begin{aligned}\sum_{k>t} w_k \,\mathsf{P}\left(|S_k| \geq \frac{\varphi_k}{\theta\psi(t)}\right) &\leq \theta^\eta \psi^\eta(t) \sum_{k>t} w_k \frac{\mathsf{E}\,|S_k|^\eta}{\varphi_k^\eta} \\ &\leq \theta^\eta \psi^\eta(t) \left[\sup_{i\geq 1} \mathsf{E}\left|\frac{S_i}{b_i}\right|^\eta\right] \sum_{k>t} \frac{w_k}{\psi_k^\eta},\end{aligned}$$

since the necessary moment exists for a stable distribution (see (A.27)). Now, from the uniform convergence of moments of normalized sums (cf. (A.35)) and Lemma A.14,

$$\frac{1}{U(\theta\psi(t))} \sum_{k>t} w_k \,\mathsf{P}\left(|S_k| \geq \frac{\varphi_k}{\theta\psi(t)}\right) \leq \operatorname{const} \theta^\eta \frac{tw(t)}{W(t)} \frac{U(\psi(t))}{U(\theta\psi(t))},$$

whence (11.32) and (11.31) follow in view of Lemma A.20 and (11.22).

Note that, by Theorem 2.6.2 in Ibragimov and Linnik [187], if X is attracted to a normal law, then either (A.26) below holds or the variance exists. Thus it remains to consider the latter case, i.e. $\mathsf{E}\,X^2 < \infty$. Then, according to the central limit theorem, $b(n) = c\sqrt{n}$, $c > 0$, $n \geq 1$. Using the Chebyshev inequality, we obtain by Lemma A.14 that

$$\begin{aligned}\frac{1}{U(\theta\psi(t))} \sum_{k>t} w_k \,\mathsf{P}\left(|S_k| \geq \frac{\varphi_k}{\theta\psi(t)}\right) &\leq \mathsf{E}\,X^2 \cdot \frac{\theta^2\psi^2(t)}{U(\theta\psi(t))} \sum_{k>t} \frac{kw_k}{\varphi_k^2} \\ &\leq D\theta^2 \cdot \frac{t\psi^2(t)}{\varphi^2(t)} \cdot \frac{tw(t)}{W(t)} \cdot \frac{U(\psi(t))}{U(\theta\psi(t))}.\end{aligned}$$

Since $\frac{t\psi^2(t)}{\varphi^2(t)} = c^{-2}$, $\frac{tw(t)}{W(t)} \to r+1$ and $\frac{U(\psi(t))}{U(\theta\psi(t))} \to \theta^{-\beta}$, as $t \to \infty$, by Lemma A.20 and (11.22). So, relations (11.30) and (11.29) also hold for the case when $\mathsf{E}\,X^2 < \infty$.

Step 5 Fix η such that $\beta < \eta < \alpha$. Then (11.29), if $\alpha < 2$, and (11.31), if $\alpha = 2$, together with (11.24) and (11.23) imply

$$\limsup_{\varepsilon\downarrow 0} \frac{1}{U(1/\varepsilon)} \left| \sum_{k=1}^\infty w_k \,\mathsf{P}(|S_k| \geq \varepsilon\varphi_k) - \sum_{k=1}^\infty w_k \,\mathsf{P}(|Z_\alpha| \geq \varepsilon\psi_k) \right| \leq C\theta^{\eta-\beta},$$

for all $0 < \theta < 1$ and some constant $C > 0$ not depending on θ. Since θ is arbitrary,

$$\lim_{\varepsilon\downarrow 0} \frac{1}{U(1/\varepsilon)} \sum_{k=1}^{\infty} w_k \,\mathsf{P}(|S_k| \geq \varepsilon\varphi_k) = \lim_{\varepsilon\downarrow 0} \frac{1}{U(1/\varepsilon)} \sum_{k=1}^{\infty} w_k \,\mathsf{P}(|Z_\alpha| \geq \varepsilon\psi_k). \tag{11.33}$$

The last step of the proof is to find the asymptotic of the series based on the tails of the distribution function of Z_α, i.e.

$$\lim_{\varepsilon\downarrow 0} \frac{1}{U(1/\varepsilon)} \sum_{k\geq 1} w_k \,\mathsf{P}(|Z_\alpha| \geq \varepsilon\psi_k) = \mathsf{E}\,|Z_\alpha|^{(r+1)\alpha p/(\alpha-p)}. \tag{11.34}$$

Relation (11.34) is an obvious corollary of Lemma A.1 below, with $\gamma = \frac{1}{p} - \frac{1}{\alpha}$. Indeed, since $(r+1)\alpha p/(\alpha - p) < \alpha$ for $\alpha > p(r+2)$, condition (A.2) is satisfied. One also needs to check condition (A.4), but this follows from (A.24), if $r > -1$, or from Lemma A.17, if $r = -1$. □

Remark 11.15 The proof in Spătaru [342] for $w(t) = 1/t$ and $p = 1$ ends at proving relation (11.33). Moreover, the limit on the right-hand side is evaluated only for the special case of $b_k = k^{1/\alpha}$, that is, for the case of normal attraction. Lemma A.1 below is given to fill these gaps in the proof of Spătaru [342], even for the more general case of $b(x) = x^{1/\alpha}h(x)$, $h \in \mathcal{RV}_0$. Note also that the proof of Lemma A.1 is simpler than the one given in [342] for the special case of $w(t) = 1/t$ and $\varphi(t) = t$.

We show how the proof can successfully be completed in the case of $w(t) = 1/t$ and $p = 1$. Fix $0 < \nu < 1 - 1/\alpha$. Recall the notation G for the distribution function of the random variable Z_α and put $\overline{G}(x) = \mathsf{P}(|Z_\alpha| \geq x)$. For $\nu > 0$, we have $k^{-\nu} \leq h(k) \leq k^{\nu}$ if k is sufficiently large, say for $k \geq k_0$. Thus

$$\sum_{k=k_0}^{\infty} \frac{1}{k}\overline{G}\left(\varepsilon k^{1+\nu-1/\alpha}\right) \leq \sum_{k=k_0}^{\infty} \frac{1}{k}\overline{G}(\varepsilon k/b_k) \leq \sum_{k=k_0}^{\infty} \frac{1}{k}\overline{G}\left(\varepsilon k^{1-\nu-1/\alpha}\right). \tag{11.35}$$

Let $v_1 = 1 + \nu - 1/\alpha$ and $v_2 = 1 - \nu - 1/\alpha$. Using Lemma A.1, with the function $w(t) = 1/t$, we get

$$\sum_{k=1}^{\infty} \frac{1}{k}\overline{G}\left(\varepsilon k^{v_1}\right) \sim \ln\left(\varepsilon^{-1/v_1}\right) \quad \text{and} \quad \sum_{k=1}^{\infty} \frac{1}{k}\overline{G}\left(\varepsilon k^{v_2}\right) \sim \ln\left(\varepsilon^{-1/v_2}\right).$$

Therefore,

$$\frac{1}{v_1} \leq \liminf_{\varepsilon\downarrow 0} \frac{1}{-\ln(\varepsilon)} \sum_{k=1}^{\infty} \frac{1}{k}\overline{G}(\varepsilon k/b_k) \leq \limsup_{\varepsilon\downarrow 0} \frac{1}{-\ln(\varepsilon)} \sum_{k=1}^{\infty} \frac{1}{k}\overline{G}(\varepsilon k/b_k) \leq \frac{1}{v_2}.$$

Passing to the limit as $\nu \to 0$, relation (11.34) is proved, for $w(x) = 1/x$ and $p = 1$, in view of the bounds (11.35).

It is worth mentioning, however, that one needs a more sophisticated argument to obtain the asymptotic of the series in (11.1) for a general function w.

11.4 Comments

This chapter is based on the works of Buldygin et al. [68] and Indlekofer and Klesov [193, 195, 196]. Similar results are obtained by Rozovskiĭ [308, 309].

Section 11.1 The notion of *complete convergence* for a sequence of random variables has been introduced by Hsu and Robbins [185].

The equivalence of the strong law of large numbers for independent, identically distributed random variables for the whole sequence and its geometrical subsequence was observed quite long ago (see Prokhorov [296]). The idea in its final form was realized by Martikaĭnen and Petrov [263], Volodin and Nagaev [362], Buldygin [55].

The limiting distribution of the estimator $\varepsilon^2 \hat{Q}(\varepsilon)$ as $\varepsilon \downarrow 0$ was found by Kao [203]. Several other functionals constructed from a sequence of independent, identically distributed random variables have been studied. For example, the last time when $S_n \geq \varepsilon n$ was considered by Robbins et al. [303] and the maximal excess of S_n over εn was studied by Chow and Lai [86] (see Gafurov and Slastnikov [140] for an early overview of this topic). Note that $Q(\varepsilon) < \infty$ is equivalent to $\mathsf{E}\,\hat{Q}(\varepsilon) < \infty$. Necessary and sufficient conditions for $\mathsf{E}\,\hat{Q}(\varepsilon) < \infty$ were found by Hsu–Robbins [185] and Erdős [120, 121] for the series (c_1), Spitzer [343] for the series (c_2), and Baum and Katz [28] for the series (c_3) (the Spitzer series (c_2) was treated in detail in [28] as well). A more general case of regularly varying weights w_k in (11.1) was considered by Heyde and Rohatgi [181]. The finiteness of higher moments of $\hat{Q}(\varepsilon)$ was studied by Slivka and Severo [336], Stratton [350], Griffiths and Wright [155], Klebaner [220], Fisher et al. [131], and Klesov and Stadtmüller [228].

The precise asymptotic as $\varepsilon \downarrow 0$ was found by Heyde [180] for the Hsu–Robbins series (c_1), by Chen [81] for the Baum–Katz series (c_3), by Chow and Lai [87] for the Spitzer series (c_2), and by Klesov and Stadtmüller [228] for the Hsu–Robbins series (c_4). Some related results for the Spitzer series are obtained by Spătaru [342], Gut and Spătaru [161], Gut and Steinebach [163].

Appendix A
Some Auxiliary Results

A.1 Introduction

This chapter contains some auxiliary results which are also of independent interest, and some of them seem to be new.

In Sect. A.2, we find the asymptotic behavior of a series related to $Q(\varepsilon)$ from (11.1). In fact, the main result of Sect. A.2 describes the asymptotic behavior of the series corresponding to the stable limit law. Note that this result is valid even for a wider class of distributions.

In Sect. A.3, we provide some large deviation bounds for distributions attracted to a stable law. Known results of this type assert that $\mathsf{P}(|S_n| \geq x_n b_n)$ and $n\,\mathsf{P}(|X| \geq x_n b_n)$ have the same asymptotic behavior as $n \to \infty$, where $\{b_n\}_{n\geq 1}$ is the normalizing sequence in the weak convergence to a stable limit distribution and where $\{x_n\}_{n\geq 1}$ is an arbitrary sequence tending to infinity, that is, $x_n \to \infty$ as $n \to \infty$. Our bounds hold for all $x \geq 0$. Note also that our estimates are uniform in x, but we do not prove an asymptotic equivalence for the probabilities of large deviations.

In Sect. A.4, we discuss some key properties of slowly varying functions used in the proof of the main result.

Some necessary properties of stable distributions and of distributions attracted to stable laws are collected in Sect. A.5. We prove a known result about the normalizing and centering sequences in the weak convergence to the stable limit law, which is convenient in many particular cases. Another result of this section provides the existence of negative moments for stable distributions.

In Sect. A.6, we state a result of Aljančić et al. [9] on the asymptotic behavior of integrals with slowly varying functions.

Finally, in Sect. A.7, a new proof of a theorem of Parameswaran [290] is given, but for sums here instead of integrals.

V. V. Buldygin et al., *Pseudo-Regularly Varying Functions and Generalized Renewal Processes*, Probability Theory and Stochastic Modelling 91, https://doi.org/10.1007/978-3-319-99537-3

A.2 The Asymptotic Behavior of Series of Weighted Tails of Distributions

Given a random variable Z, a number $\varepsilon > 0$, and functions $w \in \mathcal{RV}_r$ and $\psi \in \mathcal{RV}_\gamma$, put

$$\mathcal{Q}(\varepsilon) = \sum_{k=1}^{\infty} w_k \, \mathsf{P}(|Z| \geq \varepsilon \psi_k), \tag{A.1}$$

where $w_k = w(k)$ and $\psi_k = \psi(k)$.

Lemma A.1 *Let $w \in \mathcal{RV}_r$, $r \geq -1$, and let the series $\sum w_k$ diverge. Let W be as defined in (11.12). Consider a positive, continuous, increasing function ψ such that $\psi \in \mathcal{RV}_\gamma$, for $\gamma > 0$. Assume that Z is a random variable such that*

$$\mathsf{E}\,|Z|^\eta < \infty, \quad \text{for some} \quad \eta > \frac{r+1}{\gamma}. \tag{A.2}$$

Let f be a density of the distribution of $|Z|$. Then

$$\mathcal{Q}(\varepsilon) < \infty \quad \text{for all} \quad \varepsilon > 0. \tag{A.3}$$

In addition, if

$$\mathsf{E}\,|Z|^\eta < \infty, \quad \text{for some} \quad \eta < \frac{r+1}{\gamma}, \tag{A.4}$$

then

$$\lim_{\varepsilon \downarrow 0} \frac{\mathcal{Q}(\varepsilon)}{U\,(1/\varepsilon)} = \mathsf{E}\,|Z|^{(r+1)/\gamma}, \tag{A.5}$$

where the function U is as defined in (11.18) and where ψ^{-1} is the inverse function of ψ.

Note that conditions (A.2) and (A.4) are equivalent if $r \neq -1$. On the other hand, they are different if $r = -1$, since each of them describes a different behavior of the distribution function of the random variable Z, namely, condition (A.2) means that a certain positive moment exists, while condition (A.4) says that a negative moment exists.

Remark A.2 Lemma A.1 is used in the proof of Theorem 11.1 for an α-stable random variable Z_α and for $\psi(x) = \varphi(x)/b(x)$, where $\varphi \in \mathcal{RV}_{1/p}$ appears in the large deviation probabilities in the series of (11.1) and where $b \in \mathcal{RV}_{1/\alpha}$ provides the normalizations in the attraction of the sums S_n to the random variable Z_α.

Proof of Lemma A.1 If η is such that condition (A.2) holds, then Chebyshev's inequality implies

$$\mathcal{Q}(\varepsilon) \le \frac{\mathsf{E}\,|Z|^\eta}{\varepsilon^\eta} \sum_{k=1}^{\infty} \frac{w_k}{\psi_k^\eta},$$

whence (A.3) follows, since $w\psi^{-\eta} \in \mathcal{RV}_{r-\eta\gamma}$ and $r - \eta\gamma < -1$ in view of condition (A.2) (see Lemma A.9).

To prove (A.5), we represent the sum as

$$\sum_{k=1}^{\infty} w_k \, \mathsf{P}(|Z| \ge \varepsilon\psi_k) = \sum_{k=1}^{\infty} w_k \sum_{j=k}^{\infty} \int_{I_j(\varepsilon)} f(t)\,dt,$$

where $I_j(\varepsilon) = [\varepsilon\psi_j, \varepsilon\psi_{j+1})$. Changing the order of summation we get

$$\begin{aligned}
\sum_{k=1}^{\infty} w_k \, \mathsf{P}(|Z| \ge \varepsilon\psi_k) &= \sum_{j=1}^{\infty} W(j) \int_{I_j(\varepsilon)} f(t)\,dt \\
&= \sum_{j=1}^{\infty} \int_{I_j(\varepsilon)} W\left(\psi^{-1}(t/\varepsilon)\right) f(t)\,dt \\
&\quad + \sum_{j=1}^{\infty} \int_{I_j(\varepsilon)} \left[W(j) - W\left(\psi^{-1}(t/\varepsilon)\right)\right] f(t)\,dt \\
&\equiv \mathcal{Q}_1(\varepsilon) + \mathcal{Q}_2(\varepsilon).
\end{aligned}$$

Now we show that

$$\mathcal{Q}_1(\varepsilon) \sim U(1/\varepsilon)\,\mathsf{E}\,|Z|^{(r+1)/\gamma} \qquad \text{as } \varepsilon \downarrow 0, \tag{A.6}$$

$$Q_2(\varepsilon) = o\,(U(1/\varepsilon)) \qquad \text{as } \varepsilon \downarrow 0. \tag{A.7}$$

In the proof below, we apply Theorem A.18 (i) and A.19 (i) together with conditions (A.2) and (A.4). To prove (A.6) note that

$$\mathcal{Q}_1(\varepsilon) = \int_{\varepsilon\psi_1}^{\infty} U(t/\varepsilon)\, f(t)\,dt.$$

The rest of the proof differs for the cases $\psi_1 = 0$ and $\psi_1 > 0$.

Case $\psi_1 = 0$ Fix $B > 0$. Then

$$\mathcal{Q}_1(\varepsilon) = \int_0^B U(t/\varepsilon)\, f(t)\,dt + \int_B^{\infty} U(t/\varepsilon)\, f(t)\,dt.$$

The asymptotic behavior of the terms on the right-hand side above has been obtained in Theorems A.18 (i) and A.19 (i), with $x = 1/\varepsilon$ and $\nu = (r+1)/\gamma$. Thus (A.6) follows from conditions (A.2) and (A.4), i.e.

$$\int_0^B U(t/\varepsilon) f(t)\,dt \sim U(1/\varepsilon) \int_0^B t^{(r+1)/\gamma} f(t)\,dt, \qquad \varepsilon \downarrow 0,$$

$$\int_B^\infty U(t/\varepsilon) f(t)\,dt \sim U(1/\varepsilon) \int_B^\infty t^{(r+1)/\gamma} f(t)\,dt, \qquad \varepsilon \downarrow 0.$$

Case $\psi_1 > 0$ Fix $A > 0$. Then

$$\int_A^\infty U(t/\varepsilon) f(t)\,dt \le \int_{\varepsilon\psi_1}^\infty U(t/\varepsilon) f(t)\,dt,$$

for $0 < \varepsilon < A/\psi_1$. Thus

$$\begin{aligned}
&\liminf_{\varepsilon\downarrow 0} \frac{1}{U(1/\varepsilon)} \int_{\varepsilon\psi_1}^\infty U(t/\varepsilon) f(t)\,dt \\
&\quad \ge \lim_{\varepsilon\downarrow 0} \frac{1}{U(1/\varepsilon)} \int_A^\infty U(t/\varepsilon) f(t)\,dt = \int_A^\infty t^{(r+1)/\gamma} f(t)\,dt,
\end{aligned}$$

in view of condition (A.2) and Theorem A.18 (i), with $x = 1/\varepsilon$. Since $A > 0$ is arbitrary, we conclude that

$$\liminf_{\varepsilon\downarrow 0} \frac{1}{U(1/\varepsilon)} \int_{\varepsilon\psi_1}^\infty U(t/\varepsilon) f(t)\,dt \ge \int_0^\infty t^{(r+1)/\gamma} f(t)\,dt = \mathsf{E}\,|Z|^{(r+1)/\gamma}. \tag{A.8}$$

For any given $B > 0$,

$$\int_{\varepsilon\psi_1}^\infty U(t/\varepsilon) f(t)\,dt \le \int_0^B U(t/\varepsilon) f(t)\,dt + \int_B^\infty U(t/\varepsilon) f(t)\,dt,$$

whence

$$\begin{aligned}
&\limsup_{\varepsilon\downarrow 0} \frac{1}{U(1/\varepsilon)} \int_{\varepsilon\psi_1}^\infty U(t/\varepsilon) f(t)\,dt \\
&\quad \le \lim_{\varepsilon\downarrow 0} \frac{1}{U(1/\varepsilon)} \int_0^B U(t/\varepsilon) f(t)\,dt \\
&\qquad + \lim_{\varepsilon\downarrow 0} \frac{1}{U(1/\varepsilon)} \int_B^\infty U(t/\varepsilon) f(t)\,dt \\
&\quad = \int_0^B t^{(r+1)/\gamma} f(t)\,dt + \int_B^\infty t^{(r+1)/\gamma} f(t)\,dt = \mathsf{E}\,|Z|^{(r+1)/\gamma}.
\end{aligned} \tag{A.9}$$

Thus (A.6) follows from (A.8) and (A.9) in the case when $\psi_1 > 0$.

For the proof of (A.7), we put $u(x) = w\left(\psi^{-1}(x)\right)$. Note that $u \in \mathcal{RV}_{r/\gamma}$. Then

$$|\mathfrak{Q}_2(\varepsilon)| \le \sum_{j=1}^{\infty} (W(j+1) - W(j)) \int_{I_j(\varepsilon)} f(t)\,dt \le \sum_{j=1}^{\infty} w(j+1) \int_{I_j(\varepsilon)} f(t)\,dt,$$

since the function $W \circ \psi^{-1}$ is nondecreasing. According to Theorem A.7, there exists a constant $C > 0$ such that

$$|\mathfrak{Q}_2(\varepsilon)| \le C \sum_{j=1}^{\infty} \int_{I_j(\varepsilon)} u\left(t/\varepsilon\right) f(t)\,dt = C \int_{\varepsilon\psi_1}^{\infty} u\left(t/\varepsilon\right) f(t)\,dt.$$

By the same reasoning as above, with u instead of U, we obtain

$$\int_{\varepsilon\psi_1}^{\infty} u\left(t/\varepsilon\right) f(t)\,dt \sim u\left(1/\varepsilon\right) \mathsf{E}\,|Z|^{(r+1)/\gamma} \qquad \varepsilon \downarrow 0,$$

whence (A.7) results from Lemma A.20. Now equality (A.5) follows from relations (A.6) and (A.7). □

Remark A.3 The condition $\mathsf{E}\,U(|Z|) < \infty$ is, in fact, necessary and sufficient for the validity of $Q(\varepsilon) < \infty$, for all $\varepsilon > 0$. If ψ is not necessarily increasing, one may choose an increasing function ψ_1 such that $\psi_1 \sim \psi$, and then reformulate the necessary and sufficient convergence condition in terms of this function, i.e. $\mathsf{E}\,W\left(\psi_1^{-1}(|Z|)\right) < \infty$.

A.3 Large Deviations in the Case of Attraction to a Stable Law

In this section, we prove a result on the probabilities of large deviations for sums of independent, identically distributed random variables, which are attracted to a stable distribution of index $0 < \alpha < 2$.

Our uniform estimate given by (A.12) differs from the usual form of a large deviation result for heavy tailed distributions. One of the popular ways to express this property is

$$\lim_{n\to\infty} \sup_{x \ge t_n} \frac{\mathsf{P}(|S_n| \ge x)}{n\,\mathsf{P}(|X| \ge x)} = 1, \tag{A.10}$$

where $\{t_n\}_{n\ge1}$ is a sequence such that

$$\frac{S_n}{t_n} \xrightarrow{\mathsf{P}} 0 \tag{A.11}$$

(see Remark A.6 below). Clearly (A.12) below provides a bound for the whole axis, while (A.10) describes precise asymptotics for large arguments (see also Remark A.6 below).

Throughout this section we denote by $\{b_n\}_{n\ge 1}$ the normalizing sequence in the weak convergence to the stable limit law.

Proposition A.4 *Assume that* $0 < \alpha < 2$. *Let* X, $\{X_n\}_{n\ge 1}$ *be independent, identically distributed random variables such that* $X \in \mathcal{DA}(\alpha, \{b_n\}, \{a_n\})$. *If* $\alpha = 1$, *then we additionally assume that* X *has a symmetric distribution. We also assume that* $\mathsf{E}\, X = 0$, *if* $\alpha > 1$. *Then*

$$\sup_{x\ge 0}\sup_{n\ge 1}\frac{\mathsf{P}(|S_n|\ge xb_n)}{n\,\mathsf{P}(|X|\ge xb_n)} < \infty. \tag{A.12}$$

One can drop the assumption of symmetry of the distribution in the case when $\alpha = 1$, if S_n in (A.12) is replaced by $S_n - a_n b_n$.

Corollary A.5 *Let* $\alpha = 1$ *and* $X \in \mathcal{DA}(1, \{b_n\}, \{a_n\})$. *Then*

$$\sup_{x\ge 0}\ \sup_{n\ge 1}\frac{\mathsf{P}(|S_n - a_n b_n|\ge xb_n)}{n\,\mathsf{P}(|X|\ge xb_n)} < \infty.$$

Remark A.6 The theory of large deviations is a huge area and many authors have studied this topic in detail. One of the popular ways to express a large deviation property is written in (A.10). To get an impression of why this kind of large deviation principle does not imply (A.12) we note that, for any $m \ge 1$,

$$\begin{aligned}\sup_{x\ge 0}\sup_{n\ge 1}\frac{\mathsf{P}(|S_n|\ge xb_n)}{n\,\mathsf{P}(|X|\ge xb_n)} &\ge \sup_{x\ge 0}\frac{\mathsf{P}(|S_m|\ge xb_m)}{m\,\mathsf{P}(|X|\ge xb_m)} = \sup_{x\ge 0}\frac{\mathsf{P}(|S_m|\ge x)}{m\,\mathsf{P}(|X|\ge x)}\\ &= \max\left\{\sup_{x\le t_m}\frac{\mathsf{P}(|S_m|\ge x)}{m\,\mathsf{P}(|X|\ge x)},\ \sup_{x\ge t_m}\frac{\mathsf{P}(|S_m|\ge x)}{m\,\mathsf{P}(|X|\ge x)}\right\}.\end{aligned}$$

The second term on the right-hand side is bounded in m in view of (A.10). To prove that the first term on the right-hand side is bounded in m one can try to proceed as follows:

$$\sup_{x\le t_m}\frac{\mathsf{P}(|S_m|\ge x)}{m\,\mathsf{P}(|X|\ge x)} \le \frac{1}{m\,\mathsf{P}(|X|\ge t_m)}.$$

Unfortunately $m\,\mathsf{P}(|X|\ge t_m) \to 0$ as $m \to \infty$ in view of the necessary condition for the weak law of large numbers (A.11) and thus the right-hand side is unbounded. Therefore there is no simple way to extract (A.12) from (A.10).

Proof of Proposition A.4 We shall make use of some ideas of Heyde [178], where a similar result has been obtained in the case of attraction to a nondegenerate stable distribution of index α, if $\alpha \ne 1$ and $\alpha \ne 2$.

Let $b_0 = \inf_{n\geq 1} b_n$. Without loss of generality assume that $b_0 > 0$. Fix $x_0 > 0$. Then, according to (A.30),

$$\sup_{0\leq x\leq x_0}\sup_{n\geq 1}\frac{\mathsf{P}(|S_n|\geq xb_n)}{n\,\mathsf{P}(|X|\geq xb_n)}\leq \sup_{n\geq 1}\frac{1}{n\,\mathsf{P}(|X|\geq x_0 b_n)}<\infty.$$

This means that one can restrict the consideration to the case when $x \geq x_0$. The exact value of x_0 will be chosen below.

For the sake of simplicity, we may assume that $a_n = 0, n \geq 1$. In fact, by (11.13), a_n is in general involved in the attraction to Z_α, but Lemma A.16 allows us to switch to the case of $a_n = 0, n \geq 1$, if $\alpha \neq 1$. For $\alpha = 1$, we can do so as well, since the random variable X has a symmetric distribution.

Let $1/2 < \gamma < 1$. Fix $x \geq x_0$ and put $z = x^\gamma$. To choose x_0, let $\eta_1 > 0$, $\eta_2 > 0$, and $\eta_3 > 0$ be such that $(1-\gamma)\eta_1 + \gamma\eta_2 < (2\gamma - 1)\alpha$ and $\eta_3 < 2 - \alpha$. According to Lemma A.10, with $\ell = g$, where the function g is as defined in (A.26), there are numbers x_{1*}, x_{2*}, x_{3*} for which (A.20) is valid, with $\eta = \eta_1, \eta_2, \eta_3$, respectively, and with some arbitrary $D > 1$. Now, choose $x_0 = \frac{1}{b_0}\max\{x_{1*}^{1/\gamma}, x_{2*}^{1/\gamma}, x_{3*}^{1/\gamma}\}$. Then $zb_n \geq \max\{x_{1*}, x_{2*}, x_{3*}\}$, for all $n \geq 1$, and one may assume that $b_n \geq \max\{x_{1*}, x_{2*}, x_{3*}\}$, for all $n \geq 1$.

Fix $n \geq 1$ and put

$$X_{kn} = X_k I(|X_k| < zb_n), \qquad S_{nn} = \sum_{k\leq n} X_{kn},$$

for $1 \leq k \leq n$. Similar to Heyde [178], we have the following inclusion of random events, i.e.

$$\begin{aligned}\{|S_{nn}|\geq xb_n\} \subseteq &\left\{\exists k\leq n\colon |X_k|\geq \frac{xb_n}{2}\right\}\\ &\cup\left\{\exists k_1<k_2\leq n\colon |X_{k_1}|\geq zb_n, |X_{k_2}|\geq zb_n\right\}\\ &\cup\left\{|S_{nn}|\geq\frac{xb_n}{2}\right\},\end{aligned}$$

whence

$$\begin{aligned}\mathsf{P}(|S_n|\geq xb_n)\leq\ & n\,\mathsf{P}\left(|X|\geq\frac{xb_n}{2}\right)+n^2\left[\mathsf{P}(|X|\geq zb_n)\right]^2\\ &+\mathsf{P}\left(|S_{nn}|\geq\frac{xb_n}{2}\right).\end{aligned}\tag{A.13}$$

Each term on the right-hand side of (A.13) will be considered separately.

Step 1 Let $b = \min_{n\geq 1} b_n$. Then (A.26) implies that

$$\frac{\mathsf{P}(|X| \geq xb_n/2)}{\mathsf{P}(|X| \geq xb_n)} = 2^\alpha \frac{g(xb_n/2)}{g(xb_n)} \leq 2^\alpha \sup_{t\geq b} \frac{g(t/2)}{g(t)} < \infty,$$

by Theorem A.7, since g is a slowly varying function. Thus

$$\sup_{x\geq 1} \sup_{n\geq 1} \frac{\mathsf{P}(|X| \geq xb_n/2)}{\mathsf{P}(|X| \geq xb_n)} < \infty.$$

Step 2 We use again (A.26) and conclude that

$$\frac{n\left[\mathsf{P}(|X| \geq zb_n)\right]^2}{\mathsf{P}(|X| \geq xb_n)} = n \cdot \frac{[g(zb_n)]^2}{(zb_n)^{2\alpha}} \cdot \frac{(xb_n)^\alpha}{g(xb_n)} = \frac{ng(b_n)}{b_n^\alpha} \cdot \frac{[g(zb_n)]^2}{g(b_n)g(xb_n)} \cdot \frac{x^\alpha}{z^{2\alpha}}. \tag{A.14}$$

By Lemma A.15 (iii), the term $ng(b_n)/b_n^\alpha$ is bounded. For the second factor on the right-hand side of (A.14), we apply Lemma A.10, with $\eta = \eta_1$ and then with $\eta = \eta_2$, and get

$$\frac{g(zb_n)}{g(xb_n)} \leq D\left(\frac{x}{z}\right)^{\eta_1}, \qquad \frac{g(zb_n)}{g(b_n)} \leq Dz^{\eta_2}.$$

Therefore,

$$\frac{n\left[\mathsf{P}(|X| \geq zb_n)\right]^2}{\mathsf{P}(|X| \geq xb_n)} \leq \text{const}\, \frac{z^{-\eta_1+\eta_2-2\alpha}}{x^{-\eta_1-\alpha}} = \text{const}\, x^{\gamma(-\eta_1+\eta_2-2\alpha)+\alpha+\eta_1}.$$

Since $\gamma(-\eta_1 + \eta_2 - 2\alpha) + \alpha + \eta_1 < 0$, we conclude that

$$\sup_{x\geq 1} \sup_{n\geq 1} \frac{n\left[\mathsf{P}(|X| \geq zb_n)\right]^2}{\mathsf{P}(|X| \geq xb_n)} < \infty. \tag{A.15}$$

Step 3 Consider the third term on the right-hand side of (A.13). By Chebyshev's inequality,

$$\mathsf{P}\left(|S_{nn}| \geq \frac{xb_n}{2}\right) \leq \frac{4}{x^2b_n^2}\left[n\,\mathsf{E}\,X_{1n}^2 + n^2\,(\mathsf{E}\,X_{1n})^2\right]. \tag{A.16}$$

On integrating by parts we have $\mathsf{E}\,X_{1n}^2 \leq 2\int_0^{zb_n} x\,\mathsf{P}(|X| \geq x)\,dx$. Thus

$$\frac{\mathsf{E}\,X_{1n}^2}{x^2b_n^2\,\mathsf{P}(|X| \geq xb_n)} \leq 2\frac{\int_0^{zb_n} x\,\mathsf{P}(|X| \geq x)\,dx}{z^2b_n^2\,\mathsf{P}(|X| \geq zb_n)} \frac{\mathsf{P}(|X| \geq zb_n)}{\mathsf{P}(|X| \geq xb_n)}\left(\frac{z}{x}\right)^2. \tag{A.17}$$

The first factor is bounded in view of Lemma A.11, with $\lambda = \alpha - 1$, since $\mathsf{P}(|X| \geq x)$ is a regularly varying function (see (A.26)) and $\alpha < 2$.

The other factors on the right-hand side of (A.17) can be estimated as follows:

$$\frac{\mathsf{P}(|X| \geq zb_n)}{\mathsf{P}(|X| \geq xb_n)} \cdot \left(\frac{z}{x}\right)^2 = \frac{g(zb_n)}{g(xb_n)} \cdot \left(\frac{z}{x}\right)^{2-\alpha} \leq D\left(\frac{z}{x}\right)^{2-\alpha-\eta_3}.$$

Hence

$$\sup_{x \geq 1} \sup_{n \geq 1} \frac{\mathsf{E}\, X_{1n}^2}{x^2 b_n^2 \, \mathsf{P}(|X| \geq xb_n)} < \infty.$$

Next we consider the term $n^2\,(\mathsf{E}\, X_{1n})^2$ in (A.16).

Case $0 < \alpha < 1$ On integrating by parts, we obtain

$$|\,\mathsf{E}\, X_{1n}| \leq E|X_{1n}| \leq \int_0^{zb_n} \mathsf{P}(|X_1| \geq x)\, dx.$$

Thus (A.26) and Lemma A.11, with $\lambda = \alpha$, imply that

$$\begin{aligned}\frac{n\,(\mathsf{E}\, X_{1n})^2}{x^2 b_n^2\, \mathsf{P}(|X_1| \geq xb_n)} &\leq \frac{n}{x^2 b_n^2\, \mathsf{P}(|X_1| \geq xb_n)} \left[\int_0^{zb_n} \mathsf{P}(|X_1| \geq x)\, dx\right]^2 \\ &\leq \text{const}\, \frac{n}{x^2 b_n^2\, \mathsf{P}(|X_1| \geq xb_n)} \left[zb_n\, \mathsf{P}(|X_1| \geq zb_n\right]^2 \\ &= \text{const}\, \frac{n\left[\mathsf{P}(|X_1| \geq zb_n\right]^2}{\mathsf{P}(|X_1| \geq xb_n)} \cdot \left(\frac{z}{x}\right)^2.\end{aligned}$$

The first factor is bounded in view of (A.15). Since $z \leq x$, we get

$$\sup_{x \geq 1} \sup_{n \geq 1} \frac{n\,(\mathsf{E}\, X_{1n})^2}{x^2 b_n\, \mathsf{P}(|X| \geq xb_n)} < \infty. \qquad \text{(A.18)}$$

Case $\alpha = 1$ Here, (A.18) is obvious, since the distribution of the random variable X is symmetric.

Case $1 < \alpha < 2$ Recall that $\mathsf{E}\, X = 0$. An integration by parts yields

$$|\,\mathsf{E}\, X_{1n}| = \int_{zb_n}^{\infty} \mathsf{P}(|X| \geq x)\, dx + zb_n(\mathsf{P}(|X| \geq zb_n).$$

Thus

$$\frac{n\,(\mathsf{E}\, X_{1n})^2}{x^2 b_n\, \mathsf{P}(|X| \geq xb_n)} = \left[\frac{\int_{zb_n}^{\infty} \mathsf{P}(|X| \geq x)\, dx}{zb_n\, \mathsf{P}(|X| \geq zb_n)} + 1\right]^2 \cdot \frac{n\left[\mathsf{P}(|X| \geq zb_n)\right]^2}{\mathsf{P}(|X| \geq xb_n)} \cdot \left(\frac{z}{x}\right)^2.$$

The first factor is bounded, in view of Lemma A.12, and the boundedness of the second factor follows from (A.15). Since $z \le x$, this implies (A.18).

On combining the above bounds, the proof of (A.12) can be completed. □

Proof of Corollary A.5 Denote by ξ^s and h_ξ the symmetrized version and characteristic function, respectively, of the random variable ξ. Recall that $\xi^s = \xi_1 - \xi_2$, where ξ_1 and ξ_2 are independent copies of ξ. If $X \in \mathcal{DA}(1, \{b_n\}, \{a_n\})$, then there exists a 1-stable random variable Z_1 such that

$$h_{S_n}\left(\frac{t}{b_n}\right) e^{-ita_n} \to h_{Z_1}(t), \qquad t \in \mathbf{R}^1.$$

Thus

$$\left|h_{S_n}\left(\frac{t}{b_n}\right)\right|^2 \to \left|h_{Z_1}(t)\right|^2, \qquad t \in \mathbf{R}^1.$$

Note that $|h_{S_n}|^2$ and $|h_{Z_1}|^2$ are characteristic functions of the random variables S_n^s and Z_1^s. Therefore, $S_n^s/b_n \xrightarrow{d} Z_1^s$. Since $|h_{Z_1}|^2(t) = e^{-2c|t|}$, by (A.21), Z_1^s is a 1-stable random variable, that is, $X^s \in \mathcal{DA}(1, \{b_n\}, \{0_n\})$. Now we apply Proposition A.4 with $\alpha = 1$ and get

$$\sup_{x \ge 0} \sup_{n \ge 1} \frac{\mathsf{P}(|S_n^s| \ge xb_n)}{n\,\mathsf{P}(|X_1^s| \ge xb_n)} < \infty. \tag{A.19}$$

Then we use the symmetrization inequalities

$$\frac{\mathsf{P}(|S_n - \mu_n| \ge xb_n)}{4\,\mathsf{P}(|X_1| \ge xb_n/2)} \le \frac{\mathsf{P}(|S_n^s| \ge xb_n)}{n\,\mathsf{P}(|X_1^s| \ge xb_n)},$$

where μ_n is a median of S_n. We show that there exists a constant M for which $|\mu_n/b_n - a_n| \le M$ for all $n \ge 1$, whence

$$\mathsf{P}(|S_n - a_n b_n| \ge xb_n) \le \mathsf{P}(|S_n - \mu_n| \ge (x - M)b_n).$$

Thus

$$\sup_{x \ge M} \sup_{n \ge 1} \frac{\mathsf{P}(|S_n - a_n b_n| \ge xb_n)}{n\,\mathsf{P}(|X_1| \ge xb_n)}$$
$$\le \text{const} \sup_{x \ge M} \sup_{n \ge 1} \frac{\mathsf{P}(|S_n^s|) \ge (x - M)b_n)}{n\,\mathsf{P}(|X_1^s| \ge (x - M)b_n)} \frac{\mathsf{P}(|X_1^s| \ge (x - M)b_n)}{\mathsf{P}(|X_1^s| \ge 2xb_n)} < \infty,$$

by (A.19) and (A.26). Finally,

$$\sup_{0\le x\le M}\sup_{n\ge 1}\frac{\mathsf{P}(|S_n-a_nb_n|\ge xb_n)}{n\,\mathsf{P}(|X_1|\ge xb_n)}\le\sup_{n\ge 1}\frac{1}{n\,\mathsf{P}(|X_1|\ge Mb_n)}<\infty,$$

according to (A.26) and Lemma A.15 (iii).

It remains to prove that the sequence $x_n=\mu_n/b_n-a_n$ is bounded. This can be done via contradiction. If $\limsup x_n=\infty$, then there exists a subsequence $\{n_k\}$ such that $x_{n_k}\to\infty$ as $k\to\infty$. Let x_0 be such that $\mathsf{P}(Z_1\ge x_0)<\frac{1}{2}$ and choose k_0 such that $x_{n_k}\ge x_0$ for all $k\ge k_0$. Thus, for $k\ge k_0$,

$$\mathsf{P}(S_{n_k}\ge\mu_{n_k})=\mathsf{P}\left(\frac{S_{n_k}}{b_{n_k}}-a_{n_k}\ge x_{n_k}\right)\le\mathsf{P}\left(\frac{S_{n_k}}{b_{n_k}}-a_{n_k}\ge x_0\right)\to\mathsf{P}(Z_1\ge x_0),$$

since the distribution function of the random variable Z_1 is continuous (see (A.22)). This is a contradiction, since $\mathsf{P}(S_{n_k}\ge\mu_{n_k})\ge\frac{1}{2}$, which proves that $\limsup x_n<\infty$.

If $\liminf x_n=-\infty$, an analogous reasoning can be used. Choose x_0 such that $\mathsf{P}(Z_1\le x_0)<\frac{1}{2}$ and then find k_0 for which $x_{n_k}<x_0$ for all $k\ge k_0$. Thus, for $k\ge k_0$,

$$\mathsf{P}(S_{n_k}\le\mu_{n_k})=\mathsf{P}\left(\frac{S_{n_k}}{b_{n_k}}-a_{n_k}\le x_{n_k}\right)\le\mathsf{P}\left(\frac{S_{n_k}}{b_{n_k}}-a_{n_k}\le x_0\right)\to\mathsf{P}(Z_1\le x_0),$$

in view of (A.22). This is a contradiction again, since $\mathsf{P}(S_{n_k}\le\mu_{n_k})\ge\frac{1}{2}$, and thus the sequence $\{x_n\}$ is indeed bounded. □

A.4 Some Auxiliary Results on Slowly Varying Functions

Recall that a function ℓ is called *slowly varying* (in the Karamata sense) if it is positive for all arguments, finite on every bounded interval, and if

$$\lim_{t\to\infty}\frac{\ell(ct)}{\ell(t)}=1,\qquad\text{for all } c>1$$

(recall Definition 3.3). We further assume that the function ℓ is measurable.

A thorough treatise of slowly and regularly varying functions can be found in Seneta [324] and Bingham et al. [41]. Below we only provide a collection of simple properties to which we refer in our proofs.

One of the fundamental results concerning regularly varying functions is the uniform convergence theorem. Below is the statement of this result for slowly varying functions (see also Theorem 3.47, where the more general case of PRV-functions is studied).

Theorem A.7 *If ℓ is a measurable slowly varying function, then*

$$\lim_{t\to\infty} \sup_{c\in[a,b]} \frac{\ell(ct)}{\ell(t)} = 1$$

for any interval $[a, b]$ with $0 < a < b < \infty$.

Most of the properties of slowly varying functions, which are used here for our considerations, can easily be derived from the following result on the existence of a monotone version.

Proposition A.8 *Let ℓ be a measurable slowly varying function. Then, given an arbitrary $\eta > 0$, there exist slowly varying functions ℓ_1 and ℓ_2 such that*

(i) $x^\eta \ell_1(x)$ is nondecreasing in $[x_0, \infty)$ for some $x_0 \geq 0$,
(ii) $x^{-\eta}\ell_2(x)$ is nonincreasing in $[x_0, \infty)$ for some $x_0 \geq 0$,
(iii) $\ell \sim \ell_1$ and $\ell \sim \ell_2$,

where the symbol "$u \sim v$" stands for the asymptotic equivalence of functions u and v, that is, for $u(t)/v(t) \to 1$ as $t \to \infty$.

The following property of slowly varying functions follows immediately from Proposition A.8.

Lemma A.9 *Let ℓ be a measurable slowly varying function. Then, for all $\eta > 0$,*

$$\ell(x) = o\left(x^\eta\right), \qquad x \to \infty.$$

If $f \in \mathcal{RV}_a$, $a < -1$, then the series

$$\sum_{n=1}^{\infty} f(n)\ell(n)$$

converges.

Another useful corollary of Proposition A.8 provides the so-called Potter bounds. A similar result holds in the general case of PRV-functions (see Theorem 3.59).

Lemma A.10 *Let ℓ be a measurable slowly varying function. Given some arbitrary constants $D > 1$ and $\eta > 0$, there exists a number $x_* > 0$ such that*

$$\frac{1}{D}\left(\frac{x_2}{x_1}\right)^{-\eta} \leq \frac{\ell(x_1)}{\ell(x_2)} \leq D\left(\frac{x_2}{x_1}\right)^{\eta} \tag{A.20}$$

for all $x_* \leq x_1 \leq x_2$.

Proposition A.8 implies the following bounds for integrals of slowly varying functions.

Lemma A.11 *Let ℓ be a measurable slowly varying function and let $\lambda < 1$. If ℓ is uniformly bounded away from zero in $[0, x_0]$ for some $x_0 > 0$, then*

$$\int_0^T \frac{\ell(x)}{x^\lambda} dx \asymp \frac{\ell(T)}{T^{\lambda-1}}$$

for all sufficiently large $T > 0$.

Recall that the symbol $f \asymp g$ means that there are two constants $0 < c_1 < c_2 < \infty$ such that

$$c_1 f(x) \leq g(x) \leq c_2 f(x) \quad \text{for all} \quad x.$$

(The constants c_1 and c_2 depend on the functions f and g and are universal in x.)

Lemma A.12 *Let ℓ be a measurable slowly varying function and let $\lambda > 1$. Then*

$$\int_T^\infty \frac{\ell(x)}{x^\lambda} dx \asymp \frac{\ell(T)}{T^{\lambda-1}}.$$

A more advanced technique allows us to obtain more precise asymptotics of the integrals in both lemmas under the same assumptions (see Theorems A.18 and A.19).

On the other hand, neither Lemma A.11 nor Lemma A.12 holds for $\lambda = 1$ (for a corresponding result, see Parameswaran [290]). An analogue of Lemma A.11 for $\lambda = 1$, with sums instead of integrals, is obtained in Sect. A.7.

The proof of Lemma A.11 can be modified to obtain similar bounds for sums (instead of integrals).

Lemma A.13 *Let ℓ be a slowly varying function and let $\lambda < 1$. Then*

$$\sum_{k=1}^m \frac{\ell(k)}{k^\lambda} \asymp \frac{\ell(m)}{m^{\lambda-1}}$$

for all $m \geq 1$.

Lemma A.14 *Let ℓ be a slowly varying function and let $\lambda > 1$. Then*

$$\sum_{k=m}^\infty \frac{\ell(k)}{k^\lambda} \asymp \frac{\ell(m)}{m^{\lambda-1}}$$

for all $m \geq 1$.

The main idea for the proof of the latter four lemmas is demonstrated in the following proof of Lemma A.11.

Proof of Lemma A.11 First we obtain upper bounds. Let $\eta > 0$ be such that $\lambda + \eta < 1$. Choose a function ℓ_1 according to Proposition A.8, that is, $\ell_1(x)x^\eta$ is a nondecreasing function. Now, find an x_1 such that $\ell(x) \le 2\ell_1(x)$ and $\ell_1(x) \le 2\ell(x)$ for all $x \ge x_1$. Put $x_* = \max\{x_0, x_1\}$. The rest of the proof is easy for $T \ge x_*$. Indeed,

$$\int_0^T \frac{\ell(x)}{x^\lambda}\,dx = \int_0^{x_*} \frac{\ell(x)}{x^\lambda}\,dx + \int_{x_*}^T \frac{\ell(x)}{x^\lambda}\,dx = o\left(\ell(T)T^{1-\lambda}\right) + \int_{x_*}^T \frac{\ell(x)}{x^\lambda}\,dx,$$

since $\ell(T)T^{1-\lambda} \to \infty$, by Proposition A.8. The second term can be estimated as

$$\begin{aligned}\int_{x_*}^T \frac{\ell(x)}{x^\lambda}\,dx &= \int_{x_*}^T \frac{\ell(x)}{\ell_1(x)} \cdot \ell_1(x)x^\eta \frac{1}{x^{\lambda+\eta}}\,dx \le 2\ell_1(T)T^\eta \frac{T^{1-\lambda-\eta}}{1-\lambda-\eta}\\ &= O\left(\ell(T)T^{1-\lambda}\right).\end{aligned}$$

The proof of the lower bound is even simpler. Let $\eta > 0$. Find a function ℓ_2 such that $\ell_2(x)x^{-\eta}$ is nonincreasing. Then choose a number x_1 such that $\ell(x) \le 2\ell_2(x)$ and $\ell_2(x) \le 2\ell(x)$ for all $x \ge x_1$. Put $x_* = \max\{x_0, x_1\}$. Then, for $T \ge x_*$,

$$\begin{aligned}\int_0^T \frac{\ell(x)}{x^\lambda}\,dx &\ge \int_{x_*}^T \frac{\ell(x)}{x^\lambda}\,dx = \int_{x_*}^T \frac{\ell(x)}{\ell_2(x)}\ell_2(x)x^{-\eta} \cdot \frac{1}{x^{\lambda-\eta}}\,dx\\ &\ge \frac{1}{2}\ell_2(T)T^{-\eta}\frac{T^{1+\eta-\lambda} - x_*^{1+\eta-\lambda}}{1+\eta-\lambda} = O\left(\ell(T)T^{1-\lambda}\right),\end{aligned}$$

which completes the proof. □

A.5 Distributions Attracted to Stable Laws

Denote by Z_α a stable random variable with index α, $0 < \alpha \le 2$, and let G_α and h_α be their distribution function and characteristic function, respectively. The following canonical representation of the characteristic function h_α is well known, i.e.

$$h_\alpha(t) = \exp\left\{iat - c|t|^\alpha\left(1 - i\beta\,\mathrm{sign}(t)\omega_\alpha(t)\right)\right\}, \tag{A.21}$$

where a is a real number, c and β are real numbers such that $c \ge 0$ and $|\beta| \le 1$, and

$$\mathrm{sign}(t) = \begin{cases} -1, & t < 0,\\ 0, & t = 0,\\ 1, & t > 0,\end{cases} \qquad \omega_\alpha(t) = \begin{cases} \tan\frac{\pi\alpha}{2}, & \alpha \ne 1,\\ -\frac{2}{\pi}\ln|t|, & \alpha = 1.\end{cases}$$

The case $c = 0$ corresponds to a degenerate distribution function G_α and will not be considered below. Therefore, throughout this section we assume that $c > 0$.

Then

$$G_\alpha \text{ is a continuous function} \tag{A.22}$$

(see Gnedenko and Kolmogorov [151], p. 183). Moreover, the distribution function G_α is infinitely often differentiable in $\mathbf{R}$. Further, there exists a finite constant $C_\alpha \geq 0$ (with $C_\alpha > 0$ if $\alpha \neq 2$) such that

$$\lim_{x\to\infty} x^\alpha \, \mathsf{P}(|Z_\alpha| \geq x) = C_\alpha \tag{A.23}$$

(see Gnedenko and Kolmogorov [151], p. 182). This, for example, implies that

$$\mathsf{E}\,|Z_\alpha|^\eta < \infty \tag{A.24}$$

for all $0 < \eta < \alpha$.

Recall that $X \in \mathcal{DA}\,(\alpha, \{c_n\}, \{a_n\})$ means that the weak convergence in (11.13) holds. It is well-known that the normalizing sequence $\{c_n\}_{n\geq 1}$ is of the form

$$c_n = n^{1/\alpha} f(n), \qquad \text{where } f \text{ is a slowly varying function} \tag{A.25}$$

(see Theorem 2.1.1 in Ibragimov and Linnik [187]), while the tail of the distribution function of the random variable X satisfies the following condition

$$\mathsf{P}(|X| \geq x) = \frac{g(x)}{x^\alpha}, \qquad \text{with a slowly varying function } g. \tag{A.26}$$

This implies, for example, that for an arbitrary $0 < \eta < \alpha$,

$$\mathsf{E}\,|X|^\eta < \infty \tag{A.27}$$

(see Gnedenko and Kolmogorov [151], p. 179). The functions f in (A.25) and g in (A.26) are related to each other, since

$$\lim_{n\to\infty} n\,\mathsf{P}(|X| \geq c_n) = \lim_{n\to\infty} \frac{g\left(n^{1/\alpha} f(n)\right)}{f^\alpha(n)} = K_\alpha, \tag{A.28}$$

where K_α is a finite constant that may differ from the constant C_α in (A.23) (see (9) and (10) on p. 176 in Gnedenko and Kolmogorov [151]).

Below we provide further properties of distributions attracted to nondegenerate stable laws. We also list some properties of the centering and normalizing sequences involved in the weak convergence of (11.13).

Lemma A.15 *Let $\{S_n, n \geq 1\}$ be partial sums of independent, identically distributed random variables $\{X_n, n \geq 1\}$ and let X be a copy of the random variable X_1. Assume that $X \in \mathcal{DA}(\alpha, \{c_n\}, \{a_n\})$. Then, given a function $\varphi \in \mathcal{RV}_{1/p}$, with $0 < p < \alpha$, there exist a nondegenerate α-stable random variable Z_α and a real sequence $\{b_n, n \geq 1\}$ such that*

$$\frac{S_n}{b_n} - a_n \xrightarrow{d} Z_\alpha, \qquad n \to \infty, \tag{A.29}$$

where $\xrightarrow{d}$ denotes weak convergence and

(i) $b_n = b(n)$, $b(x) = x^{1/\alpha} h(x)$, *with a continuous, slowly varying function h,*
(ii) *the function $\varphi(x)/b(x)$ is increasing,*
(iii) $\lim_{n\to\infty} n\,\mathsf{P}(|X| \geq b_n) = K_\alpha$, *where the constant K_α is as defined in (A.28) (with $K_\alpha > 0$ if $\alpha < 2$).*

Property (iii) implies that

$$\lim_{x\to\infty} x\,\mathsf{P}(|X| \geq \theta b(x)) = \frac{K_\alpha}{\theta^\alpha}, \tag{A.30}$$

for all $\theta > 0$.

Lemma A.16 *Let X, $\{X_n, n \geq 1\}$ be independent, identically distributed random variables with distribution function F. Assume that $X \in \mathcal{DA}(\alpha, \{b_n\}, \{a_n\})$. Then the weak convergence in (A.29) holds with*

$$a_n = \frac{n}{b_n} \int_{|x|<b_n} x\, dF(x). \tag{A.31}$$

Moreover,

(A) *if $\alpha < 1$, then*

$$X \in \mathcal{DA}(\alpha, \{b_n\}, \{0_n\})\,; \tag{A.32}$$

(B) *relation (A.32) holds for $1 < \alpha \leq 2$ as well, if $\mathsf{E}\,X = 0$; in the latter case, the limiting random variable Z_α in (A.32) is such that $\mathsf{E}\,Z_\alpha = 0$.*

Lemma A.16 is known (see, for example, the translator's remark in Gnedenko and Kolmogorov [151], p. 175, or Theorem 3 in §5 of Chapter XVII in Feller [129]). For the sake of completeness, we give a new proof here based on some properties of slowly varying functions.

Proof of Lemma A.15 Let f be the slowly varying function as given in (A.25). Let $\widehat{f}$ be an asymptotically equivalent continuous version of f satisfying $\widehat{f}(n) = f(n)$, $n \geq 1$. Such a version exists in view of the statement at the end of Section 1.4 in

Seneta [324]. By Slutsky's theorem (see, e.g., Theorem 1 in Chow and Teicher [89], p. 249) the weak convergence in (11.13) together with (A.25) yield

$$\frac{S_n}{n^{1/\alpha}\widehat{f}(n)} - a_n \xrightarrow{d} Z_\alpha, \qquad n \to \infty.$$

Now we apply statement 4° of Section 1.5 in Seneta [324], with $\gamma = (\alpha - p)/\alpha p$, to conclude that there exists a slowly varying function h such that $h \sim \widehat{f}$ and $n^{(\alpha-p)/\alpha p}/h(n)$ is increasing. Put $b_n = n^{1/\alpha}h(n)$. Then $b_n \sim c_n$, so that, by Slutsky's theorem, the weak convergence in (A.29) retains. Hence property (ii) is obvious. Moreover, from (A.26),

$$n\,\mathsf{P}(|X| \geq b_n) = n\,\mathsf{P}(|X| \geq c_n)\frac{g(b_n)}{g(c_n)}\frac{c_n^\alpha}{b_n^\alpha},$$

hence property (iii) is proved by recalling (A.28), since g is slowly varying and $b_n \sim c_n$. □

Proof of Lemma A.16 The constants a_n can be obtained from a general result in the theory of weak convergence for sums of independent random variables in series schemes (see, e.g., Theorem 4 in Chapter 25 of Gnedenko and Kolmogorov [151]), that is,

$$a_n = \sum_{k=1}^{k_n} \int_{|x|<\tau} x\, dF_{nk}(x) - \gamma_n(\tau),$$

where $\tau > 0$ is an arbitrary fixed number for which $\pm\tau$ are continuity points of the spectral function H in the Lévy representation of the limit distribution function and where $\gamma_n(\tau)$ is a convergent sequence. In the case of partial sums of independent, identically distributed random variables, we have $k_n = n$ and $F_{nk}(x) = \mathsf{P}(X < xb_n)$, $k \leq n$, whence

$$a_n = \frac{n}{b_n} \int_{|x|<\tau b_n} x\, dF(x) - \gamma_n(\tau).$$

The limiting behavior of $\gamma_n(\tau)$ is also known, namely,

$$\lim_{n\to\infty} \gamma_n(\tau) = a + \int_{|x|<\tau} x\, dH(x) + \int_{|x|\geq\tau} \frac{dH(x)}{x},$$

where a is the constant in the canonical representation (A.21) and where H is the spectral function of G_α (see, e.g., relation (9) in Chapter 3 of Gnedenko and Kolmogorov [151]). Note that the spectral function H of G_α is continuous, that is, one can take, for example, $\tau = 1$. To prove (A.31), we choose $\tau = 1$ and apply Slutsky's theorem once more.

Turning to the proof of (A) we first show that the limit

$$\lim_{n\to\infty} \frac{n}{b_n} \int_{|x|<b_n} x\, dF(x) \tag{A.33}$$

exists if $\alpha < 1$. Put $\overline{F} = 1 - F$. On integrating by parts, we get

$$\begin{aligned}\int_{0\le x<b_n} x\, dF(x) &= -\int_{0<x<b_n} x\, d\overline{F}(x) = b_n\overline{F}(b_n) + \int_{0<x<b_n} \overline{F}(x)\, dx \\ &= b_n\overline{F}(b_n) + b_n \int_{0<y<1} \overline{F}(yb_n)\, dy.\end{aligned}$$

By Theorem 5, Chapter 35, in Gnedenko and Kolmogorov [151], we have $\overline{F} \in \mathcal{RV}_{-\alpha}$, whence

$$\frac{n}{b_n} \int_{0\le x<b_n} x\, dF(x) = n\overline{F}(b_n) + \frac{n}{b_n^{\alpha}} \int_{0<y<1} \frac{g_1(yb_n)}{y^{\alpha}}\, dy, \tag{A.34}$$

with some slowly varying function g_1. Now we use Theorem A.19, with $U = g_1$, $\nu = 0$, and $f(t) = t^{-\alpha}$, to prove that

$$\int_{0<y<1} \frac{g_1(yb_n)}{y^{\alpha}}\, dy \sim g_1(b_n) \int_0^1 \frac{dy}{y^{\alpha}}.$$

The limit of $n\overline{F}(b_n)$ in (A.34), as $n \to \infty$, exists in view of relation (10) in Chapter 35 of Gnedenko and Kolmogorov [151]. Thus Lemma A.15 together with property (iii) implies that the total limit in (A.34) also exists. An analogous argument proves that the limit

$$\lim_{n\to\infty} \frac{n}{b_n} \int_{-b_n<x<0} x\, dF(x)$$

exists, too. Note, however, that the proof of this property is given for F instead of $\overline{F}$ and uses relation (9) instead of (10), Chapter 35, in Gnedenko and Kolmogorov [151]. In doing so, another slowly varying function comes in instead of g_1. Combining these results, we prove (A.33), whence (A.32) follows, for $\alpha < 1$.

To prove (B), we show that (A.33) holds, for $\alpha > 1$, as well. Note that

$$\frac{n}{b_n} \int_{|x|<b_n} x\, dF(x) = -\frac{n}{b_n} \int_{|x|\ge b_n} x\, dF(x).$$

We follow the same lines of proof as above, but now for $\alpha > 1$. The areas of the integration are $x \ge b_n$ and $x \le -b_n$ instead of $0 \le x < b_n$ and $-b_n < x < 0$

above. Another difference is that we use Theorem A.18 instead of Theorem A.19. As a consequence, the integral $\int_1^\infty t^{-\alpha}\,dt$ appears in the limit instead of $\int_0^1 t^{-\alpha}\,dt$.

Finally, Lemma 5.2.2 in Ibragimov and Linnik [187] implies that, for $0 < \eta < \alpha$,

$$\sup_{n\geq 1} \mathsf{E}\left[\frac{|S_n|}{b_n}\right]^\eta < \infty, \tag{A.35}$$

if (A.32) holds. Using this result for $\eta = 1$ together with Corollary 7 of Chow and Teicher [89], p. 254, we conclude that

$$\mathsf{E}\left(\frac{S_n}{b_n}\right) \to \mathsf{E}\,Z_\alpha, \qquad n \to \infty,$$

whence $\mathsf{E}\,Z_\alpha = 0$ follows. This completes the proof. □

The final result of this section states an assertion on the existence of moments (of negative order) for stable distributions. This result has been used to apply Lemma A.1 in the proof of Theorem 11.1, for the case $r = -1$.

Lemma A.17 *Let $0 < \alpha \leq 2$ and g_α be the density of some α-stable distribution function. Then*

$$\int_{-1}^1 |x|^{-\delta} g_\alpha(x)\,dx < \infty, \tag{A.36}$$

for all $0 \leq \delta < 1$.

Proof of Lemma A.17 Note that (A.36) is obvious if $c = 0$ in the representation (A.21). Thus we consider the case when $c > 0$. Now, (A.21) implies that $|h_\alpha(t)| = \exp\{-c|t|^\alpha\}$ and, in addition, h_α is absolutely integrable on $\mathbf{R}$. Moreover, the function

$$h_\alpha(t)\int_{-1}^1 |x|^{-\delta} e^{-itx}\,dx$$

is also absolutely integrable on $\mathbf{R}$, since $0 \leq \delta < 1$. It is well-known that

$$g_\alpha(x) = \frac{1}{2\pi}\int_{-\infty}^\infty e^{-itx} h_\alpha(t)\,dt.$$

Thus

$$\begin{aligned}
2\pi \int_{-1}^1 |x|^{-\delta} g_\alpha(x)\,dx &= \int_{-1}^1 |x|^{-\delta} \int_{-\infty}^\infty e^{-itx} h_\alpha(t)\,dt\,dx \\
&= \int_{-\infty}^\infty \left[h_\alpha(t)\int_{-1}^1 |x|^{-\delta} e^{-itx}\,dx\right] dt < \infty,
\end{aligned}$$

completing the proof. □

A.6 The Asymptotic Behavior of Integrals and Sums with Slowly Varying Functions

The proof of Theorem 11.1 is based on the following obvious generalization of two results of Aljančić et al. [9], where the case of slowly varying functions U is considered, that is, the special case $\nu = 0$.

Theorem A.18 *Let f be a real-valued function and $U \in \mathcal{RV}_\nu$. Assume that the Lebesgue integral*

$$\int_A^\infty t^\eta f(t)\, dt \tag{A.37}$$

is well-defined, for some $\eta \geq \nu$ and $A > 0$. Then the integral

$$\int_A^\infty U(xt) f(t)\, dt$$

is also well-defined if either

(i) $\eta > \nu$, *or*
(ii) $\eta = \nu$ *and the function $U(t)/t^\nu$ is nonincreasing on (t_0, ∞), for some $t_0 > 0$.*

In each of these cases,

$$\int_A^\infty U(xt) f(t)\, dt \sim U(x) \int_A^\infty t^\nu f(t)\, dt \qquad \text{as } x \to \infty.$$

Theorem A.19 *Let f be a real-valued function and $U \in \mathcal{RV}_\nu$. Assume that the Lebesgue integral*

$$\int_0^B t^\eta f(t)\, dt \tag{A.38}$$

is well defined, for some $\eta \leq \nu$ and $B > 0$. Then

$$\int_0^B U(xt) f(t)\, dt \sim U(x) \int_0^B t^\nu f(t)\, dt \qquad \text{as } x \to \infty,$$

if either

(i) $\eta < \nu$, *or*
(ii) $\eta = \nu$ *and the function $U(t)/t^\nu$ is nonincreasing on (t_0, ∞), for some $t_0 > 0$.*

A.7 The Parameswaran Lemma for Sums

The following result is classical for the case when $r > -1$ and for integrals instead of sums (the case $r = -1$ has been treated in Parameswaran [290], Lemma 1).

Lemma A.20 *Let $w \in \mathcal{RV}_r$, $r \geq -1$, and let the function W be as defined in (11.12). Then $W \in \mathcal{RV}_{r+1}$ and, moreover,*

$$\lim_{t\to\infty} \frac{tw(t)}{W(t)} = r + 1. \tag{A.39}$$

Note that Lemma A.20 asserts that W is a slowly varying function if $r = -1$.

The proof in Parameswaran [290] has been given for integrals, but we need such a result for sums. Even more important is that the proof in [290] uses the l'Hospital rule for differentiable W, that is, for continuous w. In our case, however, W is a step function defined by (11.12). The Parameswaran method can be modified by using a continuous version w^* of w such that $w^*(t) \sim w(t)$, as $t \to \infty$, and by applying the uniform convergence theorem for slowly varying functions. For convenience, we provide below a new proof of Parameswaran's lemma based on an elementary argument.

As usual, the symbol $[x]$ denotes the integer part of a real number x.

Lemma A.21 *Let $w \in \mathcal{RV}_{-1}$ and let the function W be defined by equality (11.12). Then*

(i) $W([cn]) \sim W(n)$ as $n \to \infty$, for all $c > 1$, that is, W is a slowly varying function;
(ii) $nw_n = o(W(n))$ as $n \to \infty$, that is, relation (A.39) holds for $r = -1$ as well.

Proof of Lemma A.21 First we prove (ii). Fix $D > 0$ and choose $0 < \delta < 1$ with $\ln\left(\delta^{-1}\right) \geq 3D$. Let n_1 be such that

$$\sum_{k=[\delta n]}^{n} \frac{1}{k} \geq 2D \quad \text{for all} \quad n \geq n_1.$$

Such an n_1 exists, since

$$\sum_{k=1}^{n} \frac{1}{k} = \ln(n) + E + o(1) \qquad \text{as } n \to \infty, \tag{A.40}$$

where $E = 0.577\ldots$ is a universal constant, known as the *Euler constant*. Now choose n_2 such that

$$\inf_{[\delta n]\leq k\leq n} kw_k \geq \frac{1}{2} nw_n \quad \text{for all} \quad n \geq n_2.$$

Such an n_2 exists in view of Theorem A.7. Therefore,

$$\sum_{k=1}^{n} w_k \geq \sum_{k=[\delta n]}^{n} w_k \geq \frac{1}{2} n w_n \sum_{k=[\delta n]}^{n} \frac{1}{k} \geq D n w_n$$

for $n \geq \max\{n_1, n_2\}$, whence

$$\liminf_{n\to\infty} \frac{1}{n w_n} \sum_{k=1}^{n} w_k \geq D,$$

so that property (ii) is proved, since $D > 0$ can be chosen arbitrarily large.

To prove property (i), note that

$$\frac{W([cn])}{W(n)} = 1 + \frac{1}{W(n)} \sum_{k=n+1}^{[cn]} w_k,$$

for $c > 1$. Thus, by (A.40) and Theorem A.7,

$$\sum_{k=n+1}^{[cn]} w_k \leq \sup_{n \leq k \leq [cn]} k w_k \sum_{k=n+1}^{[cn]} \frac{1}{k} \sim n w_n \ln(c),$$

as $n \to \infty$. Therefore, for all fixed $c > 1$,

$$1 \leq \frac{W([cn])}{W(n)} \leq 1 + O\left(\frac{n w_n}{W(n)}\right),$$

whence (i) follows in view of (ii). □

A.8 Comments

Section A.3 Relation (A.10) is proved by Cline and Hsing [91] for the more general case of subexponential distributions. The Cline and Hsing [91] result generalizes a theorem of Heyde [178] proved for an attraction to stable laws, stating that

$$\lim_{n\to\infty} \frac{\mathsf{P}(|S_n| \geq x_n b_n)}{n\,\mathsf{P}(|X| \geq x_n b_n)} = 1 \tag{A.41}$$

for any sequence $\{x_n\}_{n\geq 1}$ such that $x_n \to \infty$ as $n \to \infty$. Related results are obtained by A.V. Nagaev [280, 281], S.V. Nagaev [282], Vonogradov [361], and Mikosch and Nagaev [273]. A similar result for dominated variation distributions (distribution functions being monotone ORV functions in our language) is proved by Baltrunas [22]. Several extensions are known for dependent random variables (see, for example, Liu [252]). Some applications of large deviations in insurance and finance are discussed by Klüppelberg and Mikosch [232], Mikosch and Nagaev [274], and Tang et al. [356].

An overview of earlier results on this topic is given in Chapter 8 of Embrechts et al. [119] (see also Christoph and Wolf [84], Saulis and Statulevičius [319], and Samorodnitsky and Taqqu [317]). Further references can be found in Borovkov and Borovkov [48] and Denisov et al. [101] (see, in addition, Foss et al. [132]).

Section A.4 The proof of Proposition A.8 can be found in Seneta [324] (Section 1.4, statement 4°). It also follows from the more general Theorem 3.79, which is valid for POV-functions.

Section A.5 See the discussion in Hall [170] and the last remark in the Introduction of the monograph of Zolotarev [375] concerning the choice of signs and constants in (A.21) as well as in the definition of $\omega_\alpha(t)$. A short and elegant proof of (A.26) and an explicit determination of the constants C_α and K_α in (A.23) and (A.28) can be found in Samorodnitsky and Taqqu [317].

Section A.6 Theorems A.18 and A.19 are obvious generalizations of two results of Aljančić et al. [9], where the case of slowly varying functions U is considered, that is, the special case $\nu = 0$ (see also Seneta [324]).

References

1. A. Abay, *Renewal theorems for random walks in multidimensional time*, Mathematica Slovaca **49** (1999), no. 3, 371–380.
2. N.H. Abel, *Méthode générale pour trouver des fonctions d'une seule quantité variable lorsqu'une propriété de ces fonctions est exprimée par une équation entre deux variables*, Mag. Naturvidenskab. **1** (1823), no. 2, 1–10.
3. A. de Acosta, *A new proof of the Hartman–Wintner law of the iterated logarithm*, Ann. Probab. **11** (1983), no. 2, 270–276.
4. J. Aczél, *Lectures on Functional Equations and their Applications*, Academic Press, New York–San Francisco–London, 1966.
5. J. Aczél, *The state of the second part of Hilbert's fifth problem*, Bull. Amer. Math. Soc. **20** (1989), no. 2, 153–163.
6. R. Agarwal, S. Hristova, and D. O'Regan, *Stability with respect to initial time difference for generalized delay differential equations*, Electronic J. Diff. Equ. **49** (2015), 1–19.
7. V. Aguiar and I. Guedes, *Shannon entropy, Fisher information and uncertainty relations for log-periodic oscillators*, Phys. A **423** (2015), 72–79.
8. S. Aljančić and D. Arandelović, *O-regularly varying functions*, Publ. Inst. Math. (Beograd) (N.S.) **22 (36)** (1977), 5–22.
9. S. Aljančić, R. Bojanić, and M. Tomić, *Sur la valeur asymptotique d'une classe des intégrales définies*, Acad. Serbe Sci. Publ. Inst. Math. 7 (1954), 81–94.
10. S. Aljančić and J. Karamata, *Fonctions á comportement regulier et l'integrale de Frullani*, Recueil des Travaux de l'Academie Serbe des Sciences **L** (1956), no. 5, 239–248.
11. G. Alsmeyer, *Erneuerungstheorie. Analyse stochastischer Regenerationsschemata,* Teubner, Stuttgart, 1991.
12. G. Alsmeyer, *Some relations between harmonic renewal measures and certain first passage times*, Statistics & Probability Letters **12** (1991), 19–27.
13. G. Alsmeyer, *On the Markov renewal theorem*, Stochastic Process. Appl. **50** (1994), no. 1, 37–56.
14. G. Alsmeyer and A. Gut, *Limit theorems for stopped functionals of Markov renewal processes*, Ann. Inst. Math. Statist. **51** (1999), no. 2, 369–382.
15. J.A.D. Appleby, G. Berkolaiko, and A. Rodkina, *Non-exponential stability and decay rates in nonlinear stochastic homogeneous difference equations, in* Discrete dynamics and difference equations, World Sci. Publ., Hackensack, NJ, 2010, pp. 155–162.
16. J.A.D. Appleby, J.P. Gleeson, and A. Rodkina, *On asymptotic stability and instability with respect to a fading stochastic perturbation*, Applicable Analysis **8** (2009), no. 4, 579–603.

V. V. Buldygin et al., *Pseudo-Regularly Varying Functions and Generalized Renewal Processes*, Probability Theory and Stochastic Modelling 91, https://doi.org/10.1007/978-3-319-99537-3

17. J.A.D. Appleby and H. Wu, *Solutions of stochastic differential equations obeying the law of the iterated logarithm, with applications to financial markets*, Electronic J. Probab. **14** (2009), no. 33, 912–959.
18. D. Arandelović, *O-regular variation and uniform convergence*, Publ. Inst. Math. (Beograd) (N.S.) **48 (62)** (1990), 25–40.
19. V.G. Avakumović, *Über einen O-Inversionssatz*, Bull. Int. Acad. Youg. Sci. **29–30** (1936), 107–117.
20. V.G. Avakumović, *Sur une extension de la condition de convergence des theoremes inverses de sommabilité*, C. R. Acad. Sci. **200** (1935), 1515–1517.
21. A.A. Balkema, J.L. Geluk, and L. de Haan, *An extension of Karamata's Tauberian theorem and its connection with complimentary convex functions*, Quarterly J. Math. **30** (1979), no. 2, 385–416.
22. A. Baltrunas, *On the asymptotics of one-sided large deviation probabilities*, Lithuanian Math. J. **35** (1995), no. 1, 14–22.
23. S. Banach, *Sur l'equation fonctionnelle* $f(x+y) = f(x) + f(y)$, Fund. Math. **1** (1920), 123–124.
24. V.S. Barbu and N. Limnios, *Semi-Markov Chains and Hidden Semi-Markov Models Toward Applications: Their use in Reliability and DNA Analysis*, Springer-Verlag, New York, 2008.
25. A. Barchielli and A. Paganoni, *On the asymptotic behaviour of some stochastic differential equations for quantum states*, Infin. Dimens. Anal. Quantum Probab. Relat. Top. **6** (2003), no. 2, 223–243.
26. N.K. Bari and S.B. Stechkin, *Best approximation and differential properties of two conjugate functions*, Trudy Mosk. Mat. Obshch. **5** (1956), 483–522. (Russian)
27. F. Barra, N. Chernov, and T. Gilbert, *Log-periodic drift oscillations in self-similar billiards*, Nonlinearity **20** (2007), no. 11, 2539–2549.
28. L.E. Baum and M. Katz, *Convergence rates in the law of large numbers*, Trans. Amer. Math. Soc. **120** (1965), no. 1, 108–123.
29. G. Berkolaiko and A. Rodkina, *Almost sure convergence of solutions to nonhomogeneous stochastic difference equation*, J. Difference Equ. Appl. **12** (2006), no. 6, 535–553.
30. S.M. Berman, *Sojourns and extremes of a diffusion process on a fixed interval*, Adv. Appl. Probab. **14** (1982), no. 4, 811–832.
31. S.M. Berman, *Spectral conditions for sojourn and extreme value limit theorems for Gaussian processes*, Stochastic Process. Appl. **39** (1991), no. 2, 201–220.
32. S.M. Berman, *The tail of the convolution of densities and its application to a model of HIV-latency time*, Ann. Appl. Probab. **2** (1992), no. 2, 481–502.
33. S.M. Berman, *Sojourns And Extremes of Stochastic Processes*, Wadsworth & Brooks/Cole Advanced Books & Software, Pacific Grove, California, 1992.
34. S.N. Bernšteĭn, *On majorants of finite or quasi-finite growth*, Doklady Akad. Nauk SSSR (N.S.) **65** (1949), 117–120. (Russian)
35. J. Bertoin, *Subordinators: Examples and Applications*, Springer-Verlag, Berlin–Heidelberg–New York, 2004.
36. J. Bertoin and M.-E. Caballero, *On the rate of growth of subordinators with slowly varying Laplace exponent*, Séminaire de probabilités (Strasbourg) **29** (1995), 125–132.
37. N.H. Bingham, *Limit theorems for occupation times of Markov processes*, Z. Wahrscheinlichkeitstheorie verw. Gebiete **17** (1971), 1–22.
38. N.H. Bingham, *Regular variation and probability: The early years*, J. Comp. Appl. Math. **200** (2007), no. 1, 357–363.
39. N.H. Bingham and C.M. Goldie, *Extensions of regular variation. I. Uniformity and quantifiers*, Proc. London Math. Soc. **44** (1982), no. 3, 473–496.
40. N.H. Bingham and C.M. Goldie, *Extensions of regular variation. II. Representations and indices*, Proc. London Math. Soc. **44** (1982), no. 3, 497–534.
41. N.H. Bingham, C.M. Goldie, and J.L. Teugels, *Regular Variation*, Cambridge University Press, Cambridge, 1987.

42. N.H. Bingham and A.J. Ostaszewski, *Beurling slow and regular variation*, Trans. London Math. Soc. **1** (2014), no. 1, 29–56.
43. H. Bohr, *Fastperiodische Funktionen*, Springer-Verlag, Berlin, 1932.
44. R. Bojanić and J. Karamata, *On slowly varying functions and asymptotic relations*, Math. Research Center Tech. Report 432, Madiso, Wisconsin, 1963.
45. R. Bojanić and E. Seneta, *Slowly varying functions and asymptotic relations*, J. Math. Anal. Appl. **34** (1971), 302–315.
46. R. Bojanić and E. Seneta, *A unified theory of regularly varying sequences*, Math. Z. **134** (1973), 91–106.
47. A.A. Borovkov, *Probability Theory*, Springer, London, 2013.
48. A.A. Borovkov and K.A. Borovkov, *Asymptotic Analysis of Random Walks. Heavy Tailed Distributions*, Encyclopedia of mathematics and its applications, 118, Cambridge University Press, Cambridge, 2008.
49. A.A. Borovkov and K.A. Borovkov, *Analogues of the Blackwell theorem for weighted renewal functions*, Sibirsk. Mat. Zh. **55** (2014), no. 4, 724–743; English transl. in Sib. Math. J. **55** (2014), no. 4, 589–605.
50. V. Božin and D. Djurčić, *A proof of an Aljančić hypothesis on O-regularly varying sequences*, Publ. Inst. Math. Nouvelle Ser. **62 (76)** (1997), 46–52.
51. D.S. Brée and N.L. Joseph, *Testing for financial crashes using the log-periodic power law model*, Int. Review Financial Anal. **30** (2013), 287–297.
52. D. Brigo and F. Mercurio, *Interest Rate Models. Theory and Practice*, second edition, Springer-Verlag, Berlin, 2006.
53. I. Brissaud, *Is the evolution of jazz described by a log-periodic law?*, Math. Sci. Hum. Math. Soc. Sci. **178** (2007), 41–50.
54. N.G. de Bruijn, *Pairs of slowly oscillating functions occurring in asymptotic problems concerning the Laplace transform*, Nieuw Arch. Wisk. **7** (1959), no. 3, 20–26.
55. V.V. Buldygin, *Strong laws of large numbers and convergence to zero of Gaussian sequences*, Teor. Imovirnost. Matem. Statist. **19** (1978), 33–41; English transl. in Theory Probab. Math. Statist. **19** (1978), 33–41.
56. V.V. Buldygin, K.-H. Indlekofer, O.I. Klesov, and J.G. Steinebach, *Asymptotics of renewal processes: some recent developments*, Ann. Univ. Sci. Budapest, Sect. Comp. **28** (2008), 107–139.
57. V.V. Buldygin, O.I. Klesov, and J.G. Steinebach, *Properties of a subclass of Avakumović functions and their generalized inverses*, Preprint Philipps-Universität, vol. 85, Marburg, 2001.
58. V.V. Buldygin, O.I. Klesov, and J.G. Steinebach, *On factorization representations for Avakumović–Karamata functions with nondegenerate groups of regular points*, Preprint Philipps-Universität, vol. 89, Marburg, 2002.
59. V.V. Buldygin, O.I. Klesov, and J.G. Steinebach, *Properties of a subclass of Avakumović functions and their generalized inverses*, Ukrain. Matem. Zh. **54** (2002), no. 2, 149–169 (Russian); English transl. in Ukrain. Math. J. 54 (2002), no. 2, 179–206.
60. V.V. Buldygin, O.I. Klesov, and J.G. Steinebach, *On factorization representations for Avakumović–Karamata functions with nondegenerate groups of regular points*, Anal. Math. **30** (2004), no. 3, 161–192.
61. V.V. Buldygin, O.I. Klesov, and J.G. Steinebach, *On some properties of asymptotically quasi-inverse functions and their applications. I*, Teor. Imov. Mat. Stat. **70** (2004), 9–25 (Ukrainian); English transl. in Theory Probab. Math. Statist. **70** (2005), 11–28.
62. V.V. Buldygin, O.I. Klesov, and J.G. Steinebach, *On some properties of asymptotically quasi-inverse functions and their applications. II*, Teor. Imovirnost. Matem. Statist. **71** (2004), 33–48 (Ukrainian); English transl. in Theory Probab. Math. Statist. **71** (2005), 37–52.

63. V.V. Buldygin, O.I. Klesov, and J.G. Steinebach, *The PRV property of functions and the asymptotic behavior of solutions of stochastic differential equations*, Teor. Imov. Mat. Stat. **72** (2005), 10–23 (Ukrainian); English transl. in Theory Probab. Math. Statist. **72** (2006), 11–25.
64. V.V. Buldygin, O.I. Klesov, and J.G. Steinebach, *PRV property and the asymptotic behaviour of solutions of stochastic differential equations*, Theory Stoch. Process. **11 (27)** (2005), no. 3–4, 42–57.
65. V.V. Buldygin, O.I. Klesov, and J.G. Steinebach, *On some extensions of Karamata's theory and their applications*, Publ. Inst. Math. (Beograd) (N. S.) **80 (94)** (2006), 59–96.
66. V.V. Buldygin, O.I. Klesov, and J.G. Steinebach, *On some properties of asymptotically quasi-inverse functions*, Teor. Imovirnost. Matem. Statist. **77** (2007), 13–27 (Ukrainian); English transl. in Theory Probab. Math. Statist. **77** (2008), 15–30.
67. V.V. Buldygin, O.I. Klesov, and J.G. Steinebach, *PRV property and the φ-asymptotic behavior of solutions of stochastic differential equations*, Liet. Mat. Rink. **77** (2007), no. 4, 445–465 (Ukrainian); English transl. in Lithuanian Math. J. **77** (2007), no. 4, 361–378.
68. V.V. Buldygin, O.I. Klesov, and J.G. Steinebach, *Precise asymptotics over a small parameter for a series of large deviation probabilities*, Theory Stoch. Process. **13 (29)** (2007), no. 1–2, 44–56.
69. V.V. Buldygin, O.I. Klesov, and J.G. Steinebach, *On the convergence to infinity of positive increasing functions*, Ukrain. Matem. Zh. **62** (2010), no. 10, 1299–1308 (Ukrainian); English transl. in Ukrain. Math. J. **62** (2010), no. 10, 1507–1518.
70. V.V. Buldygin, O.I. Klesov, and J.G. Steinebach, *Asymptotic properties of absolutely continuous functions and strong laws of large numbers for renewal processes*, Teor. Imovirnost. Matem. Statist. **87** (2012), 1–11 (Ukrainian); English transl. in Theory Probab. Math. Statist. **87** (2013), 1–12.
71. V.V. Buldygin, O.I. Klesov, J. Steinebach, and O.A. Timoshenko, *On the φ-asymptotic behaviour of solutions of stochastic differential equations*, Theory Stoch. Process. **14** (2008), no. 1, 11–29.
72. V.V. Buldygin and V.V. Pavlenkov, *A generalization of Karamata's theorem on the asymptotic behavior of integrals*, Teor. Imovirnost. Matem. Statist. **81** (2009), 13–24 (Ukrainian); English transl. in Theory Probab. Math. Statist. **81** (2010), 15–26.
73. V.V. Buldygin and V.V. Pavlenkov, *Karamata theorem for regularly log-periodic functions*, Ukrain. Matem. Zh. **64** (2012), no. 11, 1443–1463 (Russian); English transl. in Ukrain. Math. J. **64** (2013), no. 11, 1635–1657.
74. V.V. Buldygin and O.A. Tymoshenko, *On the asymptotic stability of stochastic differential equations*, Naukovi Visti NTUU "KPI" **64** (2007), 126–129. (Ukrainian)
75. V.V. Buldygin and O.A. Tymoshenko, *The exact order of growth of solutions of stochastic differential equations*, Naukovi Visti NTUU "KPI" **64** (2008), 127–132. (Ukrainian)
76. V.V. Buldygin and O.A. Tymoshenko, *On the exact order of growth of solutions of stochastic differential equations with time-dependent coefficients*, Theory Stoch. Process. **16** (2010), no. 2, 12–22.
77. S. Cambanis, *On the path absolute continuity of second order processes*, Ann. Probab. **3** (1975), no. 6, 1050–1054.
78. R.A. Carmona and M.R. Tehranchi, *Interest Rate Models: An Infinite Dimensional Stochastic Analysis Perspective*, Springer-Verlag, Berlin, 2006.
79. A.L. Cauchy, *Cours d'analyse de l'E'cole Polytechnique. Analyse algebrique*, V, Paris, 1821.
80. A. Chakak and L. Imlahi, *Multivariate probability integral transformation: application to maximum likelihood estimation*, Revista de la Real Academia de Ciencias Exactas, Físicas y Naturales, Serie A Matemáticas **95** (2001), no. 2, 201–212.
81. R. Chen, *A remark on the tail probability of a distribution*, J. Multivariate Anal. **8** (1978), no. 2, 328–333.
82. C.-P. Chen and L. Chen, *Asymptotic behavior of trigonometric series with O-regularly varying quasimonotone coefficients*, J. Math. Anal. Appl. **250** (2000), 13–26.

83. C. Chicone, *Stability theory of ordinary differential equations, in* Mathematics of Complexity and Dynamical Systems, Springer, New York, 2011, pp. 13–26.
84. G. Christoph and W. Wolf, *Convergence Theorems with a Stable Limit Law*, Akademie-Verlag, Berlin, 1992.
85. P.-L. Chow and R.Z. Khas'minskiĭ, *Almost sure explosion of solutions to stochastic differential equations*, Stochastic Process. Appl. **124** (2014), no. 1, 639–645.
86. Y.S. Chow and T.L. Lai, *Some one-sided theorems on the tail distribution of samples with applications to the last time and largest excess of boundary crossings*, Trans. Amer. Math. Soc. **208**(1975), 51–72.
87. Y.S. Chow and T.L. Lai, *Paley-type inequalities and convergence rates related to the law of large numbers and extended renewal theory*, Z. Wahrscheinlichkeitstheorie verw. Gebiete **45** (1978), no. 1, 1–19.
88. Y.S. Chow and H. Robbins, *On sums of independent random variables with infinite moments and "fair" games*, Proc. Nat. Acad. Sci. U.S.A. **47** (1961), no. 3, 330–335.
89. Y.S. Chow and H. Teicher, *Probability Theory. Independence, Interchangeability, Martingales*, Springer-Verlag, Berlin–Heidel-berg–New York, 1978.
90. D.B.H. Cline, *Intermediate regular and Π-variation*, Proc. London Math. Soc. **68** (1994), no. 3, 594–616.
91. D. Cline and T. Hsing, *Large deviation probabilities for sums and maxima of random variables with heavy or subexponential tails*, Preprint, Texas A&M University, 1991.
92. D.B.H. Cline and G. Samorodnitsky, *Subexponentiality of the product of independent random variables*, Stoch. Process. Appl. **49** (1994), no. 1, 75–98.
93. D.R. Cox, *Renewal Theory*, Wiley, New York, 1962.
94. D.R. Cox and P.A.W. Lewis, *The Statistical Analysis of Series of Events*, Wiley, New York, 1966.
95. J.C. Cox, J.E. Ingersoll, and S.A. Ross, *A theory of the term structure of interest rates*, Econometrica **53** (1985), 385–407.
96. I. Csiczár and P. Erdös, *On the function* $g(t) = \limsup_{x\to\infty}(f(x+t) - f(x))$, Magyar Tud. Akad. Kut. Int. Közl. A **9** (1964), 603–606.
97. M. Csörgő and P. Révész, *Strong Approximations in Probability and Statistics*, Academic Press, New York–London, 1981.
98. G. Darboux, *Sur la composition des forces en statique*, Bull. Sci. Math. **9** (1875), 281–288.
99. H. Delange, *Sur la composition des forces en statique*, Bull. Sci. Math. **9** (1875), 281–288. H. Delange, *Sur deux questions posées par M. Karamata*, Publ. Inst. Math. (Beograd) (N.S.) 7 (1954), 69–80.
100. A. Dembo and O. Zeitouni, *Large Deviations Techniques and Applications*, Jones and Bartlett, Boston, 1993.
101. D. Denisov, A.B. Dieker, and V.L. Shneer, *Large deviations for random walks under subexponentiality: the big-jump domain*, Ann. Probab. **36** (2008), no. 5, 1946–1991.
102. A. Di Crescenzo, B. Martinucci and S. Zacks, *Compound Poisson process with a Poisson subordinator*, J. Appl. Probab. **52** (2015), no. 2, 360–374.
103. S. Ditlevsen and A. Samson, *Introduction to stochastic models in biology, in* Stochastic Biomathematical Models with Applications to Neuronal Modeling, (eds. M. Bachar, J. Batzel, and S. Ditlevsen), Springer-Verlag, Berlin, 2013, pp. 3–35.
104. D. Djurčić, *O-regularly varying functions and some asymptotic relations*, Publ. Inst. Math. (Beograd) (N.S.) **61 (75)** (1997), 44–52.
105. D. Djurčić, *O-regularly varying functions and strong asymptotic equivalence*, J. Math. Anal. Appl. **220** (1998), no. 2, 451–461.
106. D. Djurčić, V. Konjokrad, and R.M. Nilolić, *O-regularly varying functions and strong asymptotic equivalence*, Filomat **26** (2012), no. 5, 1075–1080.
107. D. Djurčić, R.M. Nilolić, and A. Torgašev, *The weak and strong asymptotic equivalence relations and the generalizws inverse*, Lithuanian Math. J. **51** (2011), no. 4, 472–476.
108. D. Djurčić and A. Torgašev, *Strong asymptotic equivalence and inversion of functions in the class Kc* , J. Math. Anal. Appl. **255** (2001), no. 2, 383–390.

109. D. Djurčić and A. Torgašev, *Some asymptotic relations for the generalized inverse*, J. Math. Anal. Appl. **335** (2007), no. 2, 1397–1402.
110. D. Djurčić, A. Torgašev, and S. Ješić, *The strong asymptotic equivalence and the generalized inverse*, Siberian Math. J. **49** (2008), no. 4, 628–636.
111. J.L. Doob, *Renewal theory from the point of view of the theory of probability*, Trans. Amer. Math. Soc. **63** (1948), no. 3, 422–438.
112. P. Doukhan, O.I. Klesov, A.G. Pakes, and J.G. Steinebach, *Limit theorems for record counts and times in the F^α-scheme*, Extremes **16** (2013), 147–171.
113. P. Doukhan, O.I. Klesov, and J.G. Steinebach, *Strong Law of Large Numbers in an F^α - scheme, in* Mathematical Statistics and Limit Theorems, Festschrift in Honour of Paul Deheuvels (eds. M. Hallin, D.M. Mason, D. Pfeifer, J.G. Steinebach), Springer International Publishing Switzerlan, Heidelberg–New York–Dordrecht–London, 2015, pp. 287–303.
114. D. Drasin and E. Seneta, *A generalization of slowly varying functions*, Proc. Amer. Math. Soc. **96** (1986), no. 3, 470–472.
115. D. Drasin and D.F. Shea, *Convolution inequalities, regular variation and exceptional sets*, J. Anal. Math. **29** (1976), 232–293.
116. N. Elez and D. Djurčić, *Rapid variability and Karamata's integral theorem*, Filomat **28** (2014), no. 3, 487–492.
117. K.-J. Engel and R. Nagel, *One-Parameter Semigroups for Linear Evolution Equations*, Springer-Verlag, New York, 2000.
118. P. Embrechts and C.M. Goldie, *Comparing the tail of an infinitely divisible distribution with integrals of its Lévy measure*, Ann. Probab. **9** (1981), no. 3, 468–81.
119. P. Embrechts, C. Klüppelberg, and T. Mikosch, *Modelling Extremal Events*, Springer, Berlin, 1997.
120. P. Erdös, *On a theorem of Hsu and Robbins*, Ann. Math. Statist. **20** (1949), no. 2, 286–291.
121. P. Erdös, *Remark on my paper "On a theorem of Hsu and Robbins"*, Ann. Math. Statist. **21** (1950), no. 1, 138.
122. P. Erdös, W. Feller, and H. Pollard, *A property of power series with positive coefficients*, Bull. Amer. Math. Soc. **55** (1949), no. 2, 201–204.
123. K.B. Erickson, *Recurrence sets of normed random walk in R^d*, Ann. Probab. **4** (1976), no. 5, 802–828.
124. I. Fazekas and O.I. Klesov, *A general approach to the strong laws of large numbers*, Teor. Veroyatnost. i Primenen. **45** (2000), no. 3, 568–583 (Ukrainian); English transl. in Theory Probab. Appl. **45** (2002), no. 3, 436–449.
125. W. Feller, *On the integral equation of renewal theory*, Ann. Math. Statist. **12** (1941), no. 3, 243–267.
126. W. Feller, *A limit theorem for random variables with infinite moments*, Amer. J. Math. **68** (1946), no. 2, 257–262.
127. W. Feller, *On regular variation and local limit theorems*, Proceeding of the 5th Berkely Symposium on Mathematical Statistics and Probability, 1965–1966 **2** (1967), no. 1, 373–388.
128. W. Feller,*One-sided analogues of Karamata's regular variation*, Enseignement Math. **15** (1969), no. 2, 107–121.
129. W. Feller, *An Introduction to Probability Theory and Its Applications*, 2nd ed., Wiley, New York, 1971.
130. C. Feng, H. Wang, M. Xin, and J. Kowalski, *A note on generalized inverses of distribution function and quantile transformation*, Appl. Math. **3** (2012), 2098–2100.
131. E. Fisher, M. Berman, N. Vowels, and C. Wilson, *Excursions of a normal random walk above a boundary*, Statist. Probab. Lett. **48** (2000), 141–151.
132. S. Foss, D. Korshunov, and S. Zachary, *An Introduction to Heavy-Tailed and Subexponential Distributions*, second edition, Springer Series in Operations Research and Financial Engineering, Springer, New York, 2013.
133. R. Fraiman and B. Pateiro-López, *Functional quantiles*, in Recent advances in functional data analysis and related topics, Physica-Verlag/Springer, Heidelberg, 2011, pp. 123–129.

134. A. Friedman, *Limit behavior of solutions of stochastic differential equations*, Trans. Amer. Math. Soc. **170** (1972), 359–384.
135. A. Friedman, *Stochastic Differential Equations And Applications*, Academic Press, London–New York, 1975.
136. A. Friedman and M. Pinsky, *Behavior of solutions of linear stochastic differential systems*, Trans. Amer. Math. Soc. **181** (1973), 1–22.
137. B.E. Fristedt and W.E. Pruitt, *Lower functions for increasing random walks and subordinators*, Z. Wahrscheinlichkeitstheorie verw. Gebiete **18** (1971), no. 3, 167–182.
138. A. Frolov, A. Martikainen, and J. Steinebach, *Limit theorems for maxima of sums and renewal processes*, Zap. Nauchn. Sem. S.-Peterburg. Otdel. Mat. Inst. Steklov. (POMI), Veroyatn. i Stat. **278** (2001), 261–274, 314–315 (Russian); English transl. in J. Math. Sci. (N. Y.) **118** (2003), no. 6, 5658–5666.
139. J.P. Gabriel, *Martingales with a countable filtering index set*, Ann. Probab. **5** (1977), no. 6, 888–898.
140. M.U. Gafurov and A.D. Slastnikov, *Some problems of the exit of a random walk beyond a curvilinear boundary and large deviations*, Teor. Veroyatnost. i Primenen. **32** (1987), no. 2, 327–348; English transl. in Theory Probab. Appl. **32** (1987), no. 2, 299–321.
141. J. Galambos, *The Asymptotic Theory of Extreme Order Statistics*, Wiley, New York, 1978.
142. J. Galambos, K.-H. Indlekofer, and I. Kátai, *A renewal theorem for random walks in multidimensional time*, Trans. Amer. Math. Soc. **300** (1987), no. 2, 759–769.
143. J. Galambos and I. Kátai, *A note on random walks in multidimensional time*, Math. Proc. Cambridge Philos. Soc. **99** (1986), no. 1, 163–170.
144. J. Galambos and I. Kátai, *Some remarks on random walks in multidimensional time, in* Proc. 5th Pannonian Sympos. Math. Statist. (Visegrád, Hungary, 1985), Reidel, Dordrecht, 1988, pp. 65–74.
145. J. Galambos and E. Seneta, *Regularly varying sequences*, Proc. Amer. Math. Soc. **41** (1973), 110–116.
146. V.F. Gaposhkin, *Criteria for the strong law of large numbers for some classes of second-order stationary processes and homogeneous random fields*, Teor. Veroyatnost. i Primenen. **22** (1977), no. 2, 295–319 (Russian); English transl. in Theory Probab. Appl. **22** (1978), no. 2, 286–310.
147. A.O. Gel'fond and Yu.V. Linnik, *Elementary Methods in the Analytic Theory of Numbers*, Fizmatlit, Moscow, 1962; English transl. Pergamon Press, New York, 1966.
148. J.L. Geluk and L. de Haan, *Regular variation, Extensions and Tauberian Theorems*, CWI Tract 40, Amsredam, 1987.
149. I.I. Gihman and A.V. Skorohod, *Stochastic Differential Equations*, "Naukova Dumka", Kiev, 1968 (Russian); English transl. Springer-Verlag, Berlin–Heidelberg–New York, 1972.
150. P.W. Glynn and W. Whitt, *Limit theorems for cumulative processes*, Stochastic Process. Appl. **47** (1993), no. 2, 299–314.
151. B.V. Gnedenko and A.N. Kolmogorov, *Limit Distributions for Sums of Independent Random Variables*, "Gostehizdat", Moscow–Leningrad, 1949 (Russian); English transl. Addison-Wesley, Cambridge, Massachussetts, 1954.
152. J. Grandell, *Doubly Stochastic Poisson Processes*, Springer-Verlag, Berlin–Heidelberg–New York, 1976.
153. P. Greenwood, E. Omey, and J.L. Teugels, *Harmonic renewal measures*, Z. Wahrsch. Verw. Gebiete **59** (1982), no. 3, 391–409.
154. W. Greub, *Linear Algebra*, 4th edn., Springer-Verlag , Berlin-Heidelberg-New York, 1975.
155. G.N. Griffiths and F.T. Wright, *Moments of the number of deviations of sums of independent identically distributed random variables*, Sankhyā, Series A **37** (1975), no. 3, 452–455.

156. I.V. Grinevich and Yu.S. Khokhlov, *The domains of attraction of semistable laws*, Teor. Veroyatnost. i Primenen. **40** (1995), no. 2, 417–422 (Russian); English transl. in Theory Probab. Appl. **40** (1996), no. 2, 361–366.
157. Q. Guo, X. Mao, and R. Yue, *Almost sure exponential stability of stochastic differential delay equations*, SIAM J. Control Optim. **54** (2016), no. 4, 1919–1933.
158. A. Gut, *Strong laws for independent identically distributed random variables indexed by a Sector*, Ann. Probab. **11** (1983), no. 3, 569–577.
159. A. Gut, *Stopped Random Walks. Limit Theorems and Applications*, second edition, Springer, New York, 2009.
160. A. Gut, O. Klesov, and J. Steinebach, *Equivalences in strong limit theorems for renewal counting processes*, Statist. Probab. Lett. **35** (1997), no. 4, 381–394.
161. A. Gut and A. Spătaru, *Precise asymptotics in the Baum-Katz and Davis laws of large Numbers*, J. Math. Anal. Appl. **248** (2000), no. 1, 233–246.
162. A. Gut and J. Steinebach, *Convergence rates and precise asymptotics for renewal counting processes and some first passage times*, Fields Inst. Comm. **44** (2004), 205–227.
163. A. Gut and J. Steinebach, *Convergence rates in precise asymptotics*, J. Math. Anal. Appl. **390** (2012), no. 1, 1–14.
164. L. de Haan, *On regular variation and its application to the weak convergence of sample extremes*, Math. Centre Tracts 32, Amsterdam, 1970.
165. L. de Haan, *On functions derived from regularly varying functions*, J. Austral. Math. Soc., Serie A **23** (1977), 431–438.
166. L. de Haan, *Extreme value statistics, in* Extreme Value Theory and Applications, (eds. J. Galambos, J. Lechner, and E. Simiu), Proceedings of the Conference on Extreme Value Theory and Applications, Volume 1 Gaithersburg Maryland, 1994, pp. 93–122.
167. L. de Haan and A. Ferreira, *Extreme Value Theory. An Introduction*, Springer, New York, 2006.
168. L. de Haan and U. Stadtmüller, *Dominated variation and related concepts and Tauberian theorems for Laplace transforms*, J. Math. Anal. Appl. **108** (1985), no. 2, 344–365.
169. C. Hagwood, *A renewal theorem in multidimensional time*, Australian J. Statist. **31** (1989), no. 1, 130–137.
170. P. Hall, *A comedy of errors: the canonical form for a stable characteristic function*, Bull. London Math. Soc. **13** (1981), no. 1, 23–27.
171. G. Hamel, *Eine Basis aller Zahlen und die unstetigen Lösungen der Funktionalgleichung* $f(x+y) = f(x) + f(y)$, Math. Ann. **60** (1905), 459–462.
172. G.H. Hardy, *A theorem concerning trigonometrical series*, J. London Math. Soc. **3** (1928), 12–13.
173. G.H. Hardy, *Some theorems concerning trigonometrical series of a special type*, Proc. London Math. Soc. **32** (1931), 441–448.
174. G.H. Hardy and E.M. Wright, *An Introduction to the Theory of Numbers*, 5th edn., The Clarendon Press, Oxford University Press, New York, 1979.
175. P. Hartman and A. Wintner, *On the law of the iterated logarithm*, Amer. J. Math. **63** (1941), no. 1, 169–176.
176. D.R. Heath-Brown, *Mean values of the zeta-function and divisor problems, in* Recent Progress in Analytic Number Theory, Academic Press, London–New York, 1981, pp. 115–119.
177. D. Henderson and P. Plaschko, *Stochastic Differential Equations in Science and Engineering*, World Scientific, Singapore, 2006.
178. C.C. Heyde, *On large deviation probabilities in the case of attraction to a non-normal stable Law*, Sankhyā **A30** (1968), no. 3, 253–258.
179. C.C. Heyde, *Some properties of metrics in a study on convergence to normality*, Z. Wahrscheinlichkeitstheorie verw. Gebiete **11** (1969), no. 3, 181–192.
180. C.C. Heyde, *A supplement to the strong law of large numbers*, J. Appl. Probab. **12** (1975), no. 1, 173–175.
181. C.C. Heyde and V.K. Rohatgi, *A pair of complementary theorems on convergence rates in the law of large numbers*, Proc. Camb. Phil. Soc. **63** (1967), no. 1, 73–82.

182. R. Higgins, *On a problem in the theory of sequences*, Elemente Math. (1974), 37–39.
183. D. Hilbert, *Mathematical problems, Lecture delivered before the International Congress of Mathematicians at Paris in 1900*, Bull. Amer. Math. Soc. **8** (1902), 437–479.
184. L. Horváth, *Strong approximation of extended renewal processes*, Ann. Probab. **12** (1984), no. 4, 1149–1166.
185. P.L. Hsu and H. Robbins, *Complete convergence and the law of large numbers*, Proc. Nat. Acad. Sci U.S.A. **33** (1947), no. 2, 25–31.
186. H. Hult and F. Lindskog, *On regular variation for infinitely divisible random vectors and additive processes*, Adv. Appl. Probab. **38** (2006), no. 1, 134–148.
187. I.A. Ibragimov and Yu.V. Linnik, *Independent and Stationary Sequences of Random Variables*, Wolters-Noordhoff, Groningen, 1971.
188. D.I. Iglehart and W. Whitt, *The equivalence of functional central limit theorems for counting processes and associated partial sums*, Ann. Math. Statist. **42** (1971), no. 4, 1372–1378.
189. N. Ikeda and S. Watanabe, *Stochastic Differential Equations and Diffusion Processes*, North-Holland Mathematical Library, 24, North-Holland Publishing Co.; Kodansha, Ltd., Amsterdam; Tokyo, 1989.
190. P. Imkeller, I. Pavlyuchenko, and T. Wetzel, *First exit times for Lévy-driven diffusions with exponentially light jumps*, Ann. Probab. **37** (2009), no. 2, 530–564.
191. K.-H. Indlekofer, I. Kátai, and O.I. Klesov, *Renewal theorems for some weighted renewal functions*, Ann. Univ. Sci. Budapest., Sect. Comp. **34** (2011), 179–194.
192. K.-H. Indlekofer and O.I. Klesov, *Strong law of large numbers for multiple sums whose indices belong to a sector with function boundaries*, Teor. Veroyatnost. i Primenen. **52** (2007), no. 4, 803–810; English transl. in Theory Probab. Appl. **52** (2008), no. 4, 711–719.
193. K.-H. Indlekofer and O.I. Klesov, *The complete convergence in the strong law of large numbers for double sums indexed by a sector with function boundaries*, Teor. Imovir. Mat. Stat. **68** (2003), 44–48 (Ukrainian); English transl. in Theory Probab. Math. Statist. **68** (2004), 49–53.
194. K.-H. Indlekofer and O.I. Klesov, *The asymptotic behavior of the renewal process constructed from a random walk with a restricted multidimensional time domain*, Ann. Univ. Sci. Budapest, Sect. Comp. **24** (2004), 209–221.
195. K.-H. Indlekofer and O.I. Klesov, *The asymptotic behavior over a small parameter of a series of large deviation probabilities weighted with the Dirichlet divisors function*, Funct. Approx. Comment. Math. **35** (2006), no. 1, 117–131.
196. K.-H. Indlekofer and O.I. Klesov, *Strong law of large numbers for multiple sums whose indices belong to a sector with function boundaries*, Teor. Veroyatnost. i Primenen. **52** (2007), no. 4, 803–810 (Russian); English transl. in Theory Probab. Appl. **52** (2008), no. 4, 711–719.
197. B.G. Ivanoff and E. Merzbach, *What is a multi-parameter renewal process?*, Stochastics **78** (2006), no. 6, 411–441.
198. S. Jansche, *O-regularly varying functions in approximation theory*, J. Inequal. Appl. **1** (1997), no. 3, 253–274.
199. S. Janson, *Renewal theory for m-dependent variables*, Ann. Probab. **11** (1983), no. 3, 558–568.
200. S. Janson, S. M'Baye, and P. Protter, *Absolutely continuous compensators*, Int. J. Theor. Appl. Finance **14** (2011), no. 3, 335–351.
201. P.R. Jelenković and A.A. Lazar, *Asymptotic results for multiplexing subexponential on-off processes*, Adv. Appl. Probab. **31** (1999), 394–421.
202. N.G. van Kampen, *Stochastic Processes in Physics and Chemistry*, 3rd edition, Elsevier, Amsterdam, 2003.
203. Chung-siung Kao, *On the time and the excess of linear boundary crossings of sample sums*, Ann. Statist. **6** (1978), no. 1, 191–199.
204. J. Karamata, *Sur certains "Tauberian theorems" de M.M.Hardy et Littlewood*, Mathematica (Cluj) **3** (1930), 33–48.
205. J. Karamata, *Sur un mode de croissance ret'gulie're des fonctions*, Mathematica (Cluj) **4** (1930), 38–53.

206. J. Karamata, *Sur un mode de croissance régulière. Theéorèmes fondamentaux*, Bull. Soc. Math. France **61** (1933), 55–62.
207. J. Karamata, *Bemerkung über die vorstehende Arbeit des Herrn Avakumović, mit näherer Betrachtung einer Klasse von Funktionen, welche bei den Inversionssätzen vorkommen*, Bull. Int. Acad. Youg. Sci. **29–30** (1936), 117–123.
208. I. Kátai, *Personal communication*, 2009.
209. M. Katz, *The probability in the tail of a distribution*, Ann. Math. Statist. **34** (1963), no. 1, 312–318.
210. M.V. Keldysh, *On a Tauberian theorem*, Proc. Steklov Math. Inst. **38** (1951), 77–86. (Russian)
211. G. Keller, G. Kersting, and U. Rösler, *On the asymptotic behaviour of solutions of stochastic differential equations*, Z. Wahrscheinlichkeitstheorie verw. Gebiete **68** (1984), no. 2, 163–189.
212. G. Keller, G. Kersting, and U. Rösler, *The asymptotic behaviour of discrete time stochastic growth processes*, Ann. Probab. **15** (1987), no. 1, 305–343.
213. G. Keller, G. Kersting, and U. Rösler, *On the asymptotic behaviour of first passage times for discussions*, Probab. Theory Relat. Fields **77** (1988), 379–395.
214. G. Kersting, *Asymptotic properties of solutions of multi-dimensional stochastic differential Equations*, Probab. Theory Relat. Fields **82** (1989), no. 2, 187–211.
215. H. Kesten, *The limit points of a normalized random walk*, Ann. Math. Statist. **41** (1970), no. 4, 1173–1205.
216. H. Kesten, *Random difference equations and renewal theory for the product of random Matrices*, Acta Mathematica **131** (1973), 207–248.
217. R.Z. Khas'minskiĭ, *Necessary and sufficient conditions for the asymptotic stability of linear stochastic systems*, Teor. Verojatnost. i Primenen. **12** (1967), no. 1, 167–172; English transl. in Theor. Probability Appl. **12** (1967), no. 1, 144–147.
218. R.Z. Khas'minskiĭ, *Stochastic stability of differential equations*, "Nauka", Moscow, 1969 (Russian); English transl. Sijthoff & Noordhoff, Alphen aan den Rijn-Germantown, 1980.
219. K. Khorshidian, *Strong law of large numbers for nonlinear semi-Markov reward processes*, Asian J. Math. Stat. **3** (2010), no. 4, 310–315.
220. F.C. Klebaner, *Expected number of excursions above curved boundarie by a random walk*, Bull. Austral. Math. Soc. **41** (1990), 207–213.
221. F.C. Klebaner, *Stochastic differential equations and generalized Gamma distributions*, Ann. Probab. 7 (1989), no. 1, 178–188.
222. E.P. Klement, R. Mesiar, and E. Pap, *Quasi- and pseudo-inverses of monotone functions, and the construction of t-norms*, Fuzzy Sets and Systems **104** (1999), 3–13.
223. E.P. Klement, R. Mesiar, and E. Pap, *Triangular Norms*, Springer, New York, 2000.
224. O.I. Klesov, *A renewal theorem for a random walk with multidimensional time*, Ukrain. Matem. Zh. **43** (1991), no. 9, 1161–1167 (Russian); English transl. in Ukrain. Math. J. **43** (1991), no. 9, 1089–1094.
225. O.I. Klesov, *Limit Theorems for Multi-Indexed Sums of Random Variables*, Springer, Berlin–Heidelberg–New York, 2014.
226. O. Klesov, A. Rosalsky, and A. Volodin, *On the almost sure growth rate of sums of lower negatively dependent nonnegative random variables*, Statist. Probab. Lett. **71** (2005), no. 2, 193–202.
227. O. Klesov, Z. Rychlik, and J. Steinebach, *Strong limit theorems for general renewal processes*, Probab. Math. Statist. **21** (2001), no. 2, 329–349.
228. O. Klesov and U. Stadtmüler, *Existence of moments of a counting process and convergence in multidimensional time*, Adv. in Appl. Probab. **48** (2016), no. A, 181–201.
229. O.I. Klesov and J. Steinebach, *Asymptotic behavior of renewal processes defined by random walks with multidimensional time*, Teor. Imovirnost. Matem. Statist. **56** (1997), 105–111 (Ukrainian); English transl. in Theory Probab. Math. Statist. **56** (1998), 107–113.
230. O.I. Klesov and J. Steinebach, *The asymptotic behavior of the renewal function constructed from a random walk in multidimensional time with restricted time domain*, Ann. Univ. Sci. Budapest, Sect. Comp. **22** (2003), 181–192.

231. O.I. Klesov and O.A. Timoshenko, *Unbounded solutions of stochastic differential equations with time-dependent coefficients*, Annales Univ. Sci. Budapest. Sect. Comp. **41** (2013), 25–35.
232. C. Klüppelberg and T. Mikosch, *Large deviations of heavy-tailed random sums with applications in insurance and finance*, J. Appl. Probab. **34** (1997), no. 2, 293–308.
233. G. Kolesnik, *On the estimation of multiple exponential sums, in* Recent Progress in Analytic Number Theory, Academic Press, London-New York, 1981, pp. 231–246.
234. A. Kolmogoroff, *Sur la loi forte des grands nombres*, C. R. Acad. Sci. Paris **191** (1930), 910–912.
235. B.I. Korenblyum, *On the asymptotic behavior of Laplace integrals near the boundary of a region of convergence*, Dokl. Akad. Nauk. USSR (N.S.) **104** (1955), 173–176. (Russian)
236. J. Korevaar, T. van Aardenne-Ehrenfest, and N.G. de Bruijn, *A note on slowly oscillating functions*, Nieuw Arch. Wiskunde **23** (1949), 77–86.
237. J. Korevaar, *Tauberian Theorem. A Century of Developments*, Springer-Verlag, Berlin-Heidelberg, 2004.
238. I.N. Kovalenko, *Queueing theory*, Progress in Science: Probability theory. Mathematical statistics. Theoretical cybernetics, 5–109, Akad. Nauk SSSR Vsesojuz. Inst. Naucn. i Tehn. Informacii, Moscow, 1970. (Russian)
239. M.A. Kouritzin and A.J. Heunes, *A law of the iterated logarithm for stochastic processes defined by differential equations with small parameter*, Ann. Probab. **22** (1994), no. 2, 659–579.
240. F. Kozin, *On almost sure asymptotic sample properties of diffusion processes defined by stochastic differential equation*, J. Math. Kyoto Univ. **4** (1964/1965), 515–528.
241. D. Kramkov and W. Schachermayer, *The asymptotic elasticity of utility functions and optimal investment in incomplete markets*, Ann. Appl. Probab. **9** (1999), no. 3, 904–950.
242. M.A. Krasnoselskiĭ and Ya.B. Rutickiĭ, *Convex Functions and Orlicz Spaces*, "Nauka", Moscow, 1958 (Russian); English transl. Noordhoff, Groningen, 1961.
243. E. Krätzel, *Lattice Points*, Mathematics and its Applications (East European Series), 33, Kluwer Academic Publishers Group, Dordrecht, 1988.
244. A.P. Krenevich, *Asymptotic equivalence of solutions of linear Itô stochastic systems*, Ukrain. Matem. Zh. **58** (2006), no. 10, 1368–1384 (Ukrainian); English transl. in Ukrain. Math. J. **58** (2006), no. 10, 1552–1569.
245. A.P. Krenevich, *Asymptotic equivalence of solutions of non-linear Itô stochastic systems*, Nonlinear oscillations **9** (2006), no. 2, 213–220. (Ukrainian)
246. M. Kuczma, *Functional Equations in a Single Variable*, PWN, Warszawa, 1968.
247. M. Kuczma, *An Introduction to the Theory of Functional Equations and Inequalities*, Birkhäuser, Basel–Boston–Berlin, 2009.
248. M.R. Leadbetter and H. Rootzen, *Extremal theory for stochastic processes*, Ann. Probab. **16** (1988), no. 2, 431–478.
249. H.R. Lerche, *Boundary Crossing of Brownian Motion: Its Relation to the Law of the Iterated Logarithm and to Sequential Analysis*, Lecture Notes in Statistics 40, Springer-Verlag, New York, 1986.
250. G. Lindgren, *Stationary Stochastic Processes. Theory and Applications*, CRC Press, Boca Raton, FL, 2013.
251. L. Lipsky, *Queueing Theory: A Linear Algebraic Approach*, Springer-Verlas, New York, 2008.
252. Li Liu, *Precise large deviations for dependent random variables with heavy tails*, Statist. Probab. Lett. **79 (2009)**, 1290–1298.
253. A.J. Lotka, *A contribution to the theory of self-renewing aggregates, with special reference to industrial replacement*, Ann. Math. Statist. **10** (1939), no. 1, 1–25.
254. M. Maejima and T. Mori, *Some renewal theorems for random walks in multidimensional time*, Math. Proc. Cambridge Philos. Soc. **95** (1984), no. 1, 149–154.
255. S.Ya. Makhno, *The law of iterated logarithm for solutions of stochastic differential equations*, Ukrain. Matem. Zh. **48** (1996), no. 5, 650–655; English transl. in Ukrain. Math. J. **48** (1996), no. 5, 725–732.

256. E. Manstavicius, *Natural divisors and the Brownian motion*, J. Theor. Nombres Bordeaux **8** (1996), no. 1, 159–171.
257. X. Mao, *Stochastic Differential Equations and Applications*, 2nd edition, Woodhead Publishing, Oxford–Cambridge–New Delhi, 2010.
258. X. Mao, *Exponential Stability of Stochastic Differential Equations*, Marcel Dekker, Inc., New York, 1994.
259. R.A. Maller, *A note on Karamata's generalised regular variation*, J. Austral. Math. Soc. **A24** (1977), no. 4, 417–424.
260. J. Marcinkiewicz and A. Zygmund, *Sur les fonctions indépendantes*, Fund. Math. **29** (1937), 60–90.
261. V. Marić, *Regular Variation and Differential Equations*, Springer-Verlag, Berlin–Heidelberg–New York, 2000.
262. A.I. Martikainen, *A converse to the law of the iterated logarithm for a random walk*, Teor. Veroyatnost. i Primenen. **25** (1980), no. 2, 364–366 (Russian); English transl. in Theory Probab. Appl. **25** (1980), no. 2, 361–362.
263. A.I. Martikainen and V.V. Petrov, *On necessary and sufficient conditions for the law of the iterated logarithm*, Teor. Veroyatnost. i Primenen. **22** (1977), no. 1, 18–26 (Russian); English transl. in Theory Probab. Appl. **22** (1977), no. 1, 16–23.
264. A.I. Martikainen and V.V. Petrov, *On a Feller theorem*, Teor. Veroytnost. i Primenen **25** (1980), no. 1, 194–197 (Russian); English transl. in Theory Probab. Appl. **25** (1980), no. 1, 191–193.
265. S. Matucci and P. Řehák, *Regularly varying sequences and second order difference equations*, J. Difference Equ. Appl. **14** (2008), no. 1, 17–30.
266. W. Matuszewska, *Regularly increasing functions in connection with the theory of $L^{*\varphi}$ spaces*, Studia Math. **21** (1961/1962), 317–344.
267. W. Matuszewska, *On a generalization of regularly increasing functions*, Studia Math. **24** (1964), 271–279.
268. W. Matuszewska, *A remark on my paper "Regularly increasing functions in the theory of $L^{*\varphi}$-spaces"*, Studia Math. **25** (1964/1965), 265–269.
269. R. Metzler, G. Oshanin, and S. Redner (eds.), *First-Passage Phenomena and Their Applications*, World Scientific, Singapore, 2014.
270. W. Matuszewska and W. Orlicz, *On some classes of functions with regard to their orders of growth*, Studia Math. **26** (1965), 11–24.
271. V.A. Mikhaĭlets and A.A. Murach, *Hörmander interpolation spaces and elliptic operators*, Proc. Inst. Math. Ukrainian Acad. Sci. **5** (2008), 205–226. (Russian)
272. T. Mikosch, *Regular Variation, Subexponentiality and Their Application in Probability Theory*, EURANDOM Institute, Eindhoven, 1999.
273. T. Mikosch and A.V. Nagaev, *Large deviations of heavy-tailed sums with applications in insurance*, Extremes **1** (1998), no. 1, 81–110.
274. T. Mikosch and A.V. Nagaev, *Rates in approximations to ruin probabilities for heavy-tailed distributions*, Extremes **4** (2001), no. 1, 67–78.
275. T. Mikosch and O. Wintenberger, *The cluster index of regularly varying sequences with applications to limit theory for functions of multivariate Markov chains*, Probab. Theory Relat. Fields **159** (2013), no. 1, 157–196.
276. E.F. Mishchenko and N.Kh. Rozov, *Differential Equations with Small Parameters and Relaxation Oscillations*, "Nauka", Moscow, 1975; English transl. Plenum Press, New York, 1980.
277. I.S. Molchanov, *Limit Theorems for Unions of Random Closed Sets*, Lecture Notes in Mathematics 1561, Springer-Verlag, Berlin–Heidelberg–New York, 1993.
278. T. Mori, *The strong law of large numbers when extreme terms are excluded from sums*, Z. Wahrscheinlichkeitstheorie verw. Gebiete **36** (1976), no. 3, 189–194.
279. É. Mourier, *Lois des grands nombres et théorie ergodique*, C. R. Acad. Sci. Paris **232** (1951), 923-925.

280. A.V. Nagaev, *Integral limit theorems for large deviations when Cramér's condition is not fulfilled. I*, Teor. Veroyatnost. i Primenen. **14** (1969), no. 1, 203–216; English transl. in Theory Probab. Appl. **14** (1969), no. 1, 51–64.
281. A.V. Nagaev, *Integral limit theorems for large deviations when Cramér's condition is not fulfilled. II*, Teor. Veroyatnost. i Primenen. **14** (1969), no. 2, 51–63; English transl. in Theory Probab. Appl. **14** (1969), no. 2, 193–208.
282. S.V. Nagaev, *Large deviations of sums of independent random variables*, Ann. Probab. **7** (1979), no. 5, 745–789.
283. P. Ney and S. Wainger, *The renewal theorem for a random walk in two-dimensional time*, Studia Math. **44** (1972), 71–85.
284. K.W. Ng, Q. Tang, J. Yan, and H. Yang, *Precise large deviations for sums of random variables with consistently varying tails*, J. Appl. Probab. **41** (2004), no. 1, 93–107.
285. S.Y. Novak, *Extreme Value Methods With Applications to Finance*, Chapman & Hall/CRC, Taylor & Francis Group, Boca Raton, FL, 2012.
286. A.A. Novikov and N. Kordzakhia, *Martingales and first passage times of AR(1) sequences*, Stochastics **80** (2008), no. 2–3, 197–210.
287. B. Øksendal, *Stochastic Differential Equations. An Introduction with Applications*, 6th ed., Springer-Verlag, Berlin–Heidelberg–New York, 2003.
288. E. Omey and J.L. Teugels, *Weighted renewal functions: a hierarchical approach*, Adv. Appl. Prob. **34** (2002), no. 2, 394–415.
289. E.S. Palamarchuk, *Asymptotic behavior of the solution to a linear stochastic differential equation and almost sure optimality for a controlled stochastic process,* Zhurnal Vychislitel'noi Matematiki i Matematicheskoi Fiziki **54** (2014), no. 1, 89–103; English transl. in Comput. Math. and Math. Phys. **54** (2014), no. 1, 83–96.
290. S. Parameswaran, *Partition functions whose logarithms are slowly oscillating*, Trans. Amer. Math. Soc. **100** (1961), no. 2, 217–240.
291. V.V. Petrov, *Limit Theorems of Probability Theory. Sequences of Independent Random Variables*, The Clarendon Press, Oxford University Press, New York, 1995.
292. V. Pipiras and M. Taqqu, *Long Range Dependence and Self-Similarity*, Cambridge, 2017.
293. H.S.A. Potter, *The mean values of certain Dirichlet series. II*, Proc. London Math. Soc. **47** (1940), 1–19.
294. G. Polya, *Bemerkungen über unendliche Folgen und ganze Funktionen*, Math. Ann. **88** (1923), 69–183.
295. N. Privault, *An Elementary Introduction to Stochastic Interest Rate Modeling*, second edition, World Scientific, New Jersey, 2012.
296. Yu.V. Prokhorov, *On the strong law of large numbers*, Izv. Akad. Nauk SSSR, Ser. Mat. **14** (1950), no. 6, 523–536. (Russian)
297. W.E. Pruitt, *General one-sided laws of the iterated logarithm*, Ann. Probab. 9 (1981), no. 1, 1–48.
298. R. Pyke, *Markov renewal processes: definitions and preliminary properties*, Ann. Math. Stat. **32** (1961), no. 4, 1231–1242.
299. P. Řehák, *Nonlinear Differential Equations in the Framework of Regular Variation*, A–Math–Net, Brno, 2014.
300. S. Resnick, *Extreme Values, Regular Variation, and Point Processes*, Springer-Verlag, New York, 1987.
301. S. Resnick, *On the foundations of multivariate heavy-tail analysis*, J. Appl. Probab. **41A** (2004), 191–212.
302. S. Resnick, *Heavy Tail Phenomena. Probabilistic and Statistical Modeling*, Springer-Verlag, New York, 2007.
303. H. Robbins, D. Siegmund, and J. Wendell, *The limiting distribution of the last time* $S_n \geq \varepsilon n$, Proc. Nat. Acad Sci. U.S.A. **61** (1968), 1228–1230.
304. C.V. Rodríguez-Caballero and O. Knapik, *Bayesian log-periodic model for financial crashes*, Eur. Phys. J. **B 87** (2014), no. 10, Art. 228.

305. B.A. Rogozin, *A Tauberian theorem for increasing functions of dominated variation*, Sibirsk. Mat. Zh. **43** (2002), no. 2, 442–445 (Russian); English transl. in Siberian Math. J. **43** (2002), no. 2, 353–356.
306. A. Rosalsky, *On the converse to the iterated logarithm law*, Sankhyā **A42** (1980), no. 1–2, 103–108.
307. G. Rosenkranz, *Growth models with stochastic differential equations. An example from tumor immunology*, Math. Biosciences **15** (1985), 175–186.
308. L.V. Rozovskiĭ, *On exact asymptotics in the weak law of large numbers for sums of independent random variables with a common distribution function from the domain of attraction of a stable law*, Teor. Veroyatnost. i Primenen. **48** (2003), no. 3, 589–596 (Russian); English transl. in Theory Probab. Appl. **48** (2004), no. 3, 561–568.
309. L.V. Rozovskiĭ, *On exact asymptotics in the weak law of large numbers for sums of independent random variables with a common distribution function from the domain of attraction of a stable law. II*, Teor. Veroyatnost. i Primenen. **49** (2004), no. 4, 803–813 (Russian); English transl. in Theory Probab. Appl. **49** (2005), no. 4, 724–734.
310. G. Rubino and B. Sericola, *Markov Chains and Dependability Theory*, Cambridge University Press, Cambridge, 2014.
311. L. Sacerdote and E. Smith, *Almost sure comparisons for first passage times of diffusion processes through boundaries*, Methodology Comp. Appl. Probab. **6** (2004), no. 3, 323–341.
312. P.K. Sahoo and P. Kanappan, *Introduction to Functional Equations*, Chapman & Hall/CRC, Taylor & Francis Group, Boca Raton, FL, 2011.
313. N. Samko, *On non-equilibrated almost monotonic functions of the Zygmund–Bary–Stechkin class*, Real Anal. Exchange **30** (2004/05), no. 2, 727–745.
314. A.M. Samoĭlenko and O.M. Stanzhytskyi, *Qualitative and Asymptotic Analysis of Differential Equations with Random Perturbations*, Naukova Dumka, Kyiv, 2009 (Ukrainian); English transl. World Scientific Publishing Co. Pte. Ltd., Hackensack, NJ, 2011.
315. A.M. Samoĭlenko, O.M. Stanzhyts'kyi and I.H. Novak, *On asymptotic equivalence of solutions of stochastic and ordinary equations*, Ukrain. Matem. Zh. **63** (2011), no. 8, 1103–1127 (Ukrainian); English transl. in Ukrain. Math. J. **63** (2011), no. 8, 1268–1297.
316. G. Samorodnitsky, *Long Range Dependence, Heavy Tails and Rare Events*, Copenhagen Lecture Notes, MaPhySto, 2002.
317. G. Samorodnitsky and M.S. Taqqu, *Stable Non-Gaussian Random Processes*, Chapman & Hall, 1994, 1994.
318. K.-I. Sato, *Lévy Processes and Infinitely Divisible Distributions*, Cambridge University Press, Cambridge, 1999.
319. L. Saulis and V.A. Statulevičius, *Limit Theorems for Large Distributions*, Kluwer Academic Publishers, Dordrecht–Boston–London, 1991.
320. H.-P. Scheffler, *Precise asymptotics in Spitzer and Baum–Katz's law of large numbers: the semistable case*, J. Math. Anal. Appl. **288** (2003), no. 1, 285–298.
321. R. Schmidt, *Über divergente Folgen und lineare Mittelbildungen*, Math. Z. **22** (1925), 89–152.
322. C.H. Sciadas, *Exact solutions of stochastic differential equations: Gompertz, generalized logistic and revised exponential*, Meth. Comput. Appl. Probab. **12** (2010), no. 2, 261–270.
323. E. Seneta, *An interpretation of some aspects of Karamata's theory of regular variation*, Publ. Inst. Math. Acad. Serbe Sci. **15** (1973), 111–119.
324. E. Seneta, *Regularly Varying Functions*, Springer-Verlag, Berlin–Heidelberg–New York, 1976.
325. R. Serfling, *Quantile functions for multivariate analysis: approaches and applications*, Statistica Neerlandica **56** (2002), 214–232.
326. R. Serfozo, *Travel times in queueing networks and network sojourns*, Ann. Oper. Res. **48** (1994), no. 1–4, 3–29.
327. R. Serfozo, *Basics of Applied Stochastic Processes*, Springer-Verlag, Berlin, 2009.
328. B.A. Sevast'yanov, *Renewal theory*, Itogi Nauki Tehn.; Ser. Teor. Verojatn., mat. Statist., teor. Kibernet. **11** (1974), 99-128 (Russian); English transl. in J. Soviet. Math., **4** (1975), 281–302 (1976).

329. S. Schlegel, *Ruin probabilities in perturbed risk models*, Math. Econom. **22** (1998), 93–104.
330. W. Sierpiński, *Sur l'equation fonctionnelle* $f(x+y) = f(x) + f(y)$, Fund. Math. **1** (1920), 116–122.
331. Q.-M. Shao, *Maximal inequalities for partial sums of* ρ*-mixing sequences*, Ann. Probab. **23** (1995), no. 2, 948–965.
332. M.D. Shaw and C. Yakar, *Stability criteria and slowly growing motions with initial time difference*, Probl. Nonl. Anal. Eng. Sys. **1** (2000), 50–66.
333. V.M. Shurenkov, *Ergodic Theorems and Related Problems*, "Naukova Dumka", Kiev, 1981 (Russian); English transl. VSP, Utrecht, 1998.
334. V.M. Shurenkov, *Ergodic Markov Processes*, "Nauka", Moscow, 1989. (Russian)
335. D. Siegmund, *Sequential Analysis. Tests and Confidence Intervals*, Springer–Verlag, New York, 1985.
336. J. Slivka and N.C. Severo, *On the strong law of large numbers*, Proc. Amer. Math. Soc. **24** (1970), 729–734.
337. W.L. Smith, *Renewal theory and its ramifications*, J. Roy. Stat. Soc. **B20** (1958), no. 2, 243–302.
338. R.T. Smythe, *Strong laws of large numbers for r-dimensional arrays of random variables*, Ann. Probab. **1** (1973), no. 1, 164–170.
339. R.T. Smythe, *Sums of independent random variables on partially ordered sets*, Ann. Probab. **2** (1974), no. 5, 906–917.
340. P. Solier, *Some applications of regular variation in probability and statistics*, Ediciones IVIC, Caracas, Venezuela, 2009 .
341. K. Soni and R.P. Soni, *Slowly varying functions and asymptotic behavior of a class of integral transforms. I, II, III*, J. Math. Anal. Appl. **49** (1975), 166–179, 477–495, 612–628.
342. A. Spătaru, *Precise asymptotics in Spitzer's law of large numbers*, J. Theor. Probab. **12** (1999), no. 3, 811–819.
343. F. Spitzer, *A combinatorial lemma and its application to probability theory*, Trans. Amer. Math. Soc. **82** (1956), no. 2, 323–339.
344. U. Stadtmüller and R. Trautner, *Tauberian theorems for Laplace transforms*, J. Reine Angew. Math. **311/312** (1979), 283–290.
345. U. Stadtmüller and R. Trautner, *Tauberian theorems for Laplace transforms in dimension* $D > 1$, J. Reine Angew. Math. **323** (1981), 127–138.
346. A.J. Stam, *Some theorems on harmonic renewal measures*, Stoch. Process. Appl. **39** (1991), no. 2, 277–285.
347. H. Steinhaus, *Sur les distances des points des ensembles de mesure positive*, Fund. Math. **1** (1920), no. 1, 93–104.
348. W.F. Stout, *Almost Sure Convergence*, Academic Press, New York, 1974.
349. V. Strassen, *A converse to the law of the iterated logarithm*, Z. Wahrscheinlichkeitstheorie verw. Gebiete **4** (1965/1966), no. 4, 265–268.
350. H.H. Stratton, *Moments of oscillations and ruled sums*, Ann.Math. Statist. **43** (1972), no. 3, 1012–1016.
351. A. Strauss and J.A. Yorke, *On asymptotically autonomous differential equations*, Math. Systems Theory **1** (1967), 175–182.
352. P. Sundar, *Law of the iterated logarithm for solutions of stochastic differential equations*, Stoch. Anal. Appl. **5** (1987), no. 3, 311–321.
353. K. Takaŝi, J.V. Manojlović, and J. Milošević, *Intermediate solutions of second order quasilinear ordinary differential equations in the framework of regular variation*, Appl. Math. Comput. **219** (2013), no. 15, 8178–8191.
354. K. Takaâi, J.V. Manojlović, and J. Milošević, *Intermediate solutions of fourth order quasilinear differential equations in the framework of regular variation*, Appl. Math. Comput. **248** (2014), 246–272.
355. Q. Tang, *Asymptotics for the finite time ruin probability in the renewal model with consistent variation*, Stoch. Models **20** (2004), no. 3, 281–297.

356. Q. Tang, C. Su, T. Jiang, and J. Zhang, *Large deviations for heavy-tailed random sums in compound renewal model*, Statist. Probab. Lett. **52** (2001), no. 1, 91–100.
357. A.B. Trajković and J.V. Manojlović, *Asymptotic behavior of intermediate solutions of fourth-order nonlinear differential equations with regularly varying coefficients*, Electron. J. Differential Equations (2016), Paper No. 129, 32 pp.
358. O. Vašíček, *An equilibrium characterisation of the term structure*, J. Financ. Enomics **5** (1977), 177–188.
359. A.D. Ventsel', *Rough limit theorems on large deviations for Markov stochastic processes.* I, Teor. Veroyatnost. i Primenen. **21** (1976), no. 2, 235–252; English transl. in Theory Probab. Appl. **21** (1977), no. 2, 227–242.
360. W. Vervaat, *Functional central limit theorems for processes with positive drift and their inverses*, Z. Wahrscheinlichkeitstheorie verw. Gebiete **23** (1972), no. 4, 245–253.
361. V. Vinogradov, *Refined Large Deviation Limit Theorems*, Longman, Harlow, Co-published in the United States with John Wiley & Sons, Inc., New York, 1994.
362. N.A. Volodin and S.V. Nagaev, *A remark on the strong law of large numbers*, Teor. Veroyatnost. i Primenen. **22** (1977), no. 4, 829–831 (Russian); English transl. in Theory Probab. Appl. **22** (1977), no. 4, 810–813.
363. M. Vuilleumier, *Sur le comportement asymptotique des transformations linneares des suites*, Math. Zeitsch. **98** (1967), 126–139.
364. J.G. Wang, *Law of the iterated logarithm for stochastic integrals*, Stoch. Proc. Appl. **47** (1993), 215–228.
365. I. Weissman, *A note on the Bojanić-Seneta theory of regularly varying sequences*, Math. Zeitschrift **151** (1976), 29–30.
366. M. Woodroofe, *Nonlinear Renewal Theory in Sequential Analysis*, CBMS–NFS Regional Conf. Ser. Appl. Math., vol. 39, SIAM, Philadelphia, 1982.
367. J.H. Wosnitza and J. Leker, *Can log-periodic power law structures arise from random fluctuations?*, Phys. A **401** (2014), 228–250.
368. M.I. Yadrenko, *Dirichlet's Principle and its Applications*, "Vyshcha Shkola", Kyiv, 1985. (Ukrainian)
369. Yakar, M. Çiçek, M. B. Gücen, *Practical stability, boundedness criteria and Lagrange stability of fuzzy differential systems*, Comput. Math. Appl. **64** (2012), no. 6, 2118–2127.
370. A.L. Yakymiv, *Multidimensional Tauberian theorems and their application to Bellman–Harris branching processes*, Mat. Sb. (N. S.) **115 (157)** (1981), no. 3, 463–477, 496. (Russian)
371. A.L. Yakymiv, *Asymptotics properties of state change points in a random record process,* Teor. Veroyatnost. i Primenen. **31** (1986), no. 3, 577–581 (Russian); English transl. in Theory Probab. Appl. **31** (1987), no. 3, 508–512.
372. A.L. Yakymiv, *Asymptotics of the probability of nonextinction of critical Bellman–Harris branching processes*, Trudy Mat. Inst. Steklov. **177** (1986), 177–205, 209 (Russian); English transl. in Proc. Steklov Inst. Math. **4** (1988), 189–217.
373. A.L. Yakymiv, *Probabilistic Applications of Tauberian Theorems*, Fizmatlit, Moscow, 2005 (Russian); English transl. VSP, Leiden, 2005.
374. Y. Yang, R. Leipus, and Šiaulys, *Local precise large deviations for sums of random variables with O-regularly varying densities*, Statist. Probab. Lett. **80** (2010), no. 19–20, 1559–1567.
375. V.M. Zolotarev, *One-Dimensional Stable Distributions*, "Nauka", Moscow, 1983 (Russian); English transl. American Mathematical Society, Providence, R.I., 1986.

Index

Absolutely continuous functions, 99
Admissible transformations, 108
Affine model, 345
Asymptotically
  inverse function, 236
  quasi-inverse function, 235
  quasi-inverse function, left, 243
  quasi-inverse function, right, 243
Asymptotically autonomous, 391
Asymptotically inverse function, 230
Asymptotically stable
  solution of the Cauchy problem, 303
  solution of the Cauchy problem for a non-autonomous ODE, 307
Asymptotics
  of integrals with slowly varying functions, 458
  of sums with slowly varying functions, 458
Asymptotic stability of solutions SDEs
  with respect to initial conditions, 365
Autonomous, 346

Binmore–Katz result, 8

Cauchy functional equation, 198
Cauchy problem
  for an autonomous ODE, 301
  for ODE, 351
  for SDE, 351
Characterization
  PRV-functions, 260
  SQI-functions, 259
Characterization conditions
  for DL-functions, 108
Characterization properties
  of POV-functions, 91
Characterization theorem
  for RV-functions, 156
  for WRV-functions, 157
Class
  $\mathcal{ORV}$, 59
  $\mathcal{OSV}$, 60
  $\mathcal{OURV}$, 60
  $\mathcal{PI}$, 67
  $\mathcal{POV}$, 69
  $\mathcal{PRV}$, 62
  $\mathcal{RV}$, 58
  $\mathcal{SQI}$, 67
  $\mathcal{SV}$, 58
  $\mathcal{WORV}$, 59
  $\mathcal{WOSV}$, 60
  $\mathcal{WOURV}$, 60
  $\mathcal{WPI}$, 67
  $\mathcal{WPOV}$, 69
  $\mathcal{WPRV}$, 62
  $\mathcal{WSQI}$, 67
Complete convergence, 421
Component
  log-periodic, 204
  periodic, 204
  regular, 203
  slowly varying, 203
Component of a function
  log-periodic, 162
  periodic, 162
  regularly varying, 162

V. V. Buldygin et al., *Pseudo-Regularly Varying Functions and Generalized Renewal Processes*, Probability Theory and Stochastic Modelling 91, https://doi.org/10.1007/978-3-319-99537-3

slowly varying, 162
Compound counting process, 336
Compound Poisson process with a Poisson subordinator, 343
Consistently varying, 96
Constant elasticity of variance models, 346
Cramér–Lundberg model, 395

de Haan class, 227
Density, 101
almost decreasing, 125
almost increasing, 125
almost monotone, 126
Differential equation
ordinary autonomous, 347
Dirichlet divisor function, 401
Dirichlet divisor problem, 396
Dominated variation, 95
Doubly stochastic Poisson processes, 343
Duality relation, 30

Elasticity, 111
lower bound, 113
upper bound, 113
Equivalent increasing versions
for POV functions, 92
Equivalent solutions
of SDE and ODE, 351

Factorization representation
for WOURV-functions, 186
Factorization representation of limit functions
strong, 175
weak, 172
Filters, 99
First exit moment
asymptotic behavior, 322
Function
absolutely continuous, 100
almost constant, 109
asymptotically equivalent, 70
consistently varying, 96
dominatedly varying, 95
ER, 123
with finite indices of asymptotic elasticity, 121
locally integrable, 202
O-regularly varying, 59
ORV, 59
O-slowly varying, 60
OSV, 60
O-uniformly regularly varying function, 60
OURV, 60
PI, 66
with positive finite indices of the asymptotic elasticity, 121
with positive indices of the asymptotic elasticity, 121
positively increasing, 66
positively varying, 69
POV, 69
preserving equivalence of functions, 71
preserving equivalence of sequences, 71
pseudo-regularly varying, 62
quickly varying, 58
regularly log-bounded, 205
regularly log-periodic, 161
regularly periodic, 161
regularly varying, 57
slowly oscillating, 77
slowly varying, 58
SQI, 67
sufficiently quickly increasing, 67
uniformly continuous at infinity, 190
weakly O-regularly varying, 59
weakly O-slowly varying, 60
weakly O-uniformly regularly varying, 60
weakly positively increasing, 66
weakly positively varying, 69
weakly pseudo-regularly varying, 62
weakly regularly varying, 58
weakly slowly varying, 58
weakly sufficiently quickly increasing, 67
WORV, 59
WOSV, 60
WOURV, 60
WPI, 66
WPOV, 69
WPRV, 62
WRV, 58
WSQI, 67
WSV, 58
function equation
Cauchy, 198
Hamel, 198

Generalized inverse function, 229
Generalized renewal process, 29
constructed from compound counting processes, 336
constructed from compound Poisson processes, 338
constructed from Gaussian Markovian sequences, 324

constructed from random sequences, 314
constructed from stationary sequences, 320
constructed from stochastic processes, 327
constructed from sums of independent random variables, 326
Gompertz equation, 346
Group of regular points
degenerate, 156
nondegenerate, 156

Haan class of functions, 227
Harmonic renewal function, 416

Index function, 154
Index map, 154
Index of an RV-function, 58
Index of asymptotic elasticity
essential lower, 114
essential upper, 114
lower, 113
upper, 113
Integral representation, 76
for DL-functions, 106
for ORV-functions, 76
for POV-functions, 79
for PRV-functions (ORV), 77
for PRV-functions (SV), 78
for RV-functions, 76
Integral transformations, 99
Interpolation
piece-linear, 72
Invariant function
lower, 168
upper, 168
Inverse function, 229

Karamata integral theorem, 227
Karamata's theorem
converse part, 203
direct part, 202
Karamata's theorem (converse part), 227
Karamata's theorem (direct part), 227
Kronecker's theorem, 163

Large deviations
for attraction to a stable distribution, 443
rough, 392
Lebesgue density, 101
Lemma
Parameswaran, 459

LIL
for renewal processes with linear drift, 12
for renewal processes without linear drift, 18
Limit function
essential lower, 102
essential upper, 102
index, 180
lower, 56
periodic component, 180
upper, 56
Logarithmic density, 107
Long term mean level, 345
Lyapunov stability
uniform, 310

Marcinkiewicz–Zygmund SLLN
for partial sums, 4
for renewal processes with linear drift, 11
for renewal processes without linear drift, 18
Marked point process, 395
Moderate, 419
Moment
exit moment, 314
first exit, 314
first hitting, 314

One-factor short rate model of interest rates, 345
Order of growth
of solutions of autonomous SDE, 369
ORV-class, 104
Oscillation characteristic, 162

Partial order, 398
PI-class, 104
Piecewise linear extrapolation
of a sequence, 291
Piecewise linear interpolation, 141
function, 291
Poisson process
compound, with a Poisson subordinator, 343
doubly stochastic, 343
Potter's bounds
for PRV functions, 81
POV-class, 104
Principal period, 163
PRV-class, 104

Quantile function, 229
Quasi-inverse function, 233

Random walk, 2
  with multidimensional time, 398
Regular points, 154
  group, 155
Renewal counting process, 2
Renewal function, 29, 397
  for multidimensional time, 400
  for multidimensional time, local, 400
  weighted, 416
Renewal process, 2
  alternating, 52
  constructed from a solution of SDE, 366
  cycling, 52
  with linear drift, 11
  without linear drift, 17
  for multidimensional time, 403
Renewal risk models, 395
Renewal sequence, 2
  for multidimensional time, 400
Renewal theorem, 397
  for subsets, 404
Representation
  canonical, 204
Return period, 229
RV-class, 104

Semistable law, 423
  domain of attraction, 423
Sequence(s)
  asymptotically equivalent, 71
  ORV, 291
  POV, 291
  PRV, 291
  RV, 291
  SQI, 291
Series of large deviations, 419
  Baum–Katz, 421
  Hsu–Robbins, 422
  Spitzer, 421
Set of regular points, 154
SLLN
  for partial sums, 3
  for renewal processes with linear drift, 11
  for renewal processes without linear drift, 17
Small, 419
S–O functions, 199
Sojourn time, 314
Speed of reversion, 345
SQI-class, 104
Stable law, 423
  domain of attraction, 423
Steinhaus
  theorem, 156
Stochastic differential equation
  autonomous, 346
Strong law of large numbers
  for generalized renewal process, 32
  for multi-indexed sums, 399
  for multi-indexed sums in subsets, 399
  for renewal processes in a subset, 405
Strong law of large numbers for renewal processes
  for multidimensional time, 406

Theorem
  asymptotics of the ratio of POV-functions, 260
Theorem on the asymptotic behavior of integrals
  converse, 203
  direct, 202
Time
  first-passage, 30
  last-passage, 29
  sojourn, 30

Uniform convergence theorem
  for RV-functions, 75
  for SV-functions, 75
  for WPRV-functions, 75

Value-at-risk, 229
Volatility, 345

Wiener process, 346

Kolibri 360 GmbH
Ferdinand-Jühlke-Straße 7
99095 Erfurt, Deutschland
produktsicherheit@kolibri360.de